v

ECOLOGICAL EFFECTS
OF THERMAL DISCHARGES

POLLUTION MONITORING SERIES

Advisory Editor: Professor Kenneth Mellanby

ECOLOGICAL EFFECTS OF THERMAL DISCHARGES

T. E. L. LANGFORD

Manager, Environmental Information Services, National Power, London, UK

ELSEVIER APPLIED SCIENCE
LONDON and NEW YORK

ELSEVIER APPLIED SCIENCE PUBLISHERS LTD
Crown House, Linton Road, Barking, Essex IG11 8JU, England

Sole Distributor in the USA and Canada
ELSEVIER SCIENCE PUBLISHING CO., INC.
655 Avenue of the Americas, New York, NY 10010, USA

WITH 53 TABLES AND 91 ILLUSTRATIONS

© 1990 ELSEVIER APPLIED SCIENCE PUBLISHERS LTD

British Library Cataloguing in Publication Data

Langford, T. E. (Terry E.)
Ecological effects of thermal discharges.
1. Aquatic ecosystems. Effects of waste water discharges
from cooling systems. Cooling systems. Waste water
discharges. Effects on aquatic ecosystems
I. Title II. Series
574.5′263

ISBN 1-85166-451-3

Library of Congress Cataloging in Publication Data

Langford, T. E. (Terry E.)
Ecological effects of thermal discharges/T. E. L. Langford.
  p. cm.—(Pollution monitoring series)
Includes bibliographical references.
ISBN 1-85166-451-3
1. Thermal pollution of rivers, lakes, etc.—Environmental
aspects. I. Title. II. Series.
QH545.T48L36 1990
574.5′222—dc20   89-48131
                                                            CIP

Printed in Great Britain by Galliard (Printers) Ltd, Gt. Yarmouth

TO JEANIE
FOR EVERYTHING

Preface

Pollution is essentially a biological phenomenon. There would be no point in placing controls on effluents and on the alteration of habitats if they did not protect the animals and plants in those habitats, including *Homo sapiens*. Whilst this may seem very obvious to the ecologist it is clearly not always obvious to the engineer or the lawyer who is happy to reach compromises on the composition of an effluent or the degree to which a watercourse may be dredged without considering fully what those compromises may mean to a long-established ecosystem.

For example, the fact that under some legislation the constraints on the quality of an effluent may be exceeded for a percentage of the time can make a complete nonsense of protection of the flora and fauna. Even a short exposure to high levels of a toxin or the storm sewage overflow will destroy the flora and fauna of a habitat and although the constraints may be reinstated quickly the ecosystem may take months or even years to recover.

Equally, constraints on effluents which are proven to be too restrictive and not ecologically necessary to protect the habitat can inhibit developments which have overall benefit to man and can divert valuable resources which may be used more effectively elsewhere.

It is not, therefore, sufficient merely to know the physical consequences of change or the chemical composition of an effluent. The biological effects must be measured *in situ* by the best available techniques to give statistically viable results. Such data, backed by appropriate experimentation should also be usable for predictive work and ecological assessments of new developments.

The job of the responsible ecologist whether employed by industry, controlling authorities or research organisations should be, therefore, to assess quantitatively, as far as possible, the consequences of any effluent or other environmental disturbance objectively, bearing in mind both conservation of habitats and the reasonable demands of man.

My aim for this book is to assist this process of assessment as far as thermal discharges are concerned by summarising, and reviewing critically at times, the biological studies which have been carried out in various habitats and in response to varying problems related to cooling water usage and heat disposal, particularly on the large scale as in thermal power stations.

The book concentrates on field and site studies because, although controlled laboratory experiments are essential to the establishment of the relative effects of temperature, toxins and other factors on animals and plants, there is often difficulty in applying such data to the complex hydrographical and biological conditions in a wide variety of habitats.

By discussing and summarising the hugely variable field studies and by critical review I hope that I can help the reader unravel the complexity of the effects of at least one kind of effluent.

I hope that the book will also be of great use in the future planning of industries which use cooling water and for the controlling authorities whose task will be to set realistic constraints on these industries

At the same time I have intended that the book be of use to students of pollution and ecology to try to show that the simplistic view of the ecological effects of pollution propounded by some of the media and the more politically motivated environmental movements can often be misleading.

At times in the text reference may be made to terms which are more familiar to the power engineer than to the ecologist. Good references are the UKAEA *Glossary of Atomic Terms* and the CEGB *Glossary of Power Station Terms* which are listed in the bibliography.

Terry Langford
Corrie,
Kitwalls Lane,
Milford-on-Sea,
Lymington,
Hampshire

Acknowledgements

As with other projects of this type, it would have been impossible to proceed without the help and tolerance of many friends and colleagues. First of all my gratitude goes to my longest term colleague John Whitehouse for his great forbearance and tolerance while his books have been missing from his shelves. My thanks go also to all my colleagues at National Power's Marine Biology Unit at Fawley Power Station in Hampshire and the Powergen Freshwater Biology Unit at Ratcliffe-on-Soar. I have valued their advice, criticism and the time taken in discussion about the published data. Mrs Irene Martin typed some of the manuscript and tables in her own time and my colleagues originally in the Divisional Public Relations Unit of CEGB Research and now in National Power, have been helpful with discussions and advice.

My thanks also to the librarians who have traced the more obscure references and to Mr Carl Linn and Doug Godwin who provided help with the illustrations. Richard Lea at the Water Research Centre was also extremely helpful with a literature search through Aqualine.

The late Mr R. S. A. Beauchamp, who was Head of Biology in the CEGB, encouraged me to write this and my previous book. His advice and help kept the project going.

Finally, without the considerable practical and psychological assistance of my wife, Jean, this book would never have been completed. She has, once again, managed to combine her own demanding job with the time-consuming task of typing the first draft of the manuscript, transferring the edited version to a word processor, collating the bibliography and doing all the detailed searching and checking jobs which seem endless when the final manuscript is in production.

Terry Langford

Contents

CHAPTER 1

Heat Disposal and the Sources of Thermal Discharges

1.1 HISTORICAL PERSPECTIVES

Thermal discharges are not new phenomena in aquatic systems. In the geothermally active regions of the world, hot springs and geysers have poured their effluents, at temperatures up to boiling point, into natural streams, rivers and lakes for millions of years. These natural thermal discharges alter the temperature regimes of the receiving waters over considerable areas and support their own characteristic floras and faunas (Brewer 1866; Wood 1868; Hindle 1932; Stockner 1968; Forsyth and McColl 1974; Brock 1975, 1979, 1985). Biological changes are readily visible in geothermal streams in many parts of the world. The vivid greens and blue-greens of thermophilic algae cover the substrate for hundreds of metres downstream of a boiling pool. Pliny the Elder wrote 'green plants grow in the hot springs of Padua' and these blue-green algae are still evident.

Water has been used for cooling in manufacturing processes for several thousand years, for example, in glass-making, metal working, distilling and brewing. The small volumes of waste heated water were usually discharged to the nearest stream or pond. With the beginning of the industrial revolution the volumes of water used in cooling processes increased rapidly. The larger industrial complexes producing iron and steel, pulp and paper and, later on, chemicals and petroleum products, required increasing amounts of water and produced larger and larger volumes of heated and noxious effluents which were discharged to the nearest body of water without concern for the effects on the flora and fauna, or even the welfare and comfort of the human population.

In the 18th and 19th centuries the steam engines which drove mine pumps and factory machinery used small ponds known as 'engine ponds' from which they abstracted cooling water and returned it at a higher temperature. It was in some of these ponds that naturalists first noticed the presence of exotic snails (Macan 1960, 1974), probably the earliest records of the unintentional biological effects of warm water discharges. These and other exotic species were first introduced by explorers and importers of exotic plants to botanic or country-house gardens, and subsequently became dispersed by man and other animals to flourish in the warm ponds and canals near the factories.

It was the development and growth of electricity generating stations, in which steam was used to drive the turbines, which really brought waste heat as a potential large-scale pollutant to the notice of scientists, conservationists and the pollution-control authorities. At the turn of the 20th century steam-driven electricity generators were few and far between in most countries, though small generating stations had opened in London and in the United States as early as 1882 (Henessey 1971; Electricity Council 1973; Hannah 1979). By the late 1880s electricity companies were beginning to flourish in most countries which had experienced an industrial revolution similar to that in the UK. Once established, the industry then grew with remarkable rapidity, though the rates of growth varied with each country.

In the UK for example, although there were regional networks which showed great commercial and technical progress, the country as a whole was slow to realise the potential of electricity compared with the United States and Germany. Even so, in the decade from 1892 to 1902 the installed capacity of generating plant in the UK multiplied tenfold (Fig. 1.1(a)) and this exponential increase was typical of the industrialised countries.

In the following 50 years electricity generating plant increased by a factor of almost 300 in Britain (Fig. 1.1(b)). By the 1930s the industry was the largest single user of water in most industrial regions of the world, either for raising and cooling steam in thermal power stations or for driving turbines directly in hydro-electric power schemes (Langford 1983a). The relative rates of growth and the proportions of hydro-electric and thermal generating stations in different countries depended on topography, rainfall, fuel and land availability, but as electricity demand increased, the proportion of thermal power generation also increased until hydro-electricity produced the smaller proportion in most countries (Hubbert 1971). The proportion of hydro-electricity may, however, increase slightly in some regions during the early 1990s as massive new schemes begin operating in Brazil, Canada and the USSR. Electricity production in the world has almost doubled since 1970 (Electricity Council 1987). In some countries such as China the increase has been much more rapid as the rate

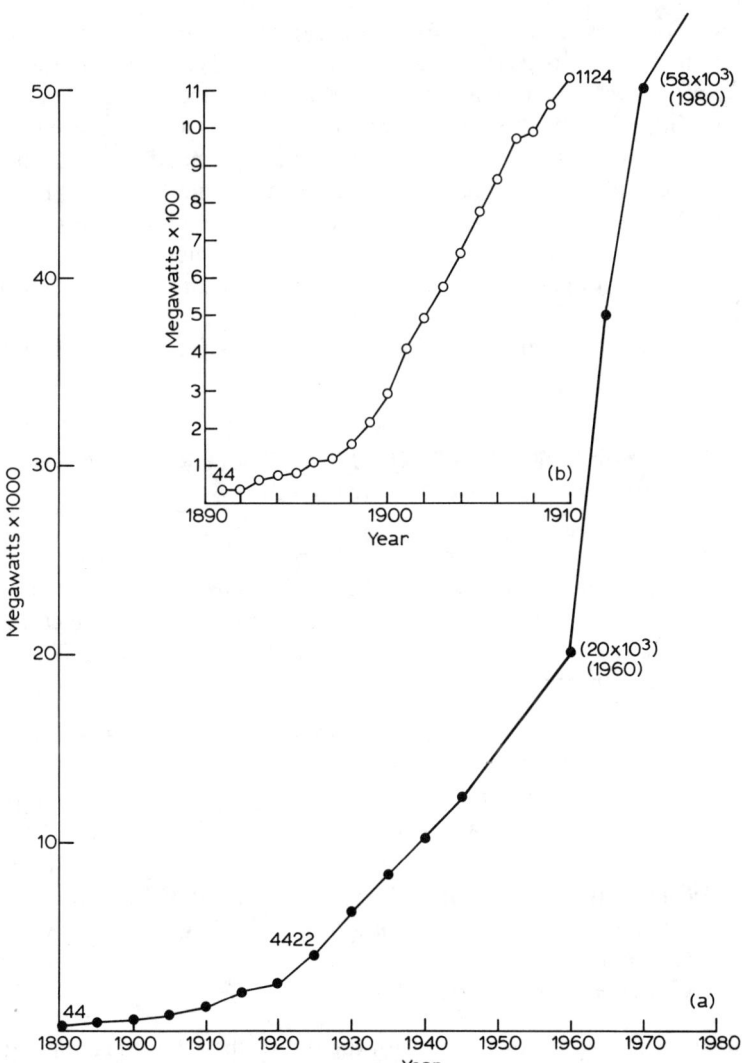

FIG. 1.1   Growth of electricity generating capacity in the UK (a) from 1890–1910, (b) from 1890–1980 (data from Hannah 1979).

of industrialisation has accelerated. In the developing countries growth may well be exponential in the next two decades.

A world recession during the late 1970s and early 1980s slowed the rate of growth of electricity generation in many developed countries but the trend began to reverse in recent years and there is an increasing demand for power in the late 1980s which will continue in the 1990s. In these countries the growth of the industry will again increase the demand for water and the output of thermal discharges (e.g. Linsley, Kraeger and Associates 1980). In most countries there has been a trend toward the building of larger power stations or concentrations of stations on smaller numbers of sites. This has led to larger volumes of cooling water being discharged from such sites. Changes in the industry in the UK could lead to a reversal in this trend at inland sites, where smaller power stations could again become economic.

1.2 MAN-MADE CAUSES OF TEMPERATURE CHANGE IN NATURAL WATERS

The largest single source of heat to most large bodies of surface water is the sun (Edinger and Geyer 1965; Parker and Krenkel 1969; Krenkel and Parker 1969a,b). There are, however, a number of means by which man's activities have changed the temperature regimes of such waters. The most significant of these are:

—forestation and deforestation (Brown 1970; Brown and Krygier 1970; Brown and Brazier 1972; Blackie *et al.* 1980);
—impoundment of running waters or lakes (Lowe-McConnell 1966; Jaske and Goebel 1967; Brown and Brazier 1972; Lavis and Smith 1972; see Ackermann *et al.* 1973; Trotzky and Gregory 1974; Ward 1974; Whitton 1975, 1984; Ward 1976; CEGB 1976; Rogers 1977; Ward and Stanford 1979a; Langford 1983a; Petts 1984; Cowx *et al.* 1987), and
—the discharge of effluents.

This book is mainly concerned with the last of these but the temperature changes associated with the other activities may match those of effluents in places (see Chapter 2).

In the naturally geothermal regions the heated water or steam is tapped to produce electricity. The effluents from these plants are discharged to surface waters (Kestin 1980).

1.3 DEFINITION OF A 'THERMAL DISCHARGE'

For the purposes of this work 'thermal discharges' are defined as those in which:

—the disposal of heat is the primary purpose;
—the heat originates from a man-made heat exchanger;
—any contaminants originate from additives, water treatments or corrosion in the system, rather than from a manufacturing process.

For the most part such thermal discharges originate from electricity generating plant, air conditioning and refrigeration units, with by far the greater proportion coming from the first of these. Substantial amounts of cooling water are also used in the paper, steel (Anon. 1988) rubber and petro-chemical industries (Table 1.1) but very few studies have been concerned with the disposal of heat alone from such industries. The liquefaction of natural gas requires the use of large quantities of cooling water (see Barnett and Hardy 1984) but plants are, as yet, much less numerous than power stations. In countries which do not have good supplies of fresh water, desalination plants are in use, usually sited on the coast. These plants, mostly in the Middle East, produce heated, highly saline effluents, which also contain heavy metal ions (see Barnett and Hardy 1984). Much of the data discussed here will however, be concerned with

TABLE 1.1
The relative use of cooling-water by major industries in the UK and USA

Category	UK (%)	USA (%)
Electricity generation	89·0	80·3
Petroleum refineries	5·4	2·5
Chemical industries	2·8	7·1
Iron and Steel	0·3	7·3
Paper, pulp and board	1·4	1·4
Food processing	0·5	0·9
Other industrial processes	0·6	0·5
Total volume	$3{\cdot}931 \times 10^6\,m^3\,day^{-1}$	$516{\cdot}3 \times 10^6\,m^3\,day^{-1}$

Other industrial processes include: brick-making, cement, gas and coke, glass, glue, paint, rubber, soap, textiles, laundries, coal-mining, leather-tanning and general engineering.
US figures: after Hocutt *et al.* 1980; UK from the Department of the Environment.

TABLE 1.2
Percentage contributions of heat to the River
Thames estuary from various sources

Source	% Total
Power stations	75
Industrial effluents	6
Sewage effluents	9
Freshwater discharges	6
Biochemical activity	4

After DSIR (1964).

thermal power stations and their cooling-water systems though, as we will see, the effects of heat alone are not always easy to distinguish from other effects.

Effluents from many industrial and domestic sources are often warmer than the waters which receive them. Table 1.2 shows the heat discharged to the River Thames from various sources during the 1950s. The greatest proportion was clearly from thermal power stations, though some 9% was from municipal sewage discharges. Parker and Krenkel (1969) also show that the run-off water from irrigation may be 5–10·5°C warmer than nearby river water in summer though the reverse may apply in winter. Generally, water from irrigation enters the receiving water through a large number of small outfalls such as land drains in contrast to the large single outfalls of power stations or other industries. As Fig. 1.2 shows, almost 25% of the total energy dissipated in the UK in 1975 went out as waste heat from power stations.

The definition of thermal discharges by single temperature criteria, for example 'at a temperature greater than 70°F (21·5°C)' which was used at one time in the United States creates complications in that in many regions natural temperatures vary such that fixed criteria may be either too high to protect an ecosystem or, in some areas, actually lower than the natural maximum.

The volume of cooling water used by the thermal power stations in the USA during the late 1960s was four times the volume of that used by all other industries combined (Table 1.1). Between 1900 and 1975 the use of water by thermal power stations increased from about 20×10^9 litre day^{-1} to 496×10^9 litre day^{-1}, a factor of almost 25. In the same period, domestic

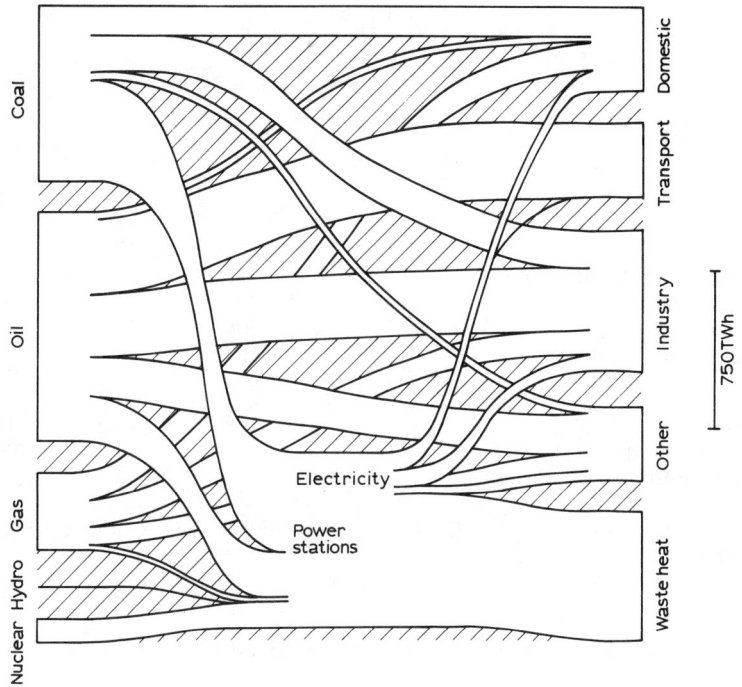

FIG. 1.2 Energy flow in the UK during 1975 (after Flowers 1976).

use rose by a factor of about 4 (United Nations 1976; Linsley, Kraeger and Associates 1980). In 1980 about 20% of the total freshwater run-off was used for power station cooling.

In Europe water use is similarly relatively high. For example, about 20% of the River Rhine passing the Dutch border has been through the cooling system of a power station (Sweers 1980). In the UK the volume of cooling water discharged from power stations accounts for over 90% of all cooling water used by industry. The generation of electricity today accounts for some 50% of the total water demand and about 75% of all industrial consumption. Similar situations exist in most industrialised countries. Since 1966 there has been a decline in the total amounts of water used for cooling at power stations in the UK mainly because of more efficient water use and recycling (Table 1.3). Evaporation has increased because of larger inland power stations using cooling towers.

TABLE 1.3
Trends in water use by UK power stations during the period from 1966 to 1981

	1966–67	*1974–75*	*1979–80*	*1980–81*
Abstracted				
Mains (for boilers,				
domestic uses)	30	38	30	28
Rivers and upper estuaries				
(cooling)	5 170	3 390	2 810	2 300
Lower estuaries and the sea				
(cooling)	8 400	10 220	9 090	7 600
(Recycled)	(6 300)	(10 850)	(12 270)	(12 140)
Total Abstracted	13 600	13 650	11 930	9 930
Evaporated				
Mains	11	14	12	11
Rivers and upper estuaries	55	97	126	126
Lower estuaries and the sea	4	2	2	2

Data supplied by the Central Electricity Generating Board—with permission.
Volumes in millions of gallons/day.

1.4 'THERMAL POLLUTION' AS AN ECOLOGICAL CONCERN

Any industrial installation requires certain basic resources, namely land,
raw materials (fuel), transport and an appropriate method of trade effluent
disposal (Modern Power Station Practice 1971; Keeney 1980). As the
electricity industry grew, the disposal of the waste heat produced in the
process became a major problem. Thus 'nearby river water for cooling
purposes became as crucial a consideration (in the 1920s in the UK) as
access to cheap railway or barge transport, or proximity to coalmines'
(Hannah 1979). Large steelworks, chemical plant and refineries suffered the
same problems.

The potential ecological effects of disposing of heat to rivers, lakes,
estuaries and the sea were predicted early in the electricity industry's history
in the UK. The Electricity Act of 1919 noted that there could be
environmental problems arising from heat disposal in rivers, but at that
time aquatic ecology was in its infancy and little was known about effects of
temperature on aquatic organisms. In 1949 a government report on river
pollution (Ministry of Health 1949) noted that, with regard to cooling water
discharges from power stations, 'during summer a rise of even 1°C may
have a very adverse effect, either directly or indirectly, on the biota'. In 1952,

a legal action in the UK known as the 'Pride of Derby' case resulted in the then British Electricity Authority having to pay some 25% of the damages and compensation for the loss of a river fishery in the River Derwent (All England Law Reports 1952; see Langford 1983a). The Court ruled that the thermal discharge from Spondon Power Station had made a 'significant contribution' to the loss of the fishery, based on the fact that 'water heated to over 100°F must be injurious to fish native to English waters'. The discharge had in fact exceeded 38°C (104°F) at times and also contained small amounts of chlorine (see Chapter 3).

This was probably the first time that heat as a separate pollutant was brought well into the public eye, at least in the UK, and it may be significant that very few research programmes dealing specifically with effects of thermal discharges were originated anywhere in the world before the early 1950s.

In 1952, biological studies of thermal discharges in rivers were in progress and studies of the effects of such discharges on oxygen and fish in rivers were being supported by the British electricity industry (e.g. BEA 1954; Ross, 1959, 1970). In 1954 the electricity industry in England opened a biological laboratory staffed to study fouling and the potential effects of thermal effluents in rivers and subsequently inaugurated the long-term biological studies of the proposed Bradwell Nuclear Power Station on the Blackwater estuary where objectors to the scheme had forecast catastrophic effects of the thermal discharge (Hawes *et al.* 1975).

In the USA, papers by Cairns (1956a,b) and reports of studies on the Delaware (Trembley 1960) began the 'thermal pollution' era. In the UK, studies of fish in relation to water temperature and thermal discharges were begun by government-funded research institutes during the mid 1950s (Alabaster 1958, 1962; see CERL 1971, 1975).

By the mid-1960s there were many research projects concentrating on thermal discharges in the UK, the United States, the USSR and in Europe and the term 'thermal pollution' was in general use. From 1960 to 1970 the literature concerned with pollution by heat grew from perhaps less than ten papers per year to several hundred (Koppe 1974). In 1969, there were some 300 projects concerned with heated discharges in the United States alone (EEI 1969). In the period between 1969 and about 1975, thermal pollution research and the associated legislation were given high priority particularly in the USA, mainly as a result of the rapid proliferation of nuclear power stations, though as far as effects on the aquatic environment are concerned, the problems of heat disposal are similar in principle whichever method of thermal generation is used.

There are many major publications dealing with the effects of temperature and the related effects of thermal discharges. They include Wurtz and Renn (1965), Warinner and Brehmer (1966), Coutant (1967, 1968a,b, 1969a,b, 1970a,b, 1971a), Kennedy and Mihursky (1967), Howells and Lauer (1969), Jensen *et al.* (1969), Krenkel and Parker (1969b), Parker and Krenkel (1969), Grimas (1970), Hechtel (1970), Kinne (1970a, 1975), Neil and Brauer (1970), Beauchamp *et al.* (1971), Nakatani *et al.* (1971), Coutant and Goodyear (1972), IAEA (1972, 1974d, 1975a, 1975c), Langford (1972, 1974, 1977, 1983a,c, 1988), AEC (1973), Clark and Brownwell (1973), Coutant and Pfuderer (1973, 1974), Parker *et al.* (1973), Gibbons and Sharitz (1974), Jensen (1974, 1976, 1978a,b), Swedish State Power Board (1974), Coutant and Talmage (1975, 1976, 1977), Grimas and Ehlin (1975), Esch and McFarlane (1976), Filatova *et al.* (1976), Koops (1976), Langford and Howells (1976), Water Research Centre (1976), CRIEPI (1977), EDF (1977, 1981), Howells (1977, 1980, 1983), Martin Marietta Corporation (1977), UNIPEDE (1977), Gerhold (1978), Hannon (1978), Khalanski (1978), Macqueen and Howells (1978), Moller (1978a), Thorp and Gibbons (1978), IUPDEE (1979), Talmage and Coutant (1980), Talmage and Opresko (1981), Cravens (1982), Gras (1982), Hesse *et al.* (1982b), Houston (1982), IAWR (1982), Jenkins and Schjodtzhansen (1982), Moller and Dahl-Madsen (1982), Cravens *et al.* (1983), Kerambrun (1983), Barnett and Hardy (1984), GESAMP (1984), Cravens and Harrelson (1985), Iwanski and Chu (1985).

From the late 1970s the number of publications declined, though the amounts of heat discharged continued to increase. Between 1989 and the year 2020 there will be an increase in the building of power stations to meet new demand and to replace the older plant. Heat disposal, especially in combination with effluents from flue-gas cleaning plants will be of prime importance, particularly if inland siting returns to favour for economic transmission reasons.

1.5 WATER USE AND HEAT DISPOSAL IN THERMAL POWER STATIONS

Predictions of the ecological consequences of thermal discharges have been hampered in the past by the lack of understanding by ecologists of the demands on the electricity industry and their effects on the design and operation of the cooling-water supply system of any individual power station and thus ultimately the amounts of water used and heat discharged. Predictions have thus been too simplistic and constraints wrongly targeted.

Much of this introductory chapter is concerned, therefore, with the ways in which water is used in the industry, particularly related to heat disposal. Chapters 2 and 3 deal with the effects of operational demands on the physical and chemical properties of discharges some of which are of more significance to ecosystems than heat. The uses of water associated with non-thermal discharges are discussed by Langford (1983a) and in Chapter 10.

1.5.1 Methods of Electricity Generation
Electricity is generated on the commercial scale by one basic process.

Water, in some form, is used to drive a turbine which in turn drives an alternator or generator which produces the electric current. There are two major types of installation or 'plant' which produce electricity on the large scale.

In a hydro-electric power station, water in a river, lake or estuary is impounded by a dam to create a hydrostatic head. The water is allowed to fall, under control, to the turbine to which it transfers its energy. The turbine drives the generator. The basic elements of any direct water power production are a flow of water from a river, stream, lake outflow or tidal water, and a potential for creating sufficient hydrostatic head. There are potential temperature effects in any relevant watercourse used for hydro-electricity production in that the water discharged from the turbine outlets (tailrace) may be at a higher or lower temperature than that of the original free-moving water, depending upon the design (see Langford 1983a). The ecological effects of impoundments and their discharge waters are well documented (e.g. Lowe-McConnell 1966; Krenkel and Parker 1969a; Ackermann *et al.* 1973; Whitton 1975, 1984; Gordon and Longhurst 1979; Ward and Stanford 1979a; Davies and Walker 1986).

A variation of the simple, or 'run-of-river' hydro-electricity scheme is the 'pumped storage scheme' where water is pumped to a high reservoir during periods of low electricity demand (e.g. at night) and then allowed to run to the lower reservoir when demand increases driving the turbines and generators in the process. Pumped storage schemes are generally used to provide electricity for peak consumer demand periods. Because of frictional and mechanical forces in the process the temperature of water leaving the turbine outlets in some such installations may be above inlet temperature (see Langford 1983a) and thus there may be some effects on the temperature of upper and lower reservoirs.

In the other major electricity generating process, known as thermal or steam generation, water is converted to steam at very high pressures and temperatures and this steam is used to drive the turbines and generators.

The water is converted to steam using a fossil fuel such as coal, oil (or in a few instances wood or peat) or alternatively nuclear fission to provide the heat. Although large volumes of water are used in boilers to create the steam (Table 1.3), by far the largest volumes of water used in thermal power stations are for cooling and condensing the steam, removing vast quantities of heat in the process. The efficiencies of the thermodynamic processes involved in converting the fuel energy to electricity, combined with physical, mechanical and other losses in the plant are such that at even the most efficient station, about twice as much heat energy is lost to the environment as is absorbed in the production of electricity. In specially designed plant, known as Combined Heat and Power plant (CHP) (Marshall 1979), these losses to the environment can be reduced by using the waste heat for heating local homes or industries. The design of these plants is, however, such that electrical efficiency may be sacrificed to provide the heat at the correct conditions for use. One of the most recent potential commercial developments in electricity generation is the Ocean Thermal Energy Conversion system which uses the temperature difference between the sea surface and deep water (about 20°C in tropical seas) to provide the heat energy. Here, however, the effluent is cold and discharged at depth (Myers and Ditmars 1985).

The alternative or renewable energies using water as a primary resource, namely wave power or tidal power, are unlikely to cause major temperature changes in the water body though there are other ecological effects originating from the physical changes. In an estuary the main effect of a tidal power scheme would be to alter the inundation regime of mud or sand flats and hence the substrate and shallow water temperature cycles (see p. 41). Near a large offshore wave-power scheme the main effect might be to reduce the physical forces of waves on the areas of shore sheltered by the structures, but temperatures would be altered very little.

1.5.2 Heat Losses from Power Stations

The basic heat cycle for producing work is the Carnot cycle, the reversible cycle for using a fluid as a driving force with the highest thermal efficiency (Keenan 1941). In the Carnot cycle heat is supplied to a fluid at the highest temperature and rejected at the lowest temperature and the fluid undergoes adiabatic expansion and compression. The second law of thermodynamics dictates that large quantities of heat are rejected in this cycle.

The steam cycle used in thermal generating plant is, however, different from the basic Carnot cycle in that heat is applied to boiler feed water converting it to steam and there is no adiabatic compression step. This cycle is more akin to the Rankine cycle (see Keenan 1941). The distinguishing contrast is between the isentropic compression of the Carnot cycle and the heat applied to the pressurised boiler water up to the critical point (Parker and Krenkel 1969).

The theoretical thermal efficiency of any heat engine is about 65% given the limitations of environmental temperatures and the properties of materials of which it is constructed. Using water as the working fluid the efficiency of the Rankine cycle is theoretically and practically much less than that of the Carnot cycle.

In the early days of thermal power stations the properties of materials and the methods of construction reduced the actual efficiencies of plant to less than 20% (Hannah 1979) but a modern fossil-fuelled generating plant is capable of thermal efficiencies of about 40%. Most power stations operating in the 1980s have thermal efficiencies of around 34–39%. This means, however, that in spite of the improvements, between 60 and 66% of the heat applied is not converted to electricity and is of necessity discharged to the environment. Cootner and Lof (1965) describe this loss as 'waste in a technological rather than an economic sense'.

Given the thermal efficiency of 40% and the normal operating conditions of a modern power station, that is steam at 550°C and $10·3 \times 10^6$ kg cm^{-2} with corresponding heat rates of 2200 kg cal kWh^{-1}, then 880 kg cal are converted to electricity energy and 1400 kg cal are rejected. This assumes a temperature of about 10°C for the cooling water intake temperature though efficiencies will vary slightly as this changes.

Modern nuclear power stations have generally higher heat rejection rates than the fossil-fuelled stations. This is because they operate at lower temperatures and pressures, i.e. 250–300°C and $4·2 \times 10^5$ kg cm^{-2}, resulting in a heat rate of about 2600 kg cal kWh^{-1}. Thus the same amount of heat is converted to electricity as in the fossil-fuelled station but about 50% more heat is rejected. The heat budgets for a representative fossil-fuelled and a nuclear plant are shown in Table 1.4.

1.5.3 Thermal Efficiency and Cooling Water Use

Power stations today are much larger than their early antecedents. Consequently the demand for water and the effluent volumes are greater at any location. In 1890, for example, power stations consisted of generators producing about 0·1 megawatts (MWe), i.e. 100 W (Hannah 1979). The British power system in the 1980s, in comparison, is mainly based on generation from 500 MWe units though fossil-fuelled and nuclear units of 600 and 660 MWe are also in operation. In Europe and the USA generating

TABLE 1.4

Typical energy balance in steam (thermal) electric power plant (figures for 1 kWh net electrical output)

Assumed overall efficiency	40%
Assumed generator efficiency	97·5%
Heat equivalent of 1 KWh	3 618 kJ
Fuel energy required	9 045 kJ
Heat losses from boiler furnace at 10% fuel use	904 kJ
Heat loss from electricity generator at 25% of generator input	92 kJ
Electrical generator output	3 618 kJ
Energy required for generator = energy output from turbine	3 710 kJ
Energy remaining in steam leaving turbine	
(removed in condenser) (conventional plant)	4 430 kJ
for nuclear plant this would be	7 102 kJ
Total cooling water use for 5·5°C rise (conventional plant)	229 litres
for nuclear plant this would be	848 litres

For higher temperature rises the volumes would be proportionately less per kWh. (Data after Cootner and Lof 1965.)

units of up to 1300 MWe have been used for some years (Electricity Council 1987).

The larger, newer power stations are more efficient than the early stations and thus require less cooling water for each megawatt of electricity generated. Figure 1.3 shows that the amounts of cooling water per unit of electricity (kWh) generated decreased by over 200% between 1910 and 1970 in the UK though the total cooling water use increased by an estimated factor of 7 over a period of 45 years, based on total generating capacities of about 4000 MWe in 1925 and about 54 000 MWe in 1970. The improvement in the rates of heat rejection in US power stations resulted in average heat losses to the environment for each unit generated being reduced by over 50% between 1925 and 1964.

A modern 2000 MWe fossil-fuelled power station requires about $65 \, m^3 \, s^{-1}$ of cooling water each day at full operation, to remove about $265 \times 10 \, kJ$ of heat. A nuclear power station of the same size may require up to 50% more because of the differences in heat rate (Table 1.4). This vital resource requirement limits the availability of inland sites in many countries particularly in those such as the UK with relatively small rivers and low summer freshwater flows. Nuclear power has expanded rapidly in many countries since the first commercial generating station opened at Berkeley in the UK in 1962 (Thompson, 1971; Howles 1984; Electricity Council 1987; FORATOM 1987). Before then nuclear generation was

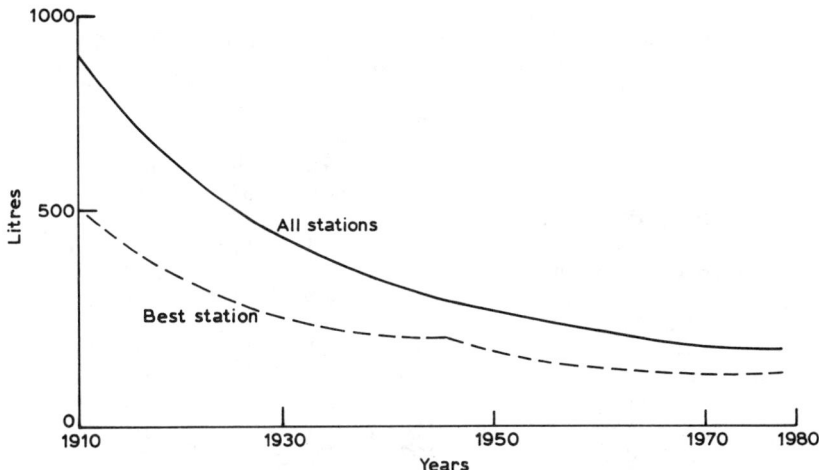

Fig. 1.3   Reduction in cooling water requirement for the production of 1 kWh (unit) of electricity in UK thermal power stations between 1910 and 1980. (Redrawn from Pipe 1972.)

practically a by-product of the military production of nuclear material. Table 1.5 shows the proportion of nuclear electricity generation in various countries from 1985 to 1988. In the latter year there were 429 nuclear reactors producing about 17% of the world's electricity. As can be seen, in several countries the proportion of nuclear power has decreased (IAEA 1989).

As far as the quality and effects of cooling-water effluents are concerned, a nuclear power station is basically the same as any other thermal power station (Langford 1983a; Ray and Skulec 1983).

Because of costs and because of long-term fuel strategies, nuclear power generation is bound to increase in most developed countries despite temporary opposition. Forecasts are for about 25% of total generation in the UK by the year 2000 and a similar amount in the USA, though political, environmental and commercial pressures may cause short-term changes in plans (Flowers 1976). In France, as Table 1.5 shows, nuclear power supplies almost 70% of the total electricity.

The increasing interest in the advanced combustion technologies and combined heat and power systems may mean that heat losses to the environment will be reduced from some plant though there are possible economic penalties for electricity generation (Fenton and Norris 1972; Marshall 1979).

Ecological Effects of Thermal Discharges

TABLE 1.5
Nuclear power's share of electricity production, 1985–88
(expressed as percentage of total electricity produced)

	1988	*1987*	*1986*	*1985*
France	69·9	69·8	69·7	64·8
Belgium	65·5	66·0	67·2	59·8
Hungary	48·9	39·2	25·8	23·6
Sweden	46·9	45·3	50·3	42·3
Korea, Republic of	46·9	53·3	43·6	23·2
Taiwan, China	41·0[a]	48·5[a]	43·8[a]	53·1[a]
Switzerland	37·4	38·3	39·2	39·8
Spain	36·1	31·2	29·4	24·0
Finland	36·0	36·6	38·4	38·2
Bulgaria	35·6	28·6	30·0	31·6
FRG	34·0	31·3	29·4	31·2
Czechoslovakia	26·7	25·9	21·1	14·6
Japan	23·4	29·1	24·7	22·7
United States	19·5	17·7	16·6	15·5
United Kingdom	19·3	17·5	18·4	19·3
Canada	16·0	15·1	14·7	12·7
USSR	12·6	11·2	10·1[a]	10·3[a]
Argentina	11·2	13·4	12·2	11·7
GDR	9·9	9·7[a]	9·7[a]	11·2[a]
South Africa	7·3	4·5	6·8	4·2
Netherlands	5·3	5·2	6·2	6·1
Yugoslavia	5·2	5·6	5·4	5·2
India	3·0	2·6	2·7	2·2
Pakistan	0·6	1·0[a]	1·8	1·0
Brazil	0·3	0·5	0·1	1·6
Italy	0·0	0·1	4·6	3·8

[a] IAEA estimates.

The cooling-water requirement of a 2000 MWe nuclear power station is, as we have seen, about 30–50% more than for an equivalent fossil-fuelled station (i.e. 85–100 m^3 s^{-1}). There are other uses of water on thermal power stations, most of which produce unheated effluents but which may contain contaminants. These may or may not be mixed with the thermal discharge for disposal, depending on the legal or operational demands at any site. Some flue-gas desulphurisation processes can produce effluents with unusually high temperatures which when mixed with cooling water discharges can raise the temperature by 3°C (see p. 75).

1.6 THE DESIGN AND SITING OF COOLING-WATER SYSTEMS

1.6.1 Heat Disposal and Site Selection

The quality and composition of the water abstracted for cooling in a power station are not as important, within limits, as its temperature. This must be low enough to provide efficient heat exchange in the steam cycle. Because thermal efficiency is partly dependent on the cooling-water intake temperature (Cootner and Lof 1965), it is also important that recirculation of heated effluent water to the water-intake is minimised (Budenholzer *et al.* 1971; Hunt 1971; Fischer *et al.* 1979; Miller and Brighouse 1984).

Conditions for rapid heat dispersal may or may not be optimal for the dispersal of contaminants (Whipple 1963; MacQueen 1980). Cooling water may be abstracted from and discharged to rivers, lakes, estuaries or coastal waters whether clean or polluted provided there is a sufficient volume available. Where sufficient natural surface water is not available, supplies may be augmented by other sources such as boreholes, sewage effluents or the supernatant water from coal-ash settlement (see Chapter 3).

Alternatively cooling-water systems may incorporate methods of minimising the use and discharge of water. In the UK even the largest rivers such as the Thames, Severn and Trent do not have sufficient all-year-round flow to provide cooling water for the simple direct transport and disposal of the waste heat from a modern 2000 MWe thermal power station (Hawes 1970). It would seem logical, therefore that all power stations in Britain should be sited on the coast or on large estuaries, but for various economic and engineering reasons only about 50% of the electricity generated originates from coastal sites. Most of the remainder use river water for cooling. The economic need for inland siting in all countries has led to the development of cooling systems for river or lake-cooled stations in which the cooling water abstracted can be cooled and recirculated, thus minimising both the volumes abstracted from and the heat loss to the receiving water. In fact since about 1960 no new power stations using river water for cooling have been built in the UK without a recirculating cooling-water system (Leason 1974).

In countries with larger rivers, e.g. the United States, the USSR and European countries bordering such rivers as the Rhine and Danube, the constraints on inland siting and the use of river water have been less rigid, though even in these countries enhanced cooling or recirculating systems have been developed for the dryer regions or as a result of environmental or political pressures.

1.6.2 The Design of Cooling-Water Systems

The criteria for the design of a cooling-water system at any power station before the 1960s were almost all those which provided the 'best' engineering or economic (or both) solutions for the efficient operation of the plant. Since the 1960s, particularly in the USA many cooling-water systems have been designed merely to reduce the amount of heat discharged to the receiving water in order to comply with environmental legislation or ecological constraints.

(i) Direct or once-through cooling

In a 'direct' or once-through system, the cooling water is pumped from the source to condensers or heat exchangers where the heat is transferred (Fig. 1.4(a)). From there the water is returned, carrying its heat load, directly to the receiving water and the heat is dispersed by radiation, conduction and convection (see Chapter 2). This method is generally the simplest and least costly to design and operate, provided there is an adequate supply of cooling water and a sufficiently large volume or area of receiving water. The amount of heat discharged to the aquatic environment is maximised, especially from a nuclear power station where there are no losses from a chimney stack (Table 1.4). Problems may arise where recirculation of effluent toward the intake occurs under various hydrological or meteorological conditions (see Chapter 2).

(ii) Recirculating cooling-water systems

These are variously known as 'closed', 'enhanced' or 'indirect' systems (Fig. 1.4(b)).

Three major types are in use throughout the world, incorporating either cooling channels, cooling ponds or cooling towers. In the first two, heat is lost mainly by evaporation from the surface of a canal or pond. In the last, various wood or concrete structures are employed to enhance the passage of air over the cooling water thus increasing heat loss to the atmosphere (McKelvey and Brooke 1959; Parker and Krenkel 1969; Cheremisinoff and Cheremisinoff 1983) (see Chapter 2).

(a) Cooling channels or canals. Cooling-water channels or canals of varying lengths are used at many power stations either for engineering (operational) or environmental reasons. At Peterborough on the River Nene in the UK a 600 m long canal carried the effluent from the small power station and discharged it some 800 m upstream of the power station intake. Siting problems dictated the use of this system which was intended to dispose of

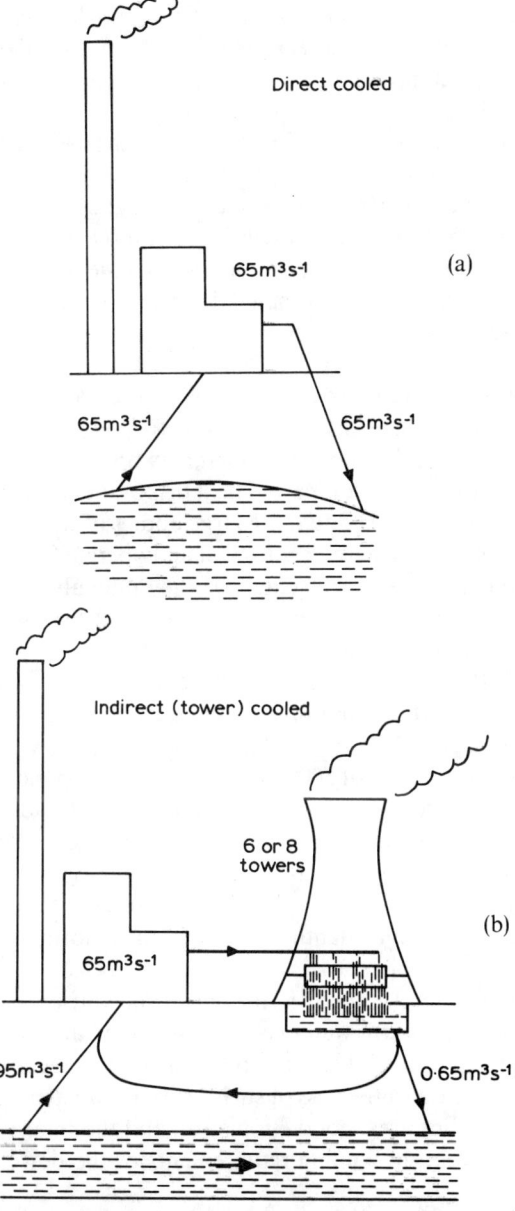

Fig. 1.4   Schematic cooling systems and flows for (a) a direct and (b) an indirect cooled 2000 MWe power station.

some of the heat before the effluent re-entered the river. However, the measured temperature loss was rarely more than 2°C and at times of low river flow water temperatures at the power station intake were much higher than expected, i.e. exceeding 27°C in summer.

At the Kingsnorth power station, in the UK, the cooling water channel is tidal and some cooling occurs through both surface phenomena and mixing before the water reaches the open estuary (Langford 1983b; Bamber and Spencer 1984). At the Turkey Point complex of power stations in Florida, environmental considerations dictated the building of a 9 km long canal in which heat was dissipated before discharge to Biscayne Bay (Thorhaug *et al.* 1978).

(b) Cooling ponds. A cooling pond may be used as a 'buffer', that is an intermediary sink, where heat can be lost before a thermal discharge enters the receiving water. Alternatively a pond may operate in a closed system as the complete abstraction source and heat disposal body. The term 'pond' is relative and may be applied to quite large bodies of water. Artificial ponds can be relatively inexpensive to construct and operate for long periods with no top-up (make-up) requirements and they may also act as a retention basin where contaminants can be detected before discharge to the receiving water. Their main disadvantage is the low heat transfer rate to the air (Brady *et al.* 1969) and thus large surface areas are required for cooling large volumes of water. For a fully closed recirculating system for a power station, about 2·5–5·0 ha MW^{-1} of generation capacity would be required. Scott (1973) gives a figure of 3000–5000 ha of shallow pond for a 1000 MWe power station in order to provide sufficient water at a low enough intake temperature for re-use in the condensers. Where ponds are used merely as a means of partial heat loss (− 2 to 3°C) before discharge to a receiving water, much smaller areas may be used. Most artificial ponds are relatively shallow and a minimum depth of about 1 m is normal (McKelvey and Brooke 1959).

Deep stratified reservoirs or natural lakes may also be used as cooling ponds and here the warm water can be discharged at the surface for most effective heat disposal while abstraction is from the deeper water which is usually colder. The effectiveness of such a system still depends, however, on an adequate surface area, good circulation and the total volume of water available in relation to cooling requirements of the power station.

(c) Spray ponds. These are systems where sprays are located above the cooling pond, or in some cases where spray nozzles are floated on the pond

surface (Kelley 1974; Majewski and Miller 1979; Miller and Brighouse 1984). Water is pumped through the sprays increasing the surface area in contact with the air and thus enhancing the rate of heat loss. Their advantage is that the surface area of a spray pond can be 20 times less than in the simple cooling pond arrangement (Scott 1973). Disadvantages include:

—limited performance because of the short time the sprayed water is in contact with air;

—nuisance from spray carried over in high winds (ASME 1975; Vogel and Hoff 1981); and

—contamination by dust or debris entering the open ponds from the atmosphere (McKelvey and Brooke 1959).

(d) Cooling towers. A cooling tower usually consists of a vertical structure of wood or concrete in which various devices are used to break up the cooling water into droplets or films as it falls from a height to a collecting pond in the tower base.

In natural draught towers, air movement is enhanced by the shape and construction of the tower (Fig. 1.5a). In assisted draught towers air movement is enhanced by motorised fans (Fig. 1.5b). There are several variations on both natural draught and forced draught towers (see McKelvey & Brooke 1959; Gurney and Cotter 1966; Parker and Krenkel 1969; Cheremisinoff and Cheremisinoff 1983).

In dry-cooling towers, the warm water is pumped through 'radiators', that is finned or honeycombed metal constructions, in which the water is not in direct contact with the air (Rossie 1974; Reymeysen *et al.* 1979). There is thus little or no evaporation loss in direct contrast to the wet-cooling towers. Problems are encountered with corrosion and blockages in these structures. The first large hyperbolic dry tower was constructed at Rugeley power station in the UK. Natural draught hyperbolic cooling towers are the most commonly used type in the UK and in other regions where atmospheric conditions are suitable. They have been incorporated into all inland British stations since about 1960 and are used entirely for cooling and recirculating water where supplies are scarce. In the United States such towers may be used merely for environmental reasons to cool effluents before discharge to the receiving water (Parker and Krenkel 1969). A modern 2000 MWe fossil-fuelled station in the UK requiring some $65 \, m^3 \, s^{-1}$ of water, usually incorporates eight hyperbolic towers. As an example of size the towers at a modern 2000 MWe power station in England (Fig. 1.5) are each 140 m high with a similar circumference for the

FIG. 1.5a   Natural draught, hyperbolic cooling towers at a UK power station.

FIG. 1.5b Low profile forced draught (fan-assisted) cooling tower.

collecting pond. About 97% of the water is continuously recirculated, 2% is evaporated from the towers and 1% is discharged to the river (Fig. 1.4(b)). Abstraction to make up these losses is, therefore, only 3% of the total requirement. The 1% discharge to the river (purge) and the resultant make-up is to prevent excessive concentration of dissolved solids in the recirculating water with its consequent problems of scale formation and blockages in pipes and condenser systems (see Chapter 3).

Assisted draught towers are not in common use in British power stations though several have been constructed in recent years and more are planned for gas-fired power stations in the 1990s. They are, however, in common use in the United States and other countries where high humidities require enhanced air movement to maintain efficient evaporative heat losses.

The main advantages of cooling towers over ponds or canals are their low use of land and high cooling efficiencies. Their main disadvantages are high capital and maintenance costs. Natural draught towers are however much less costly in maintenance than the assisted draught towers and less prone to failure. Both types of cooling tower have some environmental effect because of large aerial plumes and localised spray drift (ASME 1975).

Heat loss to the receiving water is, of course, minimised by cooling-tower systems, being roughly some 1% of that from a direct-cooled station of similar size. Thus much smaller volumes or areas of receiving water are necessary for heat disposal. The design, operation, construction and costs of cooling towers are dealt with by McKelvey and Brooke (1959), Gurney and Cotter (1966), Scott (1973) and Cheremisinoff and Cheremisinoff (1983).

TABLE 1.6
Relative costs of different cooling-water systems for power stations in the USA—based on the cost of a direct (once-through) system[a]

Cooling system	Fossil-fuelled plant	Nuclear-fuelled plant
Cooling Pond	$+0.08\ (\pm0.02)$[b]	$+0.08\ (\pm0.02)$
Wet Cooling-Towers		
Natural draught	$+0.18\ (\pm0.03)$	$0.21\ (\pm0.01)$
Mechanical draught	$+0.10\ (\pm0.07)$	$0.10\ (\pm0.08)$
Dry Cooling-Towers		
Natural draught	$+0.99\ (\pm0.16)$	$1.17\ (\pm0.27)$
Mechanical draught	$+0.81\ (\pm0.11)$	$0.93\ (\pm0.15)$

[a] Costs in millions of dollars per kWh (1970 prices).
[b] (95% confidence limits.)

TABLE 1.7
Relative, estimated, costs for various types of cooling
system at a 1100 MWe nuclear power station in the USA
(based on costs of a direct system)

System	Dollars per kW^a
Cooling-lake	+7
Wet-mechanical draught towers	+20
Wet-natural draught towers	+31
Spray canal	+24
Dry-mechanical draught towers	+187

[a] Costs include maintenance, fuel and operation over a 30 year life.

For any enhanced system of cooling at a site there are financial penalties in both construction and operation, excluding those of the extra land needed (Tables 1.6 and 1.7) (Monn *et al.* 1979).

1.7 THE DELINEATION OF THERMAL DISCHARGE EFFECTS

It is difficult to define the limits of a 'thermal discharge effect' in an ecological context. If the initial abstraction process is considered we should include the effects of the trapping and killing (impingement) of fish and invertebrates on cooling-water-intake screens (see Langford 1983a). Planktonic, epibenthic and benthic organisms may also be affected in intake areas by water currents, scour and 'entrainment'. This last applies mainly to planktonic organisms carried into and through a cooling-water system. However, as the effects of intakes generally involve no elements of increased temperature, they are not included here.

For the purposes of this volume the physical limits of actual 'thermal' effects can be regarded as:

'*from* the initial rise in temperature in the heat exchanger (see Chapter 2) to the point at which the effluent is mixed and cooled in the receiving water such that temperature rise above ambient is less than 0·5°C.'

As we will see this 'thermal plume area' is not easy to delimit in theory but in practice it may be definable for any one habitat.

This definition includes therefore all the thermal effects on entrained organisms during their subsequent passage through a cooling-water system, including channels, ponds or cooling towers. The effects of pressure changes, shear and of biocides, water-treatment chemicals and other site effluents or contaminants which may enter the discharge will be superimposed on the thermal effects and result in combined or synergistic stresses on exposed organisms (Lawler 1976; see Carrier and Hannon 1979; Yost and Uziel 1981). The major physical and chemical stresses which can be experienced within the influence of a large cooling system are described in Chapter 2.

1.8 USES AND CONSERVATION OF WASTE HEAT

For many years the potential uses of rejected heat from power stations and other industries have been investigated and many schemes have been devised and put into operation (e.g. Belter 1975). These include domestic and industrial heating, steam for industrial processes, growing and breeding fish, plants or other organisms, drying tobacco and the heating of greenhouses and soils to enhance horticultural installations. The uses and conservation of heat are discussed in Chapter 10.

CHAPTER 2

The Physical Effects of Thermal Discharges

2.1 THE PHYSICAL PROPERTIES OF THERMAL DISCHARGES

A thermal discharge is usually propelled into a receiving water in one of two general forms, namely:

—as a 'layer', that is a stream of low turbulence and velocity either at, or below the water surface;

—or as a rapid 'jet' of some form, with high turbulence and velocity, usually but not always below the water surface.

Mixing of the former with the receiving water is generally poor, and the plume remains discrete for some distance from the outfall. Most of the heat loss in a surface plume is by evaporation, but in sub-surface 'plumes', which occur in special hydrographic conditions (see p. 54), the main loss is by conduction (Table 2.1) with more heat contained in the water body.

A rapid jet mixes with the receiving water quickly after discharge and mixing may be enhanced by various outfall design, siting and constructional features, in addition to natural hydrographic characteristics.

Whichever form the discharge takes, the effluent has several physical properties which, based on existing knowledge of aquatic ecosystems, could predictably cause changes in the receiving water and hence affect the flora and fauna in its vicinity (Krenkel and Parker 1969a; Miller and Brighouse 1984).

The most obvious of these are:

—Higher temperature, causing temperature changes in the receiving water.

—Velocity which can cause changes in the directions and velocities of the water currents near the outfall, and consequently:

TABLE 2.1
Comparison of heat dispersal from surface and submerged discharges (from Miller
and Brighouse 1984)

	Surface discharge	*Submerged discharge*
1. Surface temperatures	Relatively high	Relatively low
2. Relative rate of mixing	Slow	Rapid
3. Effect on temperatures and velocities at the bottom	No change	Significant increases in both temperature and velocity
4. Rate of surface heat dissipation	Relatively high	Relatively low—more heat stored in the water body
5. Time of exposure to maximum temperature[a]	Relatively short	Relatively long
6. Relative cost	Low	High

[a] Depends on distance from condenser outlet to discharge point.

—changes in the sediments near the outfall, mostly as a result of scour or as a result of deposition where the effluent causes back eddies.
—Lower density, and thus increased buoyancy resulting in thermal and sometimes chemical stratification. This may also alter natural density-induced water currents in deeper bodies of water, and change the normal thermal regimes in lentic waters (see Hutchinson 1957; Sverdrup *et al.* 1963).

The magnitudes of these changes depend upon many factors, including plant design and operation, together with the hydrography and physiography of the receiving water (Miller and Brighouse 1984).

2.2 TEMPERATURE

2.2.1 Temperature Measurement

There have been vast quantities of temperature measurements recorded from waters in most regions of the world, mainly because temperature is one of the simplest parameters to measure accurately and ecologists have always considered it of paramount importance (Smith 1972).

In most of the earlier work on water temperature, mercury-in-glass thermometers were used for measurements. Continuous records came from similar instruments, or more usually mercury in metal casings, connected to clockwork or electrical chart recorders (see Kinne 1970b; Barnett and Hardy 1984). In recent years thermistor arrays have been deployed using micro-electronics for recording the data which are usually read directly into a computer for analysis (e.g. Burnett *et al.* 1974; Spurr and Scriven 1975).

Recently the techniques of thermal mapping and thermography have been used extensively for plotting the shape, size and behaviour of thermal plumes (Moore and James 1973; Funnell 1988).

The infra-red radiation emitted by any body is a function of its temperature and all bodies above absolute zero temperatures emit infra-red radiation. This radiation is recorded from a distance using scanners, infra-red radiation 'thermometers' and infra-red photography.

Basically, the modern thermal mapping system consists of an airborne 'camera and mapper' which is flown over the area to be surveyed and which produces a visual image of the warm-water plume (see Fig. 2.1). In the early

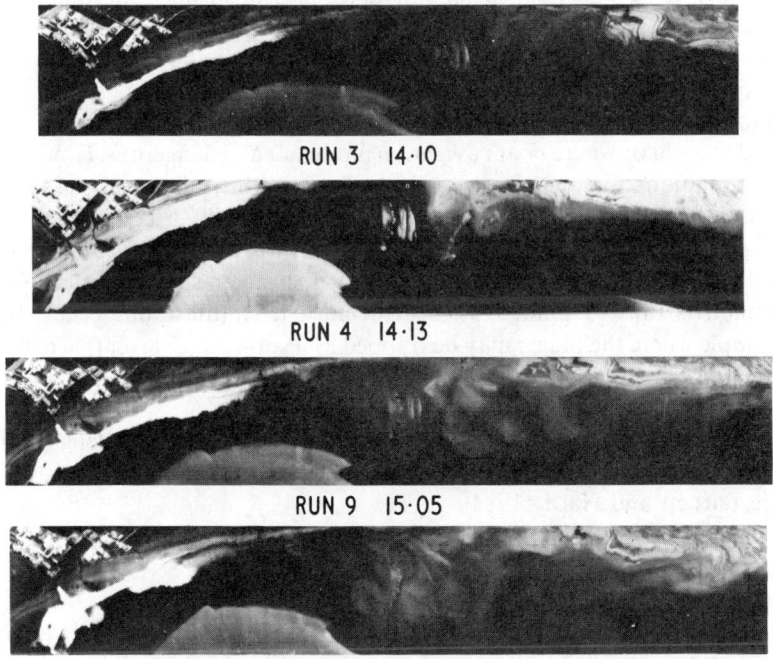

RUN 3   14·10

RUN 4   14·13

RUN 9   15·05

RUN 11   15·17   VERY NEAR LOW WATER
INFRA-RED LINE SCAN SURVEYS 8th MARCH 1967 LOW TIDE

FIG. 2.1   Infra-red thermographs of a cooling-water discharge from a coastal power station (white shows discharge).

work, images were produced in black and white with the plume showing as the lighter area. The lighter the shade, the warmer the water, as a general rule, though calibration against actual water temperatures is difficult.

In the most recent thermal mapping equipment, particularly that used in satellites, very detailed coloured thermographic images are produced. The colours can be distinguished as areas of water of different temperatures.

There are advantages and disadvantages in the use of this remote sensing compared to the more usual survey methods using water borne thermometry (e.g. Moore and James 1973; Madding *et al.* 1975; Marmer *et al.* 1975). The major advantages are, the rapid collection of data and the representation of the whole plume area. Because it is so rapid, the technique can be used to show, on film or video, very short-term variations in plume movements caused by tidal currents, winds or operational changes. Large numbers of boat-mounted or static recorders would be needed to show these total spatial variations. Continuous quantitative data are usually obtained using large numbers of automated measurements from equipment on the water surface (e.g. Burnett *et al.* 1974) but because of the rapidity of recording and the complete overall picture produced, the costs of thermal mapping are still lower than for large-scale, long-term boat surveys or automated data collection and collation. Thermal mapping can also be used at night or where boat navigation is difficult and dangerous. However, its limitations for quantitative calculations and heat-dispersal estimates dictate that some direct measurements are still usually necessary.

Its disadvantages are that it gives no vertical data, because only the surface 'temperatures' to a depth of a few micrometres are recorded. The method is thus of limited use where unusual stratifications occur, for example where the plume may be trapped as a sub-surface layer (see p. 54). Figure 2.2 shows as an example, comparison of data from boat and infrared surveys for a site on the Great Lakes.

The remote systems are reasonably accurate with abilities to detect with a precision ranging with the equipment and conditions from 0·1 to 0·7°C (see Barnett and Hardy 1984).

2.2.2 Plume Modelling

The prediction of the size and direction of a thermal plume and its temperature decay are generally made using two techniques, namely mathematical modelling and hydraulic or physical modelling. The results of such modelling are vital to the prediction of any ecological effects, apart from their importance to the siting, design and operation of a large installation such as a power station.

The mathematical modelling of thermal discharge plumes is discussed by Edinger and Geyer (1965), Ackers (1969), Parker and Krenkel (1969), Bloom *et al.* (1974), Paul and Lick (1974). MacQueen (1978), Fischer *et al.* (1979), Hauser *et al.* (1980), Ward (1982), Miller and Brighouse (1984), and in various papers in symposia (see IAEA, 1971, 1972, 1974a,b,d, 1975a,b,c; Wills, 1972; Sweers, 1974; WEC 1974) and in reviews (see Gosse, 1982; Malmgren-Hansen and Dahl-Madsen 1982; Barnett and Hardy 1984).

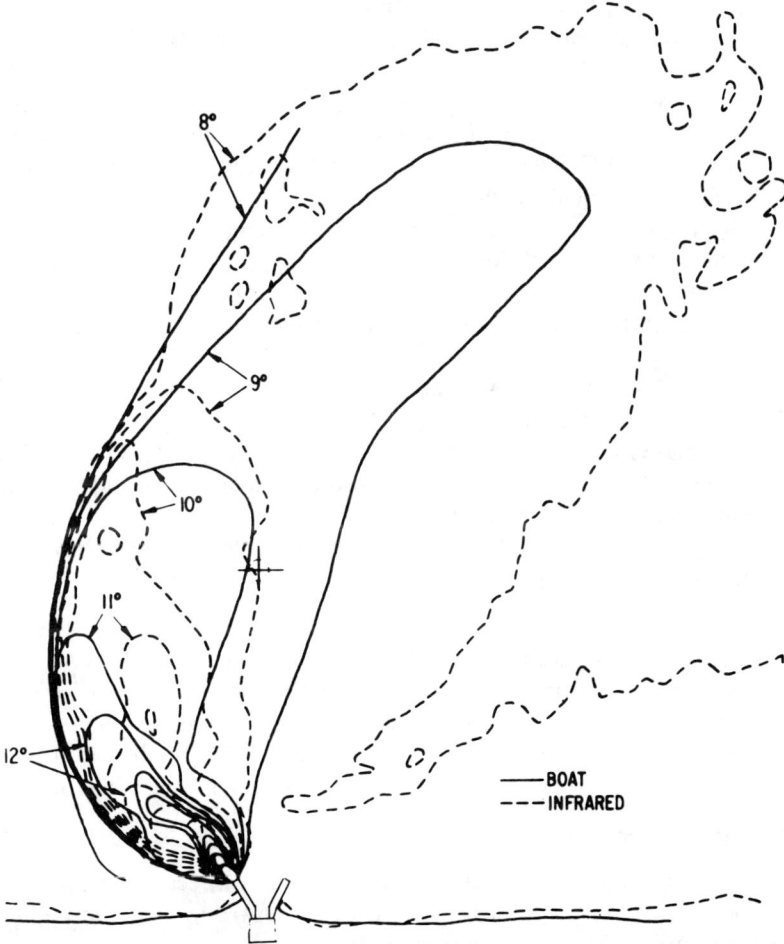

FIG. 2.2 Comparison of boat and thermal mapping surveys, of a thermal plume in a lake (from Madding *et al.* 1975).

However, Miller and Brighouse (1984) consider that 'high hopes for the development of comprehensive models have been replaced by the realisation that neither the mathematical techniques nor the available computing power are adequate for the task', and 'there is no possibility in the near future of comprehensive, reliable mathematical models being developed to simulate thermal discharges'.

Physical models are most often constructed for estuarine or tidal sites, where the currents are multi-directional and complex in both time and space, and the hydraulic characteristics are difficult to express mathematically. Effluent dispersion is usually investigated using tracer dyes under specific conditions (Fischer *et al.* 1979).

The problem with scale models is that the exaggeration in the vertical scale is so great that dispersion coefficients are different to those in the real situation (Dyer 1973). Also such factors as stratification caused by salinity differences are difficult to simulate in practice.

Whichever of the methods of plume measurements or modelling are used, the knowledge of the detailed temperature patterns and behaviour of the plume, particularly of its degree of stability in time and space, are essential to both in-situ ecological research and ecological prediction.

2.2.3 Heat Dissipation in Natural Waters

There are many accounts of heat dispersal mechanisms in water (e.g. Edinger and Geyer 1965, Edinger *et al.* 1968a,b, Brady *et al.* 1969, Parker and Krenkel 1969, Fischer *et al.* 1979) and this section draws heavily and unashamedly on these publications.

Edinger and Geyer (1965) categorised inland receiving waters into three main types for heat disposal predictions, viz.

—Cooling ponds, usually shallow (< 3·0 m) and naturally unstratified. They may be natural waters, but are more often artificial.
—Rivers and streams (running waters), with mostly unidirectional flow and little natural vertical or horizontal stratification.
—Lakes and reservoirs, usually deep and naturally thermally stratified in some seasons, with density currents also present at some times of year.

To these can be added:

—Estuaries and coastal waters, both with tidally directed currents, and which may or may not be stratified with regard to temperature and salinity.

Each has a natural temperature regime which is characteristic and discussed in detail in several authoritative works including Ruttner (1963) (rivers), Hynes (1970) (running waters), Whitton (1975, 1984) (rivers), Hutchinson (1957) (mainly lakes), Dyer (1973) (estuaries), Perkins (1974) (estuaries), Sverdrup *et al.* (1963), Kinne (1970a) (seas and coastal waters). Each has different mixing characteristics (Fischer *et al.* 1979; Miller and Brighouse 1984).

In any body of water, the main mechanisms of surface heat loss or heat transfer are evaporation, conduction and back radiation. The magnitude of these is dependent on the surface temperature of the water. For example, back radiation is proportional to the fourth power of the absolute temperature of the surface. The quantity of heat conducted from the surface is proportional to the difference between water and air temperature, and the heat loss by evaporation is proportional to the difference between saturation vapour pressure at the water surface temperature and water vapour pressure in the air. All three are surface phenomena (Edinger and Geyer 1965).

The simplest method of heat disposal is therefore to discharge an effluent to the receiving water directly, as near the surface as possible, and allow these mechanisms to remove heat and thus return the receiving water toward its original temperature. There are many interrelated factors which influence the rate at which thermal plumes disperse and heat dissipation and the complexities are discussed by many of the authors already cited (see also Barnett and Hardy 1984).

Various approaches are used to estimate the capacities of waters for heat dispersal (e.g. Krajewski *et al.* 1982). Comprehensive mathematical or physical models may also be constructed to predict plume size and behaviour (Fischer *et al.* 1979).

The natural heat budget of a water body can be calulated by balancing the heat inputs with losses. This approach was developed by Schmidt (1915) for estimating ocean evaporation rates and is still the basic primary approach. Figure 2.3 shows the main heat transfer mechanisms from the surface of a water body. The values are those for a range of lakes at different latitudes and altitudes in the USA (Parker and Krenkel 1969). As can be seen, the major heat inputs are solar and atmospheric radiation, while the main losses are through evaporation and back radiation to the atmosphere.

Although the equation shown in Fig. 2.3 is of a simple form, the derivation of values for each term relies on detailed measurements and consequently many comprehensive studies have been made of specific water bodies (see Edinger and Geyer 1965, Brady *et al.* 1969, Burnett *et al.* 1974, Jobson 1978 for examples).

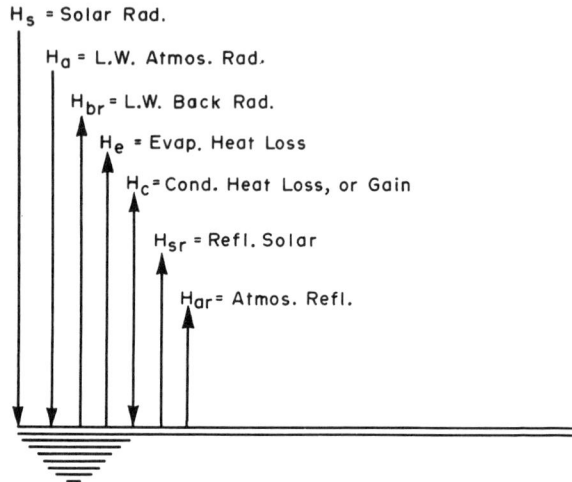

NET RATE AT WHICH HEAT CROSSES WATER SURFACE

$$\Delta H = (H_s + H_a - H_{sr} - H_{ar}) - (H_{br} \pm H_c + H_e) \; BTU \; ft^{-2} \, Day^{-1}$$

H_R

Absorbed   Radiation
Independent   of   Temp.

Temp. Dependent Terms

$$H_{br} \sim (T_s + 460)^4$$

$$H_c \sim (T_s - T_a)$$

$$H_e \sim W(e_s - e_a)$$

FIG. 2.3   Mechanisms of heat transfer across a water surface. (From Krenkel
and Parker 1969a.)

Given that the same mechanisms of heat transfer are operating in all
waters, there are detailed differences between lakes, running waters and
tidal waters which complicate the heat-balance equations and predictions
of cooling capacity. For example in cooling ponds and shallow lakes
atmospheric losses may be dominant, but in stratified waters, rivers and in
tidal waters mixing processes (conduction, dilution) and heat loss to the
surrounding water may be more significant factors than surface
evaporation (Edinger and Geyer 1965; Ackers 1969; Harleman 1969;
Hindley and Miner 1972; Spurr and Scriven 1975; Macqueen 1978; Fischer
et al. 1979). Air temperature, humidity and air movement are all significant
factors determining heat losses from water surfaces.

In some waters floating vascular plants can affect evaporation rates and smooth out the diurnal temperature regime (Dale and Gillespie 1976). Tree cover on some waters can also enhance cooling by reducing insolation (e.g. Miller, 1974).

2.2.4 Natural Water Temperatures

The surface temperatures of natural water bodies in the world vary mainly in relation to latitude and altitude. Within any one body of water, temporal and spatial variations are related to season or local weather conditions. Topographical disturbance by man, for example by impoundments or water transfers, may further complicate natural temperature regimes (see Parker and Krenkel 1969; Karadi *et al.* 1971; Ackermann *et al.* 1973; Gordon and Longhurst 1979; Ward and Stanford 1979a,b; see also Whitton 1975, 1984).

The natural temperature range for the surface waters of the world is −2°C to 100°C, from the polar seas to the hottest thermal springs. However, the temperatures of waters used for large cooling systems or industrial purposes in the world range naturally from about −2°C to about 35°C.

(i) Streams and rivers

The source of a river and its altitude are major influences in its early temperature history. Sources may be in glaciers, lakes, swamps, surface drainage or natural springs, all of which exert different thermal influences. The factors which influence the temperatures of running waters have been studied extensively in Europe, the USA, the USSR and Japan (Smith 1972) and in all these regions considerable variability has been found between different rivers and even reaches of the same river as a result of localised and upstream conditions (Edington 1966; Langford 1970; Crisp and Le Cren 1970; Walker and Lawson 1977; see Whitton 1984; Davies and Walker 1986).

Smith (1972) concluded that the volume of overall discharge is the most important single hydrological factor influencing water temperatures in any river. He also stated 'that the progressive increase in volume characteristic of all rivers in humid regions is responsible for much of the contrast in thermal response between the headwaters and lower reaches'. The relationship between river temperature and air temperature can vary with the size of the river and altitude. The modification of water temperature with time is inversely proportional to distance from the source (see Hynes 1970). The natural annual temperatures of faster-flowing streams mostly fluctuate between 0°C and 30°C in temperate and sub-tropical regions

(Macan 1958a, Mackichan 1967, Smith 1975). The larger rivers on which most industries rely for their cooling water are generally more thermally stable than the small streams.

Studies of the larger British rivers have shown maximum daily fluctuations, of about 2–3°C caused by localised weather conditions (Langford 1970), though the usual daily fluctuation was less than 1°C. British rivers are, however, very small compared with the massive rivers of, for example, the USSR and the USA where daily temperature fluctuations caused by localised influences are negligible (Smith 1972, 1975). Langford (1970) showed that in the River Severn in the English midlands the water temperature was predictable in general terms in that given values (e.g. 10°C, 15°C) tended to occur within a week or so of the same period each year with slight variations depending upon annual weather patterns. However, as a result of local and general climatic conditions the recorded temperatures could vary such that in successive years it differed by 12°C on the same June date. Over 10 years the maximum and minimum mean daily temperatures both varied by about 3°C.

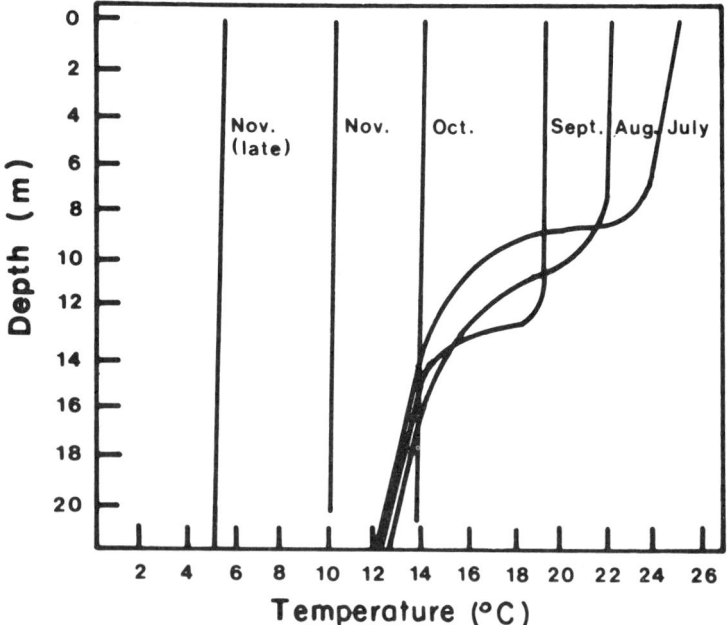

FIG. 2.4   Vertical temperature profiles through the seasons in a typical north-temperate zone lake (after Houston 1982).

Annual maxima for large rivers in temperate regions are generally below 25°C, but in the tropics temperatures exceeding 30°C and reaching 37–40°C may occur naturally in backwaters (see Langford 1983a).

In the east and south of the USA river temperatures regularly exceed 30°C in the summer (Wurtz 1969) and temperatures of over 35°C have been recorded (Profitt 1969). In the north-west of the USA, on the other hand, temperatures in the Columbia river are similar to those in the UK. In these large rivers, the mass transfer of warmer or colder water from upstream are major influences on downstream temperatures. Smith (1972) quotes references showing that snow melt water in the Mississippi could depress temperatures to well below air temperature in the lower reaches. Similar phenomena are evident in smaller rivers (Smith 1975).

Diurnal variation at any one point could also be suppressed by increases in flow. The greatest rate of temperature change in the Severn was about $0.5°C\,h^{-1}$. In small streams rates of $1–1.5°C\,h^{-1}$ have been recorded (Sprules 1947, Macan 1958a). Water currents are mostly unidirectional in rivers. There is virtually no vertical or horizontal temperature stratification and the water is generally homothermous in any one reach. Any natural lateral or vertical heterothermy is caused usually by the entry of tributaries, very low flows through deep pools or slow reaches or in backwaters where there is little flushing by the main currents (Fischer *et al.* 1979).

(ii) Lakes and reservoirs

The characteristic of deep lakes most significant to the behaviour of thermal discharges and heat dispersal is 'thermal stratification'. This phenomenon was first recorded by Simony about 1850, and Welch (1952) states that it 'is so profound and so far reaching in its influence that it forms, directly and indirectly, the substructure upon which the whole biological framework rests'. The details of seasonal changes in lakes are discussed by Welch (1952) and by Hutchinson (1957) and although there are many examples of the variations in timing and relative depths of strata, the basic seasonal pattern of stratification is similar in most stratifying lakes (Fig. 2.4). In some regions or geographical conditions stratification does not occur (Hutchinson 1957; Macan 1970). The depth of the thermocline in lakes varies with their hydrographic characteristics and even within any one lake there may be local differences (Welch 1952). In very large lakes, such as the Great Lakes of North America, or Lake Baikal, regional thermoclines may occur, for example in sheltered bays.

Diurnal variations in surface temperature are generally low, usually less than 1 or 2°C, except where solar radiation is high during the day and temperatures are low at night and the lake is strongly stratified. In such conditions the surface layer may act more like a shallow pond or stream and fluctuate over 5° or 6°C during 24 h. Shallow ponds themselves are generally unstratified, and, depending upon climate and location, can show daily fluctuations of 10°C or more (Vaas and Sachlan 1955). Natural thermal lakes are found in regions where geothermal springs occur and some of these may show considerable thermal heterogeneity where hot springs enter (McColl and Forsyth 1973).

As with other fresh waters, altitude, insolation and latitude are the main factors determining the temperature ranges in a lake (Dussart 1955).

The temperature regime in a reservoir is usually more complex than in a natural lake, especially if the reservoir is one of a chain formed by impounding a large river. Without large inflows and outflows most reservoirs would behave similarly to natural lakes of the same size, but where reservoirs provide water for a hydro-electric power station for example, and receive water from a similar installation upstream the stratification can be altered both temporally and spatially. For example water may be released from either the epilimnion, thermocline or hypolimnion or any combination of these and is generally discharged at high velocity into the downstream reservoir. The complexity of currents and stratification which may occur both upstream and downstream of a dam are well known (see Parker and Krenkel 1969, Fischer *et al.* 1979).

(iii) Tidal waters
Thermal effluents are rarely discharged more than 1 km from the shore and most commonly the outfall is on shore or within about 500 m of the shoreline. The nearshore tidal habitats can be divided into three broad categories, viz: estuaries, embayments (non-estuarine) and ocean fronts (Kinne 1970a,b). With each of these broad categories there is a wide variety of temperature regimes related to latitude and the relative influences of land and water heat balances, but there are some general characteristics of each category which can be applied for temperature predictions.

Estuaries are characterised by widely fluctuating salinities both tidally and seasonally. They may or may not be well mixed and thus may be highly stratified both thermally and as a result of salinity differences (Young *et al.* 1971, Dyer 1973, Barnes and Green 1979, Fischer *et al.* 1979, Miller and Brighouse 1984).

The temperature influences are complex. Apart from the usual heat transfer to and from the atmosphere, incoming river water and tidal inflow and outflow have significant effects. Where there are wide expanses of shore, i.e. sand or mud flats, the faster rate of heating and cooling of this land may have a considerable effect on the water temperatures at the margins (Spencer 1970).

In some estuaries, there can be an inverse thermal stratification where cold fresh water flows over warm, but more dense, saline water (Raymont and Carrie 1964; Dyer 1973; Carstens 1975).

Natural temperatures range annually from $-2°C$ to $5°C$ in polar regions, $-1°C$ to $20°C$ in temperate regions, $10-30°C$ in sub-tropical and tropical regions and $20-32°C$ in the tropics. Diurnal or tidal fluctuations vary with region and with the relative influences of oceanic and fresh water. In a small British estuary the daily range was found to be $<0.5°C$ in winter and $<1.5°C$ in summer at the surface, while in a sub-tropical estuary the ranges were $1.0°C$ and $3.0°C$. Deep water temperatures also varied but the ranges were both less, being $0.5°C$ and $0.2°C$ respectively (Spencer 1970, Dyer 1973). Water temperatures at the margins may be much more highly variable. Spencer (1970) in his study of a small British estuary showed that water flowing over wide sand and mud flats on a flooding tide either warmed in summer or cooled in winter much quicker than the open water as a result of the temperature of the substratum. In most smaller estuaries the temperature effects of the fresh water inflows are relatively small compared with those of the salt water flow and atmospheric and solar effects (Sverdrup *et al.* 1963, Kinne 1970a,b, Dyer 1973, Perkins 1974), though this may be reversed in estuaries of massive rivers such as the Amazon where freshwater flows dominate.

Embayments are categorised by some barrier which protects them from the full force of the open sea. They include lagoons and bays protected by sand bars, islands or reefs, or almost enclosed by tongues of land. They may be on the fringes of estuaries but in Kinne's (1970b) classification are not affected by fresh-water flows, except for minor run-off from land or effluents. They are not usually stratified. Surface temperatures can be very high if the bay is sheltered and if there is little tidal turbulence. In temperate regions, temperatures in such bays rarely exceed $22°C$ though in the tropics natural water temperatures in sheltered bays may exceed $33°C$ (Kolehmainen *et al.* 1975).

Ocean fronts are nearshore habitats which are exposed to the full force of the open sea. The substratum of the shores can vary widely from coarse sand to pebbles to rocks or cliffs but the water temperature always approximates that of the open sea except at the extreme margins where there may be some extra heating or cooling in calm periods. These shores are usually characterised by high energy forces, i.e. strong wave action and strong mixing processes. In the open sea or ocean there is usually marked vertical temperature stratification, with the thermocline more or less permanently at the same depth in each area though the depths vary for different seas (Sverdrup *et al.* 1963; Kinne 1970b). Extreme vertical temperature differences may be as much as 23°C in tropical waters with the coldest water in the deepest strata.

The average diurnal temperature range of surface ocean waters is 0·2–0·3°C except in shallow waters where it may reach 1°C (Sverdrup *et al.* 1963). In the Southern North Sea it is about 0·24°C at maximum while in the shallow Gulf of Bothnia in the Baltic it can be 1·9°C though here diurnal fluctuations of up to 3°C have been recorded (see Kinne 1970b). Overall ranges in open oceans are from −2°C to 30°C world wide and −2°C to 43°C in coastal waters.

There is some general predictability in the fundamental temperature patterns of major water bodies but the possibility of predicting the natural temperature accurately at any one point in any one body of water for any specific time is clearly a complex and difficult exercise. Predicting the effects of a thermal discharge on the physical characteristics of this variety of habitats is an even more difficult exercise and is usually site-specific.

2.2.5 Natural Substrate Temperatures

Many aquatic organisms live in close contact with, or actually in the substrate and therefore its temperature can be as, or more important than that of the water column. Measurements of substrate temperatures in natural waters are scarce in comparison to those of the water itself.

In streams and rivers the temperature of the substrate surface is affected by direct warming by the sun if the water is clear and shallow, but generally, substrate surface temperatures equate to those of the water above. From the few studies which have been carried out it is clear that there is a general increase in temperature stability with depth of penetration into a substrate. Thus although stream water and substrate surface temperatures may fluctuate over up to 8°C each day, some 10–20 cm into the substrate this fluctuation may be as little as 1 or 2°C (Ruttner 1963; Crisp 1990).

In stratified natural lakes and seas, deep-water substrate temperatures are relatively stable and usually at a temperature between 3·5 and 6°C depending upon the location. Even the equinoctial overturns in lakes are unlikely to change this regime.

The greatest temperature fluctuations in aquatic or semi-aquatic habitats occur on the shores of lakes or rivers with fluctuating water levels or in the intertidal zones of estuaries and coastal waters (Johnson 1965; Newell 1970; Spencer 1970; Harrison and Phizacklea 1987). Hedgpeth and Gonor (1969) concluded 'that the temperature conditions experienced by the intertidal biota are almost equally the combination of air and ocean (estuary) temperature conditions'. In fact, air temperatures may be more significant than water temperatures for some intertidal species (Newell 1970).

The natural relationship between water temperature, substrate temperature and air temperature, the depth of temperature effects in the substrate and the adaptations of organisms for a twice daily 'desiccation and inundation' pattern, may all be affected by the intrusion of a continuous discharge of heated sea water over an intertidal zone.

Natural annual temperature fluctuations exceeding $55°C$ ($-20°C$ to $+35°C$) have been recorded on some sea shores (Newell 1970). World-wide, the probable extremes are from about $-50°C$ to $+45°C$ for intertidal zones.

Daily ranges may be very wide in temperate and sub-tropical regions. Glynn (1965), for example, showed that the temperature range on a bare rock surface on a Pacific shore fluctuated by more than $15°C$ on one September day, i.e. more than twice the annual sea-water temperature range. In a nearby rock pool the daily range was $13°C$. Normal daily ranges in sea water rarely exceed $0.5–1°C$ (see p. 40). In crevices of rocks, the temperatures experienced by organisms can vary widely within very short distances as a result of weed cover, direction of the sun or height on the shore (Fig. 2.5). Harrison and Phizacklea (1987) showed that on the muddy shore of an estuary, the maximum temperature rise during inundation by the tide was about $7°C$ in autumn and the maximum fall about $10°C$ in summer.

Spencer (1970) showed daily ranges of up to $15°C$ and tidal ranges of $10°C$ in the uppermost layer of the substrate on a muddy estuarine shore. The minimum daily temperature fluctuation was $3°C$. Once the littoral substrate became covered with water, the temperature stabilised and equated with the water temperature. Even a layer of water $2–5$ cm deep suppressed the temperature fluctuation.

The surface temperature of the mud was highly variable and susceptible to rapid change as a result of local weather conditions. For example a hail shower on a sunny day caused marked lowering of temperatures on exposed areas but not where there was a cover of water. The temperature

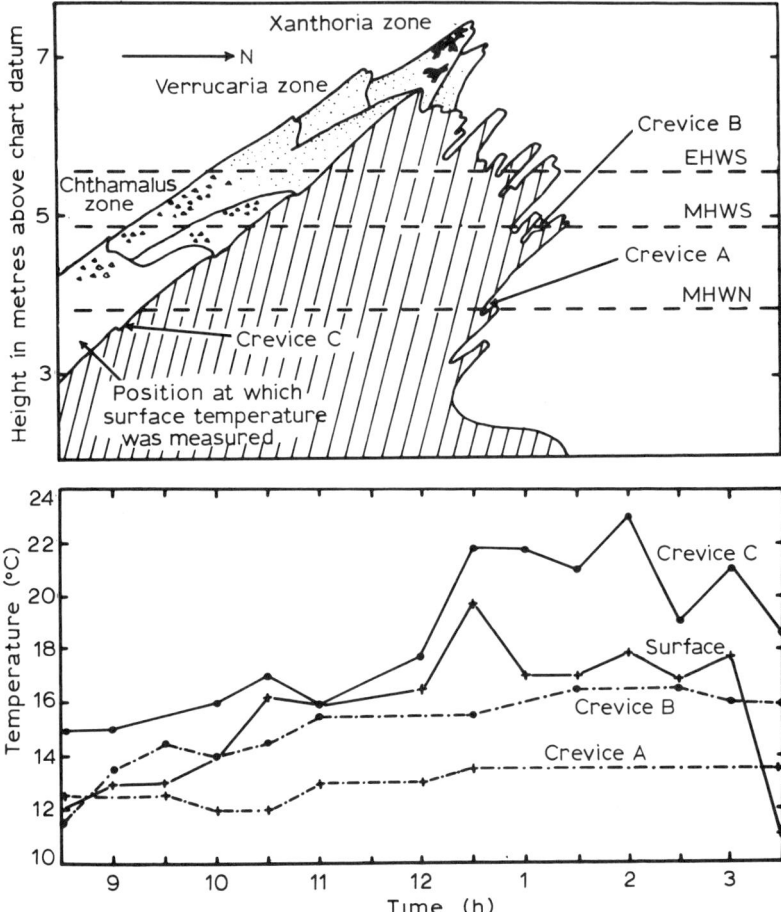

FIG. 2.5   Diagram showing the distribution of selected crevices on Wembury Reef, and the temperatures occurring in each over a complete intertidal period during June. (It will be noticed that the temperature of crevice C on the southern side of the reef exceeded that of the surface of the rocks on the southern side. This was due to the absence of cooling breezes within the crevice.) (Redrawn after Newell 1970.)

stability also increased with depth into the substrate. At 20 cm depth, the daily ranges were usually less than 1°C compared with about 3·5°C in covered areas and 8°C in exposed areas.

Johnson (1965) found similar variations in a Californian sandy beach. He showed that the daily range at 1 cm depth in the substratum was three times that at 10 cm depth and that the amplitude of the temperature change increased with elevation up to the beach. He concluded that 'diurnal cycle in intertidal sediments resembles that observed in terrestrial soils' and that 'the temperatures of surface waters (in the sea) inadequately represent the temperature regime of infaunal intertidal organisms'. Johnson (1965) also found that most animals lived deeper than 5 cm in the substrate and the temperature variation at this depth was higher in the more exposed, higher shore zones. Many animals living in the substrate have water flowing through their tubes or burrows and water temperature may be as important to these species as the substrate temperature. In many cases the temperature in the burrow is more stable than at the surface (Macintosh 1978). Records of substrate temperatures in reservoirs are scarce in the published literature but it is well known that many organisms stranded on the shores of reservoirs by drawdown are killed by heat and desiccation in summer and by extreme cold in winter (Grimas 1961; see Langford 1983a).

2.2.6 The Temperatures of Thermal Discharges

(i) Temperatures within cooling systems
Figure 2.6 shows the hydraulic gradient in a direct cooling system. The rate of temperature change is rapid as the cooling water passes through the heat

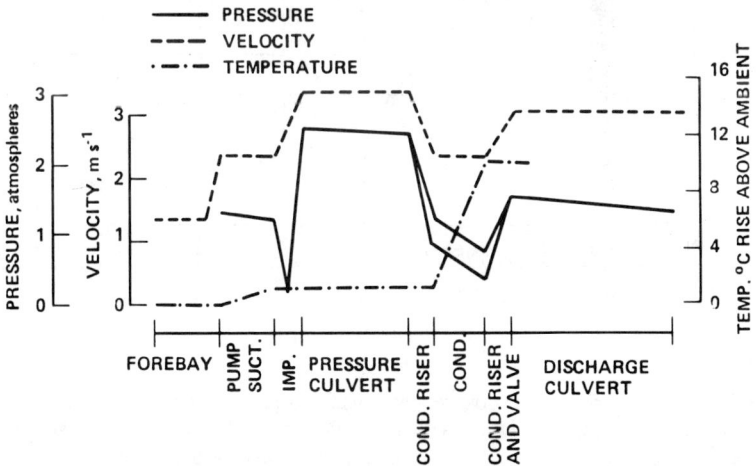

FIG. 2.6 Hydraulic gradient for a typical 'direct-cooled' cooling water system and the associated changes of pressure, velocity and temperature (from Howells and Langford 1982).

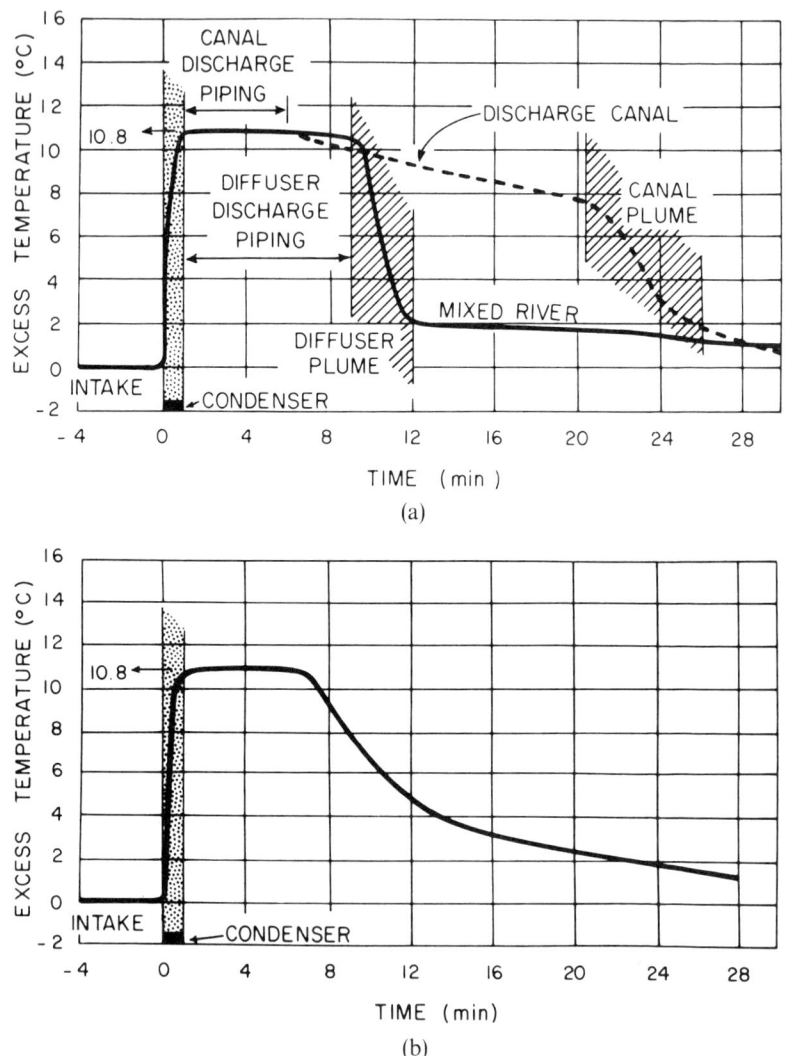

FIG. 2.7 Comparison of exposure: patterns for organisms entrained in two types of cooling-water system. Hypothetical time-courses of acute thermal shock to organisms entrained in condenser cooling water and discharged (a) by diffuser or via a discharge canal, and (b) as a jet (from Schubel *et al.* 1978).

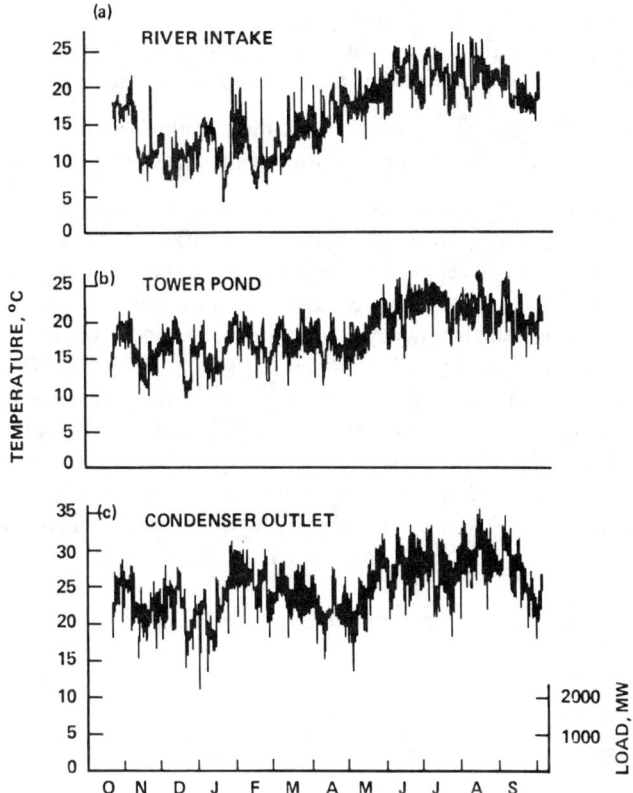

FIG. 2.8   Temperature fluctuations at various points in a tower-cooled system (from Aston *et al.* 1978).

exchangers. At most power stations the rise is between 8 and 15°C, occurring over a period of 2–3 min or less. This is far greater than in any natural habitat. The rate of fall of temperature depends on the type of discharge (Fig. 2.7a,b). In a rapidly mixing discharge, the rate of fall can also be faster than in a natural habitat. The duration of exposure can be a vital factor determining the effect of a thermal discharge, as we will see.

Figure 2.8 shows the fluctuations in temperature within a cooling tower system. At this particular site, the fluctuations in the intake water were unusually large because of a thermal discharge from a power station upstream. Within the system the rates of changes were greater than in natural waters, exceeding 10°C daily in both tower ponds and condensers.

(ii) The temperature range of discharges

The ultimate temperature of any thermal discharge, whether from a power station or any other industry depends on two major factors, viz.

—the natural ambient temperature of the cooling water, and
—the amount of heat per unit volume transferred to it.

In Chapter 1, the optimum temperature rise across a condenser for efficient cooling in a power station was given as about 10–12°C. Variations exist because of very high or very low natural ambients and inadequate volumes of water for efficient cooling, or because of variations in design.

Figure 2.9 shows the frequency distribution of temperature rises (δTs) at operating and proposed nuclear power stations in the USA. The modal temperature is about 10–11°C (18–20°F), and all stations have direct (once-through) cooling systems. Fossil-fuelled stations in the USA have similar ranges of discharge temperatures, though the amounts of heat lost per unit of electricity generated are less (see Chapter 1). In the UK and other European countries δTs are around 8–12°C under normal operating and design conditions, though at some nuclear sites δTs of 15°C may be

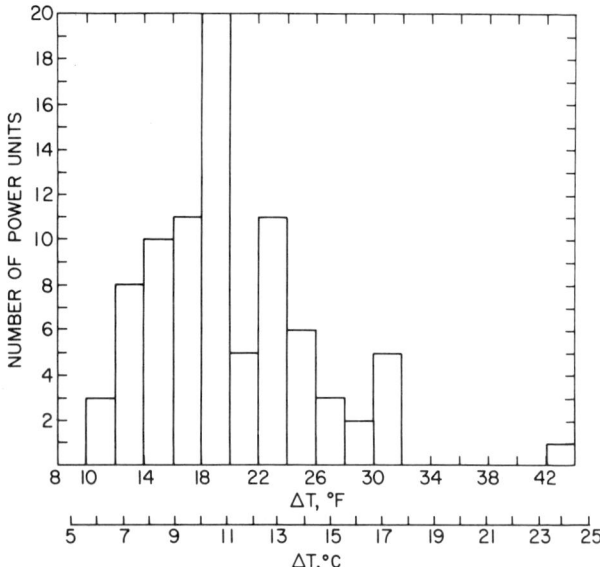

FIG. 2.9 Frequency distribution of designed temperature rises for thermal discharges in the US (power station cooling water systems: from Schubel and Marcy 1978).

attained (Spurr and Scriven 1975). These are not constant, however, at any one site and several factors can cause long- and short-term fluctuations in both volume and temperature.

At many sites throughout the world, hydrographic factors or excessive abstraction causes effluent water to recirculate back to intakes with the result that although the δTs through the system are fairly constant, the difference between true natural ambients and ultimate discharge temperatures may be higher than expected. The classic example of river recirculation in the UK was at Lincoln power station on the River Witham where during the 1950s and 1960s as a result of extremely low river flows and high abstraction, the reversal of flow in the river could result in discharge temperatures exceeding true natural ambients by over 20°C (Langford 1974). This occurred even though the power station was only of 80 MWe output with full tower cooling. Similar situations have occurred in many small rivers.

As a contrast the effluent purge water from some 2000 MWe tower-cooled power stations in the UK may show δTs of 0°C, or even -1°C or -2°C in summer if intake temperatures are already high and the atmosphere sufficiently dry to create good cooling conditions in the towers (see Langford 1983a). Some of the highest cooling-water temperatures consistently recorded from large industry and reported in the literature are those from the nuclear reactors at the Savannah River plant in South Carolina in the United States (Gibbons and Sharitz 1974; Gentry *et al.* 1975). Here, discharge temperatures have exceeded 50°C and temperatures of 35–40°C are common in the receiving habitats when the reactors are operating.

In Europe, Mesarovic (1975) recorded a temperature of 45°C in a Yugoslavian river downstream of a power station and steel mill. Some steel and textile effluents may run at temperatures up to 60°C but the volumes are generally much lower than at power stations. The temperatures of glassworks effluents are reported as 43–45°C (Bedford 1973), and desalination effluents are up to 15°C above natural temperatures. The absolute maximum discharge temperatures from power stations normally range from about 12°C in the coldest regions to 42°C in the sub-tropic and tropics (e.g. Nugent 1970; Thorhaug 1974; Kolehmainen *et al.* 1975; Kamath *et al.* 1975). In more temperate regions the maxima in summer range from about 30°C to 38°C though in extremes 40°C may be exceeded depending upon latitude and cooling system design. Details of temperatures at specific sites are discussed in the relevant biological sections.

(iii) Long-term fluctuations

The ultimate temperatures of cooling-water discharges from power stations can fluctuate seasonally as a function of natural ambient temperatures and electricity demand. The δT and volume of any one discharge may also vary with season, which is again related to demand. Latitude, and thus climate, dictates not only the maximum discharge temperature but also its timing. In the UK, for example, the demand for electricity in summer is usually lower than in winter and any power station may thus operate on half-output or less.

Maintenance work also occurs in summer in temperate regions and this may also reduce effluent volumes or temperatures because of reduced output. At some power stations, cooling-water volumes may be kept constant during such periods but because less plant is operating less heat is discharged and lower δTs occur. Alternatively the volumes used may be less and the δTs thus remain constant. In either case heat loss to the aquatic habitat is less than at full operation. Highest demand in temperate regions is usually for heating in winter, but in sub-tropical and tropical regions the greatest demand usually occurs in summer when air-conditioning plant and refrigeration are in maximum use. Thus in sub-tropical and tropical habitats the temperature effect of the discharge on the receiving water is maximised by high ambient temperatures and high power output.

In the long term, factors such as fuel costs and policy also affect discharge temperatures at any site. For example, in the UK large oil-fired power stations which were designed for continuous (base-load) operation, and ran as such in the early 1970s, now run intermittently, while many coal-fired stations which began to decline in use in the 1970s are at full base-load operation. Thus the temperature regime of the discharges has reversed at the different sites over a period of 10 years. The partial strike of miners in the UK in 1984–85 reversed this situation again for 1 year and several oil-fired power stations ran at full output. The reshaping of the industry in the UK could alter the trend toward larger discharges and, because of the commercial advantages of smaller power stations in the short term, there could be an increase in the number of smaller sites on rivers and on the coast.

At the present time nuclear power stations in many countries provide continuous base load while fossil-fuelled plant is used for higher demand (peak) period. Effluent temperatures from the nuclear sites are therefore stable for longer periods than those of fossil-fuelled sites. Other industries also show fluctuating output which will affect heat losses.

(iv) Short-term fluctuations

In a body of water with a stable temperature regime a thermal discharge may cause instability because of its own fluctuations. In contrast a continuous thermal discharge such as from a nuclear power station on base load, may stabilise temperatures in an unstable habitat, for example where it discharges across an intertidal area. Thus, instead of the normal intertidal diurnal ranges exceeding 10 or 15°C the range would be reduced to less than 1 or 2°C. Annually the range may also be reduced say from >30°C to <20°C in a temperate zone. The habitat is also usually altered in other ways. For example, the effluent inundates the substrate permanently, destroying the intertidal nature of the immediate area near the outfall. Also the effluent velocity scours away soft substrates, changing the habitat considerably.

At power stations which operate for only part of any day to meet the fluctuating demand pattern (peak loading or two shifting), the temperature of the effluent and its volume will fluctuate accordingly. For example at Fawley Power Station on the South coast of the UK, effluent temperatures can be 6°C lower at night than during the day as a result of lower output. At many sites a complete shut-down at night (i.e. two shifting) produces diel temperature fluctuations of 8–10°C in the outfall area in habitats where normal fluctuations would be less than 1 or 2°C at maximum. Responses to peak demands may result in effluents discharging intermittently for short periods, i.e. 2–4 h in any day. A more erratic pattern is shown in Fig. 2.10 for a peak-loading power station on a river. As we have seen in Fig. 2.8 the fluctuations in a tower-cooled system will also be reflected in the effluent.

The temperature of the discharge at any one site may, therefore, change considerably in both the long and short term, often unpredictably. The result may be more, or less stability depending upon demand and operating schedules.

2.2.7 Effects of Thermal Discharges on Habitat Temperatures

Whatever the volume, temperature, nature and composition of any heated effluent its ultimate ecological significance will depend on the magnitude of its effects on the receiving habitat, that is the areas 'significantly' affected by the thermal plume.

The main physical niches in any water body are:

—the surface film,
—the water column (which may or may not be homogeneous),
—the substrate surface (<5 cm),
—the deeper substrate (>5 cm).

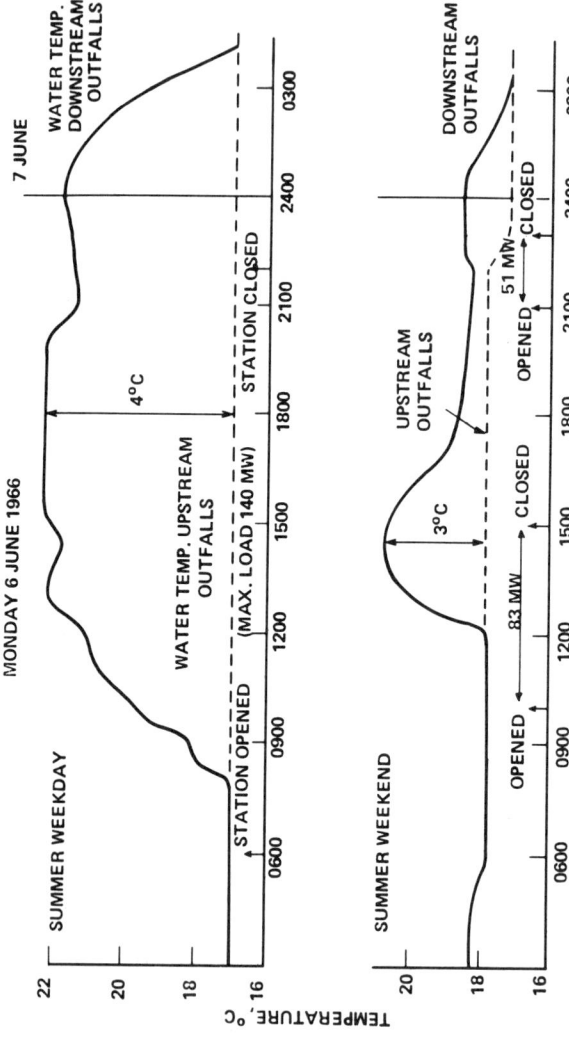

Fig. 2.10 Water temperatures upstream and downstream of the thermal discharge from a river-cooled power station operating on demand (Ironbridge 'A' Power Station on the River Severn, UK).

The two most important in ecological terms are the water column and the substrate, because the greatest proportion of the biocoenosis lives in these two zones. Most temperature measurements in thermal plumes have, however, been restricted to the water column (e.g. Gameson *et al.* 1959; Edinger *et al.* 1968; Milanov 1973; Turoboyski 1973; Burnett *et al.* 1974; Lewis 1974; Kyser *et al.* 1975; Ottendorfer 1975; Spurr and Scriven 1975; Spencer 1977; Hauser *et al.* 1980; Sweers 1980; Jarman and De Turville 1981; Masselchein and Genot 1982; Todd and Bender 1982; Masselchein 1983; Miller and Brighouse 1984).

(i) Thermal plume behaviour
One of the primary factors determining the parts of the habitat and areas which a thermal plume affects is the behaviour of the plume once it leaves an outfall. The details of plume behaviour and the forces affecting plumes are discussed at length by Miller and Brighouse (1984).

(a) In rivers. The general behaviour of a thermal plume in running water is predictable in that the main movement is downstream, whether the discharge is made on the surface or at a depth. The effect on the temperature of the water column and substrate depends on the rate and type of mixing which in turn depends on the depth of the river, its turbulence and the design of the outfall (Mangarella and van Dusen 1973; Fischer *et al.* 1979; Miller and Brighouse 1984).

Lateral stratification occurs in both shallow and deep rivers. The extent depends on outfall design and location, river flow and effluent velocity (Fig. 2.11). Thus a thermal effluent may occupy the full width and depth of the channel in small, shallow rivers for some distance downstream, while in deeper large rivers the effect will be restricted to a narrow band which may or may not be vertically as well as horizontally stratified.

In very large rivers, 'offshore' outfalls incorporating diffuser or 'pepper-pot' designs will cause the heated water to mix very rapidly, minimising both water-column and substrate effects (Jirka 1974; Eiler and Delfino 1975; Miller and Brighouse 1984). Eiler and Delfino (1975) found that maximum temperatures decreased from 6·9°C above ambient to a maximum of 1·2°C above ambient at a point 160 m downstream of the Quad Cities power station when the outfall was altered from a jet to a diffuser.

Under extreme conditions thermal plumes in rivers may move upstream (Fig. 2.11(c)). Such recirculation can cause effluent water to penetrate upstream of an intake as a result of upstream winds (Langford and Aston

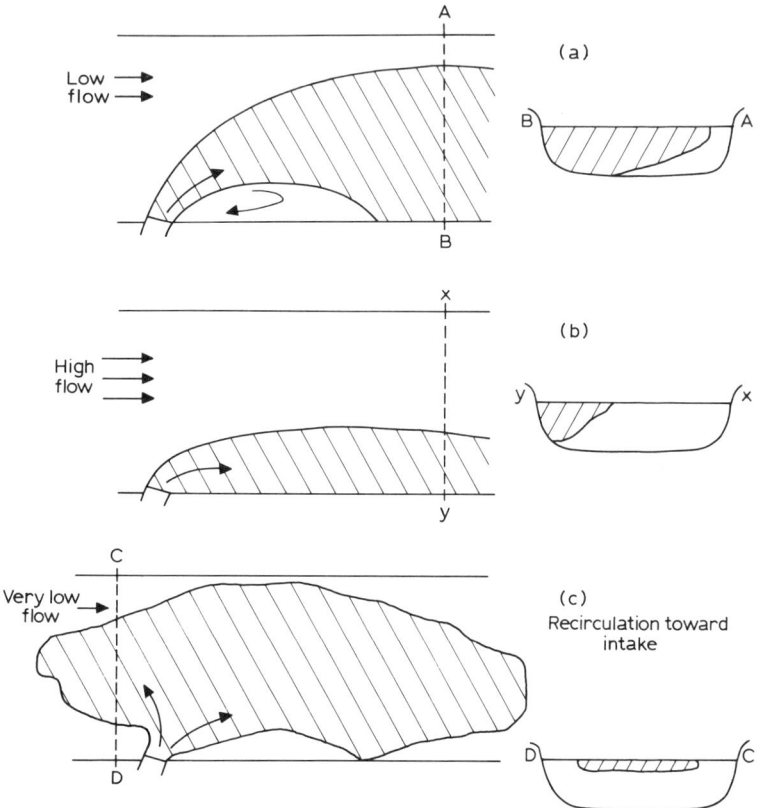

FIG. 2.11   Thermal plume patterns in a river at different rates of flow.

1972). In such instances, the warm water usually exists only as a thin surface layer. Thermal effects on substrates are maximised in shallow rivers where there is no vertical stratification and minimised in deep rivers where vertical stratification occurs. Thus where discharges occur near shallow riffle areas such as at the Martins Creek plant on the Delaware (Coutant 1962) or at Ironbridge power station on the River Severn (Langford 1970, 1971a), the surface substrates could be heated across wide areas of the river by up to 8 or 10°C above the natural ambients. In slower deeper reaches of rivers, vertical stratification restricts substrate effects to narrow marginal bands (Alabaster 1962; Mann 1965).

   The River Trent in the English Midlands is a special case in that it is used to cool some 12 000 MWe of generating plant between its source and the sea. It is probably the most heavily used river in the world for cooling water and heat disposal (Tinker 1971). Although tower cooling is used at most of the larger power stations the middle reaches of the river maintained a temperature 5 or 6°C above natural ambients during normal flows, for over 30 years (Fig. 2.12) (Trent River Authority 1964–74). The highest temperatures recorded in the Trent were 36·5–37°C downstream of Castle Donington power station in 1958 during the lowest flows for 200 years (Alabaster 1962). At the time towers were not used, for experimental reasons.

(b) In lakes and reservoirs. Natural water currents in lakes are mainly wind-generated and generally slow in relation to any effluent velocity. The depth of the discharge in relation to the natural stratification in a lake is also a significant factor determining the relative extent of water column and substrate effects.

The size, shape and direction of travel of any one plume in a lake on any day will vary mainly in relation to the wind (Burnett *et al.* 1974; Lewis 1974; Madding *et al.* 1975; Miller and Brighouse 1984). Both jet and layer discharges can be sharply stratified.

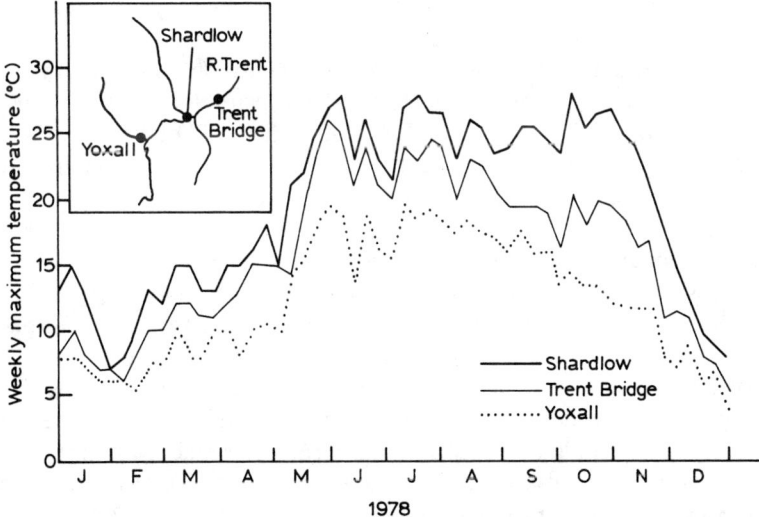

FIG. 2.12 Comparison of mean weekly water temperatures at three sites on the River Trent (UK, 1978). (Shardlow is below two major power stations using mixed cooling systems; after Lewin 1975.)

To minimise effects on summer temperatures in a lake or reservoir, cooling water may be abstracted from depths below the thermocline and discharged into the epilimnion. Thus for example intake water at 4–6°C could be warmed probably to 12–16°C and discharged to surface water which is already at 20–24°C depending upon latitude and climate.

Two potential problems arise. The hypolimnial water discharged to the epilimnion will usually be rich in nutrients or contaminants. Also, in smaller lakes there may be a breakdown of the normal stratification regime caused by a buoyant warm plume. Philbin and Philipp (1971) suggested, however, that in the Cayuga Lake in New York State (USA), the stratification pattern would be little altered by a hypolimnial abstraction and epilimnial discharge, though the lake was relatively large in relation to the volume used for the cooling water system.

(c) In estuaries. The prediction of the shape and behaviour of plumes in estuaries is complicated by tidal movements, highly variable salinity gradients, and heterogeneous current speeds and directions in addition to the other hydrographic and climatic conditions (Fischer *et al.* 1979; Miller and Brighouse 1984). In a simple, well-mixed estuary any thermal plume will stratify vertically and horizontally as in a river or lake with more or less predictable temperature decay with time and distance from the outfall. Plume shape in the short term will be again influenced by winds, tidal currents and by the outfall design and location.

In some estuaries and in special situations such as deep fjords, with freshwater inflows, discharges may be trapped between salinity strata. Thus, in the estuary of the Rivers Test and Itchen at the head of Southampton Water (UK), Raymont and Carrie (1964) found that the heated saline discharge from the Marchwood Power Station was sandwiched between the lower density fresh water on the surface and higher density cold saline water below. Similarly, heated water may be trapped at depth in Norwegian sill fjords which are strongly stratified by salinity and also have topographical restrictions to water exchange (Carstens 1975; Audunson *et al.* 1975).

In such situations where plumes do not reach the surface quickly, heat dispersal is only by conduction and convection and is very slow. The effects on substrates depend on the design of the outfall and its location. The depth of the intake in a highly stratified estuary can determine the salinity of the cooling water and this can also vary in the short term as a result of weather and variable mixing processes (Dyer 1973; Fischer *et al.* 1979).

(d) In the sea. The general behaviour of thermal plumes in the sea is predictably similar to that in lakes except that tidal and residual drift

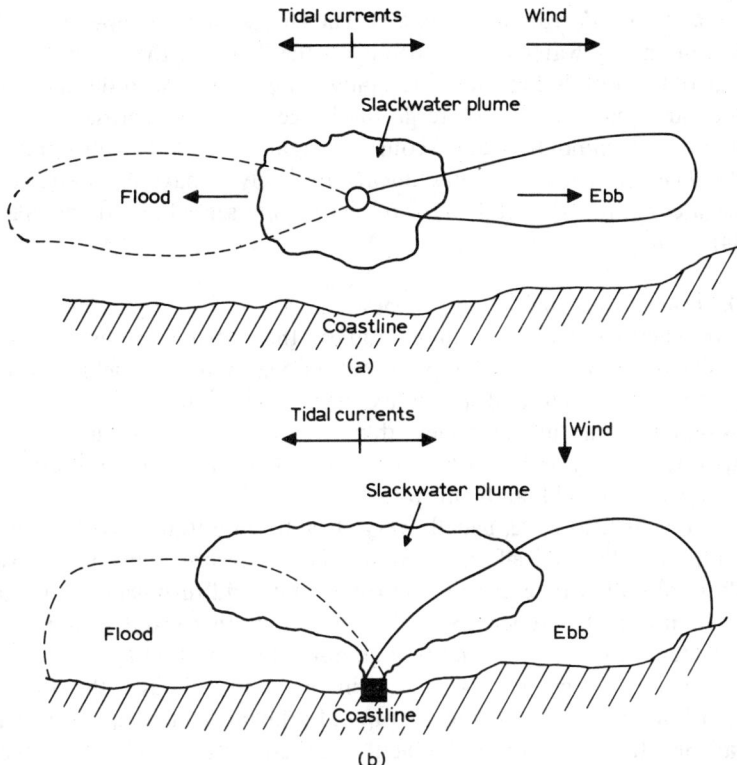

FIG. 2.13 Movements of thermal plumes in tidal waters. (a) Offshore outfall;
(b) onshore outfall.

currents may cause regular and highly predictable changes in plume
directions (Spurr and Scriven 1975) (Fig. 2.13). Wind direction and distance
offshore are further factors and determine the extent to which offshore
plumes will impinge on the shoreline as in lakes. Onshore discharges
usually inundate the shoreline as the tide floods. Stratification of the
thermal plume occurs much as in any other water but may be more readily
destroyed on open coasts by turbulent sea conditions. Around some
outfalls the buoyant layer of warm water suppresses waves in the same
manner as other buoyant substances such as oil.

The peculiarity of desalination effluents is that, although at higher
temperatures, they are negatively buoyant because of the higher salinity,
some 1·5–2·0 times that of the intake seawater (see Barnett and Hardy
1984). Temperatures of 43°C have been recorded.

(e) Effects on substrate temperatures. The effects on the temperature of a substrate in any water body depends on the depth of the water and the design of the outfall (Fig. 2.14). In shallow waters with a layer discharge the effects are maximised. Where a discharge is made across a shore, temperature, inundation and scour are significant factors affecting the biota. The heavy desalination effluents are likely to have more effects on substrates because they sink in the receiving waters (see Barnett and Hardy 1984).

2.2.8 Temperature and Mixing Zones

For predictions and modelling studies, the thermal plume is often considered as having three component effects viz. in the near-field, mid-field and far-field (Miller and Brighouse 1984). The whole plume can be considered as a 'mixing-zone', that is an area, or volume, where temperature decay and heat loss occur and there is a gradual dilution of heated water by cold 'ambient' water.

In the near-field zone, usually very close to an outfall, mixing may be minimal and thermal effects maximal. This may not apply to exposed offshore outfalls where fast tidal currents and wind turbulence can cause both mixing and heterogeneity. In the latter situation 'packets' of warm and cold water are in close contact though not truly mixed. In the mid-field a plume may still be a cohesive, definable entity though the rate of temperature decay and mixing are rapid. In the far-field, temperatures are usually less than 0·5°C above ambient and there are areas of cold, ambient water intermingled with discrete patches of the slightly warmer effluent water. These patches of residual heat may be found many miles from the original outfall.

The concept of the 'mixing zone' has been used mainly in establishing legislation, and limits on the size of the zone in relation to the receiving water have been set (Mount 1971; Majewski and Miller 1979; Neilsen and Rasmussen 1982; Miller and Brighouse 1984). As far as this volume is concerned the thermal 'plume' is more or less synonymous with the 'near' and 'mid-field' zones (see Chapter 1). The far-field effects on ecosystems have not been studied in great detail and in most waters may be considered of little ecological importance. The spatial limits of a 'mixing zone' or 'plume-area' tend to be site-specific though the size and extent of any one mixing zone is dependent on many factors including the size of the discharge in relation to the receiving water (i.e. dilution), hydrography, weather, operating conditions and outfall design and location (Fig. 2.15). The boundary of a mixing zone is defined usually in relation to the reason

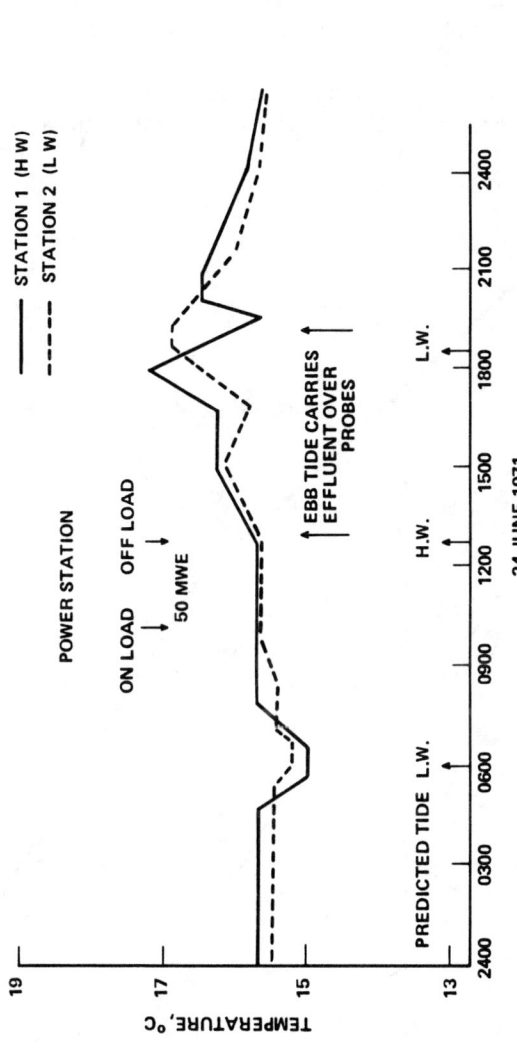

FIG. 2.14   Littoral soil substrate temperatures in the path of a tidally directed thermal discharge (from Spencer 1975).

for its definition, but for ecological purposes, volumes of water at more than 0·5°C above ambient can be regarded as within a mixing zone or plume area. The limit is, even so, more or less arbitrary as a δT of 0·5°C is probably of little ecological significance except in extreme tropical conditions.

The extent and areas of mixing zones are highly variable even at any one site. For example for a 5000 MWe nuclear power station the area of water

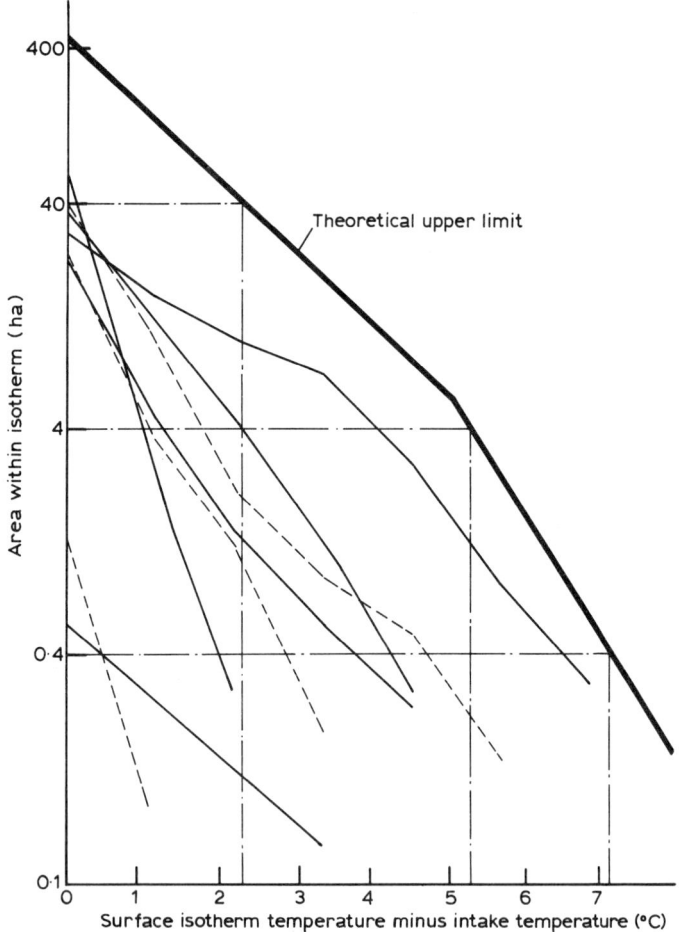

FIG. 2.15   Relationship between surface-temperature elevation and surface area affected for nine different surveys at Moro Bay Power Plant, California.

heated by 1°C or more could vary from 32 km² at an open sea coast site to 100 km² in a poorly dispersing location (IUPDEE 1979). Adams (1969) showed that for 23 power stations on the California coast, with a total capacity of 16 982 MWe, the computed mean plume area at 1·2°C above natural ambient was 1500 ha. Some 105 ha were above 5·5°C. Figure 2.16 shows the areas of water raised above 2°F (1–2°C) and 10°F (6°C) against the amount of power generated, plotted using data from US and UK power stations. Direct-cooled discharges, as might be expected, generally produce much greater plume areas than indirect or tower-cooled discharges (see also Miller and Brighouse 1984).

In lakes and tidal waters plume stratification may produce large areas of warm water but the effects on the substrate may be restricted to very small areas near outfalls. In tidal waters the temperature fluctuations in a habitat induced by the tidal movements of the plume or incursions of cold water into a discharge channel will encompass the full range from natural ambient to discharge temperature twice each day (Spencer 1975; Bamber and Spencer 1984). Temperature decay is highly variable both spatially and temporally (Fig. 2.7), again related to the various environmental and design factors.

2.3 OTHER CAUSES OF THERMAL CHANGE IN WATER BODIES

In the Chapter 1, other sources of waste heat discharges to natural waters were noted, such as sewage, agricultural drainage, urban run-off and other industries. In total these produced much less heat than thermal power station sources and temperature rises from the first three are rarely more than 2 or 3°C in the areas near outfall. It is claimed that urbanisation has increased average summer temperatures in some Long Island (US) streams by 5–8°C (Pluhowski 1970, loc. cit. Smith 1972), though winter mean values were 1·5–3°C lower than in naturally, rural area streams. Short-wave radiation and ground water flow changes were believed to be the causes of the differences.

Some localised temperature changes in small streams may be caused by thermal discharges from air conditioning and refrigeration plant discharges near building complexes but though discharge temperatures may be high (>40°C) volumes are usually very small (Brock 1975, 1978).

In the case of agricultural drainage or run-off the sources are diffuse because the points of discharge are generally small and numerous. Factory

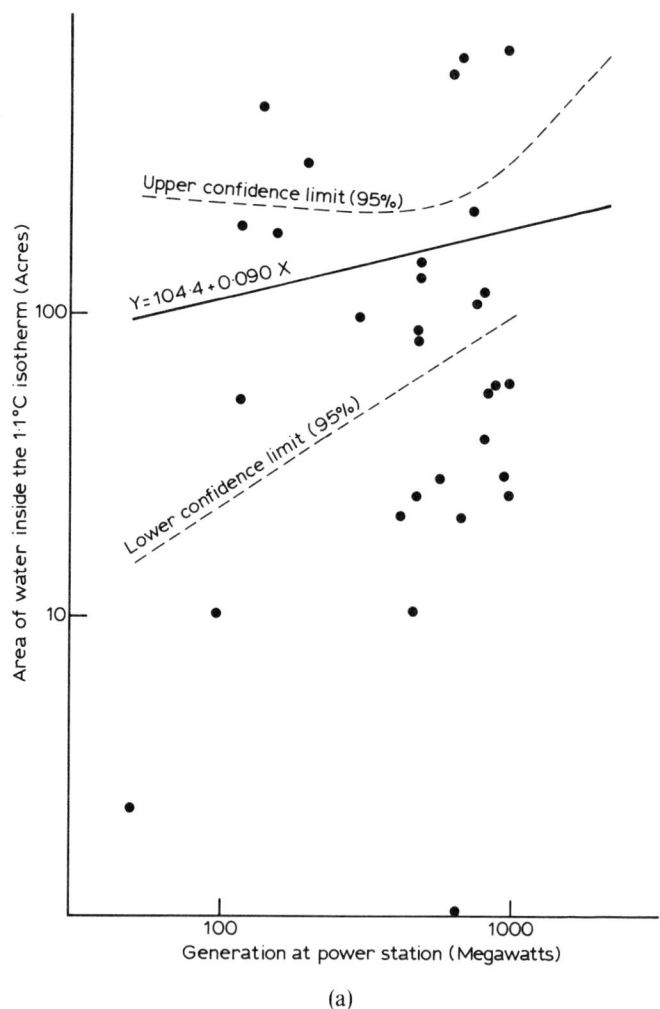

(a)

FIG. 2.16   Areas of water heated above 2°F and 10°F by thermal discharges in the US and UK plotted against maximum output of power station. (a) Areas above 2°F (1·1°C); (b) areas above 10°F (5·5°C) (after Adams, 1969). (The equations are worked in acres and °F. To convert to hectares divide by 2.47.)

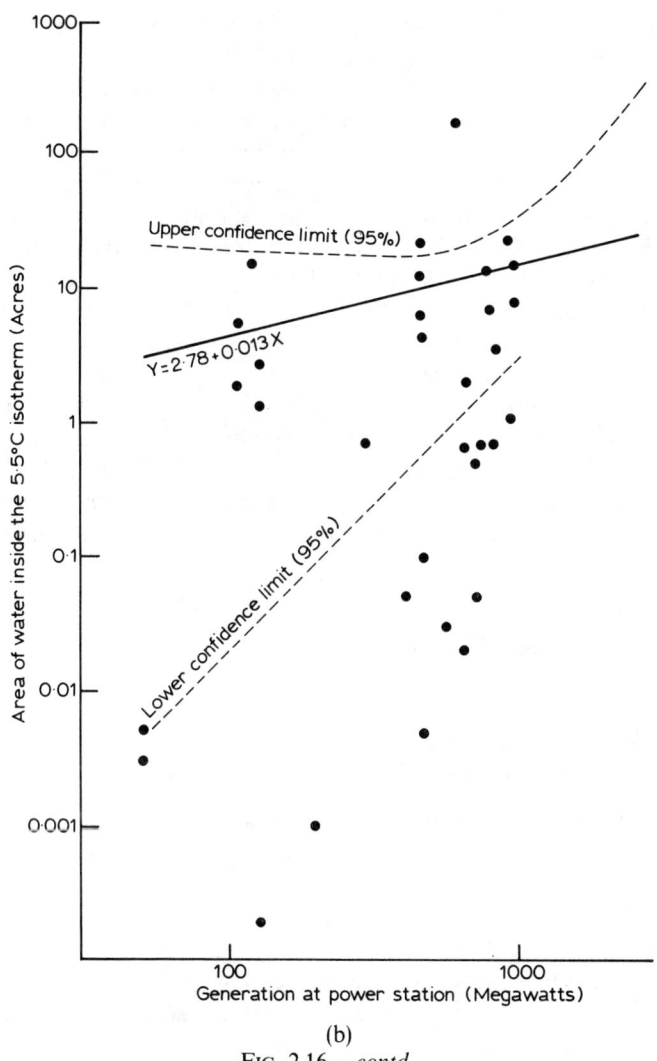

(b)

FIG. 2.16.—*contd.*

effluents, as we have seen (p. 47), may be at much higher temperatures but in most cases the chemical properties of these discharges are more significant to the ecology of the habitat than the heat, and the temperature effects are either not discernible from other polluting effects or are completely dominated by them (e.g. Mesarovic 1975; Edwards *et al.* 1984).

Thermal regimes of rivers and lakes may be totally altered by control, transfer or impoundment schemes (e.g. Major and Mignell 1966; Neel 1963; Obeng 1969; Trefethen 1972; Whitton 1975, 1984; Martin and Arneson 1978; Ward and Stanford 1979a,b; Cowx *et al.* 1987). Discharges of hypolimnial water from dams can change the temperature regime downstream by raising temperatures in winter and lowering them in summer. The extent of alteration in the downstream temperature regime depends upon the design of the dam and its use, i.e. hydro-electricity, navigation, water supply. In the case of hydro-electricity dams, short-term changes of 6–8°C may occur at irregular intervals several times each day (Pfitzer 1967, loc. cit. Ward and Stanford 1979b). Even though the diurnal and seasonal regimes may be extensively altered, the mean annual temperature may not be greatly modified.

Thermal alterations often lead to, or are associated with chemical changes which in turn can affect downstream reaches adversely (see Oglesby *et al.* 1972; Ward and Stanford 1979a; Langford 1983a).

Ward and Stanford (1979b) indicate six categories of modification to downstream temperature regimes by dams, which could be biologically significant:

—increased diurnal stability,
—increased seasonal stability,
—summer depression,
—summer elevation,
—winter elevation,
—annual pattern changes.

These categories do not include 'diurnal instability' which may be induced by the intermittent operation of a hydro-electric power station meeting peak demands only (see Langford 1983a). Although the alterations in the temperature regime may be ecologically important (Lemkuhl 1972; Ward 1974; Ward and Stanford 1979a,b; Cowx *et al.* 1987) perhaps the most significant factors, particularly near the dam outlets, are the widely fluctuating current velocities and water levels (see Ackermann *et al.* 1973; Trotzky and Gregory 1974; Ward and Stanford 1979a,b; Langford 1983a).

In upland hill streams de-forestation has led to measurable increases in maximum water temperature of up to 8·8°C as a result of direct insolation (Brown and Krygier 1967; Brown 1970; Smith 1972; Ringler and Hall 1975). The net radiation on exposed streams is up to five times greater than in shaded streams and is the main cause of temperature increase (Brown and Brazier 1972). Although all of these factors cause temperature changes in natural waters, they do not fall within the category of 'thermal discharges' defined in Chapter 1 and as such will only be considered in the following account for comparisons. Further information is to be found in Langford (1983a).

2.4 EFFECTS OF DISCHARGE VELOCITY

The water velocity changes within a cooling water system are shown in Fig. 2.6. The velocity with which a discharge enters a receiving water is related to its volume, the cross-sectional area of the outfall pipe and the hydrostatic head (Fischer *et al.* 1979). Discharge velocities of thermal effluents generally range from about 0·5 to 2·0 m s^{-1} (Miller and Brighouse 1984). The detailed mechanics of discharge velocities are probably not of ecological significance but two major effects of discharge velocity may be important.

First, the discharge velocity may be such that where the currents impinge on the bed, soft substrates will be removed or 'scoured'. Second, the 'jet' will cause water currents in the vicinity of the outfall often moving in different directions to the natural currents and also at different velocities. Turbulence usually arises from the contra-flows (see Fig. 2.11).

(i) Sediments and scour
The composition of the benthic flora and fauna in any aquatic habitat is dependent on many factors but prominent among these is the nature, stability and composition of the sediments in the substratum. The composition of the substrate sediments is, in turn, determined by the action of the water, either as currents in rivers, lakes and tidal waters, or as waves in lakes and coastal waters. In general, the faster the current or greater the wave action, the coarser are the bottom sediments (Hutchinson 1957; Macan 1963; Sverdrup *et al.* 1963; Newell 1970). Hynes (1970) discusses the sediments in rivers and concludes that in general terms the mean particle size of sediments decreases with distance downstream of the source and 'there is thus a general correlation' 'between particle size and slope'.

It is evident, therefore, that a discharge of heated effluent across a shore or in shallow water may alter the sediments in its path. Coughlan (1969) noted that the thermal discharge from Pembroke power station across a shore had removed the fine sediments and exposed the underlying clay and rock. Whitehouse and Key (1978) and Bamber and Henderson (1981) also noted that the sediments had changed since construction in the vicinity of the Bradwell power station outfall but whether this was the direct effect of the discharge velocity or of the redirection and channelling of tidal currents

TABLE 2.2
The mean current velocity of clear and muddy water
required to initiate movement along a stream bed of
various types of bottom deposit (from Hynes 1970)

Type of bed	Critical mean current velocity $(cm\,s^{-1})$	
	Clear water	Muddy water
Fine-grained clay	30	50
Sandy clay	30	50
Hard clay	60	100
Fine sand	20	30
Coarse sand	30–50	45–70
Fine gravel	60	80
Medium gravel	60–80	80–100
Coarse gravel	100–140	140–190
Angular stones	170	180

was unknown. As indications of effects of currents, Tables 2.2 and 2.3 show
the velocities required to move substrate particles. Even the slower
discharge velocities given for thermal discharges would move medium sized
particles. Outfalls designed to discharge effluent at or near the water surface
in deeper waters would not be expected to scour sediments though some
disturbance of sediments can occur during the construction phase or as a
result of changed currents around a new structure (Roessler 1971; Roessler

TABLE 2.3
The speeds of flow required to move mineral
particles of different sizes (from Hynes 1970)

Speed of flow $(cm\,s^{-1})$	Diameter of mineral particles moved (cm)
10	0·2
25	1·3
50	5·0
75	11·0
100	20·0
150	45·0
200	80·0
300	180·0

et al. 1979). Quantified effects of thermal discharges on sediments are scarce in the literature (Bamber and Spencer 1984; Bamber 1985).

(ii) Water currents

The degree of water turbulence created by a discharge is a function of its design and the size and direction of the discharge in relation to currents in the receiving water (Mangarella and Van Dusen 1973; Fischer *et al.* 1979; Miller and Brighouse 1984). In the vicinity of any outfall, local eddies and gyres may be created as the water bodies collide and mix. The degree of water turbulence is ecologically important because of its effects on the dissolved gases in the water and because of effects on the orientation and direction-finding of mobile aquatic animals (e.g. Harden-Jones 1968; Arnold 1974).

In a natural lake, water currents are generally weak except at river inflows. The creation of a strong current from a jet or layer discharge which mimics a river inflow could be highly significant to mobile animals such as fish which migrate into rivers to spawn and are unable to distinguish outfall currents from natural river currents. In coastal waters such currents could also potentially disorientate migratory fish for the same reasons. In rivers with unidirectional flow, water currents which partially counter or discharge at right angles to the natural flow could similarly cause confusion to migratory fishes (see Chapter 9).

CHAPTER 3

Effects of Thermal Discharges on Water Chemistry

3.1 FACTORS INFLUENCING THE CHEMISTRY OF THERMAL DISCHARGES

The chemical composition of any thermal effluent is a function of:

—the composition of the intake water,
—the type and design of the cooling system,
—the water treatment chemicals used, and
—the other site effluents discharged into the system.

The effects of the total discharge on the chemistry of the receiving water will, in turn, depend on the initial concentrations of the constituents and their rates of dilution, dispersal or decay and the chemical changes caused by physical factors such as turbulence or heat.

In dealing with the chemistry of cooling water it is difficult to select those constituents which are most likely to have significant effects on the flora and fauna. First, there are at least 130 chemicals which can be associated with cooling-water systems (Becker and Thatcher 1973) apart from those in the receiving water and intake water. Second, the relationships between the detailed chemical composition of natural waters and their associated biocoenoces are still not well known after many years of research. Third, there are few reliable data on the acute or chronic effects of many of the chemicals used for the treatment of cooling water on animals and plants and fourth, many water treatment preparations are proprietary and for commercial reasons their exact composition may not always be published.

There are already many reviews of the effects of contaminants such as chlorine, heavy metals, oil and its dispersants, and radioactivity on aquatic organisms, and key references will be given in the relevant sections. Here we are more concerned with the specific effects of higher temperatures and

66

other physical properties of thermal discharges *superimposed* on the effects of these contaminants and on the effects of changes in other chemical factors such as oxygen and other dissolved gases.

3.1.1 Composition of the Intake Water

Most of the cooling water used in thermal power stations is abstracted from larger bodies of surface water. Only small industries or air-conditioning plant in isolated areas utilise smaller streams, ponds or boreholes for their cooling water (Brock and Yoder 1971).

Natural surface waters vary enormously in their chemical composition in relation to geography, geology and atmospheric contributions. Welch (1952) concluded that the solute properties of water and the hydrological characteristics of catchments 'set the stage for one of nature's greatest displays of chemical diversity and chemical dynamics'. While being somewhat literary, the description aptly illustrates the problem of summarising the chemistry of surface waters in a limited space.

Accounts of the varied chemistry of surface waters in relation to their ecology are in Welch (1952), Hutchinson (1957), Ruttner (1963), Leopold *et al.* (1964), Owens and Edwards (1964), Hynes (1970), Oglesby *et al.* (1972), (fresh waters), Perkins (1974), Sverdrup *et al.* (1963), and Kinne (1970a,b) (estuaries and seas). Chemical descriptions of impounded and regulated fresh waters can be found in Hall (1971), Ackerman *et al.* (1973), Ward and Stanford (1979a), Whitton (1984) and Davies and Walker (1986).

In addition to these accounts there are vast quantities of water quality data in the published and unpublished records of pollution-control authorities and industries throughout the developed countries of the world.

Of all the surface waters, the open sea is generally regarded as the most stable chemically. However, the nearshore waters with which we are mainly concerned here, vary in their chemical composition as a result of pollution and the entry of rivers and streams.

Chemical variations in waters occur within even small geographic regions. Also, within any one body of water there are both spatial and temporal fluctuations which will influence the chemical composition of water used for cooling. Perhaps the best known spatial heterogeneity is caused by the phenomenon of 'stratification' in lentic waters (see Chapter 2). Chemical stratification tends to occur in parallel with physical or thermal stratification (Hutchinson 1957; Sverdrup *et al.* 1963). Fast-running waters generally show spatial heterogeneity only where tributaries or effluents of different water qualities enter.

Temporal variations in water chemistry also occur as a result of regular phenomena such as the well-known diurnal fluctuations of oxygen and other gaseous concentrations (see Welch 1952; Gameson and Truesdale 1959; Hutchinson 1957; Sverdrup *et al.* 1963; Owens and Edwards 1964; Owens *et al.* 1969; Perkins 1974), or irregular phenomena such as spates in rivers or intermittent pollution (Klein 1962; Leopold *et al.* 1964; Casey 1977; Casey and Farr 1982; Walling and Foster 1975).

Estuaries are perhaps the most chemically heterogeneous waters, both spatially and temporally, because of the regular tidal interchange of fresh and salt water plus the effects of climatic conditions and mixing processes (Dyer 1973; Fischer *et al.* 1979). In the open sea diurnal oxygen fluctuations can occur in the upper surface layers and very marked thermal and chemical stratification is a characteristic of most oceans. Seasonal chemical changes occur in most surface waters as a result of biological activities such as phytoplankton blooms. Many of the surface waters used for cooling are subject to pollution from many different sources and chemicals (Klein 1962; Perkins 1974; Gasparini 1982).

At some power stations and factories, treated sewage effluent (Humphris 1977), the supernatant effluent from ash-settling ponds (van Eeden 1975; Dreesen *et al.* 1977; Flook 1978; Chu and Olem 1980, 1982), or other recycled effluents are used to augment cooling water supplies (Trace Metal Data Institute 1981; Rebhun and Engel 1988) though high concentrations of nutrients and minerals can cause operational problems in the cooling system.

The initial composition of the intake water is, therefore, subject to considerable variation in both time and space. To predict the effects of a cooling system on this water requires a detailed knowledge of the patterns of variability.

3.1.2 Cooling System Design and Siting

Because of the temporal and spatial variations in the chemistry of surface water, the locations of water intakes and outfalls can be important factors determining the chemistry of cooling-water effluents and their consequent effects on a receiving water.

For example, in stratified waters, intakes sited in deep, nutrient-rich and de-oxygenated hypolimnial water produce heated, nutrient-rich effluents, which have been aerated in transit through the cooling system. Such effluents can have marked effects on the relatively nutrient-poor surface receiving waters.

An intake may also be sited on a body of water which is chemically-different to that on which the respective outfall is sited. For example, at

Fawley power station, on the south-coast of England, cooling water is abstracted from the relatively eutrophic Southampton Water and discharged to the less eutrophic Solent (Langford 1983a). On many days the effluent plume is readily visible in the receiving water because of its different colour.

Within any cooling-water system the physical processes cause chemical changes. For example in both direct and indirect systems, pressure changes of 1–2 atm through the pumps or syphons (see Fig. 2.6) alter the solubility of gases in the water, while turbulence in the system facilitates the exchange of gases between the air and water. A temperature rise of 8–10°C through the system also has effects on gas-solubility in that most common gases are less soluble at 30°C than at lower temperatures. For example, oxygen is about half as soluble at 30°C as at 5°C, while carbon dioxide is 2·5 times less soluble (see Houston 1982). A temperature rise also increases the rates of chemical and biochemical reactions (see Rose 1967; Kinne 1970a).

The differences between the effects of direct and indirect cooling systems on the chemical changes in cooling water are mainly related to aeration caused by passage through the towers, the retention time in tower ponds and concentration factors caused by evaporation and make-up proportions (see p. 24). Desalination effluents are highly saline but also contain heavy metal ions often at concentrations toxic to organisms (see Barnett and Hardy 1984).

3.1.3 Cooling-Water Treatments and Additives

The three major processes which can impair the efficiency of a cooling-water system are 'scaling', 'fouling' and 'corrosion' all of which require some form of chemical control. Timber packing in cooling towers is usually treated to prevent damage by fungal or bacterial attack (Hamer *et al.* 1961; Gurney and Cotter 1966; Cheremisinoff and Cheremisinoff 1983).

Scale formation occurs in pipes and culverts and on heat-exchange surfaces and most commonly consists of calcium carbonate (Capper 1974) or where detergents are present, calcium phosphate (Rippon 1979). Treatment is usually by concentration control or by pH control, that is, by dosing with acids. Chemical methods of removing phosphates have been used at some power stations where sewage effluents are used to augment cooling water supplies (Humphris 1977). In the UK the normal concentration factors in cooling-tower systems are between 1·3 and 2·5.

The control of pH by dosing the cooling water with sulphuric acid or in some instances hydrochloric acid, nitric acid, sulphur dioxide or chlorine (McKelvey and Brooke 1959; White 1972), prevents the precipitation of calcium salts by keeping them in solution. However at pH values lower than 7, metal surfaces may be corroded and to counteract this corrosion inhibitors are also introduced to the system. The earliest of these were mixtures of polyphosphates and chromates, or polyphosphates and persulphates which produced a thin protective film on the metal surfaces through a synergistic reaction.

Corrosion may be exacerbated or initiated by physical, chemical or biological action (Gurney and Cotter 1966; Acker *et al.* 1972; Rippon 1979). Sediments in the cooling water also erode metal surfaces allowing penetration by salt water or acidic fresh water. Even in alkaline waters, a break in the surface film of the metal allows electro-chemical action. Corrosion inhibitors have thus included a wide range of substances in addition to those mentioned above, including ferricyanides, nitrates, copper salts, chlorates, permanganates ferrous sulphate and, more recently, less toxic zinc compounds such as heterocyclic amino compounds and heptonates (Gurney and Cotter 1966; Acker *et al.* 1972; Kemp *et al.* 1973; Capper 1974; Frank 1974; Cheremisinoff and Cheremisinoff 1983). Sequestering agents and dispersants incorporated into the organo-zinc compounds maintain the zinc in solution at pH values higher than 7·0 removing the need for acid dosing. The methods of controlling fouling, scaling and corrosion may be mutually effective or antagonistic, depending on the chemistry of the cooling-water supply.

Of all the treatments, those most important to the ecological effects of cooling-water discharges are those associated with the removal and prevention of biological fouling and for this reason they are described in a separate section, as follows.

3.1.4 Biological Fouling and Control
(i) Causes of biological fouling
Although fouling is generally regarded as caused by living organisms growing and settling on surfaces (see Chapters 5, 6 and 8), many fouling deposits actually contain both organic and inorganic constituents. In practice, fouling, scaling and corrosion are very much interrelated in that living organisms produce gases which cause corrosion and the organisms themselves can gain their foothold initially by attaching to scale deposits or in the back-eddy currents caused by such deposits. In power stations both the main and auxiliary cooling circuits in the generation plant can become fouled and require treatment (Burton and Liden 1978; Chow and Kawaratani 1983; Langford 1983a; Whitehouse *et al.* 1984; Mattice 1985; Cumbie *et al.* 1985; CEGB Research 1990). Fouling occurs in both

fresh-water and salt-water cooling systems but marine fouling is the most common and best known throughout the world (Woods Hole Oceanographic Institute 1952; Ray 1959; Relini and Oliva 1972; Osborne and Lum 1978; Whitehouse *et al.* 1984). Table 3.1 shows the main fouling organisms in the US power plants.

The two most important areas for anti-fouling treatment are the heat-exchange surfaces in condensers and the surfaces within the culverts or tower circuits, where organisms can cause blockages or reduce the efficiency of heat transfer. In the condensers, bacteria, fungi or algae form organic films which trap inorganic materials to form slimes (Wood 1967). In cooling-tower circuits algae, usually blue-green (Cyanophycae) and green algae (Chlorophycae) can grow in masses on the wooden structures, at times even causing serious collapses.

In fresh waters culverts can be fouled by molluscs such as the freshwater clam (*Corbicula* sp.) in the USA (Goss and Cain 1977) or in some European waters the zebra mussel (*Dreissenia polymorpha*) (Macan 1974). In marine culverts and auxiliary cooling circuits different species of molluscs and other invertebrates can cause fouling problems but the most troublesome organism globally is the blue mussel (*Mytilus* spp.) (Beauchamp 1969b;

TABLE 3.1

Fouling organisms reported at operating power plants in the USA (from Mattice 1985)

Organism	Units (*no.*)	%
Microfouling		
Slime/algae	105	30
Slime	32	9
Algae	85	24
Total	222	63
Macrofouling		
Hydroids	25	7
Barnacles	11	3
Mussels	56	16
Clams[a]	34	10
Oysters/algae/barnacles	7	2
Total	133	37

[a] All reports from freshwater sites are assumed to be the Asiatic clam, *Corbicula fluminea.*

Jenner 1980; Khalanski and Bordet 1980; Langford 1983a; Whitehouse *et al.* 1984; CEGB Research 1990).

At thirteen of twenty-nine power stations studied in the United States, microbial slimes in condensers were the most common and important problems, though at eight sites macro-invertebrates caused serious blockages in culverts (Chow and Kawaratani 1983). The sequence of fouling colonisation on marine surfaces is described by Holmes (1970a), and the tolerances of various organisms in cooling systems are discussed in subsequent chapters of this book.

(ii) Anti-fouling treatments
By far the simplest and most widely used treatments are those using active chemicals, such as copper sulphate, mercury compounds, permanganates, phenols or oxidants for controlling slimes and algae (e.g. Lucu *et al.* 1980). For macro-invertebrates the most favoured chemicals are those based on chlorine compounds (Lamb 1972; White 1972; Whitehouse 1975, 1978; Hamilton 1978; Fava *et al.* 1985), although other oxidants such as bromine have been used (Draley 1977; see Jensen 1977; Burton and Liden 1978; Waite *et al.* 1978; Chow and Karawatani 1983; Whitehouse *et al.* 1984) (Table 3.2).

Chlorine is highly toxic, acts quickly and is relatively inexpensive (Khalanski and Bordet 1980, 1981). It is effective both for fresh-water and marine cooling-water systems and for micro- and macro-fouling but, like all other treatment chemicals, is discharged into the receiving water from the system. Because of its widespread use and its toxicity, chlorine and its derivatives are vital constituents of many thermal discharges and must be primary considerations when assessing the biological effects of such discharges.

Heat treatment as an anti-fouling procedure involves the application of unusually high water temperatures to intake culverts by reversing the cooling-water flow from the condenser outlets to the intakes. The procedure was first applied to fouling in culverts as Scottish power stations in the 1920s (Ritchie 1927). Temperatures of around 40°C are usually required to kill fouling organisms and the method is expensive and leads to intermittent discharges of effluents at higher temperatures than is usual in thermal discharges. The opposition to toxic anti-foulants has, however, led to its use at some power stations in recent years (Stock and Strachan 1977; Jenner 1982–1983), though it is expensive.

Radioactive isotopes were used to try to control the invasions of asiatic clams (*Corbicula* spp.) in the USA but disposal of the treated water was a major obstacle to the use of this method (Goss and Cain 1977).

TABLE 3.2

Alternative chlorine antifouling agents and fouling control techniques (from Chow and Kawaratani 1983)

Chemical anti-foulants	Nonchemical fouling control techniques	Detoxification of chemical anti-foulants
Oxidizing agents	Thermal	Chemical and
Bromine	Single- and	physical methods
Bromine chloride	double-pass	activated carbon
Chlorine dioxide	condenser	Aeration
Hydrogen peroxide	operations	Sodium sulphates
Iodine	(flow reversal)	and sulphites
Ozone	Heat soak	Sulphur dioxide
Potassium	Non-thermal energy	
permanganate	Gamma irradiation	
Nonoxidizing agents	Ultraviolet radiation	
Acrolein	Ultrasonic vibration	
Arsenates and	Hydraulic	
arsenites	Velocity variation	
Ammonia and	Water jet cleaning	
amines	Mechanical	
Cyano compounds	Manual cleaning	
Metals (salts)	Amertap	
Organometals	American MAN	
Phenols-chlorinated	Miscellaneous	
and phenylated	Osmotic shock	
Proprietary	Anoxic water	
formulations		
Miscellaneous		
Condenser		
biocide soak		
Anti-fouling		
coatings		

Fouling organisms can be killed by creating stagnant conditions in culverts, which involves filling them with water and leaving them unused for long periods. The fouling animals die through lack of oxygen. The stagnant water and the shells of dead molluscs still need a means of disposal and the method is expensive and not always effective. Surface paints or sacrificial surfaces can be effective on a small scale but are not widely used in large cooling-water circuits such as those at power stations.

Other chemical additives in cooling circuits include fungicides for inhibiting the decay of timber (Franco 1980), and flocculants for settling-out sediments or organic particles before they can deposit on the heat-exchange surfaces.

3.1.5 Other Contaminants

The previous section dealt with chemical contaminants deliberately introduced into cooling-water systems for some operational purpose. At any large industrial site, sewage, boiler-cleaning residues, ash-effluents, fuel-store drainage, flue-gas washings, radioactive and other wastes may be discharged separately from, or diluted in, the cooling water and, although small in volume, can introduce contaminants which may exacerbate the biological effects of the heat in the discharge (Zimmerman *et al.* 1974; see Iwanski and Chu 1985).

In many regions the majority of toxic and noxious contaminants are removed before discharge to meet legal constraints on effluents.

(i) Sewage

Once the construction phase is over, the sewage discharge from an installation employing 500 staff would be up to about $20-30 \times 10^3$ gallons $(44-66 \times 10^3$ litres) day^{-1}, i.e. approximately 25-30 gallons (50-60 litres) per worker/shift. If discharged in the cooling water of a 2000 MWe power station, the dilution factor would be of the order of 15 000 to 1. The sewage is usually treated and the effects on the chemistry of the effluent would be predictably very small.

(ii) Boiler wastes

The water used for producing the steam in large boilers is extremely pure with some 99·998% of minerals and gases removed to prevent scaling and corrosion. Treatment processes include heating, distillation or cation–anion exchange by strong acids or alkalis. The ion-exchange resins are regenerated at intervals by concentrated sodium chloride or sulphuric acid. The wastes and used regeneration solutions can be discharged separately or in the cooling water.

Boiler 'blowdown' removes any materials formed in the boiler during its operation, including scale, sediments, minerals and corrosion products such as metal ions, and these also may be discharged in the cooling water.

Boiler cleaning wastes arise from the descaling of boilers with weak acids. The resulting solutions contain high concentrations of heavy metal salts, minerals and have a low pH. Such effluents are often neutralised before discharge and metal ions may be extracted.

(iii) Fuel-store drainage and oil

The drainage from huge heaps of coal at a large power station can contain suspended solids, sulphuric acid and other acids, sodium chloride, alkalis and heavy metal ions (McGuire 1977; Nuclear Regulatory Commission 1977; WPCD 1982), with the specific composition depending upon the composition of the coal. Liquid effluent originates from rain or snow and from the natural water in the coal and is usually discharged into surface drainage separately from the cooling water though often near the outfall. There are over 40 known constituents of the leachate water from coal piles which may have some effects on receiving waters (e.g. Birge 1978).

Oil is a common constituent of industrial site drainage and the ecological effects have been well studied (e.g. Baker 1977). At most large sites oil-traps are used to remove oil from drainage water before discharge.

(iv) Flue-gas washings and desulphurisation effluents
In recent years the effects of the emission of gaseous oxides such as sulphur dioxide and nitrogen oxides from chimneys have been considered to have contributed to the acidity of rainfall and, through various mechanisms, to acidification of fresh waters in various parts of the world and there is a vast literature available (e.g. Overrein *et al.* 1982; Martin 1986; CEGB Research 1987).

Because of uncertainty about the role of power station and large industrial burners in the formation of acid precipitation and in damage to trees and crops, it has been considered necessary in several countries to install gas washing, low-nitrogen oxide burners or desulphurisation equipment to power stations to reduce emissions. Once fitted, the wastes from the desulphurisation processes need to be disposed of and this has serious implications for local surface waters. The chemistry and high temperatures of the effluents from some desulphurisation processes are well known and experience in the 1940s and 1950s in the UK highlighted the potential pollution problems (DSIR 1964; Bettelheim *et al.* 1981).

There are a number of different processes for desulphurisation (Elder and Hollingden 1976; Chu and Olem 1980, 1982; Kyte 1986) though the amounts of aqueous effluents vary with the process. In the solid treatments the disposal of wet sludge or the transport of solids using slurry and settling ponds can lead to major effluent disposal problems (Burnett and Fedyko 1978; Dvorak and Lewis 1978; Woodyard and Sonning 1978). The main constituents of aqueous effluents are suspended solids and heavy metal ions. The pH values are usually between 3 and 5. An FGD effluent from a 2000 MWe power station using an average English coal would produce an effluent with a pH of around 3·0 (Fig. 3.1) which when diluted in an estuarine cooling water discharge would result in an effluent of about 1300 mg day^{-1} (5000 million litres day^{-1}) at a pH of 5·6. Such pH values,

Ecological Effects of Thermal Discharges

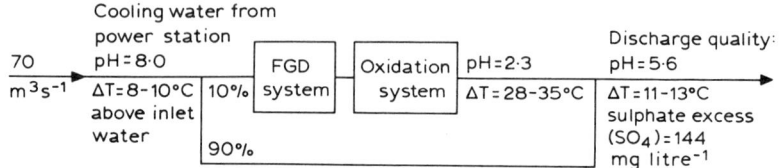

FIG. 3.1   Schematic diagram of a basic sea-water washing FGD system showing water quality through the system.

if not alleviated, could cause mortalities or physiological damage to some marine and estuarine organisms (Knutzen 1981; Bamber 1986, 1987b).

Flue-gas washing at two London power stations produced effluents containing sulphites, which although partially oxidised before discharge still caused serious oxygen depletion in the receiving water totalling about $4 \cdot 2 \, t \, day^{-1}$ (DSIR 1964). In experiments on the use of dechlorinated cooling water for aquaculture, Emberton and Turnpenny (1981) showed that sulphur dioxide, which is one of the main constituents of flue gases, bubbled into sea water caused significant reductions in the pH while neutralising the toxic effects of chlorine residuals. In addition to such chemical effects of gas-washing or other aqueous FGD effluents there are also effects on the temperature. FGD effluents may reach $50°C$ and when diluted may raise cooling-water discharge temperatures by an extra $3-4°C$.

(v) Ash-effluents

Where pulverised fuel ash is disposed of as a slurry from some power stations, the supernatant liquid from the large settling ponds is usually discharged to the nearest surface water. The volumes of water used to sluice the ash to the ponds range from about 5000 to 150 000 litres t^{-1} or about 40×10^6 litres day $^{-1}$ per 1000 MWe of coal fired plant. Major ash discharges are usually separate from the cooling water (Crecelius 1985), though they may enter the same water body close enough to have combined effects (Cherry *et al.* 1976, 1978). The qualities of ash effluents and their effects are discussed by Chu and Olem (1980, 1982). The composition of any ash is directly related to the original coal composition and can vary widely even in any one region (US Department of the Interior 1953; Watt and Thorne 1965; Brown *et al.* 1976; Dvorak and Lewis 1978). The major leachates from US coal ash are shown in Table 3.3. The concentrations of some trace metals in ash can be as high as 100 times that of the original coal. Where ash effluents are used to augment cooling-water supplies the disposal of the final effluent can cause problems because of the high levels of trace metals

TABLE 3.3
Trace elements released from fly-ash samples after a 3-day
shake test with distilled water (200 g fly ash/litre)[a]

Metal	Concentration (mg/litre)	
	Fly ash 1 (pH 11)	Fly ash 2 (pH 4)
Zinc	0·06	0·59
Cadmium	0·035	0·05
Copper	0·030	0·30
Chromium	0·21	0·04
Lead	0·75	0·25
Arsenic	<0·2	0·5
Mercury	0·000 7	<0·000 1

and either high or low pH depending upon the location (Duedall *et al.*
1985a,b).

(vi) Radioactive wastes

Small amounts of radioactive wastes are discharged in the cooling-water
effluents of both nuclear and fossil-fuelled power stations (Bertsche 1971;
Robson 1984). The dispersal of aqueous radioisotopes from nuclear
installations has been studied for many years in most countries and the data
have been analysed and published widely (e.g. Polikarpov 1966; IAEA
1969, 1971, 1972, 1974a,b, 1975a,b,c, 1976, 1979; Eisenbud 1973; see e.g.
Hunt 1985, 1989). Eisenbud (1973) states that as far as ionising radiation is
concerned 'it is not unfair to say that more is known about this subject than
is known about the effects of any other of the noxious agents which man has
introduced to his environment'.

The levels of the various nuclides in effluents vary enormously from site
to site. Monitoring has been carried out at British nuclear power stations
over almost 25 years and data exist over similar periods in other countries
(see Hunt 1985, 1989; Osterberg 1985). Table 3.4 shows data from the Calvert
Cliffs nuclear power station in the USA as a typical example though there
are differences between sites (see MAFF 1972 *et seq.*). These data refer to
direct discharges to the aquatic environment almost all of which are
discharged in cooling water and are thus greatly diluted.

The accident at Chernobyl in Russia in 1986 added small amounts of
radioactive materials to many waters in Europe and Table 3.5 shows

TABLE 3.4
Radio-nuclides released by liquid effluents from Calvert Cliffs Power Station,
Maryland, in 1980 (from Osterberg 1985)

Radio-nuclide	Activity (Bq)	Radio-nuclide	Activity (Bq)
^{3}H	1.8×10^{13}	^{110m}Ag	1.1×10^9
^{51}Cr	2.7×10^{10}	^{113}Sn	1.5×10^8
^{54}Mn	3.2×10^9	^{124}Sb	6.2×10^8
^{56}Mn	4.6×10^7	^{125}Sb	7.6×10^9
^{57}Co	3.0×10^7	^{131}I	5.8×10^9
^{58}Co	7.4×10^{10}	^{132}I	6.0×10^7
^{59}Fe	2.7×10^8	^{133}I	2.6×10^9
^{60}Co	1.6×10^{10}	^{133}Xe	5.7×10^{11}
^{65}Zn	8.9×10^8	^{133m}Xe	3.9×10^9
^{85m}Kr	6.6×10^6	^{134}Cs	3.9×10^9
^{89}Sr	5.2×10^8	^{135}I	2.6×10^8
^{90}Sr	8.5×10^8	^{135}Xe	7.8×10^9
^{91}Sr	2.3×10^7	^{135m}Xe	1.2×10^8
^{95}Nb	1.4×10^9	^{137}Cs	6.8×10^9
^{95}Zr	7.8×10^9	^{140}Ba	2.2×10^8
^{97}Zr	2.2×10^7	^{140}La	1.2×10^9
^{99}Mo	3.2×10^7	^{141}Ce	1.0×10^7
^{103}Ru	9.2×10^8	Unidentified	3.8×10^9
^{106}Ru	2.7×10^7		

predictions of exposures to organisms in the cooling water reservoirs at Chernobyl and other waters (OECD 1988).

It has been shown in experimental systems that both the rate of accumulation of radioisotopes in the flesh of animals and in plant tissues and the toxic effects of radioactivity are exacerbated by higher water temperatures (e.g. Grayum 1973; Harvey 1974).

3.2 CHEMICAL CHANGES CAUSED BY PASSAGE THROUGH COOLING WATER SYSTEMS

Chemical changes occur in cooling water as it passes through a system as result of physical processes such as heating and turbulence. These differ significantly in direct and indirect cooling systems. For example, in a direct system the water passes through with no make-up or purge, but in a tower-cooled or other recirculating system, evaporation and make-up lead to

TABLE 3.5

Calculated estimates and predictions of exposure doses for aquatic organisms (mrad/h) in the area of the radiation plume from the Chernobyl nuclear power station, 10–20 June 1986 (from USSR State Committee on the Utilization of Atomic Energy 1986)

Water body	External exposure		Internal exposure		
	From water	From bottom sediments	Aquatic plants	Plankton	Fish
Chernobyl Power Station cooling pond	10 (2)[a]	4 300 (300)	10 000 (1 000)	1 000 (100)	500 (50)
River Pripyat (buoy 204)	0·1 (0·009)	40 (3·3)	110 (15)	12 (2)	6 (0·8)
River Dnepr (Kiev, Inst. of Hydrobiology, Ukranian Academy of Sciences)	0·002 (0·000 2)	0·3 (0·025)	1·0 (0·2)	0·1 (0·015)	0·04 (0·01)
Natural background	0·0001–0·006	0·002–0·02	0·08–0·2	0·002–0·016	0·003–0·005

[a] The figures in brackets are predictions for June 1987.

concentration factors in the cooling water of between 1·3 and 2·5 in power stations and up to factors of 8 in industrial air-conditioning systems (McKelvey and Brooke 1959; Davies 1966; Milner 1984). Cooling-tower sprays and packing produce fine droplets and films which enhance aeration and gaseous exchange while pumps in both types of system cause considerable turbulence and have a similar effect. In tower ponds the warm aerated conditions also encourage bacterial action and nitrification, particularly where inlet waters are polluted with sewage (Klein 1962; Humphris 1977).

3.2.1 Dissolved Gases

Figure 3.2 shows the results of a survey of *dissolved oxygen* concentrations in cooling waters before and after passing through the condensers of

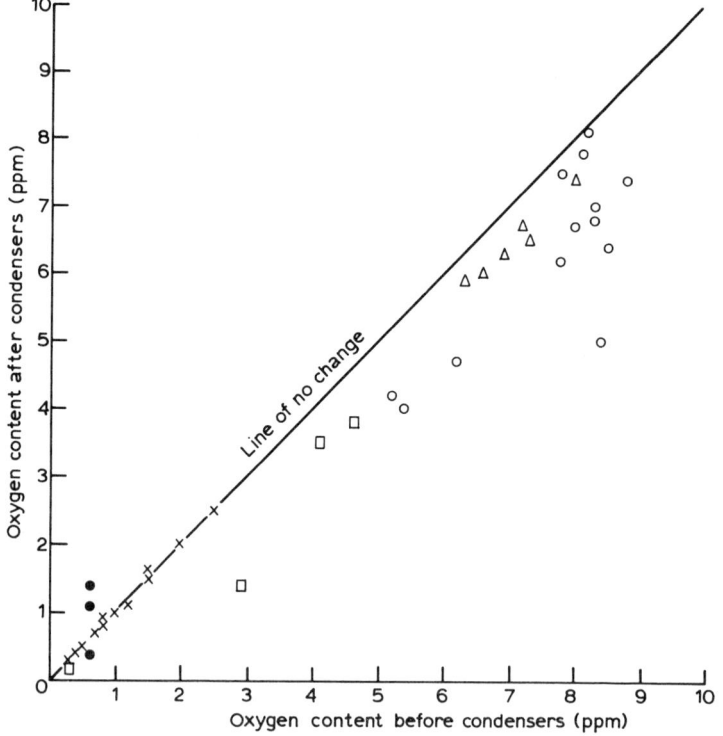

FIG. 3.2   Oxygen concentrations at condenser outlets from British Power Stations, in relation to inlet concentrations (effect of condensers only; after BEA, 1954). ×, Cliff Quay; ○, Cardiff; △, Upper Boat; ●, Keadby; □, Thornhill.

cooling systems at power stations in England and Wales during 1953 and 1954 (BEA 1954; Ross 1954). There were no marked deleterious effects on oxygen concentration where upstream concentrations were less than 3.0 mg litre^{-1}. Reductions occurred, however, where inlet concentrations exceeded 4.0 mg litre^{-1}. The reductions varied from 0.1 to 3.5 mg litre^{-1}.

The presence of a weir or cascade causes a significant increase in oxygen concentrations at an outfall (Gameson 1957; Powell 1963; Gammon 1969; Clement 1972) where the water is unsaturated and such structures were used to oxygenate the effluents from the deoxygenated gas-washing effluents at British power stations on the Thames (DSIR 1964).

Davies (1966) did not find a clear relationship between oxygen concentrations at the inlet and in the effluent at tower-cooled power stations, though Aston (1973) showed that the Drakelow power stations on the River Trent in England contributed up to 4.0 t of oxygen to the river each day from mixed tower and direct-cooling systems (Fig. 3.3). Both Aston (1973) and Gammon (1969) found higher oxygen concentrations in the recirculating water than in the inlet water in tower systems. Davies (1966) found that the effects of passage through cooling-tower packing on the oxygen content of recirculating cooling-water could be described by the equation:

$$O_2 \text{ (inlet)} = 0.57 \times O_2 \text{ (outlet)} + 42$$

where O_2, is expressed percentage saturation.

The oxygen contribution by the total benthic community to the thermal discharge area of a power station in Florida was 70–75% less than that of an unheated area. Respiration was 1.5 times less. Photosynthetic contributions in heated and unheated areas were similar in summer (Smith *et al.* 1974b). There are few published measurements of *carbon dioxide* concentrations in effluents from direct-cooled power stations though reductions would probably occur between intake and outfall if inlet concentrations were high because of turbulence and temperature. Davies (1966), Parker and Krenkel (1969), Ross and Whitehouse (1973) and Humphris (1977) note that carbon dioxide is 'scrubbed' from the water by cooling towers and Humphris (1977) showed reductions from between 20 and 40 mg litre^{-1} to around 3.0 mg litre^{-1} during passage from inlet to cooling-tower ponds. Davies (1966) concluded that the increase in pH which he found in some cooling systems was a result of half-bound carbon dioxide in bicarbonates being converted to carbonates. Reductions in the pH of cooling waters at power stations are usually a result of acid-dosing for descaling (Milner 1984) or chlorination. *Nitrogen* has a low solubility in

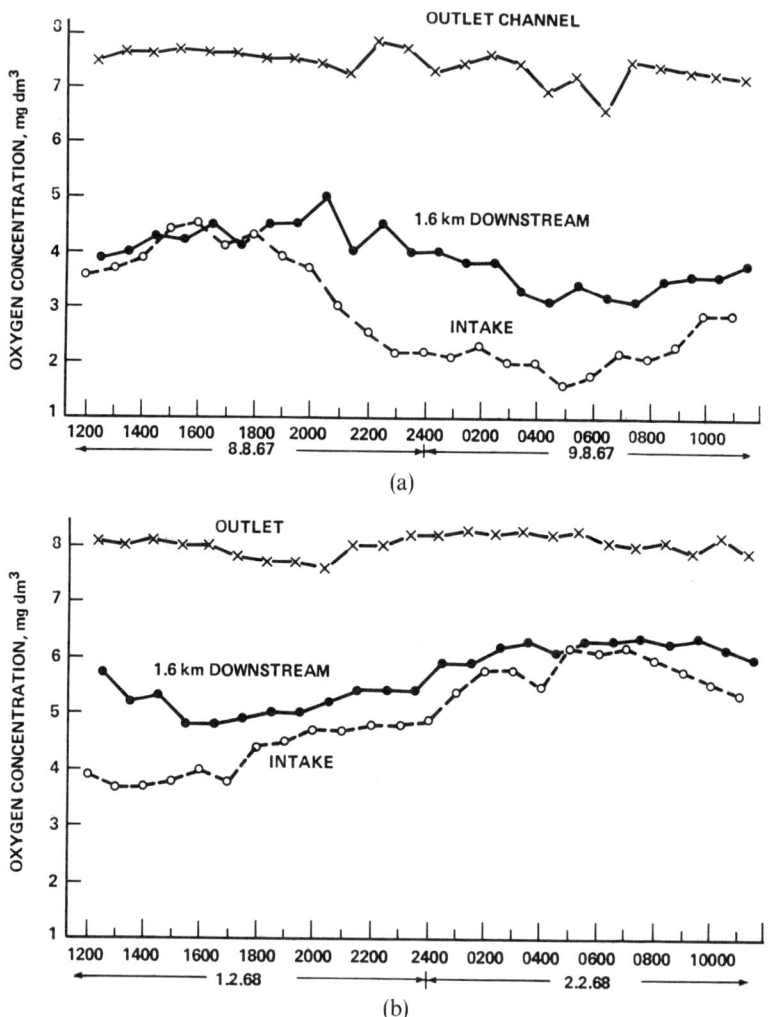

FIG. 3.3   Oxygen content of the River Trent adjacent to Drakelow power stations;
(a) 8th and 9th August 1967, (b) 1st and 2nd February 1968 (after Aston 1968b).

water and as a gas appears to be of little significance, except where high temperatures or pressures cause supersaturation in the body fluids of fish or invertebrates. Once the pressure or temperature falls the gas emerges from solution to form embolisms and the symptoms of the phenomenon known as gas-bubble disease appear in exposed organisms (Demont and Miller 1971; Adair and Hains 1974; Miller 1974; Fickeisen and Schneider 1976; Otto 1976).

The concentrations of other gases vary widely with the quality and composition of the intake water. In polluted or in hypolimnial waters *hydrogen sulphide, methane, ammonia, sulphur dioxide* or *carbon monoxide* may be present in significant amounts.

Ammonia and ammonium compounds originating from the decay of proteinaceous materials are generally 'lost' during passage through an indirect cooling-water system as a result of biological or chemical nitrification processes in cooling towers (Davies 1966: Stratton and Lee 1975). Table 3.6 shows the relationship between the concentrations of ammonia in make-up water with those in the circulating water in cooling-tower systems. Chlorination also reduces concentrations of ammonia through the formation of chloramines (Brown and Aston 1975). One unusual result was recorded in the cooling tower of a US installation, where it was found that the tower water was scrubbing ammonia from the air near an animal stockyard producing higher concentrations in the purge water than in the intake water (Stratton and Lee 1975).

Nitrification, that is the oxidation of ammonia and ammonium compounds to nitrates (Table 3·6), is enhanced in the warm aerated water

TABLE 3.6
Regression coefficients for the relationship between ammonia and nitrate levels in cooling waters, compared with those in intake waters at four UK power stations (after Davies 1966)

Station	Substance	Regression equation	Confidence limits	Range
G	Ammonia	$0.79Cm$	± 1.1	0·1–6·48 as N
	Nitrate	$0.82Cm + 1.35$	± 9.9	1·2–9·6 as N
M	Ammonia	not significant	—	—
	Nitrate	$-2.65Cm - 14.5$	± 52	11–33 as NO_3
	Nitrite	$0.37Cm + 0.44$	± 6.0	0·4–1·5 as NO_2
R	Nitrate	$1.14Cm + 16.7$	± 55	11–89 as NO_3
	Nitrite	$0.76Cm + 0.82$	± 0.48	0·16–0·46 as NO_2
B	Ammonia	$-0.04Cm + 0.97$	± 0.05	0·1–1·8 as NH_3

circulating in a tower system. The conditions favour the metabolism of nitrifying bacteria, particularly where organic pollution is prevalent in the intake water (Humphris 1977).

3.2.2 Concentration Factors, pH and Solids in Recirculating Systems

Davies (1966) produced a series of regression equations relating the concentrations of various constituents in make-up, circulating waters and effluents of power station cooling-water systems. Concentration factors varied from 1·1 to 1·3.

pH values in cooling tower purge waters were found by Davies (1966) to range from 0·3 to 0·9 units higher than predicted from known concentration factors and this was attributed to the removal of dissolved carbon dioxide.

Table 3.7 shows the regression equations for the relationship between the hardness of water at intakes and in the circulating water of cooling-tower systems. Precipitation of mineral salts, mainly calcium, occurred at the station with the lowest slope value (i.e Station C). Acid treatment was necessary to prevent this causing blockages in the system.

There was no consistent correlation between the Biochemical Oxygen Demand or Oxygen Absorbed values at the intakes and outfalls of these power stations (Davies 1966) and this also applied to the solids concentrations. In cooling-tower systems suspended solids tend to settle

TABLE 3.7

Regression coefficients for the relationship between various types of hardness in cooling water, compared with those of intake waters at six UK power stations (after Davies 1966)

Station	Substance	Regression equation	Confidence limits	Range as $CaCO_3$
B	Total hardness	$1·22Cm + 30$	± 56	339–467
G	Total hardness	$0·59Cm + 170$	± 68	320–422
	Temporary hardness	$1·13Cm - 23$	± 37	163–243
	Permanent hardness	$0·89Cm + 31$	± 38	95–229
	Calcium hardness	$0·78Cm + 89$	± 42	298–380
	Magnesium hardness	$0·26Cm + 27$	± 15	10–50
M	Calcium hardness	$1·98Cm - 82$	± 88	226–295
R	Calcium hardness	not significant	—	850–1 750
C	Temporary hardness	$0·18Cm + 36$	± 30	78–176
W	Total hardness	$1·33Cm + 15$	± 83	210–1 900
	Temporary hardness	$0·67Cm + 115$	± 96	125–300

Cm = concentration in ppm in intake water.

out in the ponds beneath the towers, and Davies estimated that about 30% of the solids originating from the inlet water settled out this way. At Croydon power station, however, the sewage effluent used as tower make-up increased solids in the circulating water by up to a factor of 7·5 (Humphris 1977).

In direct-cooled systems, suspended solids may increase in the effluent as a result of the flushing out of fouling or scaling deposits. At the Turkey Point power station in Florida suspended solid concentrations were high enough to blanket the area near the outfall with a layer of silt (Roessler 1971).

3.2.3 Chlorination and its Residues in Cooling Systems
Of all the constituents of thermal discharges, chlorine will be seen during the subsequent chapters to be the most significant biologically, not only because it is the most widely used biocide, but because it is clear that in many cases the ecological effects of many thermal effluents are mainly the result of chlorination rather than heat. There are now very many publications dealing with the chemistry and ecological consequences of the chlorination of cooling water and other industrial and sewage effluents, e.g. Beauchamp (1969a), White (1972), Block and Helz (1976), Coughlan and Whitehouse (1977), Jensen (1977), Jolley *et al.* (1978a,b,c, 1980, 1983a,b, 1985), Schubel and Marcy (1978); Morgan (1980), Hall *et al.* (1981), Jolley and Carpenter, (1983a,b), Langford (1983a), Helz *et al.* (1984).

Chlorine was first used as a disinfectant in the 19th century and as a water treatment for potable supplies in the early 20th century. Its use as an anti-fouling agent for power-station cooling-water systems began in the mid-1920s in both the USA and in the UK. Originally, continuous treatment was used but was regarded as uneconomic and replaced by intermittent dosing. At the present time chlorine in some form is the predominant anti-fouling chemical used in electricity generating stations throughout the world. In the USA more than 25 000 t of chlorine are used annually in power stations. The respective figure in the UK is about 10 000 t.

There are three major methods of application of chlorine, viz.

(a) Chlorine gas is dissolved in water and pumped to the application point, usually at the intake screens.
(b) Hypochlorite solution is dosed directly from tanks to the application point. This method is most commonly used in smaller installations or in emergencies at power stations where the normal equipment has broken down.

(c)   Chlorine gas is produced by an electrolytic process from sea water
      on site and the resulting solution applied at the intakes. This method
      is increasing in use as it is inherently safer and easier to control than
      the first method.

The complex chemistry of chlorination is beyond the scope of this volume
but there are a large number of reviews, collected papers and bibliographies
which have already been noted, e.g. White (1972), Hall *et al.* (1981) and
Jolley *et al.* (1978a,b,c, 1980, 1983a,b, 1985) are recommended (Fig. 3.4).
These also include reviews of the main methods of chlorine measurement.

In fresh waters the chemical reactions of chlorine are relatively simple,
initially forming hypochlorous (HOCl) acid which then dissociates to form
hydrogen (H^+) and (OCl^-) ions. In most natural waters the presence of
ammonia and ammonium compounds lead to the formation of chloramines and the extent and rates of the reactions are dependent on the
relative concentrations of the constituents, temperature and pH.

The reactions of chlorine in sea water are extremely complex and not yet
fully understood (e.g. Block *et al.* 1977; Carpenter and Smith 1978;
Goldman *et al.* 1978; Hartwig and Valentine, 1983; Jolley 1985). The
complexities arise from the serial reactions of the chlorine with other

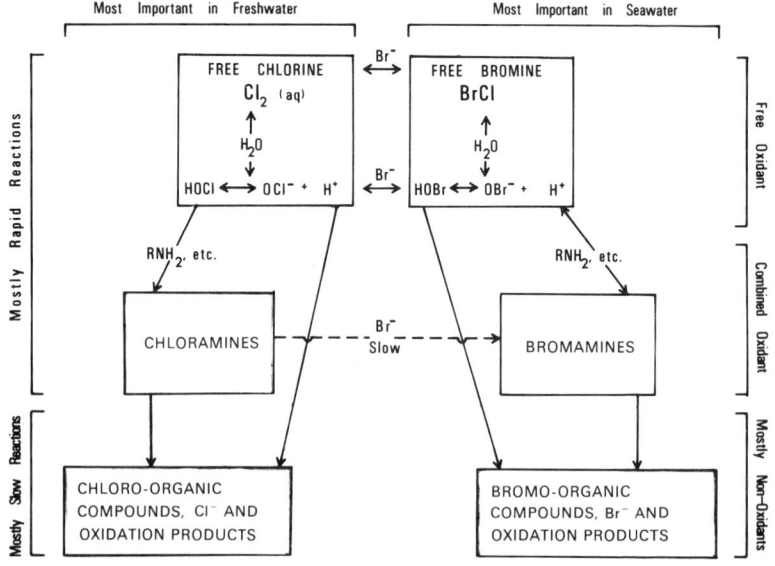

FIG. 3.4   Diagram of reaction paths of chlorine in freshwater and sea-water (after
Hall *et al.* 1981).

oxidants such as bromine and iodine naturally present in sea water, and subsequently with organic compounds (Katz 1977; Wong and Davidson 1977; Sugam and Helz 1980; Wong 1980a,b; Grove *et al.* 1985). Figure 3.4 shows some of the possible reaction pathways in both fresh and salt water.

It is traditional to distinguish between 'free chlorine' and 'combined chlorine' in waste waters. The former is chlorine not combined with other substances, the latter describes the by-products of the chemical reactions of chlorine and other substances, for example ammonia, to form chloramines. 'Total residual chlorine' (TRC) is the sum of these two. In sea water the residuals are generally expressed as 'Total residual oxidants' (TRO) which includes the bromine and iodine compounds formed in the reactions. (When chlorine is dosed to natural waters, the measured values of TRC are usually less than might be predicted from the dose.) The difference is known as the 'chlorine demand' (Helz *et al.* 1983), and is related to the composition of the water and the temperature (Wong 1980b). Its measurement is also affected by time in that some reactions may be immediate while others occur over varying finite periods (Cole 1977; Hall *et al.* 1981; Grove 1983; Helz *et al.* 1983; Morris and Isaac 1983; Seaman *et al.* 1983). There is thus a period of 'chlorine decay' over which the amount of measurable TRC decreases more or less exponentially, even in a static body of water with no dilution. The eventual products of this decay are most probably chlorides of various elements.

The chlorine demand of water to be used for cooling largely determines the effective dose and is determined during the planning and design stages of cooling-water systems. At most power stations throughout the world, chlorine is dosed such that measured residuals (TRO/TRC) immediately after application are between 0·5 and 10·0 mg litre^{-1} depending upon the composition of the water. In Britain and the USA dose rates are generally less than 1·0 mg litre^{-1} and the required residual at the inlets to the condensers between 0·2 and 0·5 mg litre^{-1}. Doses may need to exceed 1·0 mg litre^{-1} if algae are to be controlled in cooling-tower systems (Cole 1977; Cheremisinoff and Cheremisinoff 1983). Dosing procedures vary from intermittent (for example for about 30 min every 8 h) to fully continuous, depending upon the problem and site (Fig. 3.5).

Intermittent chlorination is most widely used in fresh-water systems where the control of slimes in condensers is the major objective. Continuous, low-level chlorination is more common at marine or estuarine sites, usually to prevent the settling of sessile organisms in spring and summer. In winter when there is less probability of organisms settling, intermittent dosing may be used at these sites.

At marine power stations in the UK, continuous dosing is at

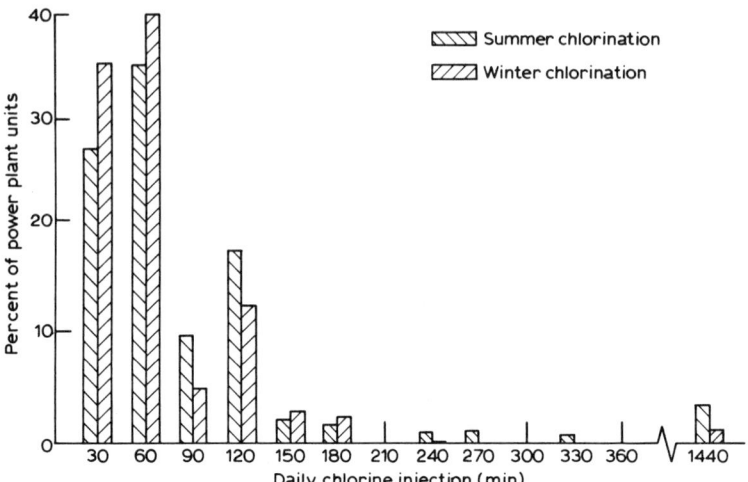

FIG. 3.5   Distribution of power plants injecting chlorine into cooling water for up to a given time each day. Data are for a subsample of existing power plant units, but reasonably represent the industry with respect to cooling water type and plant design (from Mattice 1985).

approximately $0.5\,\mathrm{mg\ litre^{-1}}$ and there is a general requirement by the industry to produce effluents which contain no free chlorine at outfalls and a minimum TRO.

Figure 3.6 shows the chlorine concentrations in the cooling-water circuits of two power stations using cooling towers over the period following dosing. Free chlorine was low or absent after passage through the condensers but the total residuals (TRO) exceeded $0.8\,\mathrm{mg\ litre^{-1}}$ at the British power station and $0.6\,\mathrm{mg\ litre^{-1}}$ at the American power station. Three hours after dosing TRO concentrations were negligible in both systems. Clearly, however, there were short periods immediately following dosing when the chlorine concentrations in the purge effluent from the systems could contain concentrations up to about 0.4 or $0.5\,\mathrm{mg\ litre^{-1}}$ of chlorine residuals.

Chlorine and chloramines are volatilised to the air in cooling towers (Lutz and Merle 1982; Holzwarth *et al.* 1984a,b), the former by about 2%, the latter almost entirely and very rapidly as the water passes through the tower. Typically, some 2% of free chlorine is lost to the purge effluent and 88% to the chlorine demand of the water.

Studies on the biological effects of other biocides are infrequent in the literature (Wackenuth and Levine 1977; Bongers *et al.* 1978; Chow and

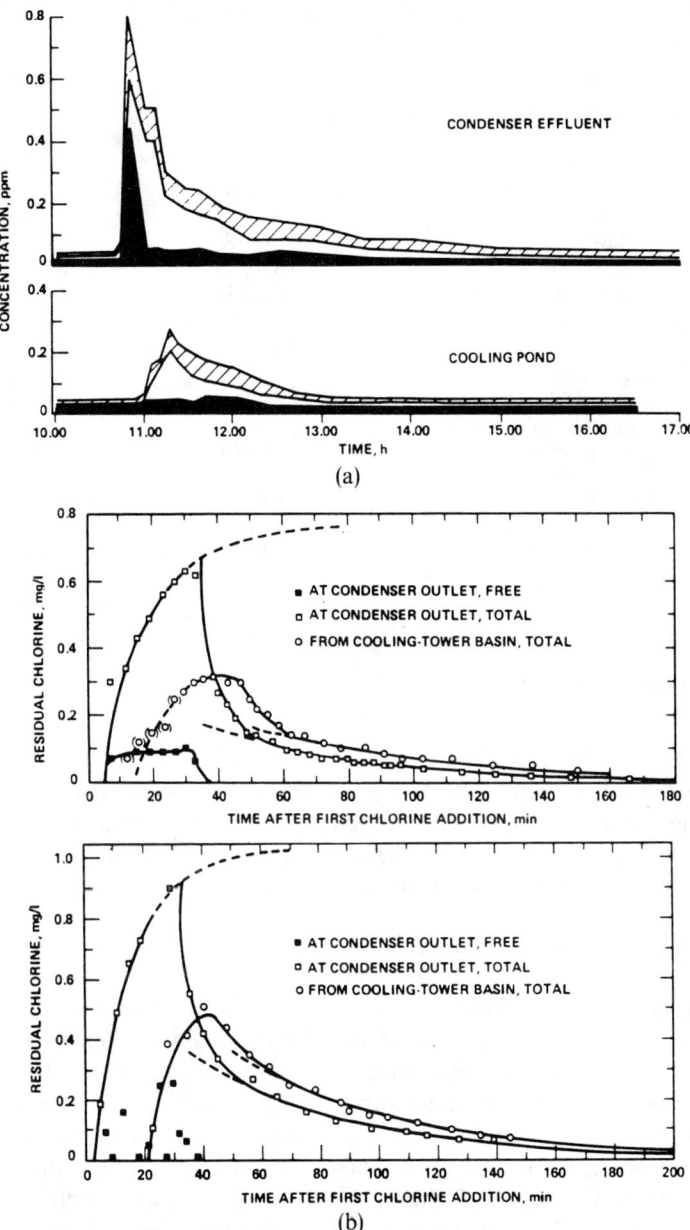

FIG. 3.6   (a) The concentrations of free chlorine (■), monochloramine (□) and dichloramine (▨) in the cooling water of a UK power station (after Brown and Aston, 1975); (b) chlorine and residual concentrations in two cooling water circuits using towers (after Dickson *et al.* 1974).

TABLE 3.8
Trace metal concentrate ions in circulating cooling water at 11 sites in the USA

Element[a]	Inter-water (range)	Circulating water (range)
Total Phosphorus (mg litre^{-1})	0·01–0·80	0·25–6·01
Zinc (μg litre^{-1})	<0·05–3 430	0·7–15 740
Copper (μg litre^{-1})	1·0–102	6·0–2 200
Chromium (μg litre^{-1})	1·0–17·0	1·0–1 400
Lead (μg litre^{-1})	0·0–0·02	0·0–38
Manganese (μg litre^{-1})	?	50–150
Nickel (μg litre^{-1})	?	0·6–24·3
Cadmium (μg litre^{-1})	?	0·3–2·9
Mercury (μg litre^{-1})	<0·3–30·0	0·4–84
ICPB Standards[b]	*Effluent*	*Lake Michigan*
Zinc[a]	1·0	1·0
Copper[a]	1·0	0·02
Chromium[b]	0·3	0·05
Lead[b]	0·1	0·05
Mercury[b]	0·5	0·5

[a] Standards in μg litre^{-1}.
[b] Standards in mg litre^{-1}.

Kawaratani 1983). Dechlorination of effluents has been carried out by various means to reduce the biological effects of residuals. These methods include photochemical sterilisers, sulphur dioxide, sodium thiosulphate and physical aeration (e.g. McCauley and Scott 1960; Nash 1969; Armstrong and Scott 1974; Constable 1979).

3.2.4 Heavy Metals in Cooling-Water Systems

In cooling waters from metal finishing or processing industries, wastes from the process itself may be present, causing very high metal concentrations (Klein 1962; Lindberg and Hutchinson 1987). In cooling towers or other recirculating systems the concentration factors may increase concentrations of metals in the make-up water by 30–250%, though settlement or flocculation may in turn reduce such concentrations. Most heavy metal ions originate from the treatments and additives rather than from corrosion in the system. Table 3.8 shows levels of heavy metals in the cooling waters of eleven systems with circulating volumes ranging from 30 litre s^{-1} to 45 000 litre s^{-1}. All the concentrations in the system were higher than in the inlet water mainly as a result of treatment chemicals (Stratton and Lee 1975).

3.3 EFFECTS ON RECEIVING WATERS

Most of the data on the chemical effects of thermal discharges on receiving waters come from reports with limited circulation (see Becker *et al.* 1979a,b), or from the studies of specific subjects such as oxygen or chlorine already described. The effects of any discharge will to some extent be site-specific but there are clear patterns which are useful for predictive studies.

3.3.1 Oxygen and Other Dissolved Gases

Dissolved oxygen in natural waters mainly originates from atmospheric exchange at the surface and from the photosynthesis of aquatic plants (Owens *et al.* 1969). Sea-water typically contains about 15% less dissolved oxygen than fresh water at the same temperature. Under normal and quiescent conditions the oxygen saturation value is inversely related to water temperature and therefore heated effluents should predictably cause oxygen depletion in any receiving water (Hynes 1960). This is not typical in practice, however, mainly because of the effects of turbulence in the system as we have already seen (see p. 80).

Aston (1973) found that in the polluted reach of the River Trent at Drakelow in the English Midlands the oxygen contributed to the river by the nearby power stations caused sustained increases up to 1·6 km downstream of the outfall (Fig. 3.5) and Koppe (1974) showed a similar effect on the Ruhr. However, subsequent work (Alabaster *et al.* 1971), on the Trent, while confirming Aston's results at Drakelow, concluded that the raised temperatures below the next power station downstream, resulted in slightly reduced oxygen concentrations.

Figure 3.7 shows the oxygen concentrations upstream and downstream of British power station outfalls in the 1950s. From the pooled data for all samples the relationship between upstream and downstream concentrations could be expressed by the equation

$$O_2 \text{ (outfall)} = 0.78 \times O_2 \text{ (inlet)} + 0.82$$

(expressed in parts per million of O_2).

Using this equation the calculated mean oxygen concentrations and percentage saturations would be as shown in Table 3.9.

In the Thames estuary power station effluents also exacerbated lowered oxygen concentrations though their contribution was relatively minor compared with those of other effluents (DSIR 1964).

In Laguna Lake, in New York State, oxygen concentrations below saturation in the effluent from the Merulco power station were inversely

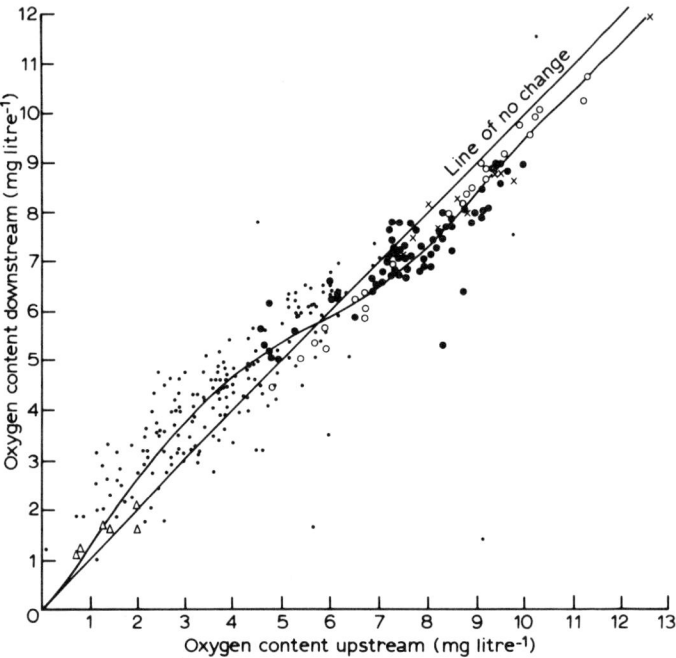

FIG. 3.7   Oxygen concentrations downstream of thermal discharges in relation to upstream values—data from British rivers (line drawn by eye; after BEA 1954). ×, Earley; △, Cliff Quay; ○, North Wilford; ●, Clydes Mill Spondon.

related to temperature though the overall fluctuations ranged from supersaturation to very low values as a result of both photosynthesis and the effluent (Cabral *et al.* 1976). The effects of natural oxygen stratification combined with a hypolimnial intake were shown by Keith and Perrin (loc. cit. Becker *et al.* 1979a,b). Oxygen levels at a cooling-water outfall ranged from 2·8 to 3·9 mg litre^{-1} while the natural lake surface values were 6·6–7·7 mg litre^{-1}. Outfall values coincided closely with those in the hypolimnion of the lake though there was some increase after passage through the system.

Where heated water is discharged to an already stratified lake the oxygen stratification is predictably altered, though there are few studies of such sites before and after development. Such data that are available indicate that the areas of effect are small (McNeely and Pearson 1974; Gallup and Hickman 1975; Todd and Bender 1982; Eloranta 1983).

TABLE 3.9

Effects of passage through cooling-water systems on oxygen concentrations and saturation in river water—predicted from surveys in UK power stations, assuming a ΔT of approximately 6·6°C on a natural temperature of 18·5°C (after BEA 1954)

Concentrations in ppm		Percentage saturations	
Upstream	Downstream (% change)	Upstream	Downstream
1·0	1·6 (+60)	11	18
3·0	3·16 (+5)	33	36
5·0	4·72 (−5)	54	54
7·0	6·28 (−10)	76	72
9·0	7·84 (−13)	98	89

Data from marine sites are scarce in the literature. Adams (1969) concluded that effects at Californian coastal sites were small and variable. Sandstom (1985) found that percentage saturation was higher in the effluent at Forsmath station than at the intake. The difference was mainly because of the increase in temperature. As a general observation, from all the published data it seems that in all polluted waters with oxygen concentrations below 4 mg litre^{-1} the presence of a power station, preferably incorporating cooling towers, could cause considerable oxygenation in the receiving water. At higher concentrations, however, the raised temperatures may cause reductions (e.g. Jacobsen 1976; Loftus 1978).

Parker and Krenkel (1969) suggested that the capacity of fresh waters to assimilate wastes would be adversely affected by heat, and an effluent could, under certain conditions, add the equivalent of almost 30% to the BOD loading. These effects would be compensated for, at least in summer, by increased photosynthesis, provided that turbidity, and hence light penetration, were not affected.

There are no published data on the concentrations of carbon dioxide, nitrogen and other gases in receiving waters of cooling-water discharges though some inferences may be drawn from their concentrations in cooling systems (see p. 84).

3.3.2 Chlorine and its Residues in Receiving Waters

Because there is a significant time factor affecting the decay of chlorine, the concentrations in any effluent will depend on the time taken for the cooling water to travel between the dosing point and the outfall, in addition to

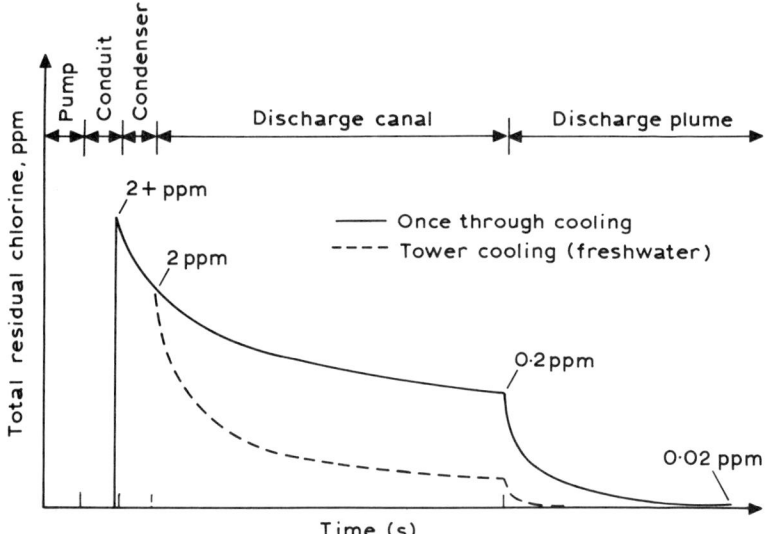

FIG. 3.8   Schematic of chlorine decay in cooling system (after Tetra-Tech Inc. 1980).

factors such as chlorine demand, temperature, turbulence and dilution (Beauchamp 1967, 1969a,b; Dickson *et al.* 1974; Wong and Davidson 1977; Hergott *et al.* 1978; Lee 1979; Crane *et al.* 1980; Haag and Lietzke 1980; Helz *et al.* 1980; Sugam and Helz 1980; Wong 1980b; see Hall *et al.* 1981 (Fig. 3.8)). Clearly, intermittent dosing will produce pulses of TRC at an outfall (Plumb *et al.* 1980), while continuous dosing will produce a more constant effluent concentration though the values may fluctuate as a result of changes in the intake water and within the system.

Figure 3.9 illustrates decay curves through time at two points in a river downstream of a discharge of cooling water dosed intermittently with chlorine. The concentration of TRC in the plume decreased from about 0·58 mg litre^{-1} at a point 9 m below the outfall to 0·14 mg litre^{-1} at 90 m. The peak occurred at the outfall some 30–40 min after dosing (Dickson *et al.* 1974). The TRC concentrations at the point of dosing ranged from 0·6 to 1·3 mg litre^{-1} and free chlorine concentrations ranged from 0 to 0·35 mg litre^{-1}. The Carbo power plant, at which the measurements were made, incorporates cooling towers in the circuit. At all three sampling sites in the river the peak was very short-lived, typically less than 15–20 min with a slower decay at the lower concentrations. In a once-through (direct) cooling system the peaks would be similarly ephemeral though there would not be

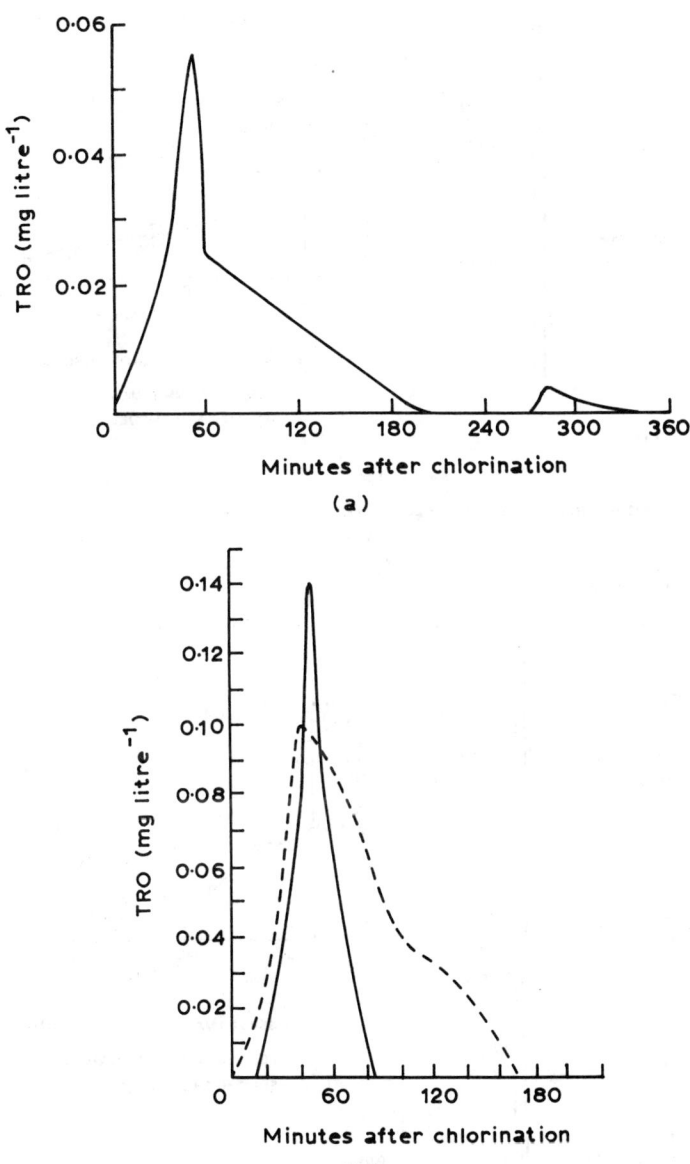

FIG. 3.9   Total residual oxidants (TRO) in the discharge plume of a river-cooled power station, at two sampling points: (a) 8 m downstream of outfall; (b) 90 m downstream of outfall.

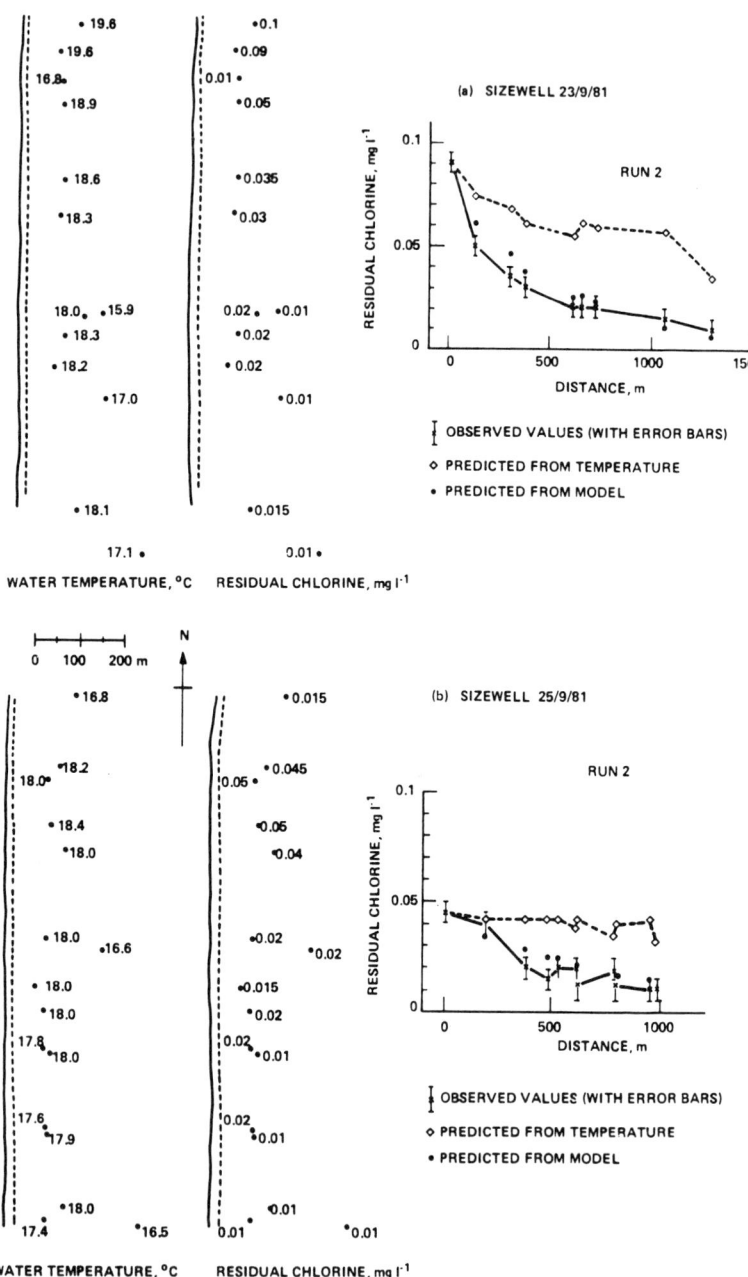

FIG. 3.10 Temperature and residual chlorine distribution and decay curves for validation of model—Sizewell discharge 1981 (from Coughlan and Davis 1983).

such a delay in the system. Figure 3.10 compares predicted and measured TRC concentrations at a site.

In marine and estuarine habitats a number of studies have illustrated chlorine decay in thermal plumes. Spencer (1982) found effluent concentrations of TROs reaching 0.35 mg litre^{-1} in the effluent canal from the Kingsnorth power station on the Medway estuary. By a point 400 m downstream there was generally a decay factor of between 5 and 6. Davis and Coughlan (1983) in one of the most comprehensive studies of chlorination at power stations, produced models of TRO decay at various sites around the British coast. They showed that decay rate was exponential and very predictable under experimental conditions at each site, though the model was significantly different from any based simply on the reductions of temperature in a plume. They stated that in receiving waters 'the most significant decreases in chlorine concentration arise from dilution bringing in "chlorine demand"'. A model based on temperature would almost invariably underestimate the rate of chlorine decay. Table 3.10 shows the variations in decay constants and chlorine demand with temperature from experiments. Based on studies of the effects of chlorination on the productivity of entrained phytoplankton and on the derivation of the decay curve 'breakpoint' the British sites were categorised as 'open coast' or 'estuarine' types. At the latter, the instantaneous chlorine demand was less than at the former (Coughlan and Davis 1983; Davis and Coughlan 1983).

TABLE 3.10

Variation of instantaneous chlorine demand (C_{id}) and decay constant (k) with temperature[a]. C_{id} is the difference between the 'chlorine' concentration determined at 1 min after NaOCl addition to sea water, and the concentration yielded by the same addition to distilled water (2·5 mg/litre) (from Davis and Coughlan 1983)

Water temperature (°C)	Observed chlorine concentration (mg/litre)	Instantaneous chlorine demand (mg/litre)	Decay constant (k)
0	2·4	0·1	0·012
2·5–4·0	2·2	0·3	0·011
10·0–10·5	2·0	0·5	0·020
18·0	1·7	0·8	0·027
18·5	1·6	0·9	0·029
25·0–26·5	1·8	0·7	0·042
32·0–33·0	1·5	1·0	0·042
Distilled water	2·5		

[a] Estimated activation energy of rate determining reaction = 3 kJ/mol.

Two studies were used to validate the models. Figure 3.10 shows typical results for the effluent plume at Sizewell nuclear power station on the east coast of England. There was a 50% reduction in the TRC concentrations at distances of between 150 m and 400 m from the cooling-water outfall. There is a very clear agreement between the predicted and observed values (Coughlan and Davis 1985). This is not, however, always true, especially where the water is polluted (see Carpenter and Smith 1978) and complex halogen compounds may be formed (Bean *et al.* 1983).

Surveys of chlorine concentrations have been carried out at a number of outfalls to both fresh and saline waters and although there are site-specific characteristics resulting from the different chemical and hydrographic factors, the basic patterns of decay follow similar trends (Hostgaard-Jensen *et al.* 1977; Grieve *et al.* 1978; Jolley *et al.* 1978b; Lietzke 1978; Jolley and Carpenter 1983a). Clearly, the effects of chlorination can be readily detected chemically in cooling water plumes with total chlorine residuals typically between 0.1 and 0.6 mg litre^{-1} at outfalls. Three factors need to be considered which will have relevance to the biological effects of these residuals, namely:

—the toxicity of most chemicals increases with temperature,
—decay rates of residuals are related to temperature,
—in a thermal plume, chlorine residuals are likely to be related to thermal stratification.

One of the more recent concerns is the accumulation of stable halogenated organic compounds in sediments and in organisms. There are many such compounds but trihalomethanes and chlorinated phenols have been most studied (e.g. Bean *et al.* 1985; Grove *et al.* 1985; Punzi and Patel 1985).

3.3.3 Heavy Metals and Other Contaminants

(i) Metals

Generally the concentrations of heavy metals in power station cooling-water discharges originating from corrosion are below the levels of detection by normal methods (Romeril 1972). However Roosenburg (1969) found sufficiently high levels of copper in a US power station effluent to cause physiological changes in oysters and Martin *et al.* (1977) recorded concentrations lethal to abalone (*Haliotis* spp.) in cooling water from a Californian coastal power station. In Long Island Sound (USA), the contribution of particulate copper and zinc from a power station was estimated as 0.5 g litre^{-1} compared with 1.0 g litre^{-1} of other dissolved metals (Waslenchuk 1983). Surveys around power stations on Southampton Water in the UK showed no detectable levels in the cooling-water effluents (Romeril 1974).

Rothwell (1971) suggested that high levels of heavy metals found in Trawsfynnydd lake originated from pipework corrosion in the nuclear power station which used the acidic lake water for cooling, but subsequent calculations showed that natural sources were responsible, and that the metals had leached from soils and rocks in the catchment.

In the Savannah River drainage area selenium concentrations reached 0.107 mg litre^{-1} as a result of ash effluents discharged near the thermal discharges (Cherry *et al.* 1976).

(ii) Flue gas desulphurisation effluents
The effluents resulting from the desulphurisation of coal where wet scrubbing methods are used are generally extremely noxious and unacceptable in most surface waters without considerable treatment (Elder and Hollingden 1976; Duedall *et al.* 1985a). The gas-washing effluents discharged to the River Thames from three power stations in the 1940s and 1950s contained dissolved sulphite which caused serious oxygen depletion in the river. The pH was less than 5 and there were quantities of heavy metal ions present (DSIR 1964).

To date very little is known about the effects of sea-water gas-washing effluents on marine organisms (see Chapter 10).

(iii) Radioactivity
The dispersal and effects of aqueous radioactive materials has been discussed in many publications including several major symposia (e.g. IAEA 1969, 1971, 1975a,b,c) and by Kamath *et al.* (1971), Nelson *et al.* (1972), Nelson and Evans (1973), Dunster (1978), Pentreath (1980) and Carr and Blackley (1986). Both Polikarpov (1966) and Eisenbud (1973) give excellent introductory accounts of radioactivity and radio-nuclides in marine habitats. Klein (1962) quotes a value of $10^{-5} \mu C$ (37×10^{3} Bq) as the normal maximum concentration of natural radioactivity in surface waters. Of the man-made fission products in water those from fallout far exceed those from nuclear power installations (see Osterberg 1985).

The biological consequences of radio-nuclides are mainly related to their effects on *Homo sapiens* in most studies, and most of the discharge limits and standards are set with these consequences in view. For example in the UK standards are based on the 'Derived Working Limit' (DWL) which is the 'maximum rate of release of any nucleide that will not exceed the maximum permissible dose to nearby humans' (see Preston 1974). In the vicinity of nuclear installations this dose is based on the human group most exposed to radioactivity from the most concentrated source, for example the group consuming organisms with high rates of accumulation.

Okamoto (1980) (loc. cit. Osterberg 1985) considers that the dose of radioactivity received by man from coal-fired power stations through the marine food chain could be up to 800 times that from nuclear power in Japan.

The literature on the accumulation and potential effects of radioactivity on organisms is vast as has been indicated. Here the main concern is the effects of higher water temperatures on both the rates of accumulation and biological consequences of radio-nuclides in the vicinity of cooling water discharges and the discussion in subsequent chapters is of necessity limited to that area. Accidental releases to water have been rare (Preston 1974). The most recent, through atmospheric releases at Chernobyl did not cause excessive contamination of the surrounding water (Emmings 1990).

CHAPTER 4

Experimental Studies and the Predicted Biological Effects of Thermal Discharges

4.1 POTENTIAL EFFECTS OF THERMAL DISCHARGES

It should be clear from the previous chapters that the potential ecological effects of any thermal discharge cannot be regarded simply as a direct function of the increased water temperatures.

The major potential effects on ecosystems are from:

—The entrainment of animals and plants into and through cooling-water systems and the long- and short-term effects, within the system, of heat, in conjunction with contaminants, turbulence and other physical and mechanical stresses.

—The exposure of organisms in receiving waters to unnaturally high water temperatures in conjunction with low concentrations of biocides and other contaminants, chemical changes in the water caused by physical processes (e.g. aeration caused by turbulence), and unusual water currents, scour or siltation.

In-situ studies of operating thermal discharges or cooling water systems can rarely distinguish the separate effects of all the constituents. Thus for devising the most effective and economic strategies for the prevention or alleviation of any biological effects, the relative magnitude of the effects of constituents need to be quantified. This can be through the use of controlled field experiments, careful in-situ studies of operating systems or laboratory scale experiments, or some combination of these. Such data can also be used to assist the prediction of the consequences of proposed new developments though there are limitations to the extrapolation of experimental data to the field.

There is already a vast amount of literature on the effects of both physical and chemical influences on aquatic organisms and major works are noted in the References and in the text. It is already well known that alterations in habitats, including those caused by pollution, produce generalised responses at individual, population and community level (Hynes 1960; Hawkes 1962; Perkins 1974; Nilsen and Kallquist 1975; Hellawell 1978, 1986; Phillips 1980). Further, sub-lethal temperature and chemical changes induce a number of 'generic' physiological and behavioural responses which follow similar patterns irrespective of the stress or stimulus (Jones 1964; Alabaster and Lloyd 1980; Hellawell 1986).

For example, gradual increases in both chlorine concentration and temperature will stimulate increased metabolic activity in animals, followed by an activity peak prior to torpor and death at higher levels. The values which cause the various phases tend to be specific to the organism and this knowledge is therefore vital to the protection of species in altered habitats. As another example, exposure to heat or sub-lethal concentrations of some chemicals during hatching can cause similar morphological or meristic changes in the early stages of fish (see p. 120).

In this chapter we will be discussing mainly the tolerance of organisms to the various known constituents of thermal discharges for the purposes of:

—Assessing the relative effects of these influences in any discharge and particularly the effects of heat.
—Comparing the effects of effluents predicted from experimental studies with those found *in situ* as described in subsequent chapters.

The three major components which are considered as dominating the potential effects of a thermal discharge are temperature, biocides (mainly chlorine) and water currents. The effects of other contaminants are considered in relation to these three, particularly where they may be acting in combination or synergistically.

4.2 TEMPERATURE

4.2.1 Tolerance in Relation to Natural Habitats

Living organisms are found over almost the whole temperature range of natural surface waters from $-2°C$ to almost $100°C$ (e.g. Mellanby 1939, 1940a; Hynes 1970; Kinne 1970a; Brock 1975, 1978, 1985). As a broad generalisation, among the poikilotherms, temperature tolerance decreases with physiological and morphological complexity (Table 4.1). Within taxa, however, temperature tolerance can vary with individuals and even within the lifetime of any one individual. Thus the concept of temperature

TABLE 4.1
Upper temperature limits for aquatic organisms. Data from studies
of geothermal waters (from Brock 1975)

Group	Temperature (°C)
Animals	
Fish and other aquatic vertebrates	38
Insects	45–50
Ostracods (crustaceans)	49–50
Protozoa	50
Plants	
Vascular plants	45
Mosses	50
Eucaryotic algae	56
Fungi	60
Procaryotic micro-organisms	
Blue-green algae	70–73
Photosynthetic bacteria	70–73
Non-photosynthetic bacteria	>99

tolerance for any organism is complex and tolerance ranges for various functions are very specific (Fig. 4.1).

Essentially, however, most animals and plants can survive over a genetically predetermined temperature range, which can be characteristic of the species. This range can be modified by many factors, but for any species the ultimate upper and lower lethal temperatures vary little from those determined genetically.

Organisms can be broadly categorised as:

—Cold stenotherms: those organisms with narrow tolerance ranges (e.g. ±10°C) in arctic regions.

—Warm stenotherms: those organisms with narrow tolerance ranges in warm regions (±10°C) in the tropics.

—Eurytherms: those species with wide tolerance ranges (more than ±30°C), e.g. in temperate or sub-tropical regions. This category may be further sub-divided with respect to distributions within any region.

The survival and life processes of most aquatic organisms are dependent to a great extent on water temperature, though many species have developed behavioural and physiological strategies for survival in apparently hostile natural conditions, for example the shores of arctic or tropical lakes or the intertidal zones of temperate seas (see Chapter 2).

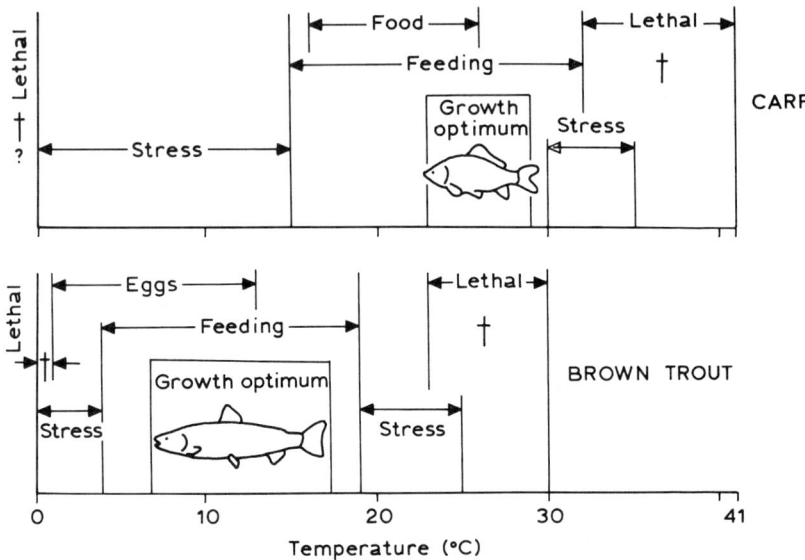

FIG. 4.1   Comparison of thermal requirements of carp (*Cyprinus carpio*) and
brown trout (*Salmo trutta*) (after Elliott 1980).

It is safe to assume that any natural body of surface water will contain a
biocoenosis comprising species tolerant of and adapted to its normal
temperature extremes. However, in most habitats there are species which
are not permanent members of the community, either because they are at
the limits of their tolerance ranges or because they are transients passing
through the habitat.

Most of the data on the temperature tolerances of organisms in natural
habitats have come from studies of their distribution and from the detailed
studies of natural geothermal habitats (Brock 1975, 1978; Castenholz and
Wickstrom 1975) (Table 4.1). Natural tolerances of the major taxa are
discussed in the subsequent chapters.

4.2.2 Effects of Temperature Changes

(i) Classification of effects
Fry (1967) produced probably the clearest classification of temperature
effects on organisms. He described three main categories, viz.

—Lethal effects: high or low temperatures which will kill an organism
   within a finite time, usually less than its normal life span.

—Controlling effects: sub-lethal temperatures which affect physiological or biochemical processes, such as growth, metabolic rate or reproduction.

—Directive effects: which cause behavioural responses, movements or migrations.

A fourth category can be distinguished which probably includes influences of all three but is worth considering separately, viz.

—Indirect effects: caused by the action of temperature on some other factor, e.g. oxygen or toxins, which in turn affects organisms. Indirect biotic effects may also be predicted because of effects of temperature on food species, competitors, predators or parasites (e.g. Gibbons, 1976).

(ii) Lethal temperature
The concept of lethal temperature, that is, of heat or cold (Block 1974; Wolters and Coutant 1976), causing the death of an organism, is perhaps the simplest concept on first consideration. However, there are many interrelating factors which can influence lethal temperatures. Some of these relate to the methods used to apply the stress, that is 'external' factors, others to the state of the organism itself, that is 'internal' factors which can also include behavioural responses. These factors are of extreme importance to the survival of organisms *in situ* when exposed to any stress, including temperature stress.

The factors which can affect the lethal temperature at any one time are:

—rate of change in temperature (up or down);
—duration of exposure;
—acclimatisation, that is, the previous temperature history of the organism;
—life history stage of the organism;
—physiological state of the organism, that is, effects of other stresses, either natural or applied;
—adaptive strategies, for example behavioural or physiological modifications.

(a) Lethal temperature assessment. The classical methods of assessment can be divided into two main categories, namely: exposure of the organisms either as individuals or groups to test temperatures by a direct transfer from another temperature in which they have been maintained for a given time. The period of exposure after which a proportion or all of the organisms die

is used as a criterion of lethality. This method produces the following values: LT_{xy}, where LT is the temperature causing death, x is the time of exposure, and y is the proportion of organisms dead after that exposure.

For example, the $LT_{24,50}$ for any species is the temperature at which 50% of the exposed group died after 24 h exposure.

A variation of this is to express the value as a time, for example, $TL_{30,50}$ which is the time taken for 50% of the organisms to die when exposed to 30°C.

The second method usually involves exposure to a gradual change in temperature until death. Both methods have been used extensively for assessing lethal temperatures and are suitable for statistical treatments (Fry 1967).

In recent work the criterion of 'effect' has been modified from death to some behavioural or physiological change known to preclude death. The 'critical thermal maximum' (CTM) is the temperature at which there is some observable but usually reversible behavioural or physiological change, for example loss of movement or of orientation. This can be assessed by either of the above methods but is generally more applicable to the latter, e.g. (Cowles and Bogert 1944; Gonzalez 1974; Sylvester 1975a; Everich and Gonzalez 1977; Becker and Genoway 1979; see Houston 1982). The CTM can also be affected by the rate of temperature change (Cox 1974) and the size of the organism.

Lethal temperatures of fish can be estimated from data on optimum growth and preferred temperatures using the regression models produced by Jobling (1981).

The general equation is

$$y = ax + b$$

The various relationships are as follows,

    Growth optimum (x)
    Lethal temperature (y)

$$y = 0.76x + 13.81 \qquad (r = 0.866)$$

    Final preference (x)
    Lethal temperature (y)

$$y = 0.66x + 16.43 \qquad (r = 0.880)$$

    Growth optimum (x)
    Final preference (y)

$$y = 1.05x - 0.53 \qquad (r = 0.937)$$

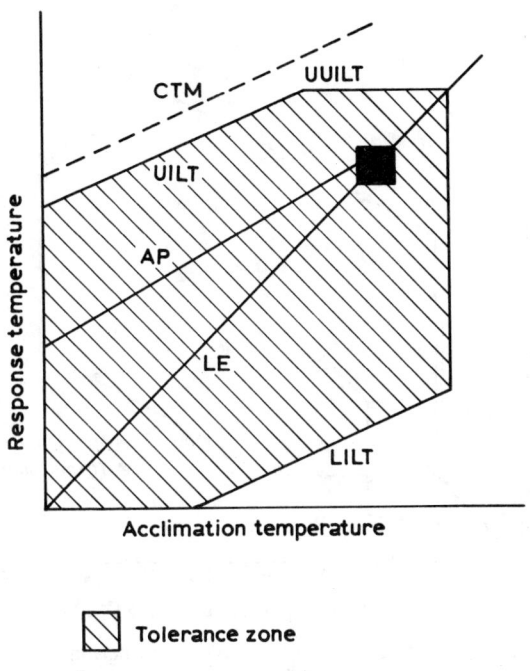

Tolerance zone

Zone of final preferendum

FIG. 4.2 The temperature relations of fish (CTM, critical thermal maximum; UILT, upper incipient lethal temperature; LILT, lower incipient lethal temperature; UUILT, ultimate upper incipient lethal temperature; AP, acute thermal preferendum; LE, line of equality; after Jobling 1981).

The lethal temperature data used mostly originated from LT_{50}s at set times. The terminology and definitions most frequently used in lethal temperature assessments are shown in Fig. 4.2. Species in the same habitat can have widely different tolerances (Elliott 1980), though they are rarely as extreme as shown in Fig. 4.3, which compares the temperature tolerances of two species from contrasting habitats.

Various methods have been used to provide lethal temperature data. For example Fry and his co-authors (1946) acclimatised fish to predetermined temperatures by raising the temperature by 1°F per day and then exposing them to test temperatures for 14 h. The highest temperature at which all survived, the lowest at which all died and the 50% survival value were all recorded.

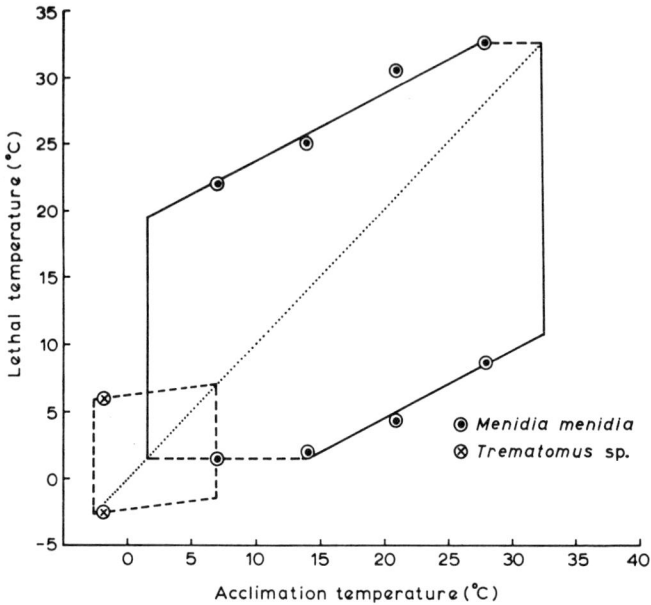

FIG. 4.3   Temperature tolerances of a cold-stenothermal fish and a warm-eurythermal fish from different regions (after Brett 1970).

Macan (1963) observed that the temperature causing 100% mortality differed from that causing 50% mortality by only about 2–3°C in most cases.

The UILT and LILT (Fig. 4.2) are those at which 50% of any population should survive indefinitely. Within the hatched area is the 'zone of tolerance' where any acclimatized individual should survive indefinitely. Death is usually rapid once the CTM is exceeded (Fry 1967).

The size of the tolerance zone is a useful measure of the degree of eurythermy of any species. For example, Fig. 4.3 illustrates clearly the extreme differences in the genetically determined tolerances of the cold-water stenotherm from the antarctic and the warm-water eurytherm from the Atlantic. Similar polygons can be constructed for all organisms.

(b) Factors affecting lethal temperature

Rate of change and duration of exposure. Sudden changes in temperature do not necessarily cause immediate death provided that both temperatures are within the tolerance limits (Hoss *et al.* 1974; Widdows 1976;

Schubel *et al.* 1978; Hartwell and Hoss 1979; Burton *et al.* 1980–1981). However, temperature changes which expose organisms suddenly to temperatures outside their UUILT or ULILT (Fig. 4.2) or the CTM for the given acclimatisation temperature are likely to kill organisms rapidly depending on the final temperature reached. Even if exposed to potentially lethal temperatures, the survival of an organism may depend upon the duration of exposure (see Houston 1982).

The duration of exposure has long been known as an important factor determining the effects of both temperature and toxic substances on fish, and recent studies have shown similar trends for other organisms (e.g. Brungs 1971; Hair 1971; Mountford *et al.* 1974; Ginn *et al.* 1976; Schubel *et al.* 1978; Ito 1980). The need for data to predict the effects of 'entrainment' exposures (i.e. for periods from 2 min to 1 h) on organisms passing through power stations has led to massive amounts of experimental work over the past 10 years (e.g. Frank 1974; Schubel 1974; Hettler and Clements 1978; see Schubel *et al.* 1978; Lawler, Matusky and Skelly 1979: Patten 1980; Talmage and Ospresko 1981; Chung and Strawn 1982; Itzkowitz *et al.* 1983). As we saw in Chapter 2, the times of exposure of organisms to temperature changes will vary with the design of the cooling-water system. Thus the calculation of survival for species passing through systems requires somewhat more than the simple time and temperature exposure data.

Figure 4.4 shows the method of estimating the potential effect of temperature on the mortality of organisms entrained in a cooling system based on physical data and the time–temperature tolerances of the organism. At a temperature of 30°C the 50% mortality time for the chosen organism is > 120 min. As it takes only 5 min to pass through the cooling system, the temperature rise is therefore unlikely to cause a 50% mortality. The prediction could also be adjusted to take CTM values into account.

The relevant equations for predicting 50% mortality are as follows:

$$\log t = a + b \ (\text{temp.})$$

where t is the time of exposure in minutes and the temperature is in °C, and a and b are the regression coefficients (Schubel *et al.* 1978).

By setting the equation differently the conditions for survival can be assessed, e.g.

$$1 \geq t/10a + b(\text{temp.})$$

If the result is equal to or less than 1, the organism is likely to survive.

The protection of species can be predicted better if the CTM values are incorporated into the equation (see Schubel *et al.* 1978). Also if the exposure

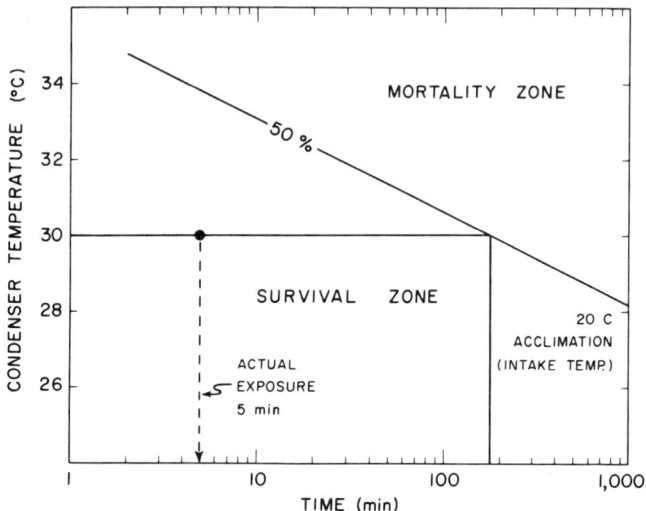

Fig. 4.4 Example of how a thermal resistance curve can be used to predict whether or not mortality from thermal stresses will be expected for entrained organisms. (Consider a plant with an intake temperature of 20°C, a ΔT of 10°C, and a transit time of 5 min. Assume the organisms are acclimated to 20°C and that cooling takes place instantaneously.) (After Schubel *et al.* 1978.)

to decreasing temperature needs to be assessed the equation can be modified to account for the incremental losses through time, until the temperature falls below the lethal threshold. The effects of prolonged exposure in thermal discharge canals or cooling ponds can also be estimated using the sum of time–temperature exposures at reducing temperatures such as would be experienced in a plume. The equation would read

$$1 \ll \text{Time}_1/10^{(a+b)}; \quad \text{Temp.}_1 + \text{Time}_2/10^{(a+b)}\text{Temp.}_2$$
$$+ \text{Time}_n/10^{(a+b)}\text{Temp.}_n$$

where $\text{Time}_{1,2,n}$ are those for which organisms are exposed to temperatures 1, 2 and n in set increments during passage through the cooling system. Results less than unity again indicate survival, if temperature is the only factor to consider (see Chapter 3).

Acclimatisation and acclimation. Both terms are commonly used in the literature and can be regarded as essentially synonymous. If there is a distinction, it is that the former is generally applied to the phenotypic adaptation of an organism to its natural environment (Mellanby 1939; 1940a,b,c; Precht *et al.* 1973; Kaya 1978), while the latter applies to the normal experimental technique of keeping organisms in predetermined conditions prior to testing.

The classic relationship between acclimation and lethal temperature is shown in Fig. 4.2. Four characteristics of acclimation are significant to the effects of a thermal discharge, particularly for sessile organisms or those in enclosed waters, namely:

—There is a consistent direct relationship between acclimation temperature and lethal temperature (or CTM) both upward and downward until the UUILT or ULILT are reached. After this point the acclimation temperature is equal to the lethal temperature (Fry *et al.* 1946; Schroeder and Callaghan 1981; see Houston 1982).

—Acclimation to higher temperature tolerance is generally accompanied by decreased tolerance to lower temperatures (Brett 1970; Coutant *et al.* 1976; Shafland and Pestrak 1982).

—Natural acclimation is mainly, though not exclusively, a seasonal phenomenon in any one species in any one habitat (Fry 1967). Thus as a general rule, aquatic organisms will tolerate higher temperatures in summer than in winter provided that the UUILT is not exceeded.

—temperature tolerance in the same species can differ over its geographical range such that the warmer the habitat the higher the UUILT (Fry 1967; Matthews 1986).

Acclimation occurs relatively rapidly in fish, usually faster than 1°C in 24 h (Fry 1967), but such a rate of acclimation is still much slower than the rate at which experimental temperatures may be raised. For this reason many of the experimental data may be regarded as conservative. It is also difficult to assess the point beyond which many organisms will not recover when temperatures are being raised rapidly. The CTM or death point of sessile organisms can only be assessed properly therefore by measuring metabolic processes or observing moving parts such as cilia.

Acclimation also affects the time of survival of organisms exposed to thermal shock and other short-term exposures. Organisms acclimated to fluctuating temperatures in short-term cycles, generally adapt and attain increased tolerance to the median values (Widdows 1976; Houston 1982), and, in various studies, have shown increased tolerances of 50–60%. Acclimation to fluctuating temperatures also appears to increase resistance to short-term exposures to potentially lethal temperatures.

Physiological state. The presence of another physiological stress clearly affects the temperature tolerance of organisms. For example, Parasitised snails (*Gonobiasis* sp.) have been shown to be more susceptible to chemical

and thermal stress than uninfested individuals (Cairns *et al.* 1975). Coho
salmon infected with the bacterium *Aeromonas* sp. suffered higher
mortalities at higher temperatures than uninfected fish (Groberg *et al.*
1978). The reverse was, however, true for those fish infected with bacterial
kidney disease (Sanders *et al.* 1978).

It is, in fact, well known that combinations of stresses are more lethal
than single stresses and that certain combinations are truly synergistic
(Jones 1964; McKenney and Dean 1974; Cairns *et al.* 1975, 1978). Thus in
polluted waters it is predictable that temperature tolerance may be reduced
when compared with clean waters. Also fatigued animals are often less
tolerant to external stresses (e.g. Schneider *et al.* 1974).

Temperature tolerance can also be affected by salinity (Lewis 1966;
Alabaster 1967; Brett 1970; Hartnoll 1978; Cotter *et al.* 1982; Yagi and
Ceccaldi 1984), oxygen and salinity combined (McLeese 1956; Tagatz 1961;
Almatar 1984), water hardness (Murphy *et al.* 1976) or physical factors such
as pressure (Knight-Jones and Morgan 1966). Clearly there are points
where one factor is dominant, i.e. at lethal levels (e.g. Fig. 4.5).

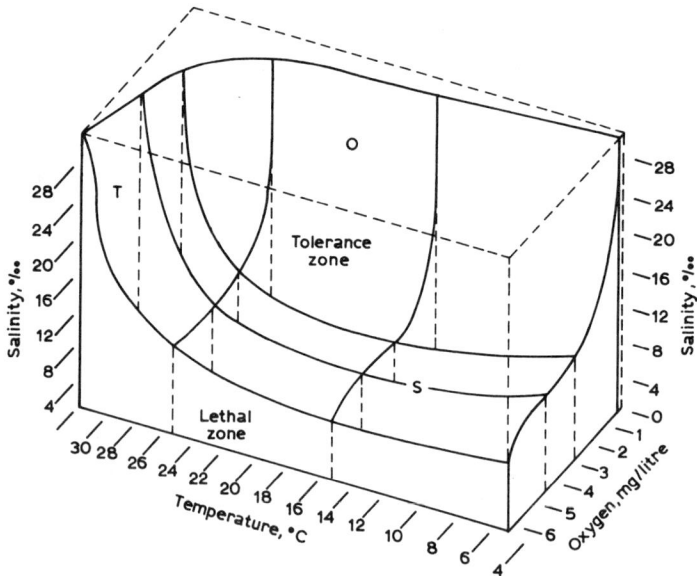

FIG. 4.5   Diagram of the boundary of lethal conditions for lobsters for various
combinations of temperature, salinity, and oxygen. (T is the region in which
temperature alone acts as a lethal factor, S is the region in which salinity alone acts
as a lethal factor and O is the region in which oxygen alone acts as a lethal factor.)
(Adapted from McLeese 1956.)

In aquatic plants desiccation, pressure, light and salinity can all influence temperature tolerance (Gessner 1970; Kinne 1970a; see Precht *et al.* 1973). *Behaviour and internal temperature control.* Most mobile aquatic species show behavioural responses to temperature change (McCauley 1977b; see Hocutt *et al.* 1980) and this is one of the most important assets for survival in heated waters. Fish, for example, can detect temperature changes as small as 0·05°C and selection is controlled by the neuronal system (Muller and Fry 1976; Crawshaw 1977; Stauffer 1980). Crawshaw stated that 'two important events are initiated by changing water temperatures' namely:

—transmission of peripheral information to the central nervous system, and
—thermal exchange with the altered environment.

The body temperature of most fish follows water temperature very closely but the rate at which any individual equilibrates depends on body shape and size and factors such as gill ventilation, blood flow, activity and water movement (e.g. Spigarelli *et al.* 1977; see Houston 1982). Temperature perception occurs in the early stages of the life history and is evident in larval fish very soon after hatching.

Clearly in the smaller organisms, body size will be a major factor determining the rate of equilibration, and as most invertebrates and plants are relatively small, the rate will be rapid, and for most species can be regarded as almost instantaneous. The time lag between heat being applied and the internal organs reaching higher temperatures can be vital to the survival of organisms which can detect and avoid adverse conditions (e.g. Spigarelli *et al.* 1974; Dizon *et al.* 1977; Spigarelli *et al.* 1983). There has been a variety of experimental apparatus used to establish the behavioural responses of both fish and invertebrates, ranging from the outdoor wooden channels used in the British studies in the 1950s (Alabaster 1958; Alabaster and Downing 1966) to the complex 16-chambered 'selectatron' used by Reynolds and Thompson (1974), and electronic shuttle boxes in which the fish themselves are able to control the temperature (McCauley 1977a; Reynolds 1977). In most cases heat has been supplied by immersion heaters, heated blocks or heating lamps.

Because of the variety of experimental techniques the results for the same species have tended to differ. For example the preferred temperatures for rainbow trout (*Salmo gairdneri*) shown by experiments in a horizontal trough differed markedly from those shown by experiments in a vertical column. However, Stauffer (1980) considered that there is ample evidence to show that fish can detect and avoid adverse temperatures and also select areas of temperature which, for some reason, suit their behavioural or physiological requirements. This also appears to apply to mobile invertebrates (Hocutt *et al.* 1980). Species from the same macro-habitat show differences in temperature preferences which may reflect natural niche separation (Ehrlich *et al.* 1979; Reynolds and Casterlin 1979; Spigarelli and Thommes 1979; Shrode *et al.* 1982).

Many factors affect temperature selection and avoidance in any one species, including acclimation, season, habitat, photoperiod, diet, hormone levels, age, salinity, toxins, pathogens, social rank, growth and size (e.g. Beitinger and Magnuson 1975; Beitinger 1976; McCauley and Huggins 1976; Otto *et al.* 1976: Stuntz and Magnuson 1976; Gray *et al.* 1977; Hokanson 1977; McCauley 1977b; McCauley *et al.* 1977; Richards and Ibara 1977; Beitinger and Magnuson 1979; Ehrlich *et al.* 1979; Hestagen 1979; McCauley and Casselman 1981; Ingersoll and Claussen 1984) though Stauffer (1980) suggests that some of the results prior to 1980 are ambiguous. He does conclude, however, that the relationship between acclimation temperature and the final preferred temperature is best expressed by the quadratic equation:

$$P = \alpha A + \alpha A_2 + \beta$$

where P = preferred temperature and A = acclimation temperature in °C. α and β are constants.

Excellent reviews of behaviour in relation to temperature are available (e.g. Crawshaw 1977; Richards *et al.* 1977; Hocutt *et al.* 1980; Cherry and Cairns 1982).

Predictably 'upper avoidance temperature' is higher than the preferred temperature but lower than the upper lethal temperature (Stauffer 1980).

The data on temperature preference and related behaviour by invertebrates are scarce (Hynes 1970; Kinne 1970c). Nymphs of some Odonata exhibited some temperature preference when placed in heated troughs (Martin and Gentry 1974; Gentry *et al.* 1975). Avoidance temperatures of 34–35°C were reported for blue crabs (*Callinectes sapidus*) and 15–20°C for the American lobster (*H. americanus*) (see Hocutt *et al.* 1980).

Sessile invertebrates clearly have little chance of avoiding or selecting temperatures though Newell (1970) notes references to temperature preferences of intertidal species which could explain their distribution in the habitat. Many species show ingenious mechanisms for controlling body temperatures and for avoiding heat death and desiccation in intertidal habitats (Kinne 1970c; Newell 1970).

The abilities to detect, select and avoid adverse temperature conditions are collectively known as 'behavioural thermo-regulation' and are vital to the survival of species in both natural and altered habitats. Internal temperature control is a feature of several species of fish (e.g. Smith and Rhodes 1983). For example some tuna fishes may have muscle temperatures up to 8°C higher than the surrounding water though the duration of such high temperatures is not known (Brett 1970). In contrast Brett (1970) also refers to the large sunfish (*Mola mola*) which can have body temperatures 2–4°C lower than those of the sea surface. In any situation mobile animals may enter water at above their lethal temperature provided that they remain for very short periods and do not allow the body temperature to reach lethal levels (see p. 295). Some reptiles and amphibians show both behavioural and physiological thermo-regulation but this is more common in terrestrial than aquatic forms (e.g. Mellanby 1940b; Fry 1967; Precht *et al.* 1973).

Among aquatic species, the alligator has been well studied. The preferred temperature for *Alligator mississippiensis* is reported as 32–35°C and lethal temperatures as 38–39°C (Colbert *et al.* 1946, *loc. cit.* Glassman and Bennett 1978).

Size and stage in the life history. Many of the lower organisms, e.g. bacteria, fungi, algae, protozoa and plants have resting stages such as spores, cysts or seeds which are extremely resistant to stresses, including extremes of temperature or desiccation over long periods (Christophersen 1973). In contrast, many fishes and larger invertebrates are more vulnerable to stress as larvae or in the juvenile stages than as adults (Schubel *et al.* 1977, 1978).

Prediction of lethality. It is clear from the above that the concept of lethality is by no means simple and that unless a change in temperature is likely to exceed the criteria of UUILT, or ULILT, by a wide margin for all life-history stages and without any possibility of behavioural thermoregulation the prediction of the direct lethal effects of temperature in a thermal discharge may be a complex problem. It is also clear that lethal temperature data from simple experiments are not satisfactory for assessing effects of a discharge on an ecosystem.

(iii) Controlling effects

(a) At cellular level. The causes of heat death are based on the effects of temperature on the cells and metabolic processes (see Stauffer 1980), though the precise mechanisms are still debated. The fact that some individuals survive after short exposures to apparently lethal temperatures suggests

that the cell and tissue damage may be reversible, provided that a certain threshold of exposure or dose is not exceeded. On the other hand, although there may be some recovery, some tissue mechanisms may yet be irreversibly damaged (Szybalski 1967; see Stauffer 1980; Houston 1982).

Early predictions of the effects of thermal discharges were based on the predictions of rises in metabolic rates and the subsequent effects on the physiology of exposed organisms. The effects of temperature on metabolic processes have been described using the equations derived from the temperature coefficient or Q_{10} value, which is the ratio of the rate of a process at a given temperature to the comparable rate $10°C$ lower (Farrel and Rose 1967). The equation is:

$$Q_{10} = Kt + 10/K_t$$

where K is the velocity constant of the reaction and t is the temperature.

The rate of a reaction doubles for each $10°C$ rise in temperature. The Arrhenius equation which is used to calculate rate of biochemical reactions is:

$$K = A\,e^{-E/RT}$$

where K is the velocity constant, T is the temperature in degrees Kelvin and A is a derived constant.

The calculations of metabolic rates from the equations are not simple, however, and there are differences between 'in vitro' and 'in vivo' estimates which are not always easily explained (Farrel and Rose 1967; see Precht *et al.* 1973). Kinne (1970b) is critical of the Q_{10} coefficient as a predictive ecological tool mainly because 'biological processes are known not always to fit the chemical laws'.

Metabolism and activity. Most of the effects of temperature on the body processes of organisms follow a similar model in that they increase with temperature to an optimum after which there is a rapid decline to the death point. The main processes for which data are available are oxygen consumption, growth and activity and the reproductive functions, namely, gonad maturation, spawning, hatching and development. The maintenance of any population in a habitat depends on the organism maintaining all metabolic processes successfully.

The rate of oxygen consumption in animals or respiration and photosynthesis in plants are the main criteria used to measure metabolic rates. Figure 4.6 shows the relationship between temperature and oxygen consumption for a fish at different levels of activity. The patterns for other organisms are similar (Fry 1967; Brett 1970; Garside 1970; Gessner 1970; Kinne 1970a,b; Oppenheimer 1970).

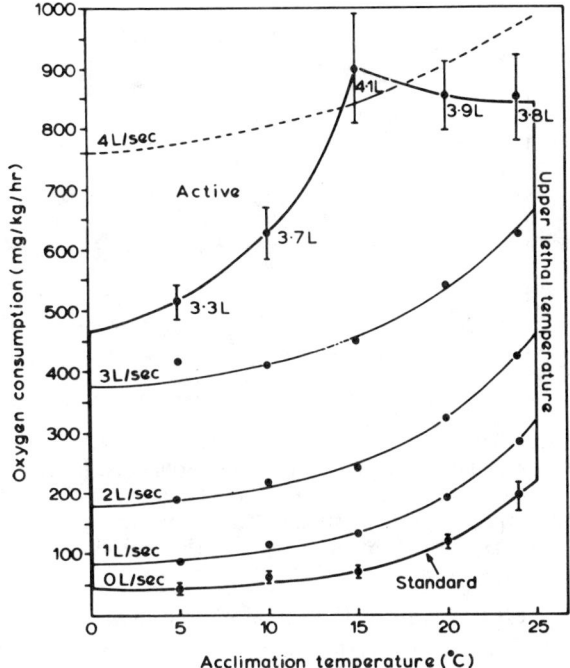

FIG. 4.6 Relation between oxygen consumption and temperature at various swimming speeds for yearling sockeye salmon (18 cm long; 50 g). (Standard and active rates are indicated as mean ± 2 S.E. The 60 min 'critical speed' accompanying each active rate is shown. Swimming speeds of 1–3 lengths L/s were obtained by interpolation; those at 0 L/s were obtained by extrapolation.) (From Kinne 1970.)

Fry (1967) described three levels of oxygen consumption, viz.

—Standard: the minimum rate of a resting organism.
—Routine: the mean rate measured during normal activity with no unusual outside stimuli.
—Active: the maximum steady rate under forced activity (or caused by an abnormal outside stimulus).

The active rate for sockeye salmon (*Onchorhynchus nerka*) is approximately 12 times the standard rate (Brett 1964). Routine rates for other fishes are 2–4 times the standard rates.

'Activity', as used here refers to movement of some kind, either total mobility or the action of some body structure such as cilia. In most animals the basic models relating activity to temperature are similar. As an

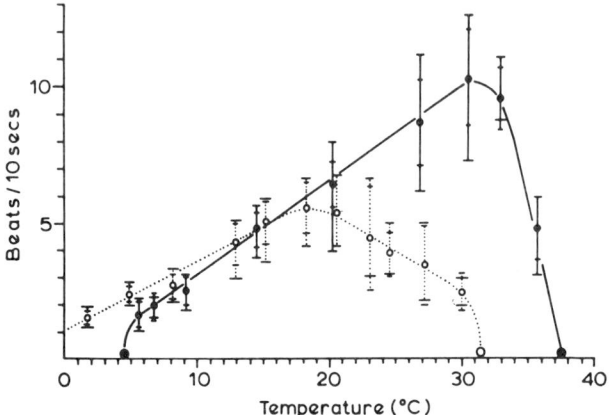

FIG. 4.7   Cirral activity as a function of temperature in the cirripedes *Balanus balanoides* (dotted line) and *Chthamalus stellatus* (solid line). (Small circles indicate mean rate at each temperature, large cross lines the range of observations and the small cross lines the standard deviation on each side of the mean. The large circles indicate absence of cirral beating.) (From Kinne 1970.)

example of a sessile organism, Fig. 4.7 shows the cirral activity of barnacles in relation to temperature. Although the optima vary with species the model is similar in principle for all groups of animals.

Growth rates and size. Provided that no other criteria are limiting, for example food supply or oxygen, growth rate is closely related to temperature. The range for efficient growth is generally narrower than for other metabolic processes (Fig. 4.1). Fluctuating temperatures affect the growth optima in various ways (Hokanson *et al.* 1977; Biette and Geen 1979). The relationship between preferred and optimal growth temperatures were described previously (see p. 106). Light and daylength are also significant growth factors for plants and animals and their effects are closely related to temperature (e.g. Gessner 1970; Precht *et al.* 1973; Clark *et al.* 1981; Boehlert 1982). Salinity changes also affect growth at set temperatures depending upon the extent to which the salinity exceeds the normal range for the individual (Hoonbeek *et al.* 1982; see Houston 1982) and the effects of other factors such as the availability of food (Peters *et al.* 1976).

Food supply is of course a vital factor. At temperatures which increase metabolic rates outside the normal rate for that season, a lack of food will result in starvation and death (Wurtsbaugh and Davis 1977; Allen and Wootton 1982). As an example, the maintenance ration for sockeye salmon,

(*Onchorhynchus nerka* at 1°C is less than 0·5% day^{-1} but increases to 1·45% day^{-1} at higher temperatures (see Brett 1970, 1979; Houston 1982).

The rate of food conversion is related to temperature though the optimum temperature for this is often lower than the optimum growth temperature (Fry 1967: Brett 1970). This is because the exponentially increased maintenance requirements at the higher temperature reduce the energy available for growth.

The ultimate size of any individual can depend on many factors, among which is its geographical location and the temperature regime. Kinne (1970c) notes that many marine species attain larger body sizes in colder than in warmer regions. Some aquatic insects which overwinter as aquatic stages and emerge from early summer onwards are larger as adults than those individuals which grow in summer from early eggs (Macan 1957, 1958b, 1963; Langford 1975). Similar phenomena also occur in the estuarine amphipod *Gammarus duebeni* and the hydroid *Cordylophora caspia* under experimental conditions (see Kinne 1970c). The effect is related to the onset of sexual maturity in that delayed maturity in colder conditions allows somatic rather than gonadial growth.

Feeding and digestion. The rates at which bacteria absorb nutrients and at which plants and animals feed and digest, increases with increasing metabolic rate and therefore with increasing temperature given suitable conditions (see Farrel and Rose 1967; Oppenheimer 1970; Precht *et al.* 1973). For example, the maximum meal size taken by fish was found to increase with temperature to an optimum and then decline (Hathaway 1927; Peters *et al.* 1974). Board (1972) showed that the feeding, and hence tunnelling rates of wood-boring molluscs in experimental systems were clearly related to water temperature.

Digestion rates follow a similar pattern. For example the hydranths of the marine hydroid *Clava multicornis* digested food 2–2·5 times faster as temperatures rose from 12°C to 25°C (see Kinne 1970c). Experiments with fish have produced significant relationships between digestion rates and temperature, mostly expressed by simple logarithmic regression equations (see Brett 1979; Peters *et al.* 1974; Elliott 1976, 1980; Hofer *et al.* 1982).

At lower temperatures there can be a delay period between ingestion and the onset of digestion begins (Peters *et al.* 1974). The food material can also affect digestion rates. In experiments it was found that the roach (*Rutilus rutilus*) digested vegetable material quicker than animal material at all temperatures (Hofer *et al.* 1982).

Reproduction and early development. Reproduction and growth are clearly related to temperature in most organisms (e.g. see Kinne 1970a,b;

Masse *et al.* 1978; Morawska 1984). Under experimental conditions micro-organisms show increased rates of reproduction up to their optimal temperatures provided that all other conditions are suitable (Farrel and Rose 1967; Oppenheimer 1970). Aquatic plants produce more gametes at higher temperatures, given the correct daylength (Langridge and McWilliam 1967: Gessner 1970; see Precht *et al.* 1973).

The temperature range at which reproduction takes place in both invertebrates and vertebrates is generally lower than that over which growth occurs (see Fig. 4.1), except for unusual cases such as the oyster drill (*Urosalpinx cinerea*) where reproduction occurs at higher temperatures than the optima for growth and feeding (see Kinne 1970c).

In general given that the organism is physiologically prepared and that all other conditions are suitable the time of reproduction in temperate and sub-tropical aquatic animals depends on temperature (Horoszewicz 1983). In the tropics species are not so dependent on temperature though many do show a periodicity in reproduction probably related to the optimisation of space and food.

The rate of development of eggs of aquatic poikilotherms after fertilisation is dependent upon temperature under normal conditions (EIFAC 1968b; De Sylva 1969; Brooke 1975; Sastry, 1976, 1978; Guerin and Reys 1978; Guma'a 1978; Sarvala 1979).

Alderdice and Velsen (1978) described the relationship between the incubation time of fertilised eggs and temperature. The simplest equation is:

$$y = k/x$$

where y is the development time in days, k is a constant and x is the water temperature in $°C$.

More complex relationships have been described, though this equation gives satisfactory approximations for most species. In euryhaline organisms the combination of temperature and salinity can influence hatching success and development significantly and the size of larvae at hatching (e.g. Schubel and Auld 1974; Schubel and Koo 1976; Schubel *et al.* 1977; Hrs-Brneko 1978; Olla and Samet 1978; see Houston 1982).

Meristic characters and morphology. Meristic changes in fish can be caused by exposure to higher temperatures during embryonic development (EIFAC 1968b; Garside 1970; Coutant and Talmage 1976, 1977; see Langford 1983a). Carp (*Cyprinus carpio*) and roach (*Rutilus rutilus*) showed over 20% of deformed individuals when hatched at 20° C. Brooke (1975) also found that abnormal fry of Coregonids were more abundant after hatching at 10°C than at 4°C. Changes in the numbers of gill rakers, vertebrae, pyloric caecae fins and fin-rays have also been caused by temperature changes and other ecological variables (Moodie 1985).

(iv) Directive effects (behaviour and movements)
Many studies have shown that mobile organisms select or avoid appropriate temperatures in water (see Fry 1967; Kinne 1970a,c; Meldrim and Gift 1971; Martin and Gentry 1974; Reutter and Herdendorf 1974; Cherry *et al.* 1975, 1977a,b; Olla *et al.* 1975; Ott and Forward 1976; Reynolds and Casterlin 1976, 1978; Stuntz and Magnuson 1976; Coutant 1977; Medvick and Miller 1979; Peterson and Sutterlin 1979; Hocutt *et al.* 1980; Houston 1982). As we have seen already (p. 113) temperature selection and avoidance are likely to be important factors determining the distribution and tolerances of animals around thermal discharges both in the long and short terms (Stauffer *et al.* 1980). In natural habitats the mass migrations and diel vertical migrations of some organisms are also thought to be related to temperature (Harden-Jones 1968; Kinne 1970b,c; Casterlin and Reynolds 1982). The drifting behaviour of invertebrates in streams and rivers can be triggered by sudden increases in temperature (Wojtalik and Waters 1970). Short-term alterations in the thermal regime of a water body can also change the thermal preferences of an organism acclimatised to the natural regime (Calhoun *et al.* 1982).

Behaviour other than simple movements or migrations are related to temperature. For example eels (*Anguilla anguilla*) and yellow bullheads (*Ictalurus natalis*) both showed increased aggression at higher temperatures under experimental conditions (Nyman 1972; McLarney *et al.* 1974; Knights 1987). The preferred temperature for some species is also affected by the presence of dominant individuals (Power and Todd 1976).

(v) Indirect effects, combined effects and synergisms
The effect of the elimination or attraction of a plant or animal species in relation to an outfall area is likely to have an effect on other species, particularly if it is a food species or a predator. In experiments, thermal shocks caused mobile animals to be more susceptible to predation (Deacutis 1978). Also, rapid cooling ($>2°C\,min^{-1}$) made small channel catfish (*Ictalurus punctatus*) more susceptible to predation than slower rates (Coutant 1973; Coutant *et al.* 1976), presumably because of changes in their behavioural responses, though in these experiments the size of catfish may also have been significant. Similar effects have been found with salmonids (Sylvester 1972). The combined and synergistic effects of temperature with

other stresses are well known (e.g. Klein 1962; Jones 1964; Hawkes 1969; Hiryama and Hirano 1970; Stewart *et al.* 1972; Duever and Abernethy 1974; Vernberg and Vernberg 1974; Burton *et al.* 1976, 1980; Cairns *et al.* 1976a,b, 1978; Vernberg 1978). In general higher temperatures exacerbate the effects of poisons, though in the case of phenols and ammonia, higher temperatures over a part of the tolerance range do not do so. The potential synergistic effects of temperature and chlorine are of considerable importance to thermal discharge effects *in situ*.

4.3 EFFECTS OF CHEMICAL CONTAMINANTS

There are many groups of organisms which can tolerate unusual chemical conditions in natural waters, for example thermal springs (Brock 1978), organically polluted rivers and saline inland waters (Hynes 1960, 1970; Perkins 1974) and waters polluted by metals (Hynes 1960; Klein 1962; Perkins 1974; Mance 1987). The various biotic indices used to assess the effects of pollution on communities, are generally based on the differential tolerances of the various organisms (Hellawell 1986).

4.3.1 Chlorine and Related Biocides
All the biocides used in cooling-water systems whether chlorine and its derivatives, or specific fungicides, algicides, herbicides or molluscicides, are, by definition, potentially toxic to organisms either in general or to the target groups. As with temperature, the effects of any biocide vary with taxon, duration of exposure, with environmental factors such as salinity, oxygen, or the presence of other pollutants and with biological factors such as age and size of the organism or physiological state. At sub-lethal concentrations most biocides can affect physiological and behavioural systems and, like temperature changes, produce controlling and directive effects.

Microcosm studies with established marine communities and chlorine have shown that there are also clear community effects (e.g. Sanders and Ryther 1980). Over 24 months, continuous dosing with chlorine at 50 ppb caused reductions in species-richness in large-scale micro-cosms, but neither intermittent dosing at up to 50 ppb nor continuous dosing at 10 ppb had any effect on the composition of the communities (Vanderhorst *et al.* 1983).

(i) Lethal effects
The estimated threshold acute and chronic concentrations of chlorine for freshwater and marine organisms, expressed as μg litre^{-1} TRO, were

deduced in a model by Mattice and Zittel (1976). Several aspects of the model have been criticised (Seegert and Boġardus 1980; Turner and Thayer 1980; Hall *et al.* 1981). Modified models and regression equations, shown in Fig. 4.8, suggest that the critical values from the Mattice and Zittel model may have been too high to protect a few species. The criteria set for surface waters in the USA were $2 \mu g$ litre^{-1} for salmonid fish and $10 \mu g$ litre^{-1} for other fresh-water and marine organisms but such single criteria do not take account of specific conditions at sites, chlorine decay, duration of exposure and of complex chemical reactions (Larson *et al.* 1977; Hall *et al.* 1981; Mitchell and Cech 1983).

For some species the residual by-products of chlorination, for example chloramines, may be more toxic than free chlorine (see Evins 1975; Morgan and Carpenter 1978; Roberts and Gleeson 1978; Scott *et al.* 1980; Thomas *et al.* 1980; Venkataramiah *et al.* 1983), but the reverse may be true for other species (Capuzzo *et al.* 1977a). Turner and Chu (1983) derived simple regression models to describe the relationships between chlorine, its residuals, other environmental factors, duration of exposure and toxicity. Table 4.2 shows the LC_{50} values derived from the analysis. Clearly, free chlorine is much more toxic than combined residuals. Combined toxic effects are shown in Table 4.3, in relation to other parameters.

Intermittent chlorination can be less harmful than constant chlorination at lower concentrations because of the short exposure times, provided that

TABLE 4.2

Comparisons of the toxicity of chlorine and its residuals (after Turner and Chu 1983)

	LC_{50} (mg litre^{-1})	
	24 h	*48 h*
Fresh water[a]		
Free residual chlorine (FRC)[b]	0·06	0·04
Combined residual chlorine (CRC)[c]	0·47	0·37
Total residual chlorine (TRC)[d]	0·14	0·10
Sea water[a]		
Total residual oxidants (TRO)[e]	0·20	0·16

[a] Regression equation: n = observations; r = correlation coefficient.
[b] Log concentration = 1·15–0·75 log duration ($n = 88$; $r = 0·58$).
[c] Log concentration = 0·71–0·33 log duration ($n = 64$; $r = 0·52$).
[d] Log concentration = 0·96–0·57 log duration ($n = 438$; $r = 0·53$).
[e] Log concentration = 0·31–0·32 log duration ($n = 109$; $r = 0·39$).

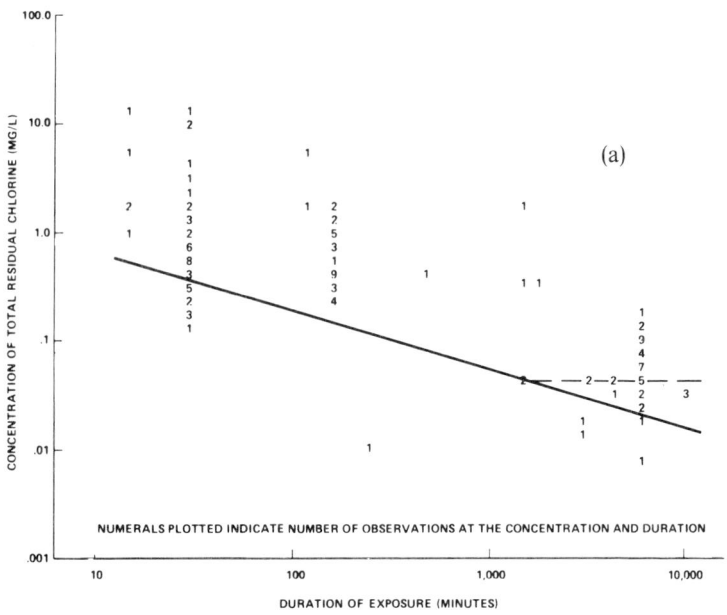

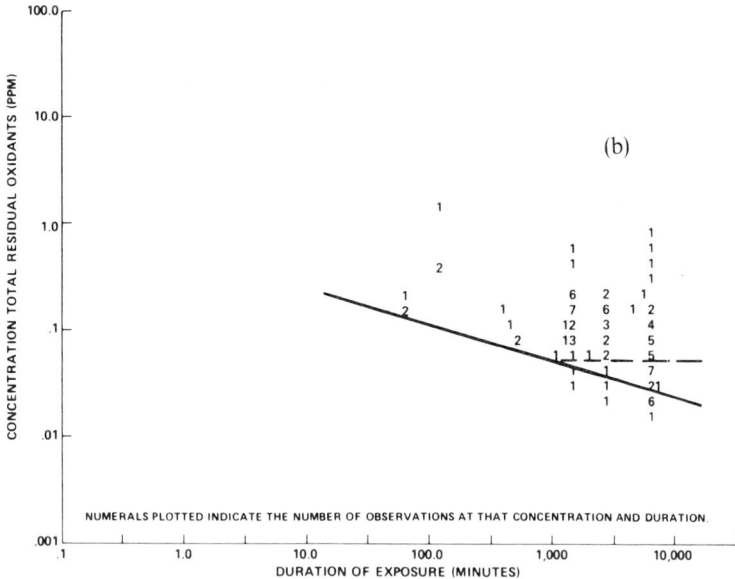

Fig. 4.8 Alternative models for effects of chlorine on aquatic organisms (cf. Mattice and Zittel 1976). (a) Alternative freshwater model, log (conc) = 0·81 − 0·55 log (duration); (b) alternative marine–estuarine model, log (conc.) = 0·10 ± 34 log (duration) (after Turner and Thayer 1980).

TABLE 4.3

Comparison of calculated LC_{50} values derived from simple and multiple regression analysis (after Turner and Chu 1983)

Model[a] (*range of secondary independent variable in data base*)	LC_{50} (*mg litre^{-1}*)	
	24 h	*48 h*
Simple TRC[b]	0·14	0·10
TRC with pH[c] (pH range: 6·5–8·9)	0·17–0·70	0·13–0·56
TRC with alkalinity[d] (Alkalinity range: 42·3–318 mg/litre)	0·21–0·76	0·16–0·57
TRC with hardness[e] (hardness range: 45·3–467 mg $CaCO_3$/litre)	0·25–0·27	0·23
TRC with temperature[f] (temperature range, 4–32°C)	0·06–0·39	0·04–0·27
Simple TRO[g]	0·20	0·16
TRO with salinity[h] (salinity range, 6·5–35‰)	0·15–0·29	0·13–0·24

[a] Regression equation: n = observations; p = probability of F-test (MS regression/MS deviation).

[b] Log concentration = 0·96–(0·57) log duration; $n = 438$; $p < 0.001$.

[c] Log concentration = -1.49–(0·31) log duration + (0·26) pH; $n = 117$; $p < 0.001$.

[d] Log concentration = 0·54–(0·41) log duration + (0·002) alkalinity; $n = 37$; $0.005 < p < 0.01$.

[e] Log concentration = 0·03–(0·19) log duration–(0·000 04) hardness; $n = 27$; $0.10 < p < 0.25$.

[f] Log concentration = 1·38–(0·53) log duration–(0·3) temperature; $n = 434$; $p < 0.001$.

[g] Log concentration = 0·31–(0·32) log duration; $n = 109$; $p < 0.001$.

[h] Log concentration = 0·28–(0·24) log duration–(0·01) salinity; $n = 91$; $p < 0.001$.

concentrations are not immediately lethal or that the total dose is not lethal (e.g. Dickson *et al.* 1974, 1977; Brooks and Seegert 1977, 1978; Seegert *et al.*, 1977; Larson and Schlesinger 1978; Brooks and Liptak 1979). Temperature and other stresses clearly can affect the tolerance of species to chlorine (e.g. Waugh 1964: EIFAC 1973; Groth 1975; Gibson *et al.* 1976; Thatcher *et al.* 1976; Cairns 1977; Heath 1977, 1978; Middaugh *et al.* 1977, 1978; Truchan 1977; Scott and Middaugh 1978; Capuzzo 1979a; Burton *et al.* 1978; Hall *et al.* 1979, 1981; Margrey *et al.* 1981; Vanderhorst 1982; Hall *et al.* 1983; see Langford 1983a; Fava and Seegert 1983; Murray 1983; Scott *et al.* 1985) but increased sensitivity to chlorine is most evident at temperatures near lethal values (Turner and Chu 1983). These last authors concluded that the response was not linear and varied with species. On average, invertebrates

Ecological Effects of Thermal Discharges

TABLE 4.4

Comparisons of the toxicity of chlorine residuals to vertebrates and invertebrates (after Turner and Chu 1983)

Analysis	Equation (concentration = $mg\,litre^{-1}$; duration = min)	Derived LC_{50} values	
		24 h exposure	48 h exposure
Marine–estuarine			
Vertebrate species	log concentration TRO = 0·12–0·28 log duration ($n = 69$; $F = 16·6$; $p < 0·001$; $r = 0·45$)	0·17 mg TRO litre^{-1}	0·14 mg TRO litre^{-1}
Invertebrate species	log concentration TRO = 1·09–0·51 log duration ($n = 40$; $F = 9·7$; $p < 0·005$; $r = 0·45$)	0·30 mg TRO litre^{-1}	0·21 mg TRO litre^{-1}
Freshwater			
Vertebrate species	log concentration TRC = 0·75–0·43 log duration ($n = 138$; $F = 166·1$; $p < 0·001$; $r = 0·74$)	0·25 mg TRC litre^{-1}	0·18 mg TRC litre^{-1}
Invertebrate species	log concentration TRC = 1·10–0·63 log duration ($n = 300$; $F = 45·1$; $p < 0·001$; $r = 0·36$)	0·13 mg TRC litre^{-1}	0·08 mg TRC litre^{-1}

were more tolerant to chlorine than vertebrates in saline waters but the reverse was true for fresh waters (Table 4.4). There is no ready explanation for this. The relative toxicities of selected residual chlorine compounds for one species are shown in Table 4.5.

Dechlorination of effluents has been attempted by various methods (see Chapter 10), and most are effective at decreasing the toxicity of chlorinated water.

(ii) Controlling effects

Chlorine doses, both constant and intermittent, at below lethal levels can affect histopathology and physiological processes, and experimental data exist for most groups of aquatic organisms (e.g. McLean, 1972; Brungs 1973, 1977; Bass and Heath 1977; Bass *et al.* 1977; Morgan and Prince 1977, 1978; Zeitoun 1977, 1978; Block *et al.* 1978; Larrick *et al.* 1978; Hall *et al.* 1981; Dempsey 1983; Hose *et al.* 1983a,b,c; Scott 1983). Brooks and Liptak (1979) also showed that the photosynthesis of freshwater phytoplankton inhibited by chlorine, could recover from doses which produced residuals of <0.1 mg litre^{-1} TRC, though concentrations greater than 0.5 mg litre^{-1} were lethal. Similarly Hirayama and Hirano (1970a) also demonstrated the recovery of photosynthesis in cultures of *Chlamydomonas* exposed to chlorine residuals.

TABLE 4.5

The relative toxicities of chlorine compounds to one species of fish, *Gambusia affinis*[a] (after Mattice and Tsai 1983)

Compound	Chemical formula	LC_{50}[b] ($mg\ litre^{-1}$)
Dichloramine	$NHCl_2$	0·366
Hypochlorous acid	$HOCl$	0·455
Cyclohexylmonochloramine	$C_6H_{11}NHCl$	0·547
Ethylmonochloramine	C_2H_5NHCl	0·646
N-Propylmonochloramine	C_3H_7NHCl	0·673
Methylmonochloramine	CH_3NHCl	0·799
Monochloramine	NH_2Cl	1·31
N-Chloroethylglycinate	$C_2H_5COOCH_2NHCl$	1·70
Hypochlorite ions	OCl^-	2·21
Ethanolmonochloramine	$HOCH_2CH_2NHCl$	15·4
N-Chlorotrisamine	$(HOCH_2)_3CNHCl$	90·4
N-Chloroglycine	$COOHCH_2NHCl$	575

[a] Mortality assessed 48 h post-exposure.
[b] Expressed as TRC.

Chlorine also affects growth, reproductive processes and meristic characters. For example the crustacean *Gammarus tigrinus* showed inhibited growth, respiration and reproduction after exposure to both short- and long-term chlorination (Poje *et al.* 1983). Marine mussels (*Mytilus edulis*) showed reduced body condition factors when treated with 0·2–0·4 mg litre^{-1} TRC (see Whitehouse *et al.* 1984). The growth rates of plaice (*Pleuronectes platessa*) and sole (*Solea solea*) decreased with increased exposure to chlorine over about 0·02 mg litre^{-1} (Alderson 1974). The hatchability of the eggs of striped bass (*Morone labrax*) was completely suppressed by 0·21 mg litre^{-1} TRC (Middaugh *et al.* 1977). Morphological deformities also occurred in one species of fish after exposure to 0·29 mg litre^{-1} TRO for 29 weeks (Thatcher 1978). Physiological effects reported have included reductions in haemoglobin, changes in respiration and symptoms of haemolytic anaemia in fish exposed to chlorine residuals for varying experimental periods (Capuzzo 1979b; see Hall *et al.* 1981).

Of the fifteen most common chlorine residuals in water, mono-chloramine was found to be the most inhibitory to algal metabolism (Erickson and Freeman 1978). Chlorine residuals also exacerbate the sub-lethal effects of heavy metals, synergistically (Anderson 1983). Synergistic effects have also been demonstrated between temperature and both free chlorine and chloramine (Zillich 1972; Capuzzo *et al.* 1977a; Capuzzo 1979a).

(iii) Directive effects
Avoidance of chlorinated water by both fish and invertebrates has been amply demonstrated. The threshold values for avoidance vary with the dose and with the species. They are also influenced by other factors such as salinity and, more importantly temperature (Tagatz 1969; Stober and Hanson 1974; Ginn and O'Connor 1976; Bogardus *et al.* 1978; Morgan 1980; Stober *et al.* 1980; Hall *et al.* 1981, 1983b; Cherry *et al.* 1982a,b).

Sprague and Drury (1969) reported avoidance of free chlorine concentrations as low as 0·001 mg litre^{-1} by rainbow trout (*Salmo gairdneri*) though there was evidence of preference at about 0·1 mg litre^{-1}. Morgan (1980) analysed the relationship between chlorine-avoidance thresholds of fish and water temperature from various experiments and showed that the thresholds were at 0·01 mg litre^{-1} at 12°C and 0·023 mg litre^{-1} at 30°C. From Fig. 4·9 it is evident that there is considerable variation in the data.

The effects of temperature on chlorine-avoidance are still not clearly defined. Meldrim and Fava (1977) concluded that temperature preference

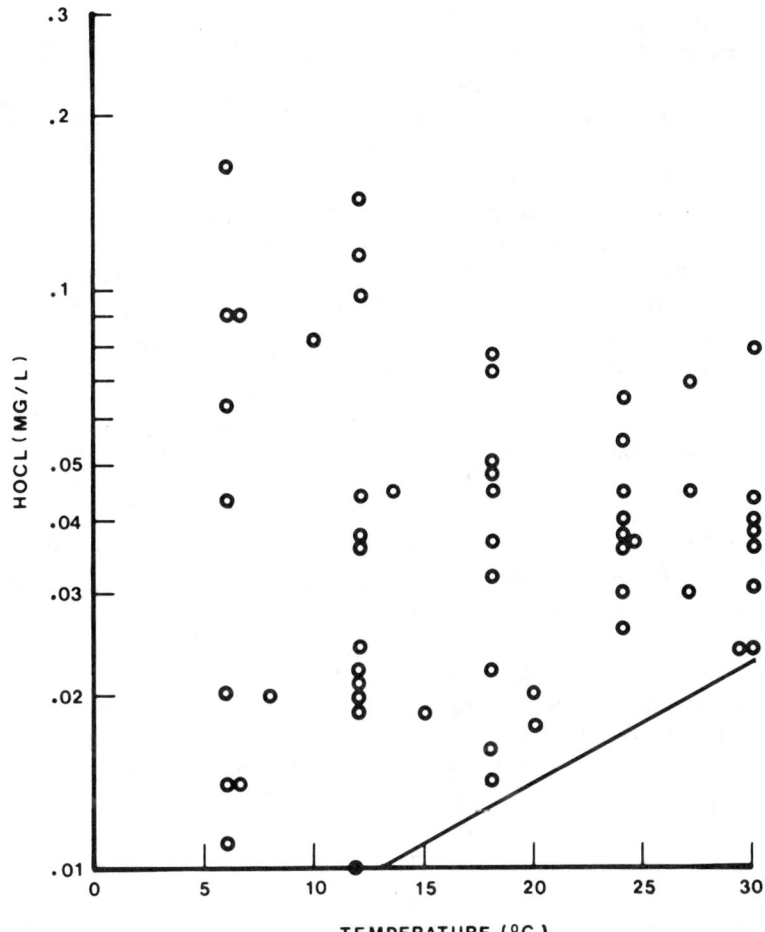

Fig. 4.9   Avoidance of freshwater, estuarine and marine fishes to HOCL as a function of temperature (after Hocutt *et al.*, 1980).

over the range 13–27°C overrode chlorine avoidance to an extent, but this was found not to be true for all species (Hall *et al.* 1983b). Higher temperatures increased the sensitivity of fish to chlorine during intermittent dosing (Giattina *et al.* 1981). Avoidance thresholds are generally specific and influenced by the proportions of different chlorine residuals in the total (Cherry *et al.* 1982b).

(iv) Long-term products and bioaccumulation

The effects of chlorination on the formation of toxic and accumulatable organic compounds such as PCBs or trihalomethanes has been of concern, particularly in countries where sewage effluents are treated (see Jolley and Carpenter 1983a,b; see Chapter 3). Experimental data are scarce (Edgren *et al.* 1979; Gibson *et al.* 1980; Hall *et al.* 1981; Scott *et al.* 1983) though a number of field studies have been carried out near cooling-water discharges (Murray 1979; Edgren *et al.* 1981). In experiments, various marine molluscs did not accumulate bromoform in large quantities and depuration occurred rapidly where any accumulation occurred (Scott *et al.* 1983). The fact that chlorination increased the bioaccumulation of nickel in fish has also been demonstrated (Anderson 1983).

4.3.2 Other Contaminants

(i) Heavy metals

The literature on the effects of heavy metals on aquatic organisms is extensive and covers many species. Two effects are of interest, namely the direct toxic effects of metals in solution, and the bioaccumulation by organisms, which can be summed through the food chain (see for example Klein 1962; Jones 1964; Bryan 1971; Romeril 1972, 1974, 1976a,b; Bryan and Hummerstone 1973; Sandholm *et al.* 1973; Perkins 1974; Morris and Bale 1975; Leland *et al.* 1977; Harrison and Rice 1978; Jones 1978; Oehme 1978; Edgren and Notter 1980; Hodson *et al.* 1980; Eisler 1982; Mance 1987). Both the toxicity of many metals in solution and the ability of organisms to accumulate metals in their tissues are exacerbated by higher temperatures and other factors (e.g. Thatcher 1974; Hoss *et al.* 1975; Fales 1978; Yongue *et al.* 1979) and thus where concentrations are small the effects of bioaccumulation may be significant, especially where human consumption is involved (Pringle *et al.* 1968; Cember *et al.* 1978; Edgren and Notter 1980).

Heavy metals can also influence behavioural thermoregulation, for example by lowering the preferred temperature of a species (Peterson 1976; Macinnes and Calabrese 1978).

(ii) Radioactivity

As we have seen in Chapter 3, small quantities of radioactive isotopes are discharged into water from both nuclear and fossil-fuelled power stations and from most other factory and sewage outfalls (see Klein 1962). The acute effects of radio-toxins are to produce cell damage through reactions with water and protein molecules. Longer term effects are most often centred on genetic materials (see Polikarpov 1966; Eisenbud 1973; IAEA 1976).

Lethal doses of ionising radiation for the range of aquatic taxa differ by almost three orders of magnitude (Polikarpov 1966; Rice and Baptist 1974; IAEA 1976). Figure 4.10 shows the ranges of recorded LD_{50}s for the various major taxa. As a general rule, the more complex the organism the more sensitive it is to radiation. Also, early developmental stages are more sensitive than older organisms. Summaries of experimental data on lethality can be found in Polikarpov (1966) and Eisenbud (1973).

The radiation doses which might cause mortalities of aquatic organisms are many orders of magnitude above those which are likely to be found in thermal discharges (see MAFF 1972 *et seq.*; Tilly *et al.* 1978; Coughtrey 1983; Goldberg *et al.* 1983). Similarly, chronic effects of exposure would be expected to be rare, though authors disagree about the levels which might cause sub-lethal effects after long-term exposures. For example Rice and Baptist (1974) quote Russian research which shows that low-level irradiation caused significant malformations of plaice (*Pleuronectes platessa*) on hatching, but similar American studies found no such effect after much higher levels of irradiation. Temperature can clearly exacerbate

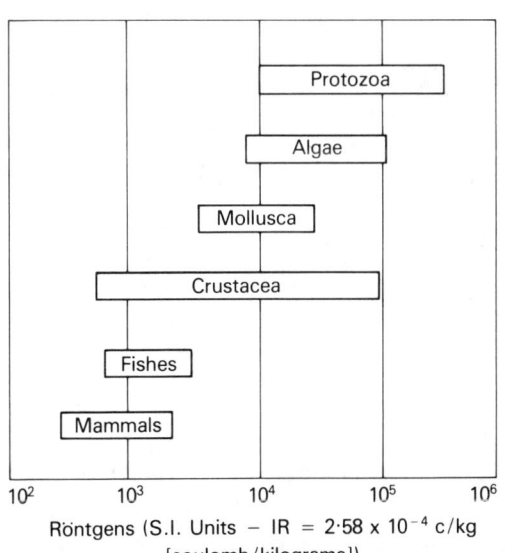

FIG. 4.10   Dose of X-rays or gamma-rays required to kill 50% of organisms (after Rice and Baptist 1974).

Ecological Effects of Thermal Discharges

TABLE 4.6
Concentration factors (CF) for various groups of aquatic
organism (after Eisenbud 1973)

(a) Marine organisms

Element	Group	CF range	Mean CF
Cs	Plants	17–240	51
	Molluscs	3–28	15
	Crustacea	0·5–26	18
	Fish	5–244	48
Sr	Plants	0·2–82	21
	Molluscs	0·1–10	1·7
	Crustacea	0·1–1·1	0·6
	Fish	0·1–1·5	0·43
Mn	Plants	2 000–20 000	5 230
	Molluscs	170–150 000	22 080
	Crustacea	600–7 500	2 270
	Fish	35–1 800	363
Co	Plants	60–1 400	553
	Molluscs	1–210	166
	Crustacea	300–4 000	1 700
	Fish	20–5 000	650
Zn	Plants	80–2 500	900
	Molluscs	2 100–330 000	47 000
	Crustacea	1 700–15 000	5 300
	Fish	280–15 500	3 400
Fe	Plants	300–6 000	2 260
	Molluscs	1 000–13 000	7 600
	Crustacea	1 000–4 000	2 000
	Fish	600–3 000	1 800
I	Plants	30–6 800	1 065
	Molluscs	20–20 000	5 010
	Crustacea	20–48	31
	Fish	3–15	10
Ce	Plants	120–4 500	1 610
	Molluscs	100–350	240
	Crustacea	5–220	88
	Fish	0·3–538	99
K	Plants	4–31	13
	Molluscs	3·5–10	8
	Crustacea	8–19	12
	Fish	6·7–34	16
Ca	Plants	1·8–31	10
	Molluscs	0·2–112	16·5
	Crustacea	0·5–250	40
	Fish	0·5–7·6	1·9
Cu	Plants	—	1 000
	Molluscs	—	286
	Fish	0·1–5	2·55

TABLE 4.6—*contd.*

Element	Group	CF range	Mean CF
Mo	Plants	12–42	23
	Molluscs	11–27	17
	Crustacea	8·9–17·3	13
	Fish	7·6–23·8	17
Mu	Plants	15–2 000	448
	Molluscs	1–3·6	2·2
	Crustacea	1–100	38
	Fish	0·4–26	6·6
Ar–Nb	Plants	170–2 900	1 119
	Molluscs	8–165	81
	Crustacea	1–100	51
	Fish	0·05–247	86

(b) Freshwater organisms

Element	Group	CF range	Mean CF
Cs	Plants	80–4 000	907
	Fish	120–22 000	3 680
Sr	Plants	80–410	200
	Fish	0·85–90	14
Mn	Plants	1 300–600 000	150 000
	Molluscs	1 100–1 600 000	~ 300 000
	Crustacea	1 700–250 000	125 000
	Fish	0·1–400	81
Co	Plants	300–30 000	6 760
	Molluscs	300–85 000	32 408
	Crustacea	—	—
	Fish	60–3 450	1 615
Zn	Plants	140–15 000	3 155
	Molluscs	30–140 000	33 544
	Crustacea	300–4 000	1 800
	Fish	10–7 600	1 744
Fe	Plants	40–45 000	6 675
	Molluscs	20–80 000	25 170
	Crustacea	60–1 800	930
	Fish	0·1–1 225	191
I	Plants	10–200	69
	Molluscs	60–1 000	320
	Crustacea	—	—
	Fish	0·5–25	9
Ce	Plants	200–35 000	3 180
	Molluscs	400–1 500	1 100
	Crustacea	300–1 000	600
	Fish	2–160	81
K	Fish	340–18 000	4 400
Ca	Plants	64–720	350
	Fish	0·5–470	70

the acute effects of irradiation (e.g. Abernethy and Watson 1976). It was found that temperature was the main factor influencing growth and morphological changes in post-larval pinfish (*Lagodon rhomboides*) exposed to various levels of irradiation though these were well above levels recorded *in situ* (Angelovic *et al.* 1973). Low level irradiation appears to stimulate growth in some algae (Polikarpov 1966). Williams and Murdoch (1973) demonstrated long-term effects of exposure on marine organisms. Growth of sponges was reduced at 0.0085 Gy h^{-1}, growth and survival of oysters and corals were only slightly reduced by dose rates of 0.0042 Gy h^{-1} and slipper shells and sea-squirts grew well at dose-rates of 0.085 and 0.017 Gy h^{-1} respectively. The exposure periods varied from 3 to 7 months.

In the shorter term, high doses of radioactivity did not affect the CTM of mosquito fish (*G. affinis.*) under experimental conditions, though some days after exposure to doses exceeding 1500 rads, thermal tolerance was reduced (Blaylock and Frank 1978). The effective doses were, however, very much higher than would be found in any discharge to the environment.

The fact that organisms accumulate radioactive elements in their tissues in the same manner as they accumulate metals is well known and has been widely studied in aquatic organisms (Rice 1963; Gutknecht 1965; Polikarpov 1966; Eisenbud 1973; Rice and Baptist 1974; see IAEA 1974c, 1975b, 1976; van Weers 1975; Hetherington and Harvey 1978). Table 4.6 shows the concentration factors for freshwater and marine taxa estimated from experiments. Two points are obvious, viz.

—There are wide variations within the major taxa, probably reflecting differences between species.
—Concentration factors are greater in freshwater than in marine organisms.

The potential concern about bioaccumulation hinges on the pathways from water to man via the organisms. Critical pathways have been studied and reported, e.g. Preston (1971), see MAFF (1972 *et seq.*); see IAEA (1975a,b,c, 1976, 1979), Hetherington (1976), Hunt (1985, 1989) and Osterberg (1985).

The influence of temperature on bioaccumulation can differ with the organism and with other environmental conditions (e.g. Cross *et al.* 1969; Wolfe and Coborn 1970). For example Harvey (1974) showed that accumulation in some algae was not affected by raised temperatures though others showed an inverse relationship between accumulation and temperature. There was little evidence of enhanced accumulation in molluscs at higher temperatures, except for isotopes of iodine (Patel *et al.*

1975). On the other hand the accumulation rates of ^{60}Co and ^{65}Zn by shrimps (*Crangon crangon*) were faster at higher temperatures though total body burdens were ultimately similar at all temperatures (van Weers 1975). In clams (*Rangia cumeata*), concentration factors of ^{137}Cs increased with temperature but decreased with salinity (Wolfe and Coborn 1970).

(iii) Acids, alkalis, FGD and ash effluents
Although the tolerance of many species to suspended solids, acids and alkalis in fresh waters is well known (Doudoroff and Katz 1950; Jones 1964; Thomson 1963; EIFAC 1965, 1968a; see Alabaster and Lloyd 1980), data on estuarine and marine organisms are scarce (Bamber 1987b). The prediction of the effects of heated FGD effluents from sea-water washing processes is, therefore, difficult as yet, though the effects of the heavy metals in such effluents is more predictable. Bamber (1987b) has shown that there are measurable sub-lethal effects of acidified sea water on at least one marine mollusc.

Langford (1983a) has discussed the effects of contaminants and ash discharges in relation to heated effluents from power stations and the factors affecting the toxicity of coal ash and leachates have been studied (e.g. Cherry *et al.* 1987). Site-specific studies are discussed in the relevant chapters of this volume. The effects of many metals leached from ash and FGD wastes have been studied in both fresh-water and marine organisms (e.g. Birge 1978; Duedall *et al.* 1985a; Osterberg 1985).

4.4 WATER CURRENTS AND SCOUR

Aquatic organisms are adapted to live in habitats where water currents usually vary within specific velocity ranges during their life histories. Apart from providing the basic requirements such as food, oxygen and waste disposal, water currents act as directive stimuli for migratory species and as agents for the dispersal of progeny. Substrate composition and hence the composition of benthic communities is also closely related to water movements (Hutchinson 1957; Macan 1963; Sverdrup *et al.* 1963; Hynes 1970; Newell 1970; Perkins 1974).

It is predictable, therefore, that changes in water currents can have effects on the animals and plants in the vicinity of large outfalls. The displacement of fine sediments from substrates will clearly affect organisms adapted to such habitats.

Water currents are not generally lethal in themselves though where

animals are trapped by water currents against solid objects such as intake
screens or where fish become exhausted mortalities can occur (e.g. Uziel and
Hannon 1979; see Langford 1983a; Turnpenny 1988).

Fish respond in several ways to water currents, such as attraction,
avoidance and variation in swimming speed (e.g. Rickhus 1975). Under
experimental conditions, fish generally respond to water currents by
heading upstream and increasing swimming speed in relation to current
speed, thereby maintaining station, until the critical speed is reached after
which the fish is carried downstream. It is predictable, therefore, that fish
may be swept away from outfalls which produce very rapid currents.

Swimming speed is closely related to temperature in most fish (Fig. 4.11),
up to a point where prolonged forced swimming at elevated temperatures

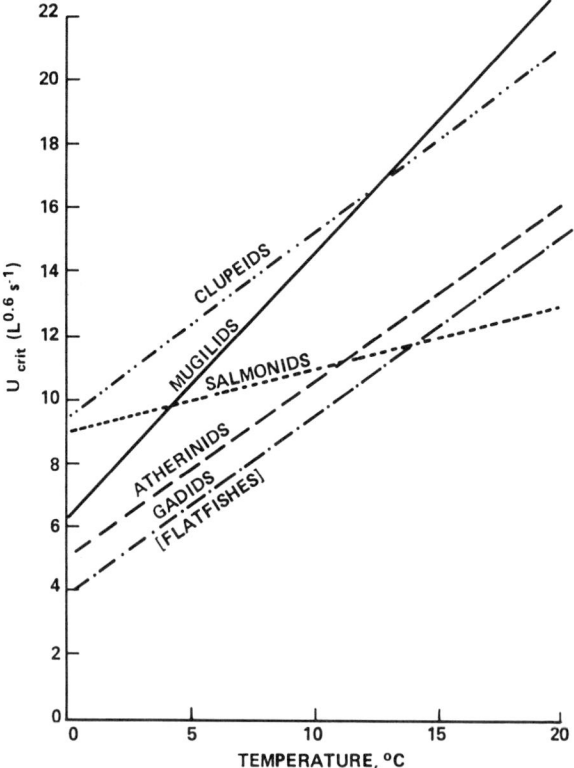

FIG. 4.11   Swimming speeds in relation to body length and water temperature for
various groups of fish (from Turnpenny 1988).

exhausts individuals (Fry 1967; Brett 1970; Hocutt 1973; Kutty and Sukumaran 1975; Rulifson 1977; Gibson 1978; Turnpenny 1984). The fact that fish exhibit temperature preferences and velocity preferences suggests that in some heated areas the two influences may be in conflict. Acceleration and response time to increased current velocities are also influenced by temperature though the relationships are not simple (Webb 1978).

Swimming performance may also be affected by other factors such as oxygen, fatigue, disease or parasitism or combinations of chemical and other stresses (Oseid and Smith 1972; Cairns *et al.* 1978; Hocutt and Edinger 1980; see Langford 1983a). The effects on mobile invertebrates would predictably be similar to those on fish. Most sessile organisms are able to withstand intermittently high current velocities once established though they may have difficulty becoming established in high velocities. Lewis (1964) showed that settlement of mussels (*Mytilus edulis*) on smooth surfaces was difficult at velocities exceeding 2.4 m s^{-1}. Established organisms could withstand velocities up to 3.9 m s^{-1} (Lewis 1964). Burton and Liden (1978) found no settlement of organisms at velocities of 1.5 m s^{-1} at Calvert Cliffs power station in the USA.

CHAPTER 5

Bacteria, Fungi and Other Heterotrophs

5.1 INTRODUCTION

Most of the bacteria, fungi and protozoa in natural waters are heterotrophic decomposers and vital to the degradation of organic materials. Fuller accounts of the micro-organisms in water are given, for example, by Zobell (1946), Oppenheimer (1963, 1970), Morita (1966), Ferguson-Wood (1967), Sykes and Skinner (1971), Mitchell (1972), Rheinmeimer (1974), Fjerdingstad (1975), Jones (1975, 1977), Skinner and Shewan (1977) and Brock (1978, 1979, 1985). Micro-organisms in the water column may be carried into a cooling water system and given suitable conditions can settle on surfaces within the system or be carried through it into the receiving water.

5.2 BACTERIA

5.2.1 In Natural Waters

Bacteria are ubiquitous in natural waters, mostly derived from forms present in soils (Fjerdingstad 1975). Most species are truly heterotrophic though a small group of chemosynthetic autotrophs can derive their energy from the oxidation of ammonia, sulphur and ferrous compounds. Others, the photosynthetic autotrophs, use a chlorophyll-like substance to photosynthesise.

The open sea has a less abundant bacterial flora than most estuaries and fresh waters. Organic pollution increases numbers dramatically, though proper treatment of sewage can reduce numbers by up to 90% (Klein 1962). Vertical stratification of bacteria is well known in lakes (Welch 1952; Hutchinson 1957; Fliermans et al. 1975), and seas (Sverdrup et al. 1963), with a tendency for densities to increase below the thermocline.

138

Numbers vary seasonally with temperature and nutrient cycles. Marine species differ from soil forms in that they form a specific community and, unlike their terrestrial counterparts, require additional sodium, halogen ions, magnesium and calcium. Bacteria are an important food source for many aquatic species, especially those which filter particulate matter from the water column or feed on decaying organic materials. Some forms are parasitic and others are the causative agents of disease and these latter have particular relevance to thermal discharges as we will see later in this chapter.

5.2.2 Temperature Tolerance
Bacteria are generally regarded as the organisms most tolerant to temperature (Brock 1975, 1978, 1979, 1985). Activity of most forms decreases sharply at temperatures above 85°C though there are records of species living at temperatures over 93°C in thermal springs (Brock 1975). Recently, bacteria have been isolated from the superheated, sulphurous waters in the abyssal depths of the Pacific (Baross and Denning 1983). These have been found to grow at 300°C with a 250°C optimum when cultured at 265 times atmospheric pressure. At 100°C they grow actively even when kept at atmospheric pressure. Psychrophilic bacteria mostly grow best at temperatures between 0 and 2°C (see Farrel and Rose 1967), though some forms are reported as metabolising at $-11°C$ (Oppenheimer 1970). Psychrophils are widespread and include the genera *Achromobacter*, *Acaligenes*, *Flavobacterium*, *Proteus*, *Pseudomonas* and *Serratia*. *Pseudomonas* is recorded as containing the largest number of psychrophilic species with *P. geniculata* being the commonest. Some *Clostridium* spp. are anaerobic psychrophils (see Farrel and Rose 1967). The normal survival temperature range for psychrophils is 0–20°C.

Mesophilic bacteria have an optimal growth range of temperature from between 20 and 45°C. These include the majority of the known bacterial species.

Thermophilic bacteria have optimum growth ranges above 45°C. Many grow best at 50–70°C. They are common in such habitats as silage, compost, cooked foods and dung, all of which are, or have been, at higher than average environmental temperatures. Aquatic forms are common in thermal springs (Brock, 1985).

As we have seen in Chapter 2 temperatures in the larger thermal discharges rarely exceed 40°C and thus optimal conditions for obligate thermophils are infrequent. Temperature rises of 10°C in arctic waters would, however, destroy the habitat of some obligate psychrophils.

5.2.3 Chemical Tolerance

Chlorine and other halogens have been used as bacteriocides for many years (Hall *et al.* 1981). Sewage discharges treated with chlorine show reductions in bacterial activity of 99·9% over crude sewage levels (Klein 1962). Copper, lead and hexavalent chromium also suppress bacterial activity at concentrations of $<0·5$ mg litre^{-1}, $1·0$ mg litre^{-1} and $0·01$ mg litre^{-1} respectively. Reductions in pH also produce similar patterns (Klein 1962). Bacteria also accumulate heavy metals and radioactive isotopes very effectively (see p. 130).

5.2.4 Bacteria in Cooling-Water Systems

Bacteria in cooling-water systems can be categorised as follows:

—Fouling types: which settle on surfaces and reduce heat exchange or clog pipework.
—Damaging types: which cause stresses, e.g. corrosion or structural damage through biochemical action.
—Nuisance or pathogenic types: which cause allergies or disease if transmitted.
—Purifying types: which are instrumental in processing the natural cycles, e.g. the nitrogen cycle, and are, for example, commonly found reducing ammoniacal compounds in cooling-tower systems (see p. 83).

There are, of course many bacteria which may not fit into these classifications and are performing other functions in a water body before being drawn into and passed through the system (see Jensen 1978a), but these four categories are of most significance here.

(i) Fouling bacteria

Solid surfaces immersed in water bodies are colonised very rapidly by bacteria and other micro-organisms as the first stages of the development of a fouling community (Holmes 1970a; Rippon 1971; Corpe 1972, 1978; Marshall 1972; Bott *et al.*, 1983; Whitehouse 1989). Gerchakov and Sallman (1978) found metal surfaces populated by bacteria within 4 h of immersion and Corpe (1972) counted up to 2×10^6 cells cm^{-2} on glass slides immersed in sea water for 1 week. The surfaces of culverts, pipes, cooling towers and condenser tubes immersed in water develop thin, visco-elastic layers consisting mainly of bacteria and fungi. These layers increase frictional resistance in pipes causing reductions in flows of up to 55% (Characklis

1978). Bacteria and fungi also grow on the main heat-exchange surfaces in condenser tubes (Rippon 1979). The absorption of fine organic and inorganic particles on to these organisms creates an insulating layer which reduces the efficiency of heat exchange. It has been estimated that this type of microbial fouling was costing the electricity industry in the USA some 400 million dollars each year (1975 prices). The development of microbial films and their relationships with the water–solid surface interface have been described and discussed by Marshall (1972) and Corpe (1978).

The succession of microbial organisms at any one site is generally predictable and in open environments, is the precursor of the total macro-fouling community (Holmes 1970a; Crisp 1965, 1972). Table 5.1 shows the stages of development of the microbial fouling layer on a solid surface (Corpe 1978). The adsorption of the organic particles accelerates bacterial and fungal growth and the exponential growth rate alters to a linear rate once the surface is covered. Limits to oxygen diffusion cause the growth rate to slow as the layer thickens.

The species of bacteria which foul cooling-water systems vary with the community of the inlet water and the conditions in the system. The colonisers require the following properties for long-term survival:

—ability to adhere to wood, concrete or metal surfaces,
—resistance to copper or other metal ions leached from tube surfaces,
—ability to form a slime and trap silt,
—high temperature tolerance and growth optima.

The walls of heat-exchanger tubes may be 10°C warmer than the water passing over them and thus temperatures can exceed 45°C even in a power

TABLE 5.1
Stages in the microfouling of solid surfaces submerged in aqueous environments
(from Corpe 1978)

a. Conditioning stage or molecular fouling of the surface. Adsorption of high molecular weight substances and low molecular weight nutrients. Micro-organisms not directly involved

b. Attraction and attachment of pioneer bacteria to the conditioned surface, chemotaxis by motile bacteria, synthesis of bridging polymers, facilitating attachment

c. Attachment and colonisation by secondary species: stalked, budding and filamentous bacteria; microalgae and various protozoa

d. Accumulative stage: adhesion of particles, dead cells and debris to a substantial microbial film

station cooling system. Thermophilic forms are at some advantage during colonisation. Many species have been isolated from cooling-water systems (McKelvey and Brooke 1959; White 1972; Rippon 1979). Within any cooling system, therefore, there is predictably a community of naturally occurring bacteria along with other micro-organisms which is resistant to heat and contaminants. Bacterial slimes have also been identified in nuclear fuel element storage and cooling ponds at several nuclear power stations. Although the slimes were basically harmless, they were found to accumulate and adsorb radioactive materials and discolour the pond water (Rippon 1971). Biocides and increased pond circulation rates controlled the problems. Chlorine is generally used to control bacterial fouling in cooling-water systems, though Whitehouse (1989) found that it was not particularly effective on fouling layers formed on cooling-tower packing.

(ii) Damaging bacteria
Bacteria have long been suspected as having involvement in the corrosion processes of metals and concrete. Some species are also involved with fungi in the rotting of wood. The corrosive processes include anaerobic electrochemical reactions and chemical changes, for example in pH, caused by bacterial metabolism (Gerchakov and Sallman 1978). Enzyme extracts alone do not cause corrosion, indicating that the living bacteria produce other catalysts during their metabolism. Acidic metabolites such as hydrogen sulphide, acetic acid and sulphuric acid produced by the metabolism of *Desulphovibrio desulphurans*, *Clostridium aceticum* and *Thiobacillus thioxidans* are obvious corrosion agents.

Once a surface is pitted, bacteria are involved in the formation of tubercles which further enhance the corrosion process. On concrete, *Thiobacillus concretivorus* can cause rapid corrosion. The high concentrations of sulphur in concrete provide an ideal substrate for sulphur and sulphate reducing forms. Optimal activity occurs at around $30°C$.

(iii) Pathogenic bacteria
Several groups of human pathogenic bacteria can survive for long periods in surface waters, e.g. *Vibrio cholerae* and *Salmonella* spp. Water borne viruses are also associated with disease though no studies of these in relation to cooling waters have been published. Fish and water-bird pathogens may also be associated with heated waters (Ross and Smith 1974; see also Chapter 9).

Power station cooling systems, particularly those involving cooling towers, have been considered as potential culture systems for several human pathogens and nuisance organisms and have been closely studied (Christensen *et al.* 1984). Faecal coliform 'and other enteric bacteria, e.g. *Escherichia coli*, *Klebsiella* spp. and *Enterobacter* spp. are common particularly where sewage effluent is used for make-up water. Adams and Lewis (1978) concluded that at the five power stations they studied, pathogens such as *Salmonella, Shigella* and *Yersinia* did not occur in sufficient numbers to cause public health problems. McDonald and Bernhard (1978) also found that faecal coliforms did not increase in number in cooling water either at direct or tower-cooled sites. Salmonella was found to be present in 97% of samples taken from the cooling system of a French power station but the densities were similar at intake and outfall (Schwartzbrod and Noel 1980). The effects of chlorination on these organisms are discussed later (see p. 144).

Legionella. This causal agent of Legionnaires disease was first identified as a serious problem in 1976 (Industrial Water Society 1981). It has since been reported from many countries in Europe, Australasia and the United States of America, particularly associated with small industrial cooling systems, air-conditioning units and domestic hot water systems (e.g. Howland and Pope 1983; Orrison *et al.*, 1983; Morton *et al.* 1986).

As an example, the bacterium was isolated from 10 out of 14 cooling-water systems serving air-conditioning units in London (Kurtz *et al.* 1982).

L. pneumophila, of which there are a number of strains, occurs widely in natural habitats and has been isolated from mud, river water and wet soils in the 'wild'. Its optimum growth temperature is around 35°C and temperatures over 60°C are lethal (Industrial Water Society 1981). There is some evidence for an association between *Legionella* and blue green algae or free-living amoebae. The bacterium may be transported for short distances in spray drift from open cooling water units but, so far, there is no evidence of such occurrences in power station water vapour or drift from large cooling towers, which have led to infections.

In the UK a detailed survey of power station cooling waters during 1985 found *Legionella* in the cooling-water systems of nine stations out of 33 using cooling towers. Of the samples taken at the nine stations, 20% contained the organism. Cell concentrations were considered to be below critical levels for infection of humans. The organism was isolated from both intake and effluent water at nine power stations in the USA and showed marked seasonal variations in numbers (Christensen *et al.* 1984). Viability and infectivity were higher in effluents from indirect (closed-cycle) systems. Contrary to expectation, the degree of infectivity was not related to the density or viability of cells and no relationship was found between these properties and the water quality.

A new species, *L. oakridgensis*, has been identified as widespread in the USA (Orrison *et al.* 1983), and ten strains of this species have been isolated. The physiological differences between this and the other seven known species of *Legionella* were confirmed.

(iv) Purifying bacteria
This group includes all those forms involved in the degradation of organic materials in surface waters and which are particularly important to the self-purification processes in polluted waters (Klein 1962). The groups most widely studied in large cooling-water systems are the nitrifying bacteria such as *Nitrobacter* and *Nitrosomonas* which oxidise ammonia to nitrites and nitrates in tower systems (Humphris and Rippon 1978). The chemical changes in cooling water caused by these organisms have been discussed in Chapter 3 (see p. 83).

5.2.5 Effects of Chlorine and Heat within Systems
Chlorination is generally considered to be an effective controller of slimes on heat exchangers (White 1972) though Rippon and Wood (1970) in fact stated that 'Chlorination is rarely effective in killing bacteria (in already-formed slimes) and as slime builds up will be even less so'. The effect of chlorination is to reduce frictional losses in fouled pipes under experimental conditions, presumably as a result of the killing and removal of slime films. Reductions in cell counts on fouled surfaces is also indicated after chlorination (see Characklis 1978; Characklis *et al.* 1980). When surface layers of slimes were removed by anti-fouling procedures the under layers grew at an accelerated rate.

As far as the pathogens are concerned, Snow and Sladek (1978) found that a concentration of $0.18\,\text{mg}\,\text{litre}^{-1}$ TRC produced significant reductions in the numbers of faecal bacteria in a river-fed power station cooling-water system. However, free chlorine concentrations of $1.2\,\text{mg}\,\text{litre}^{-1}$ and chlorinated phenol concentrations of up to $100\,\text{mg}\,\text{litre}^{-1}$ did not eliminate *Legionella* from air-conditioning systems (Kurtz *et al.* 1982). Continuous chlorine concentrations of $1–2\,\text{mg}\,\text{litre}^{-1}$ were not effective either. Berg *et al.* (1985) found that chlorine dioxide was a much more effective control than hypochlorite. There is clearly a need for more research to find a suitable biocide for controlling *Legionella* in both large and small cooling systems (Tyndall 1980; England *et al.* 1982; Fliermans *et al.* 1982).

Nitrifying bacteria are apparently more resistant to cooling water treatments than fouling bacteria though it was clear from experiments carried out at an English power station that the nitrification processes were inhibited by chlorination in a cooling tower system (Humphris 1977). Davis and Coughlan (1978) used the rate of assimilation of ^{14}C-labelled glucose to assess the effects of entrainment on the heterotrophic organisms passing through the cooling-water system of a coastal power station. They found that bacterial activity in the effluent 'virtually ceased when chlorination was in progress'. In fact gross changes in the rate of assimilation (V_{max}) indicated the leakage of small amounts of chlorine into the system which were not measurable by the DPD colorimetric methods. Temperatures of 27–30°C alone did not suppress activity.

Wood (1979) noted that 0·1 mg litre^{-1} of residual chlorine killed off at least 90% of the bacterial flora, assessed from plate counts after entrainment. The effects of passage through a cooling-water system in the absence of chlorine varied with site and with the initial bacterial community. McDonald and Bernhard (1978) found higher concentrations of coliform bacteria in the effluents from tower-cooled stations than in the intake water. The increases could be accounted for however by the normal concentration factors in such systems (see p. 84). No differences were found between intakes and outfalls at direct cooled stations.

5.2.6 Effects of Discharges in Receiving Waters
Brock (1975, 1978) in his extensive work, found that the bacteria characteristic of thermal springs also colonise heaters and steam condensate pipes of industrial and domestic installations. Temperatures in all of these often exceed 70°C.

The obligate thermophile *Thermus aquaticus*, with a tolerance range of 55–80°C is one of the most characteristic and ubiquitous species in such habitats. It is found in the heated springs of Yellowstone and for several kilometres downstream, only disappearing as the water cools below 40°C (see Brock 1975, 1978).

Thermal discharges have been found to cause changes in the abundance and populations of bacteria in a number of receiving waters.

One of the most comprehensive studies was carried out at the James Stuart generating station on the Ohio River (Miller *et al.* 1976). Outfall temperatures ranged from 8·5 to 15·6°C above ambients and the maxima and minima in the discharge canal were 40°C and 12°C respectively. Samples were taken at four stations, one upstream and three downstream of the outfall at distances of < 100 m, 1·6 m (still in the discharge canal) and

TABLE 5.2

Density differences in bacteria between the discharge canal and intake water at the James Stuart power station on the Ohio River (figures compared are in 10^3 bacteria ml^{-1}; modified from Miller *et al.* 1976)

Date	Intake temp. (°C)	Discharge temp. (°C)	Proportional change (outfall/intake)
15 April	9·5	22·5	0·84 −
29 April	13·5	24·6	1·12 +
29 May	19·5	32·0	9·57 +
23 June	27·0	40·3	1·83 +
15 July	28·0	39·5	1·97 +
12 Aug	25·0	41·0	1·44 +
5 Sept	29·5	44·0	2·7 +
30 Sept	25·0	37·0	0·37 −
11 Nov	11·6	21·9	—[a] +
16 Dec	7·0	18·0	1·64 +
6 Jan	4·5	12·0	1·69 +
27 Jan	8·0	18·0	1·7 +
24 Feb	5·8	14·0	0·67 −
17 Mar	7·2	17·8	1·08 +

[a] A nil count for ambient, 0·5 for discharge.

in the river itself about 300 m downstream of the confluence of the canal. Three techniques were used to measure changes viz. counts of stained cells, counts of viable cultures and V_{max}.

Table 5.2 shows the changes in densities of colony-forming bacteria units on cultures, in relation to natural and effluent temperatures at the upstream and 1·6 km station. The conclusions were that the fluctuations in densities of viable bacteria were not related to temperature fluctuations. The counts for the downstream station were generally higher than for the control and the numbers for both stations followed the same trends. Regression analysis showed an inverse but not significant relationship between the numbers and temperature ($F_{(1,27)} = 0.874$; $p = > 0.05$). The densities at the control station were significantly related to those at the downstream station ($F_{(1,27)} = 4.414$; $p = < 0.05$), though the actual peaks did not coincide. The fluctuations in numbers of bacteria in the river were directly related to river flow with the maxima at peak flows.

As might be expected, the V_{max} was related to the temperature with an optimum at 20–25°C. At the higher temperatures in the discharge ($> 35°C$),

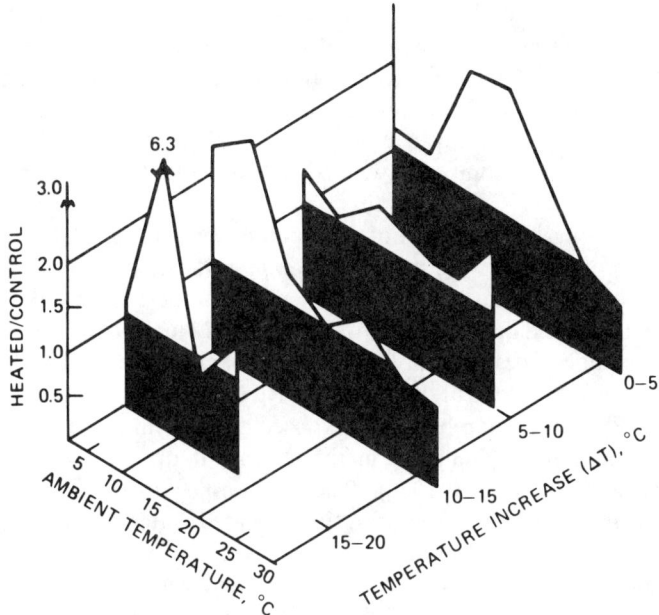

FIG. 5.1   Effects of a thermal effluent on heterotrophic uptake of glucose (V_{max}), as a function of ambient temperature and ΔT (from Miller *et al.* (1976)).

V_{max} declined sharply (Fig. 5.1). Community metabolism was found to be increased in 63% of samples from the heated water. While the increased metabolism could be related to water temperature, the increased densities in the discharge were most likely a result of mixing processes in the cooling-water system (see p. 181) rather than any effect on reproduction.

Bacteria could also be sloughed from internal surfaces in the system at some times of year. The period for which activity was reduced was April to October, when temperatures over 25°C were inhibitory. In the river plume, activity was, on average, lower than in the discharge canal, though 10% of the samples showed an increase. This study is one of the few where effects can be attributed to temperature as no biocides were used.

Fox and Moyer (1973) also found that numbers of bacteria increased in the discharge of the Crystal River power station in Florida. The authors attributed the increase to prolonged exposure to higher temperatures.

In the Connecticut River, the abundance of bacterial populations was unchanged downstream of the Connecticut Yankee nuclear power stations outfall when compared with upstream, though psychrophilic forms and marine bacteria were least abundant at summer temperatures which rose to 30°C (Rankin *et al.* 1974). Solski (1974) found a similar relative change in the Vistula and Skawinka Rivers though here the numbers in the heated reaches were more constant over a year than in the unheated reaches, i.e. around $80\,000\,\text{ml}^{-1}$ compared with $12\,000$–$400\,000\,\text{ml}^{-1}$ throughout the year.

In the cooling-water lagoon of a French power station on the Moselle River, where biocides were not in use, *Salmonella* survived after passage through the power station. There were no differences in the abundance between intake and discharge (Schwartzbrod and Noel 1980), though there were some differences in serotypes (Noel and Schwartzbrod 1981). In the heated areas of Par Pond at the Savannah River nuclear installation, *E. coli*, placed in diffusion chambers and exposed to the unheated and heated waters became more abundant in the warmer than in the cooler zones (Gorden and Fliermans 1978). The lactose-negative strains became dominant in the effluent but not in the unheated waters, indicating some temperature-induced selection (e.g. Kasweck 1978). When reactors were operating, the cooling-water discharge carried quantities of sloughed organic and bacterial material from the dense mats which were formed in the canal.

A chlorinated thermal discharge on an Indiana river, reduced total coliforms, faecal coliforms and faecal streptococci for distances of about 200–400 m downstream of the outfall (Snow and Sladek 1978). TRC values varied from 0.18 to $0.53\,\text{mg litre}^{-1}$ during intermittent chlorination.

In a stream polluted by both coal-ash and thermal effluents, the bacterial communities changed little in their composition but the assimilation rates (of glucose), were reduced by 66–86% in the ash basins. The effects of heat and solids were difficult to identify separately (Larrick *et al.* 1981). However, in another stream, heat affected the stability of the community more than increased mineral loadings caused by ash effluents (Guthrie *et al.* 1974a,b, 1978). Both caused reductions in species-diversity and in the proportions of chromagens in the population. In a natural stream (Bott 1975) found that the doubling time of both unicellular and filamentous bacteria decreased from 42 to 51 h at 0.5°C to 2.8 to 6 h at temperatures of 16.5–21°C. Running and lentic waters studied by other workers also showed increases in bacterial counts when water temperature was raised by 3–5°C, though rises of more than 5°C caused reductions (Cherry *et al.* 1974; Guthrie *et al.* 1974a; Guthrie and Cherry 1978).

As a general conclusion, therefore, it is evident that where bacteria are exposed to prolonged temperature changes, either in natural seasons or in heated waters, some alterations can occur in both community structure and abundance. Short-term temperature exposure has little effect over the range normally found in thermal discharges. The total effects of a thermal discharge will mainly depend on the concentrations of chlorine residuals present.

5.2.7 Fish and Reptile Pathogens
Several species of the bacterium genus *Aeromonas* are pathogenic to vertebrates in water. Udey (1978) found that the progress of several bacterial diseases in Pacific salmonids was a logarithmic function of temperature. In most cases mortality rates decreased with temperature. In fish, the condition known as red-sore disease is reputedly caused by a combination of *A. hydrophila* and the colonial ciliate *Epistylis* (Udey *et al.* 1975; Esch and Hazen 1978; Fliermans *et al.* 1978). The disease is responsible for large-scale mortalities of fish in the South-eastern states of the USA and has also been identified in fish and alligators which died in Florida lakes (Esch and Hazen 1978; Glassman and Bennett 1978). Temperature is believed to be a major factor in the development of the disease.

The incidence of the disease in largemough bass (*Micropterus salmoides*) was closely related to the density of *A. hydrophila* cells in the water column at two sites on Par Pond (Fig. 5.2) (Esch and Hazen 1978). Linear regression analysis showed significant relationships between cell density in month M and infection rate (as a proportion of the catch) in months M, $M+1$ and $M+2$ ($r = 0.53$; $p = 0.05$; $r = 0.62$; $p = <0.01$; $r = 0.56$; $p = <0.05$ respectively). However, there was no significant difference between the incidence of the disease at natural ambient temperatures (16%) and at 5–10°C above ambients (19%). Seasonal plots (Fig. 5.3) showed that there were lower infections of the disease in winter at natural temperatures. There was little statistical evidence of direct relationships between incidence and water temperature. Even so, the authors concluded that normal seasonal temperature fluctuation is a factor influencing the incidence of the disease. Body condition and the presence of external stresses were also significant factors. Figure 5.4 shows the hypothetical relationships between stresses and the incidence of red-sore disease. No mass mortalities were recorded at the sites studied.

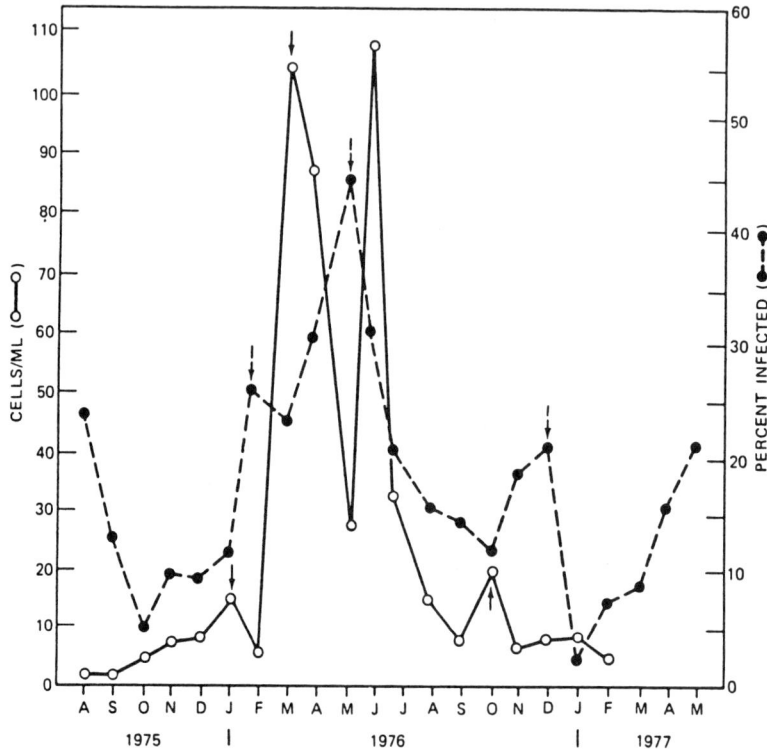

FIG. 5.2   The incidence of red-sore disease in largemouth bass in relation to the density of *Aeromonas hydrophila* cells in the water column of Par Pond (USA)—a heated lake (from Esch and Hazen 1978).

A. hydrophila had a temperature optimum of 'around 33°C', which is colder than the summer temperatures recorded in the heated areas of Par Pond, though upper lethal temperatures were 45–48°C (Fliermans *et al.* 1978; Hazen and Fliermans 1979). The highest densities of cells were recorded in the hypolimnion when the Savannah River plant nuclear reactors were operating and densities were generally higher in the pond than in unheated lakes (Fliermans *et al.* 1978). Thus it was predictable that fish in Par Pond could be more prone to the disease than fish in other waters but may also develop some immunities because of prolonged exposure.

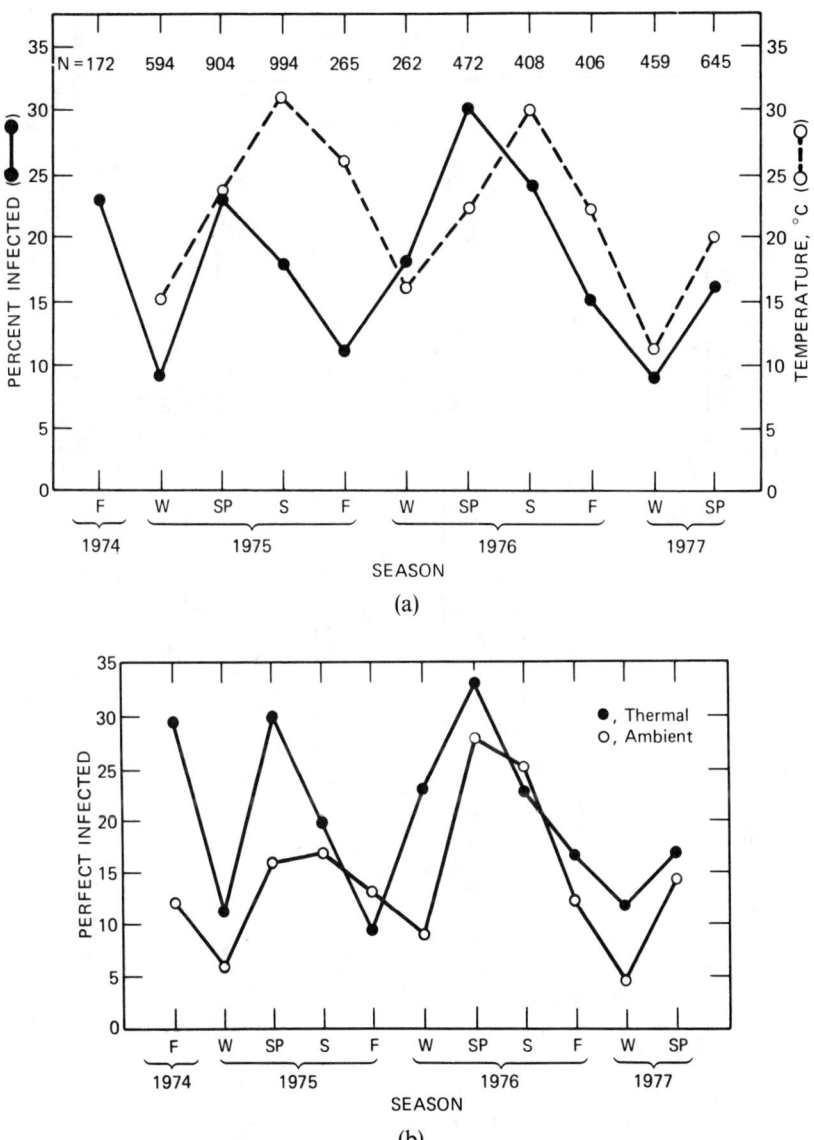

FIG. 5.3 The seasonal incidence of red sore disease in relation to water temperature in Par Pond (USA). (From Esch and Hazen 1978.)

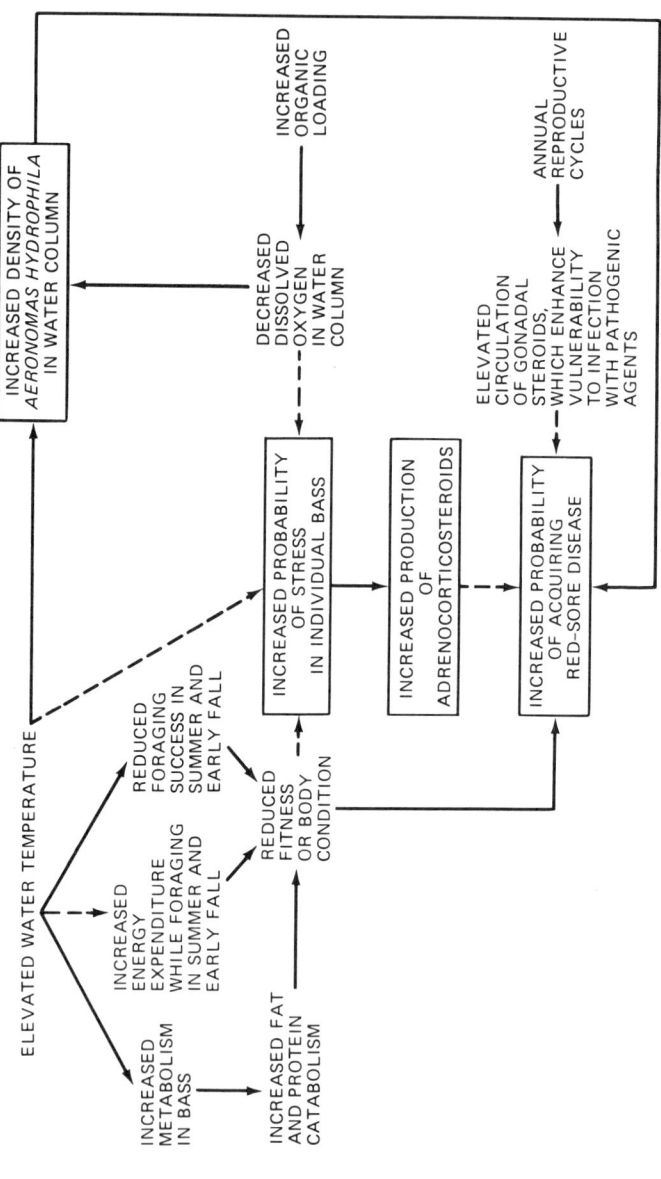

FIG. 5.4 Probable relationships among elevated temperature, increased organic loading, stress, and red-sore disease in largemouth bass. (Known relationships are shown as solid lines; speculated relationships are shown as dashed lines.) (From Esch and Hazen, 1978.)

The alligator, *Alligator mississippiensis*, found commonly in the heated and natural waters of the Savannah River swamps, is also infected with *A. hydrophila* (Glassman and Bennett 1978). In experiments, the highest infection was at about 35°C. Only small numbers of animals were tested and the presence of leeches and a blood parasite complicated the interpretation of the results, but the authors concluded that 'thermal stress' was a major influence on the susceptibility of alligators to the infection. In fact, at 35°C all four experimental animals died.

In the Columbia River, the bacterium causing Columnaris disease (*Chondrococcus columnaris*) is endemic, though it was suggested that it might focus on the heated effluents from the Hanford reactors. It was also suggested that the radioactive isotopes discharged might produce more virulent strains of the bacterium through mutation. To date, there is no evidence of this increased virulence in the wild though experimental exposure to a 10 krad (1000 Gy) dose did actually produce such mutants in experiments (Nakatani 1969; Becker 1973).

The incidence of the disease in coarse (rough) fish ranged from 6% to 9% at all sampling sites. Under natural conditions hatchery salmonids reached peak incidence and mortalities at 17–18°C and at 21°C mortalities declined. Most problems occurred where migrating fish congregated with coarse fish in overcrowded fish ladders in summer. There was no evidence of a simple relationship with temperature. It seems clear, therefore, that although warm water no doubt enhances the growth and survival of pathogens of aquatic vertebrates, there is no clear relationship between temperature *per se* and the incidence of disease.

5.3 FUNGI

5.3.1 Tolerance and Natural Distribution
The fungi found in natural waters have considerable relevance to cooling-water systems because of their involvement in fouling and in the rotting of wood (Eaton and Jones 1970; Jones 1976; see Cheremisinoff and Cheremisinoff 1983). The fungi have a similar or complementary role to that of bacteria in natural waters in that they are primary decomposers of both autochthonous and allochthonous materials and are also food material for higher organisms, notably worms and small crustaceans (Hynes 1970). In rivers more true fungi are found in clean water than in polluted water though some species are clearly associated with pollution

(Klein 1962). In both fresh and saline waters, fungi, particularly some Ascomycetes and Fungi Imperfecti, are especially associated with waterlogged timber. On sea shores several species have been found associated with thalloid algae and one species (*Phyllachorella oceanica*), occurs on Sargassum.

Naturally thermophilic fungi have been known since the late 19th century, but only six species had been described by 1963 (Farrel and Rose 1967). The temperature range for growth of thermophiles is generally between 20°C and 50°C though the highly tolerant species *Aspergillus fumigatus* will grow at both higher and lower temperatures. The temperature limits for thermophilic fungi are different for those of thermophilic bacteria (see p. 139). The responses of fungi to temperature are described and discussed by Farrel and Rose (1967) and Christophersen (1973).

Natural habitats for thermophiles are hay, silage, rotting compost and thermal springs. The role of fungi in natural and polluted waters is discussed in many publications, including Klein (1962), Ferguson-Wood (1967), Christophersen (1973), Rheinmeimer (1974), Cairns (1976) and Brock (1985).

Sewage fungus, evident in organically polluted rivers, and usually seen as white slime or fluffy streamers attached to stones or macrophytes, is a complex community of bacteria, fungi and protozoa mainly dominated by the sheathed filamentous bacterium *Spaerotilus natans* and the true, non-septate fungus *Leptomitus lacteus* (Klein 1962; Curtis 1969). In culture the optimal temperature range for growth is reported as 7–17°C and the outer limits 5 and 40°C, though other reported optima are as high as 25–37°C. There is evidence, however, that the composition of the sewage fungus community differs with location and water quality and this could account for the different temperature optima.

5.3.2 Fungi in Cooling-Water Systems

(i) Occurrence

Fungi are a constituent of the slimes which foul heat-exchange surfaces in cooling-water systems. They are instrumental in binding the fouling material together and making removal difficult. They do not respond readily to chlorination (see Cheremisinoff and Cheremisinoff 1983) and have resistant spores which allow them to colonise treated surfaces very rapidly (Jones 1972). Slime accumulation and fungal abundance are closely related in cooling systems. Fungi have also long been implicated in the deterioration of timber packing in cooling towers and have been at least partly responsible for the collapse of such timbers in the USA and UK (McKelvey and Brooke 1959; Eaton 1976). The causal species are generally among the Basidiomycetes, which may penetrate deeply into the wood or occur in the dampened surface layers. The species involved in the various types of timber rot are described by McKelvey and Brooke (1959).

Fungi isolated from fouling slimes have included the genera *Aspergillus*, *Cephalosporium*, *Paecilomyces*, *Penicillium* and *Trichoderma*. Yeasts have also been isolated from slimes in heat exchangers and cooling towers including *Torula*, *Odium* and *Monilia*. Of the 127 species of fungi isolated from cooling-tower timbers most have come from fresh water because most indirect cooled power stations and factories are sited on river or lakes. Eaton (1976) in recording species from British cooling towers, noted that one species, *Monodictys putredinis*, was common to all locations. There are generally more species found on rotting timbers in fresh water than in the sea.

(ii) Effects of biocides and heat
Chlorination at the levels found typically in large cooling-water systems, is not generally regarded as an effective controller of fungi in such systems (see references in Cheremisinoff and Cheremisinoff 1983; Rosenwig *et al.* 1983), though in heated waters intermittent dosing of more than $1{\cdot}0$ mg litre^{-1} can be effective for slime fungi (McKelvey and Brooke 1959). Ultimately, the removal of persistent fouling layers requires physical or mechanical methods. Pretreatment of timbers with copper salts, chromates or arsenates is an effective anti-fungal measure and fungicides are rarely dosed into operating cooling-water systems today.

5.3.3 Effects of Discharges on Fungi in Receiving Waters
Shearer (1971) recorded fewer species of Ascomycetes and Fungi Imperfecti in a power station discharge canal than in the intake area. Temperatures in the canal varied from about 26 to 35°C during the survey, some 6–10°C above natural ambients. Tansey and Fliermans (1978) recorded 45 species in the Par Pond system of which eight were known to be pathogenic to aquatic and terrestrial animals. Effluent temperatures from the Savannah River reactors could reach 70°C and temperatures over 50°C were common. Tansey and Fliermans sampled cold waters, heated canals and ponds, surface foam, microbial mats and air and soils near the water. Of the known pathogens, *Aspergillus fumigatus* and *Dactylaria gallopava* were considered as most important because of their high incidence and the

TABLE 5.3

Species of fungi recorded at given temperatures from habitats around the thermal discharges from the Savannah River nuclear reactors, USA (extracted from Tansey and Fliermans 1978)

Species	Water temperatures (°C)	Foam[a] (°C)
Aspergillus fumigatus	22–53	30·5–63 (p)
Dactylaria gallopava	51[b]	32·5–63 (p)
Phanerochaete chrysosporium	51[b]	— (p)
Rhizopus nigricans	—	46–56·5 (p)
Rhizopus rhizopodiformis	—	46·5–50 (p)
Emericella (Aspergillus) nidularis	—	47 (p)
Humicola languginosa	—	30·5–42·5 (p)
Mucor pusillus	—	30·5–42·5 (p)
Thilavia terricola	—	30·5–42·5
Malbranchea sulphurea	—	42·5
Thermoascus thermophilus	—	32·5
Thielavia terrestris	—	35
Mucor mehei	—	30·5–37·5 (p)

[a] Nearest water temperature.
[b] One colony only.
(p) = pathogenic or believed so.
NB One measured temperature in foam was 52°C, above water at 68°C.

severity of their infections. Both species were ubiquitous in the Par Pond habitats across almost the whole temperature range (Table 5.3). *A. fumigatus* is pathogenic, allergenic and toxigenic and 'is a frequent cause of disease in humans and other animals' (Tansey and Fliermans 1978). *D. gallopava* is the reported cause of encephalitis in turkeys and chickens in the USA and Australia and may also infect waterfowl and other wild birds. It has been isolated from poultry litter, spontaneously heating stocks of coal, geothermal soils and thermal springs. The studies in Par Pond indicated that the fungus was associated with effluents warmer than 44°C but that the primary habitats were the soil and microbial mats near the water's edge.

However, the paper by Tansey and Fliermans (1978) did not give clear temperature associations. Actual temperatures of the foam, microbial mats and soils were not given and it is misleading to use the temperature of the nearest water as this may differ significantly from that of the relevant microhabitat. It is significant that few cultures could be grown from water samples. The authors do not give any indications of the occurrence of

TABLE 5.4
Fungi recorded at the highest temperature from microbial mats
at the edge of the hot ponds near the Savannah River nuclear
reactors (from Tansey and Fliermans 1978)

Species	Nearest water temperature (range for all habitats)
Aspergillus fumigatus	21–72°C (p)
Dactylaria gallapava	21–72°C (p)
Thielavia terrestris	21–72°C
Talaromyces thermophilus	21–72°C
Rhizopus rhizopodiformis	22·4–72°C (p)
Thermoascus crustaceus	27–60°C (p)

(p) = pathogenic.

disease among either the human or other animal populations around Par
Pond which may be associated with the effects of the fungi.

Only two other species of fungi occurred in water at over 45°C
(Table 5.4). Brock's (1975) conclusion was that the upper limit for fungi was
60°C and this is in fair agreement with the work at Par Pond.

5.4 PROTOZOA

5.4.1 Tolerances and Distribution

Protozoa are ubiquitous as are the other main groups of micro-organisms.
Most species in water are heterotrophic though some groups have
photosynthetic pigments. Many species are planktonic (Zobell 1946; Klein
1962; Ferguson-Wood 1967; Rheinheimer 1974).

Protozoa have been used as biological indicators of pollution for many
years (see Hellawell 1986) but because of the difficulties of taxonomy,
sampling and classification of zones, such systems have fallen into disuse.
Upper temperature limits for the group are around 50°C in natural waters.
Some species of pseudomonads grow readily at 40°C in swimming pools
(Hoadley *et al.* 1975).

In recent years an amoeba pathogenic to humans, *Naegleria fowleri*, has
been discovered in both naturally heated waters and industrial effluents
(Shapiro *et al.* 1980). This free-living amoebo-flagellate is the causal
organism of the disease known as PAME (primary amoebic meningoencephalitis). The disease has an incubation period of 4–7 days and is fatal

TABLE 5.5

The occurrence of *Naegleria fowleri* in heated water in Belgium (from de Jonckheere *et al.* 1975)

(a) Summer 1975

Factory	Date	Place	Temp. (°C)	pH	N. fowleri[a] in 250-ml samples	On solids	Plaque forming units in 1 ml at	
							45°C	37°C
1. Metallurgical	6 Jun	Discharge	27	7·6	+	+	0	ND[b]
		On the outlet across the canal ≈200 m from discharge	23	7·8	+	+	0	ND
2. Electric power plant	29 Jul	Discharge	21	8·1	+[c] (1)	+	0	ND
		In river at discharge	24	8·1	+	ND	0	3
		Idem	24	8·4	−	ND	2	8
	19 Aug	Discharge	24	8·4	−	ND	2	10
		Idem	27	8·0	0	−	0	7
		Idem	27	8·0	0	+	3	7
3. Chemical plant	12 Aug	Brooklet	32	9·1	+	+[c] (2)	0	0
4. Metallurgical factory	12 Aug	Below discharge	27	8·2	+	−	0	0
		At discharge	31	7·8	+	+	0	0
		Above discharge	27	8·4	0	ND	0	0
5. Coking plant	19 Aug	Near discharge	26	8·3	+	ND	0	11
		Discharge	30	8·7	+	ND	0	15
6. Chemical plant	19 Aug	Discharge	31	7·9	0	ND	0	6
		Near discharge	27	7·8	+[c] (3)	ND	0	1

(b) Winter 1975

Factory	Date	Place	Temp. (°C)	pH	N. fowleri[a] in 250-ml samples	On solids	Plaque forming units in 1 ml at	
							45°C	37°C
1. Metallurgical	17 Nov	Discharge	32	8·0	+	ND	1	23
2. Metallurgical	3 Dec	Near discharge	18	7·9	+	ND	0	3
3. Metallurgical	3 Dec	Discharge	21	7·9	−	−	0	3
		Inlet	10	7·7	−	+	0	0
		≈2 km from discharge	12	7·6	0	0	0	0
		≈2 km from discharge	13	7·6	−	0	0	0
		Across the canal at discharge	18	8·1	+	−	0	1
		Discharge	20	8·4	+	0	0	1
		Upstream outlet	16	7·9	−	ND	0	0
4. Chemical plant	3 Dec	Brooklet	17	7·4	+	−	0	0

[a] 0, no amoebae; −, negative for N. fowleri but positive for other amoebae; +, positive for N. fowleri.
[b] ND, not done.
[c] Highly pathogenic for mice when inoculated intranasally. Strain nos (1), 75/14/250; (2), 75/36/5; (3), 75/46/250.

in 2–5 days. *Naegleria fowleri* is naturally thermophilic and can grow at temperatures of 37–45°C. It is intolerant of salinities over 50‰ chloride (Duma 1978). In New Zealand and California and in the UK, the organism has been recorded from hot mineral springs at 37–38°C. In Australia drinking water carried in open channels at 37°C harboured the organism and in other countries swimming pools and a warm stream were discovered to be sources.

5.4.2 Effects of Discharges on Protozoa in Receiving Waters

There are few published studies of protozoa in thermal discharge plumes. Cairns (1969) could not demonstrate differences in the protozoan communities on artificial substrates in the Potomac River upstream and downstream of a thermal discharge. In experiments, he showed that on colonised substrates exposed for periods as short as 7–8 min to sudden temperature rises of 12–30°C, the numbers of species surviving fell sharply. A steady exposure to 30°C for 24 h showed a similar effect. Once restored to natural temperatures most populations recovered their abundance and diversity over periods of 24–144 h. In similar experiments, Cairns and his co-workers demonstrated that intermittent exposure of protozoan communities to chlorine in fresh water caused reductions in diversity. The less tolerant species were destroyed by three successive doses of 1·45 mg litre^{-1} of hypochlorite within a 2-h period. Increasing the frequency of exposures decreased the lethal threshold (Cairns and Plafkin 1975). In these experimental communities 11 species showed high tolerance including *Peranema* sp., *Entosiphon* sp., and *Asidisca* sp., all known inhabitants of sewage discharges. *Stylonichia* sp., is also tolerant of a wide range of salinity and two others, *Oxytrichia* sp. and *Actinophrys* sp., are tolerant of temperatures of 38–52°C and pH values from 3·0 to 1·8. Several other species have resistant spores. Most communities recovered their diversity and abundance when left in uncontaminated water.

Pathogenic protozoa have been found in waters heated by industrial effluents (Shapiro *et al.* 1980). *Naegleria fowleri* was found in a brook near Antwerp, downstream from the heated effluent from a metal-finishing factory (De Jonckheere *et al.* 1975; De Jonckheere and van de Voorde 1977; De Jonckheere 1981). Temperatures reached 26·5°C in the brook and exceeded 30°C at the factory outfall. Table 5.5 shows the occurrence of the organism in heated waters in Belgium. It was not present consistently at any site but occurred intermittently at all sites.

In the USA pathogenic *Naegleria* have been found in some fresh-water lakes receiving effluents from power stations and air-conditioning plants

(Willaert and Stevens 1976; Stevens *et al.* 1977; Tyndall *et al.* 1978). Two of the five locations in Florida and one of eight man-made lakes in Texas produced the pathogens in samples taken from within 1000 m of thermal discharges (Stevens *et al.* 1977). Temperatures ranged from 35 to 38°C during sampling and all the *Naegleria* came from samples taken from the shoreline, probably from the sediments. Surprisingly, *Naegleria* was not found in water from cooling towers. Neither were pathogens found in unheated lakes though this does not agree with the work of others (Wellings *et al.* 1977) who found pathogenic species in other naturally warm lakes. *Naegleria* have been found in lake muds at temperatures of over 16°C but only in the water column at temperatures over 26·5°C. Natural maximum temperatures range from 24 to 32°C and thermal discharge temperatures from 29 to 41°C in Florida. In Texas and Tennessee a survey of nine heated and four unheated waters showed that only one lake, with average temperatures around 40°C near a heated discharge, contained *N. fowleri*. No pathogenic organisms were found in four sets of cooling towers (Tyndall *et al.* 1978). Pathogenic *Acanthamoeba* and *Naegleria* were isolated from cooling-water discharges at several coal-fired power stations in the USA (Shapiro *et al.* 1980) and the authors concluded that such cooling systems may be a source of some infections. This has not been

TABLE 5.6

Prevalence of pathogenic *Naegleria*[a] before and after chlorination of a cooling system (from Tyndall *et al.* 1983)

Site	Sampling date	Temp. (°C)	No. samples positive/ No. tested (water)		
			1 ml	*10 ml*	*100–250 ml*
Tower Canal	May 1981	22	0/10	2/5 (2)[b]	NT[c]
Tower Canal	Aug 1981	28	1/10 (1)	2/10 (2)	NT
Intake	Aug 1981	34	3/10 (3)	4/10 (3)	3/5 (1)
Tower Canal	Sept 1981[d]	28	0/10	0/10	0/5
Intake	Sept 1981[d]	27	0/10	0/10	0/5

[a] Pathogenic *Naegleria* as indicated by their ability to grow at 45°C, trophozoite and cyst morphology, well-defined growth front, ability to flagellate. Majority of positive cultures appeared to be pure cultures of *N. fowleri*.
[b] Number in parentheses indicates number of isolates tested and subsequently confirmed pathogenic in mice.
[c] NT = not tested.
[d] Samples taken after chlorination.

confirmed by most other workers. It is clear, however, that *Naegleria* at least can be found in both naturally and artificially heated waters but is more common at temperatures over 30°C.

The control of *Naegleria* is difficult, though chlorination clearly has some effect (De Jonckheere and Van de Voorde 1976; Chang 1978) and the organism is unlikely to survive for long periods in the chlorination regimes of most large cooling-water systems. In studies at a power station in the USA, however, chlorine concentrations exceeding 2·0 mg litre^{-1} applied for 6 h were required to kill *Naegleria* and dechlorination was necessary to make the effluent acceptable for fish survival (Tyndall *et al.* 1983; Table 5.6).

CHAPTER 6

Algae and Macrophytes

6.1 TOLERANCE AND NATURAL DISTRIBUTION

The natural temperature range inhabited by algae is from around $-40°C$ on polar sea-shores to 75°C in thermal springs (Brewer 1866; Wood 1868; Stockner 1968; Patrick 1969; Gessner 1970; Eppley 1972; Boylen and Brock 1973; Round 1973; Castenholz and Wickstrom 1975; Brock 1978). The blue-green algae (Cyanophycae) which are so prominent as brilliantly coloured mats in thermal streams, mostly stenothermal thermophils, include unicellular forms such as *Synchoccus* spp., and filamentous forms such as *Mastigocladius* spp. and *Phormidium* spp. These last have the highest known temperature tolerance of any aquatic organism bearing chlorophyll (Castenholz and Wickstrom 1975). In Japan these genera have been recorded from thermal streams at between 40 and 60°C. In the USA the range is described as 60–80°C though 75°C is probably the highest constant temperature at which they survive. The lack of grazing organisms in some thermal streams allows thick mats of blue-green algae to grow over underlying layers of bright orange filamentous bacteria (Brock 1975, 1978, 1985; Castenholz and Wickstrom 1975).

Apart from their dominance in these geothermal environments, blue-green algae also tend to increase in abundance in rivers, lakes and reservoirs as temperatures increase in summer, particularly in warmer climates (Patrick 1974). Hawkes (1969) summarised the optimum growth temperatures for the main algal groups as broadly:

Diatoms: 15–25°C
Green algae: 25–35°C
Blue-green algae: 30–40°C (excluding the extreme thermophils).

There are, of course, exceptions. For example, the diatom *Achnanthes marginulata* occurs, and presumably grows, at temperatures over 35°C, while the blue-green alga *Oscillatoria rubescens* is a cold-water stenotherm with an optimum growth range of 4–10°C. Predictably, the majority of algae are mesotherms living in a temperature range of 5–32°C, though species are known to evolve strains with different temperature tolerances within that range (Patrick 1974).

Many species inhabit intertidal areas or the drawdown zones of reservoirs where extremes of temperature and varying degrees of desiccation occur (see Price *et al.* 1980). Many marine algae have evolved resting stages or special physiological strategies to survive in such habitats (Gessner 1970; Newell 1970; Patrick 1974). Gessner (1970) lists the critical $LT_{(24, 50)}$ for some Japanese algae and shows temperature tolerances as low as -55 to -196°C for short exposures. At the other end of the scale, some marine species could tolerate long exposure to temperatures 1°C below their lethal level provided that they were eventually returned to normal temperatures. Many tropical algae were killed by 12-h exposures to between 32 and 35°C, and only a few littoral species survived 40°C. This appears to compare with the lethal values for temperate algae and 'conflicts with the idea of genetic fixation based on habitat' (see Gessner 1970).

Some freshwater algae are highly eurythermic. For example, the thermophilic species, *Synchococcus lividus*, can tolerate sudden changes from 70°C to 'room-temperature' usually 20–25°C, without harm. Diatom species have been found to tolerate fluctuations of 10–12°C in 20 s without much damage, provided that the temperatures did not exceed the lethal values. *Cladophora*, reputedly, can withstand short exposures up to 60°C (see Gessner 1970).

Hirayama and Hirano (1970a) found that both *Chlamydomonas* and *Skeletonema costatum* were resistant to short-term exposures to high temperatures, but growth was inhibited after 10-min exposures to 43°C and 35°C respectively. Photosynthesis began to decline after similar exposures to 40°C and 35°C respectively. Algae have been used as indicators of pollution or stress for many years (see Hynes 1960; McMahon and Docherty 1975; Mellanby 1980; Wilde and Tilly 1981; Shubert 1981; Hellawell 1986).

As far as the effects of cooling water and thermal discharges are concerned, algae in natural waters can be regarded as two functional groups, viz. planktonic and attached or substrate-dwelling forms. The former, whether obligate or facultative species, are vulnerable to entrainment and transit through the system. The latter, whether living on hard surfaces, other plants or substrate particles (Round 1973) are likely to be vulnerable to a discharge. In some locations normally attached algae may be swept into the water column and become facultative plankton (Pickering and Davis 1983). Some attached algae are also potential colonisers of surfaces within a cooling-water system.

Both obligate and facultative planktonic algae are entrained, the latter particularly where an intake is sited near mud flats or very shallow areas of water from which benthic species may be swept into the water column (Pickering and Davis 1983).

6.2 ALGAE WITHIN COOLING-WATER SYSTEMS

6.2.1 Fouling Algae

A number of species are sufficiently tolerant of temperature and contaminants to survive and flourish in cooling-water systems. Such fouling algae are most troublesome in cooling-tower systems, usually growing profusely on the timber packing or other surfaces where light is sufficient (McKelvey and Brooke 1959; Gurney and Cotter 1966; White 1972; Smedile and Parisi 1975). Blank (1984) described massive growths of the filamentous green algae *Cladophora glomerata* on packing timbers in power station cooling towers in the English Midlands. McKelvey and Brooke (1959) describe blue-green algae as the most troublesome, blocking pipes and sprays in towers and destroying wooden structures by their weight. Diatoms are also found in cooling systems incorporated into fouling deposits on the cooler surfaces.

Algal control in such systems is usually carried out by chlorination at doses exceeding $1 \cdot 0 \, \text{mg litre}^{-1}$ (White 1972), by herbicides or specific algicides (Blank 1984). Species growing in cooling-water systems are clearly adapted to stress and it may be that they are also more resistant to treatments than algae in natural waters. *C. glomerata* and some blue-green algae are notably tolerant to pollution (Hynes 1960).

6.2.2 Effects of Entrainment

(i) Methodology

A comprehensive review of the methodology used to study algae in cooling-water systems (Lawler, Matusky and Skelly, Engineers 1979), shows clearly that the methods and criteria for assessing effects of entrainment on the algae have varied considerably and there is little comparability from location to location.

As a general rule, 'control' samples are taken at or near the cooling-water intake or some distance upstream. Samples following transit through a system are taken at the outfall, preferably before the effluent mixes with the receiving water. At some sites samples have been taken from culverts prior to the actual outfall to prevent such mixing (Davis and Coughlan 1978, 1983).

The criteria used to describe and quantify the effects of cooling waters on algae fall into four main categories, viz.

—Abundance: usually expressed as numbers per unit volume or area.
—Biomass: usually expressed as the weight of chlorophyll 'a' per unit volume or area, or by weight.
—Viability: usually expressed as amounts of ATP or enzymes per unit volume or area.
—Primary productivity: usually expressed as the rate of assimilation of ^{14}C-labelled glucose by cells in samples held in light and dark bottles.

In some studies respiration rates have been compared at intakes and outfalls (Brook and Baker 1972).

In studies of attached algae in receiving waters the changes in community diversity and species composition are used to assess the effects of a discharge (Hellawell 1986). Sampling methods for phytoplankton have included pumps, specially designed bottles or standard hydrographic samplers and tow nets. Methods for attached algae have mainly employed artificial substrates such as glass slides or plastic strips suspended at predetermined depths in the water column.

One of the major criticisms of the methods used for sampling planktonic organisms entrained through cooling-water systems is that sampling design has not usually allowed for the innate 'patchiness' (heterogeneity) of the distribution of such organisms in the water column.

In most waters, other than shallow streams, plankton generally shows discontinuous distribution both temporally and spatially (both vertically and horizontally). In passage through a cooling system the plankton is thoroughly mixed, causing a major change in distribution irrespective of any other effects. Clearly this 'averaging' results in a more homogeneous distribution which reflects on densities, biomass and activity measurements which may be higher or lower at the outfall than at the intake, depending upon the situation (Lawler, Matusky and Skelly, Engineers 1979).

(ii) Effects of temperature within cooling systems
Site studies of entrained phytoplankton are plentiful though many are not

reported in the open literature (Lawler, Matusky and Skelly, Engineers 1979). Many of the data are summarised and reviewed in various other publications, e.g. Schubel *et al.* 1978; Yost and Uziel 1981; Langford 1983a; Barnett and Hardy 1984).

(a) Fresh waters. The relationships between outfall temperatures and rates of photosynthesis of entrained phytoplankton at various power stations in the USA are summarised in Fig. 6.1 (lakes) and Fig. 6.2 (rivers). Both groups of sites include different intake and outfall configurations (see Chapter 2) and thus the effects of stratification and temporal and other spatial discontinuities varied widely. This probably accounts for much of the variation in the data.

The overall pattern for both groups is similar in that at outfall temperatures lower than 35°C there was a widely variable effect with site, from a 50% decrease to a 40% increase in rates of photosynthesis.

Consistent decreases only occurred at temperatures above 37°C at the lake sites and above 38°C at the river sites. After exposure to temperatures over 37°C in both Lake Norman (Smith *et al.* 1974b) and in Lake Wylie (Gurtz and Weiss 1974a, b) phytoplankton activity was reduced by 60–100%. At the highest temperature there was no measurable photosynthesis.

In contrast, among the river group there was still measurable activity at the outfall of the Chesterfield power station on the James River when temperatures exceeded 45°C (Smith *et al.* 1974a). Evidence suggests, however, that this residual activity would subsequently disappear. For example, Gurtz and Weiss (1974b) showed that in the 2 h following exposure to 44·3°C any residual activity declined to zero and did not recover after 4·5 h. After exposure to temperatures around 35–38°C, a similar decline took place over 2 h but levelled out at about 40–60% of the outfall values. It is clear, however, that measurements immediately following entrainment do not necessarily reflect the full extent of effects. After entrainment in May and August the phytoplankton of Lake Wylie showed maximum growth after nutrients were added to samples. Temperatures were around the natural growth optima in the discharge at these times.

In most studies of the separate effects of temperature at operating power stations, the maximum adverse effects on entrained phytoplankton predictably occurred in high summer when natural temperatures were at their highest, standing crop was at its maximum and productivity had declined (see Koops 1972, 1975; Kreh and Derwort 1976; Dunstall 1985).

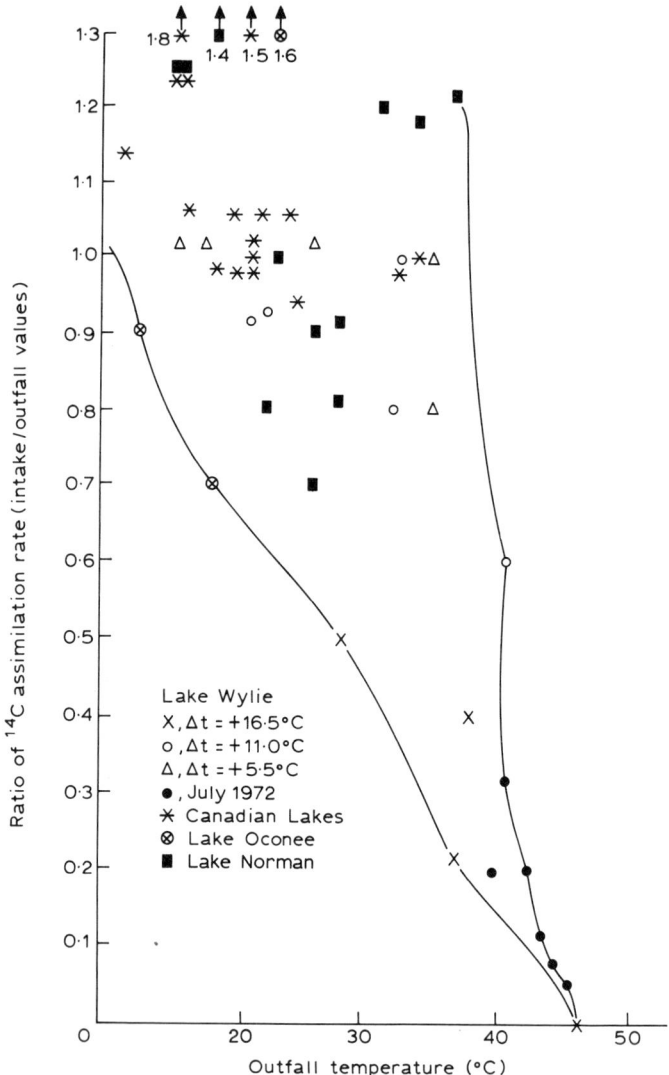

FIG. 6.1   [14]C assimilation rates of entrained phytoplankton at the outfalls of power stations using lake-water for cooling. Lake Wylie, all points are means of multiple samples; Gurtz and Weiss (1974). (See text for references.)

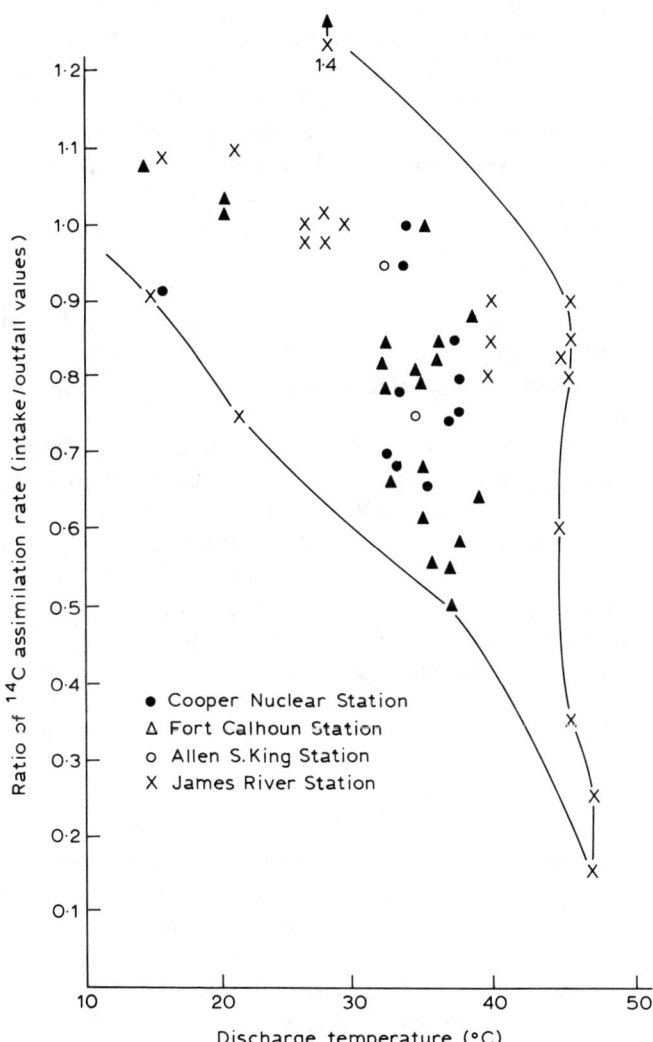

FIG. 6.2 Effects of discharge temperatures on carbon (^{14}C) assimilation of phytoplankton after entrainment (data from US rivers). (See text for references.)

Thus any stimulation of activity in winter will represent a much lower absolute value than a decline in summer and cannot be regarded as a compensatory effect. For example there was a factor of 30 times difference between the natural rates of photosynthesis in summer and winter in Lake Wylie (Gurtz and Weiss 1974a). Losses of 30–100% of the entrained phytoplankton in summer were not, therefore, balanced by the 10–20% stimulation of the winter rates in the cooling water of the power station. Experimental data were used to produce a series of significant equations relating temperature to primary productivity (Smith *et al.* 1974a). The field data indicated a less predictable system with a marked threshold temperature around 33°C.

The differences in species-composition of phytoplankton communities may also explain some of the variation between the effects at different sites. For example, in Lake Wylie, Cyanophyta and Crypophyta were most affected by passage through the cooling-water system (Gurtz and Weiss 1974b). At the Kewaunee power station on Lake Michigan, the greatest effect was on the $< 10\,\mu m$ and $> 64\,\mu m$ size fractions of the phytoplankton and the intermediate size groups suffered little damage. However, at this site, the highest temperature was 30°C at the outfall and the total reduction in phytoplankton productivity never exceeded 13%. Using oxygen metabolism as their criterion, Brook and Baker (1972) found no relationship between entrainment and metabolic rate up to the maximum outfall temperature of 36°C.

Field studies of the effects of entrainment temperatures on the abundance, biomass and viability of phytoplankton have also shown considerable variation with site (e.g. Lawler, Matusky and Skelly, Engineers 1979). For example, at the Fort Calhoun power station on the Missouri River (Reetz 1982), differences in the abundance of cells between intake and outfall ranged from a 22% decrease to a 25% increase in the period from January to December. There was no correlation with temperature. Gurtz and Weiss (1974a, b), on the other hand, suggested that in their studies decreases in abundance and species diversity were related to outfall temperature but no statistical relationship was demonstrated. Where differences in abundance are recorded they are not usually statistically significant at any site (e.g. Lawler, Matusky and Skelly, Engineers 1979).

In terms of biomass, as indicated by pigments, there are no clear relationships between temperature and the differences between intake and outfall. The greatest reductions were reported as about 33%.

The evidence indicates that irrespective of experimental data, short-term exposures to temperatures lower than 35°C do not cause significant damage to entrained freshwater algae. Long-term exposure to such temperatures is potentially harmful. At temperatures of 40°C and above even short-term exposures may be lethal.

(b) Saline waters. Figure 6.3 shows the ratios of intake to outfall productivity, as measured by the assimilation of ^{14}C, for entrained phytoplankton at eight power stations. As with the fresh-water sites, there is considerable variation at temperatures lower than 33°C. Below 25°C results vary from decreases of 40% to increases of 300–400%. European studies (Khalanski 1977; Davis and Coughlan 1978; Coughlan and Davis 1983) showed some stimulation of photosynthesis at < 23°C but decreases of around 20% at 27–28°C.

Briand (1975) suggested that losses of cells during entrainment were a result of high temperatures but it is clear from the data that there was no direct correlation and that maximum reductions occurred at temperatures in the middle and not the highest ranges. Chlorination is noted by Briand but, surprisingly, not implicated in the assessments of effects. Long exposure to high effluent temperatures can cause marked effects on entrained marine phytoplankton. Brooks (1974) showed that, in summer, assimilation rates of ^{14}C declined with increasing exposure over 1–6 h, though in cooler seasons this did not occur. At the Millstone power station, exposure to effluent temperatures of 29–34°C depressed rates by 61–84% immediately following entrainment (Peck and Warren 1978) but after longer exposure and a slight cooling in the effluent canal the rates increased by 15–30% from the lowest values. The activity of the enzyme nitrate-reductase was also depressed by the higher temperatures though not so much by chlorination.

The major effect of temperature on both fresh-water and marine phytoplankton during entrainment in cooling-water systems is clearly on the metabolic processes (see also Morgan and Stross 1969; Carpenter *et al.* 1972; Jensen *et al.* 1974; Ramade 1981) and only at temperatures exceeding about 37°C is there some evidence of longer-term damage, though even short exposures to temperatures over 40°C can cause total mortality of some species. In most site studies it is evident that entrained populations which suffer partial damage can recover in a relatively short time.

(iii) Effects of biocides in cooling systems

(a) Fresh waters. Figure 6.4 shows a clear relationship between the rates of photosynthesis, respiration and chlorine concentrations for entrained fresh-water phytoplankton. About 5–15% decreases in photosynthesis and

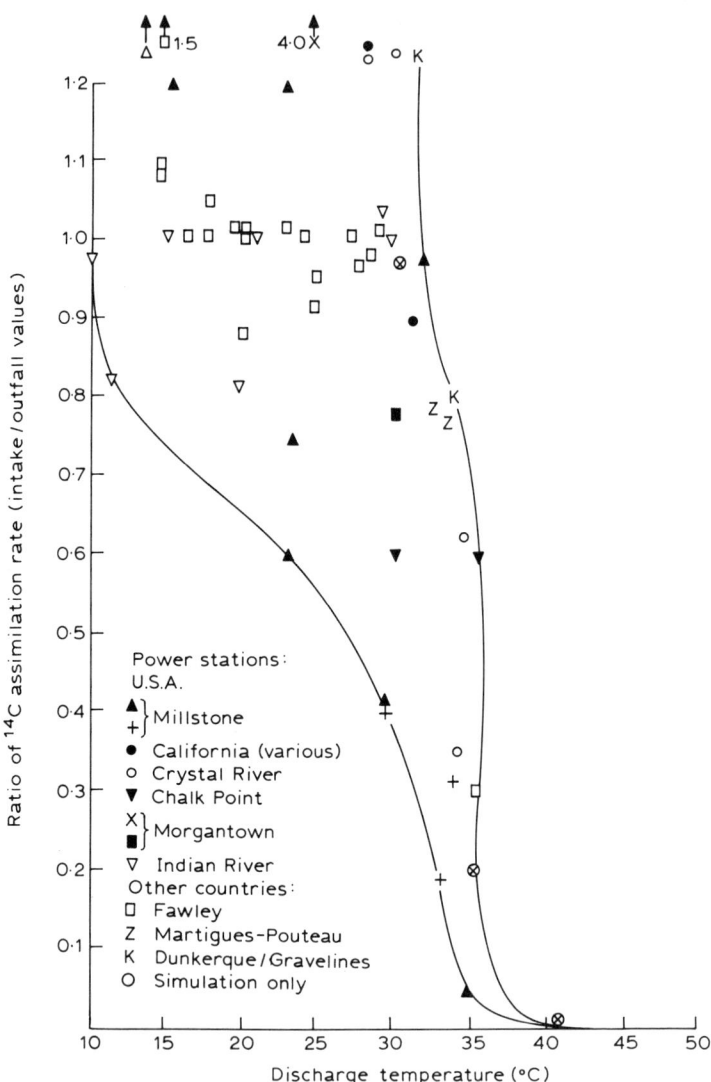

FIG. 6.3　Effects of discharge temperatures on carbon (^{14}C) assimilation by phytoplankton after entrainment (data from estuarine and coastal sites). (See text for references.)

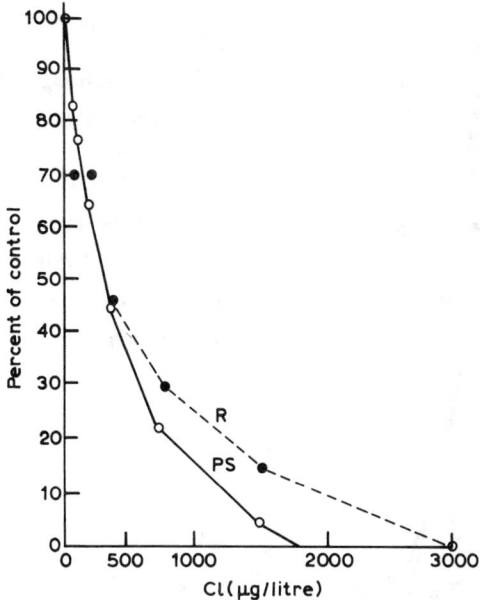

Fig. 6.4 Effect of chlorine addition on rates of photosynthesis (PS, open circles) and respiration (R, closed circles), expressed as percentage of control, in a freshwater phytoplankton community, after entrainment (after Brook and Baker 1972).

respiration were related to temperatures of 32–36°C. Chlorination produced much greater responses. The EC_{50} after only 1·5 min of exposure through the system was 0·32 mg litre^{-1} of dosed chlorine. At 2·7 mg litre^{-1} all photosynthesis and respiration ceased at the outfall. Brook and Baker (1972) found that where chlorination caused reductions in photosynthesis and respiration of phytoplankton in a cooling-tower system both were restored to a great extent following passage through the towers.

In many of the larger cooling-water systems using fresh-water sources chlorination is intermittent and thus the peak values are short-lived (see Chapter 3). Thus there are long periods when entrained organisms are not chlorinated and the total daily effect is much less than with constant chlorination.

(b) Saline waters. Models describing the relationship between chlorination and the photosynthesis of entrained marine algae have been devised from studies at coastal power stations. Khalanski (1977) produced a logarithmic

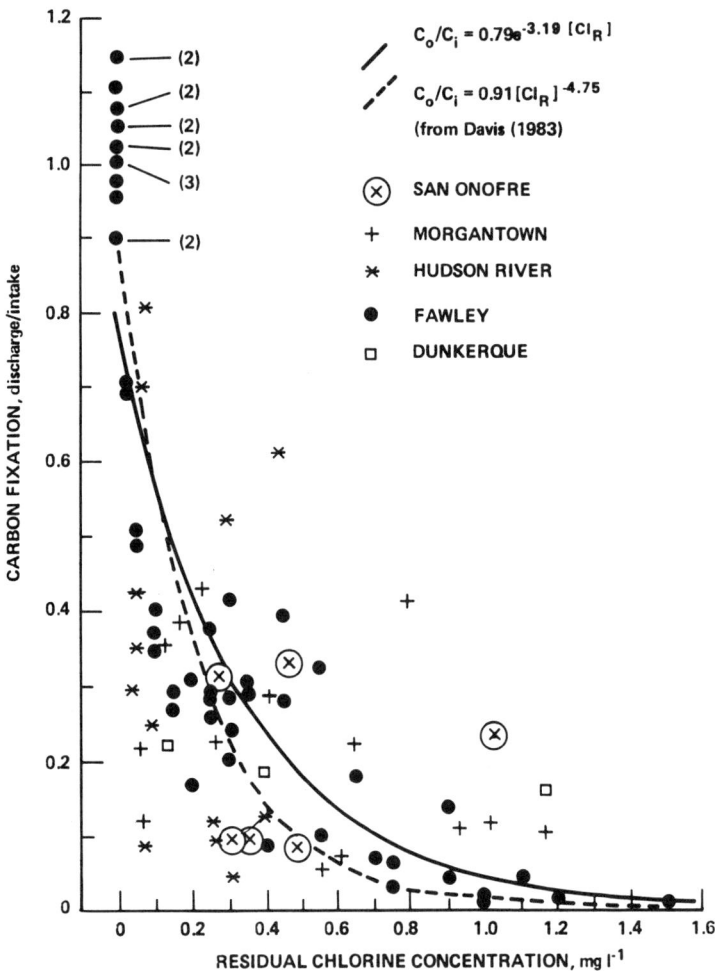

FIG. 6.5   The effect of cooling water chlorination on carbon fixation in marine phytoplankton (modified after Davis 1983a).

model from field studies which covered a range of TRC concentrations from 0 to 0.41 mg litre^{-1}. From this he concluded that the EC_{50} was 0.0002 mg litre^{-1}. The equation was:

$$\delta P\% = 94.7 + 5.35 \ln \delta Cl \, (\text{mg litre}^{-1})$$

where δP is the change in primary productivity (^{14}C assimilation rate) between intake and outfall, and δCl is the total residual chlorine concentration (TRC).

Davis and Coughlan (1978) and Coughlan and Davis (1983), in comparison, found a close exponential relationship from studies at Fawley power station (Fig. 6.5) where TRC values ranged from 0 to $1 \cdot 5$ mg litre^{-1}. The equation was:

$$C_o/C_i = 0 \cdot 91 \; (Cl_r) \, e^{-4 \cdot 75}$$

where C_o and C_i are the ^{14}C fixation rates at outfall and intake respectively, and Cl_r is the TRC value in mg litre^{-1} in the sample before incubation, as determined by the DPD colorimetric method. From this work it was concluded that the EC_{50} was $0 \cdot 14$ mg litre^{-1}, some 700 times higher than the value produced by Khalanski. There is no fully satisfactory explanation for the difference in these results. The differences in techniques were considered to be responsible for the difference in results, but exactly how these occurred is not fully clear.

In their subsequent studies, Coughlan and Davis (1983) found that there were differences between the effects of similar doses of chlorine at different power station sites and that open-coast sites apparently required greater doses of chlorine to suppress ^{14}C assimilation than estuarine sites. Thus although the relationships were still best described by exponential equations there were differences in the values for different sites (Table 6.1). Figure 6.5 also shows data from other studies in relation to the model. Although there is some variability, the results are fairly consistent, despite the widely varying field conditions and methods.

The most likely explanation for the differences between open-coast sites and estuarine sites was the difference in water quality, particularly in the concentrations of ammonia present, which are important to the chemical consequences of chlorine discharges in natural waters (Inman and Johnson 1978). Erickson and Freeman (1978), for example, showed that of 15 compounds formed during chlorination, monochloramine was the most inhibitory to phytoplankton. The method of applying chlorine, whether as a gaseous solution, hypochlorite or through electrolysis (see Chapter 3) has a negligible effect on the relationship between TRC and ^{14}C assimilation in a natural phytoplankton population (Davis and Coughlan, 1984), though hypochlorite produced a lower EC_{50} than the other two methods.

In the work at Fawley power station, Davis and Coughlan (1978) obtained the range of TRC values by manipulating the dose rates at the condenser inlets. In most other studies concentrations were measured during normal operations and chlorine residuals are quoted as either dosed

TABLE 6.1
Effect of chlorination on primary productivity of entrained phytoplankton[a] at coastal power stations in England (from Coughlan and Davis 1983)

Site and year	a	b	EC_{50}
Fawley, 1977	0·79	3·19	0·14
Fawley, 1978	0·71	2·99	0·12
Kingsnorth, 1978	0·70	3·00	0·11
Bradwell, 1979	0·90	4·21	0·14
Sizewell, 1978	0·85	1·82	0·29
Dungeness, 1980[b]	0·53	2·20	NA

[a] From the relationship:

$$P_o/P_i = ae^{-b(R.Cl)}$$

where P_o/P_i = the ratio outlet/inlet productivity,
       R.Cl = concentration of residual chlorine,
       a and b = regression coefficients.
[b] Values refer to adjusted data; EC_{50} is not given because it is extremely sensitive to error in the calculated transit time.

rates or discharge concentrations but rarely both. Thus useful comparisons with the models and other data are difficult (Fig. 6.5; see Carpenter *et al.* 1972). In some cases no chlorine concentration data are available.

For example, at two power stations in the USA chlorination to unknown levels caused reduced primary productivity by 78–100%, but no quantitative relationship could be described because of the lack of chemical data (Smith *et al.* 1974a). Outfall concentrations measured at a Californian power station as 0·02–0·04 mg litre^{-1} TCRC caused a suppression of 70–80% of primary productivity of entrained phytoplankton (Eppley *et al.* 1976). This does not fit either of the models described above, but if the actual dose rates are compared, i.e. about 1·0 mg litre^{-1}, the effect fits Davis's model well. Similarly the 65–90% suppression found by Lauer and his colleagues (Lauer *et al.* 1974) at measured TRC values of 0·23–0·5 mg litre^{-1} is within the range of the model. In every case where chlorine is used for anti-fouling purposes, it is clear that it is the major influence on the suppression of productivity in entrained phytoplankton (Flemer 1974; Flemer and Sherk 1977; Schubel and Marcy 1978; Lawler, Matusky and Skelly, Engineers 1979; Langford 1983a).

The production of ATP was reduced in entrained cells by up to 93–95% in the cooling system of a Californian power station in the presence of 0·5–

0·6 mg litre^{-1} TRC. In contrast, nitrate-reductase activity was less affected by chlorination than by heat though the enzyme is known to be heat labile (Peck and Warren 1978).

As in fresh waters, losses of chlorophyll and reductions in the numbers of cells during entrainment have been found to be associated with chlorination (Hamilton *et al.* 1970). In any cells killed, the degradation of chlorophyll 'a' will result in products such as phaeophytins. The losses in chlorophyll 'a' tend to follow a similar trend to the changes in productivity but no quantitative relationships are published in the open literature (see Lawler, Matusky and Skelly, Engineers 1979).

Changes in the abundance of phytoplankton also occur at marine sites following entrainment. Briand (1975) reported losses of 41% of cells passing through two Californian power stations. Seasonal variation was from 25% to 60%. Diatoms were apparently more affected than dinoflagellates and the community diversity was always lower at the outfalls than at the intakes. The cause of this change was the increased dominance of *Gonyaulax polydora* and *Asterionella japonica* rather than an overall loss of species. Briand related the changes to heat but chlorination also occurred at dose rates of 0·2–1·0 mg litre^{-1} giving concentrations of 0·03–0·16 mg litre^{-1} at the condenser inlets. The mechanism by which cells were lost is not described.

Davis (1983b) found that diatoms were less affected than the other phytoplankton by chlorination but that cell size of the non-diatom fraction did not influence chlorine effects.

(iv) Recovery from chlorination
Experiments have suggested that where photosynthesis is adversely affected the community will recover in a relatively short time provided that the cells are not all killed. Brooks and Liptak (1979) found that at about 0·1 mg litre^{-1} of TRC, photosynthesis was suppressed by 25% but recovered to the control rate 48 h after exposure. Further, a 30 min exposure to 0·64 mg litre^{-1} TCRC completely suppressed photosynthesis but after 24 h activity was back to about 50% of the control level. Both magnitude of effect and rate of recovery were affected by water temperature. At 2°C, recovery occurred after exposure to 1·43 mg litre^{-1} but at 10–12°C no recovery occurred after exposure to 0·5–1·4 mg litre^{-1} TRC. In these experiments the 50% suppression concentrations of TRC (EC_{50}) varied with temperature from 0·5 to 0·9 mg litre^{-1}. Similar recovery trends were also shown by marine algae (Hirayama and Hirano 1970a; Goldman and Quimby 1979).

At concentrations over 0.1 mg litre^{-1} TRC there was a loss of chlorophyll 'a' and this was also evident in some field studies. This loss was usually accompanied by an increase in phaeophytins which arose from the decomposition of chlorophylls (see Lawler, Matusky and Skelly, Engineers 1979).

In studies at a power station on Narrangasset Bay, cells did not recover after exposure to TRC concentrations of 0.55 and 0.32 mg litre^{-1} through the system and subsequent exposure to reducing levels at high temperatures in the discharge canal (Gentile *et al.* 1976).

Coughlan and Davis (1983) noted that the relationship between the percentage change in productivity and TRC concentration was constant through the year at any one site. This indicated that the effect of chlorine was not related to the different species composition and probably represented a reduction in the activity of all cells present rather than a proportional or selective mortality. However, the presence of phaeophytins and the corresponding reductions in chlorophyll found in some field studies (Gentile *et al.* 1976) clearly indicates cell mortalities, at least among some species.

It was suggested (Eppley *et al.* 1976) that the effects could be regarded as 'a killing of the phytoplankton in proportion to the chlorine concentration, followed by uninhibited growth of the few survivors'. This would not appear to be true, however, as in the experimental systems, recovery occurred over periods of 24–48 h which is too short for sufficient reproduction of cells. The effects would seem to be a rate-limiting process rather than selective mortality up to a threshold point when a combination of temperature, chlorine concentration and duration of exposure begin to cause direct mortalities.

The magnitude of any chlorination effects are clearly related to the initial dose at an intake.

(v) Effects of dechlorination
The experiments by Hirayama and Hirano (1970a) showed that the growth rates of cultured *Skeletonema costatum* did not recover after exposure to 5 and 10 min doses of chlorine even after the water was dechlorinated with sodium thiosulphate. The marine flagellate *Monochrysis lutheri* was not affected by exposure to dechlorinated water with a nutrient medium (Garey 1980).

(vi) Mechanical effects
Table 6.2 shows data from nine power station sites where there were

TABLE 6.2

Changes in Phytoplankton activity caused by passage through cooling-water systems in the absence of heat and chlorine

Power station	Conditions	Effects on organisms in transit	References (comments)
Cooper nuclear	Missouri River	(^{14}C) +3 to +10% (Chl_a) −2 to −12%	7–26 h incubation Reetz (1982)
Fort Calhoun	Missouri River	(^{14}C) 0 to +5% (Chl_a) −5 to +4%	7–26 h incubation Reetz (1982)
Allen S. King	Lake Wylie	(^{14}C) −7 to +20%	Gurtz and Weiss (1974a)
Marshall	Lake Norman	(^{14}C) +17·5 to +30%	Smith et al. (1974a)
Indian River	Estuary	(^{14}C) +10 to −25%	Brooks (1974)
Vienna plant	Estuary	No effects	Flemer (1974)
Kewaunee power plant	Lake Michigan	(^{14}C) −1 to −34% (mean (^{14}C) −8%) (mean (Chl_a −5%)	see Zeeman and Grunewald (1978)
Fawley power station	Coastal	(^{14}C) +8 to +14%	Davis (1983a)
Millstone nuclear	Coastal	(^{14}C) (outfall) +26% (Feb.) (^{14}C) (outfall) +22% (Jul.)	Peck and Warren (1978)
Martigues–Ponteau	Coastal	(^{14}C) (outfall) −16%	

differences between the phytoplankton abundance and metabolism after entrainment but where no chlorine was applied and no heating occurred. Any effects were therefore apparently related solely to mechanical stresses in the cooling system. The adverse effects were clearly much smaller than those of heat or chlorination though at one site up to 72% of diatom cells belonging to five species were reported as damaged by mechanical stresses. The increased productivity at some outfalls was most likely caused by aeration and mixing during passage.

Colonial algae and filamentous forms become fragmented as they pass through pumps and pipes and several authors have suggested that increases in cell surfaces after fragmentation could increase ^{14}C fixation (Gurtz and Weiss 1974a, b; Koops 1974; Bradford and Burns 1977).

Deep-water intakes bringing water from hypolimnia, rich in nutrients, to be discharged at the surface could also cause increased primary productivity near outfalls. Davis (1983a) suggested that both increases and decreases in ^{14}C fixation when only cold, unchlorinated water was being pumped, were caused by the changes in the spatial distribution of cells during pumping compared with the natural heterogeneity at the intake.

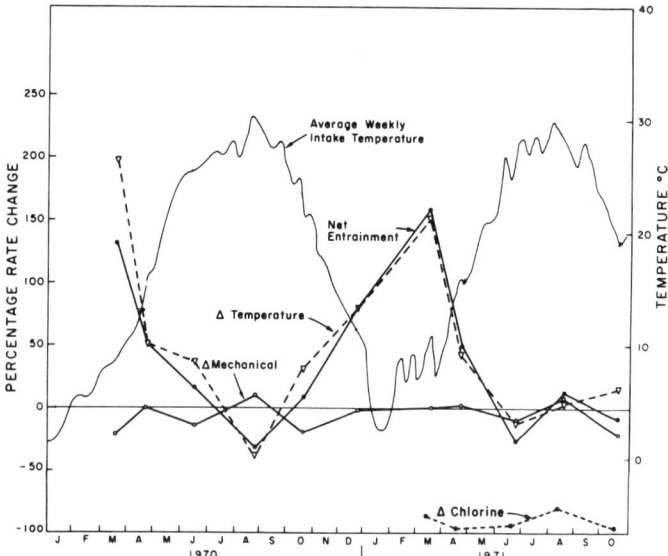

FIG. 6.6 The effects of temperature, mechanical stresses and biocides on the phytoplankton entrained by the Indian River Power Station, US (after Brooks 1974).

Large-scale reductions in the density or biomass are unlikely to be caused by heat, contaminants or mechanical shocks in a cooling-water system (Coutant 1970b; Coughlan and Whitehouse 1977). The most likely explanation is the most obvious and that is that the mixing processes in the system integrate the spatially heterogeneous concentrations of cells at the intake (Davis 1983a) and that sampling at the intake tends to be in the areas of highest density, usually near the surface.

Figure 6.6 shows the relative effects of heat, chlorine and mechanical stresses during entrainment of phytoplankton of a small tidal estuary. Clearly the greatest adverse effects were caused by heat and chlorine in summer.

6.3 EFFECTS IN RECEIVING WATERS

In most receiving waters, dilution and mixing are so rapid and the generation times of phytoplankton so short that long-term entrainment damage or discharge effects are unlikely to be significant (Goldman and Davidson 1977; see Lawler, Matusky and Skelly, Engineers 1979; Bordet 1980). One conclusion is that the most likely effect of thermal discharges on the phytoplankton of receiving waters is an increase in abundance, 'despite cross-plant (transit) reductions in density and primary productivity' (Simmons *et al.* 1974; Lawler, Matusky and Skelly, Engineers 1979). An opposing viewpoint proposes that the onset of the exponential growth phase of any species which culminates in the annual 'bloom' can be sensitive to the loss of as little as 1% of productivity. This loss would reduce the population maximum and alter its timing which in turn could imbalance the phased succession of phytoplankton and its dependent herbivores and ultimately reduce total production in the ecosystem (Cushing 1976).

Whilst this is theoretically possible, the huge sampling and analytical effort required for the empirical validation of such an effect would seem impractical given the known annual and seasonal variability in natural populations.

6.3.1 Fresh Waters
(i) Rivers
In the River Tagus, the increased density of phytoplankton at a point downstream of the Jose Carrera nuclear power station was mainly caused by an increase in the diatom *Achnanthes lanceolata*. Increased temperature and eutrophication were both suggested as causes of the algal blooms (Urbistondo *et al.* 1974).

The reach of the Connecticut River near the Haddam Neck nuclear power station is a complex hydrographical system. Here it was also found that the growth of phytoplankton increased downstream of the cooling-water discharge (Foerster *et al.* 1974). Blooms of *Scenedesmus quadricanda* and *Microcystis aeruginosa* were associated with the discharge area and the increases were measurable up to 2 km downstream.

In large impounded rivers, the major influence of the standing crop of phytoplankton is river discharge. Thus where the volumes of water used in a cooling system represent only a small proportion of the river discharge volume, the total effects of the thermal discharge are predictably small.

For example, at two nuclear power stations on the Missouri River (Reetz 1982) and at the J. M. Stuart generating station on the Ohio River (Miller *et al.* 1976), the thermal discharges represented only 4% and 15% of the river discharge respectively and although entrainment caused measurable changes in the phytoplankton passing through the system, the populations in the receiving water showed negligible effects. Similarly, at the Quad-Cities power station on the Mississippi River, the thermal discharge caused reductions in the chlorophyll 'a' distribution across the river but only a relatively small area was affected (Lawler, Matusky and Skelly, Engineers 1979).

In two European rivers, temperature rises of up to 8°C caused no changes in the composition of the populations of floating algae (Appourchaux 1952; Swale 1964).

Smith and Jensen (1974) used a model of oxygen concentrations in the James River estuary in Virginia, to predict the reductions which would result from a 100% kill of phytoplankton after entrainment in a power station cooling system. Measurements on two dates showed that the predictions were valid. Reductions were around 10–20% (0·5–1·0 mg litre^{-1}) in July and around 5% (0·2–0·5 mg litre^{-1}) in October.

(ii) Lakes
After summarising the data from studies of algae in cooling lakes in the USA, it was concluded that 'the weight of evidence revealed no significant ecosystem response that can be attributed to heated effluents per se' (Becker *et al.* 1979a). Any measurable effects were only in the immediate vicinity of the outfall at most sites. In some such areas productivity increased in cooler seasons and decreased in summer. In at least three cooling lakes, reduced productivity in heated areas was not attributable to heat alone. The fact that blue-green algae are tolerant to higher temperatures has already been noted (see p. 163) and the dominance of these organisms is regarded by

some authors as indicative of thermal stress in a community (Patrick 1969, 1974; see Becker *et al.* 1979a, b). In Lake Julian (North Carolina) discharge temperatures reached 42°C in summer and here primary productivity was reduced in the outfall area even though blue-green algae (*Oscillatoria*) were more abundant than green algae. In a heated lake in Texas, optimum temperature for algal productivity was 25°C (Welch and Ward 1978).

In Lake Arlington (Texas) *Oscillatoria* was also abundant but here the green algae maintained numerical dominance. Water temperatures were, however, some 5–10°C lower than in the former lake. In North Lake (Texas) where water temperatures in a thermal discharge reached 38°C, blue-green algae were much less abundant than predicted and the standing crop of the whole lake was much lower than those of other lakes nearby (Stuart and Stanford 1978). There was no convincing explanation for this phenomenon though the authors suggested that the low density and productivity of the phytoplankton were caused by 'nutrient limitation aggravated by power-plant caused evaporation and consequent dissolved solids accumulation'.

It is also possible that cumulative cell mortalities over a period of 10–12 years resulted in the type of effect on the exponential growth phase predicted by Cushing (1976) leading to the ultimate collapse of the community. Irrespective of their thermal tolerance, the blue-green algae are obviously tolerant of other stresses such as pollution (Hawkes 1962) and are, in fact, indicators of general, as well as thermal stress in a water body. An indirect effect of heating on phytoplankton was reported from Lake Wabamun in Canada. Here the temperature rise did not kill the organisms even though the δT could range from 8–19°C. The low abundance of phytoplankton in the discharge area was apparently caused by high macrophyte densities which reduced light penetration in summer. In winter, phytoplankton persisted unseasonally in the ice-free area comprising about 300 m^2 near the outfall (Nursall and Gallop 1971). *Oscillatoria* was abundant in autumn and was the most common of the algae in summer.

In the heated Konin Lakes in Poland, the flushing rate of water through the lake system was the major influence on the phytoplankton community because of the increased nutrients, though the thermal discharges were believed to stimulate algal growth (Hillbricht-Ilkowska and Zdanowski 1978). In Par Pond (see p. 149), higher winter standing crops and a 20% rise in respiration rates were found in the phytoplankton populations of the heated areas (Tilly 1974).

Very few studies have attempted to assess which parts of the phytoplankton community are most affected by thermal discharges. Zeeman and Grunewald (1978) measured the primary productivity of three different size fractions of cells before and after heating by the Keewanee power station on Lake Michigan. Their results were inconclusive, as all the fractions showed some effect but there was no clear difference between the size groups.

(iii) Estuaries and coastal waters
Fox and Moyer (1973) found that ^{14}C assimilation of entrained cells recovered during passage down the discharge canal, after an initial reduction caused by entrainment through a cooling system. At the outer end, some 3·2 km distant from the outfall, assimilation rates were at the level of control stations. Samples from areas in the Gulf of Mexico about 4 km from the end of the discharge canal showed assimilation rates of up to twice the control values. Water temperatures were 2–3°C above natural ambients.

At the estuarine Indian River power station (Smith *et al.* 1974a), a regression model was used to estimate rates of photosynthesis at points 2·5 km upstream and downstream of the discharge canal mouth, mostly within the areas bounded by the 0·2°C isotherm. Where natural water temperatures were around 5°C in the absence of chlorination the increases in rates of photosynthesis ranged from 3% to 260% over a δT range of 0·2–8°C. Elevated rates were predicted over areas of 2 km upstream and 2–3 km downstream. In summer when natural temperatures were around 25°C the predicted changes in photosynthesis ranged from −7% in the discharge canal to +3% at the 2°C isotherm.

Bradford and Burns (1977) reported damaged cells in the receiving water some distance from a thermal discharge outfall. Sampling damage was the only feasible explanation here. The authors concluded that with a dilution factor of 30 times each second, even a 100% kill of plankton would not be measurable in the receiving water.

Intermittent chlorination caused reductions of 80–90% of photosynthesis measured some 50 m from the vertical discharge 'boil' at the San Onofre nuclear power station on the California coast (Eppley *et al.* 1976). Although the thermal plume was theoretically free to mix, there was considerable vertical and horizontal stratification for some distance from the outfall.

In most cases where measurements have been made in receiving waters the effects of the discharge on phytoplankton have been restricted to the near-field plume, usually within a very short distance of the outfall as in fresh waters (see Lawler, Matusky and Skelly, Engineers 1979; Yost and Uziel 1981). Khalanski (1981) found that the phytoplankton activity was inhibited within 200 m of the outfalls where chlorine residuals exceeded 0.03 mg litre^{-1}.

The possibility that higher water temperatures caused by thermal discharges could enhance blooms of toxic algae has been suggested (e.g. Eng-Wilmot *et al.* 1977) but this has not been verified in the field.

6.3.2 The Significance of Entrainment Effects

Briand (1975) estimated that the equivalent of 1700 t of organic carbon was represented by the mortality of phytoplankton entrained in two Californian power stations. There was no attempt to relate this to the standing crop in the waters nearby. Davis (1983a) estimated that Fawley power station used about 2–5% of the tidal exchange in Southampton Water in the UK. Given that the average summer suppression of ^{14}C assimilation of entrained phytoplankton was about 60% the total effect on the productivity was 1·2–3% of the tidal exchange.

It is evident from the studies described so far, that entrainment in large cooling-water systems has not caused the wholesome destruction of phytoplankton communities in fresh or saline waters.

In most cases the magnitude of any effect could have been easily estimated from assuming 100% mortality and the known hydrographic data. The models produced by Brook and Baker (1972), Khalanski (1977) and Davis (1983a), though differing in their assessments of lethal values, are the basic concepts for future assessments if these are considered necessary. However, the reasons for the different lethal values require further investigation only if the criteria are applied to legal constraints. Khalanski's data applied to legislation would give criteria of orders of magnitude less than the 0.2–0.5 mg litre^{-1} used at British power stations. Davis's model (p. 174) would indicate 60–90% mortalities at these concentrations, giving some room for recovery of cells even without dilution.

6.3.3 Sampling and the Interpretation of Results

The sampling effort required to measure effects to some statistically acceptable criterion in receiving waters is immense, even in the near field. For example, in a detailed study at the Millstone power station, it was found that although 80–100% of the primary productivity was suppressed during passage through the cooling system, no significant ($p > 0.05$) differences could be detected between mixing zone and control areas (Carpenter *et al.* 1974a; Carpenter *et al.* 1974b, c). The authors concluded that 88 samples were required at each station to detect a plus or minus 5% change in productivity. Detecting a 10% change would require 22 samples at each station, a 20% change in six samples. The number of replicate samples used in surveys is not specified in most field studies. The general interpretation of field data on phytoplankton and comparisons of different site studies require some caution (see Lawler, Matusky and Skelly, Engineers 1979). Three main problems arise:

—differences in sampling methods and locations in relation to the design and configuration of the cooling system,
—differences in the methods of incubation and temperatures,
—whether the sampling intensity was adequate for correct statistical treatment, particularly for separating out such factors as spatial and temporal heterogeneity.

These points also apply to many of the studies of other taxonomic groups as we will see.

6.4 EFFECTS OF DISCHARGES ON SESSILE AND BENTHIC ALGAE

In contrast to the dynamic nature of phytoplankton populations, the benthic algae form a stable community and are much more likely to reflect longer-term changes caused by thermal or other discharges. Some of the earliest studies of thermal discharge effects used benthic algae as indicators of change (Trembley 1960; see Patrick 1969). The temperature relations of benthic algae have been reviewed by Patrick (1969, 1974), Brock (1975, 1978, 1985), Shubert (1981) and Luning (1980).

The main method of sampling is by the use of an artificial substrate, usually glass slides or strips of plastic suspended in the water column. Trembley (1960), for example, used a frame containing microscope slides, known as a pralgometer, floating just below the surface. Patrick (1969) used a similar apparatus which she described as a diatometer.

Experimental studies have indicated that temperature and chlorine can affect community structure and the physiological state of periphyton (Murray 1980; Wilde and Tilly 1981).

6.4.1 Fresh Waters
(i) Rivers
There are clear indications of changes in communities in thermal plumes

and several species found to be common to such areas. Decrease in the numbers of species, diversity and abundance of sessile micro-algae have been recorded in the discharge areas of several power stations where the control sampling sites were upstream of the intake (Stangenberg and Pawlaczyk 1960; Trembley 1960; Coste *et al.* 1978; Farrell and Tesar 1982). At the Martins Creek site (Trembley 1960) effluent temperatures of 36–42°C in summer, combined with intermittent chlorination, caused lower species diversities in the discharge area but greater total abundance because of the growths of blue-green algae. Where temperatures exceeded 34·5°C these blue-green algae were dominant. Trembley suggested that chlorination caused a reduction in the numbers of individual cells but no reduction in diversity. Higher temperatures with no chlorination, produced the classic reductions in diversity characterised by smaller numbers of species together with dominance by one or two favoured species (Southwood 1966; Hellawell 1986).

The data from a Polish river (Stangenberg and Pawlaczyk 1960), indicated that fewer species colonised slides where temperatures exceeded 30°C but here blue-green algae did not increase. Higher temperatures favoured the diatoms *Navicula cuspidata*, *N. ambigua* and *Tabellaria* spp.

In contrast, Patrick (1969) did not demonstrate changes in the benthic diatoms which could be attributed to the thermal discharge from the Dickerson power station on the Potomac River, where the highest temperatures were around 34°C. At the Porcheville power station on the Seine (Coste *et al.* 1978) effluent temperatures reached only 31°C and were typically less than 30°C. In the discharge canal *Cyclotella Kuetzingiana, C. meneghiniana, Melosira ambigua, M. granulata, M. italica, Stephanodiscus astrea, S. minutula* and *M. rantzchii* all increased in their relative abundance when compared with the intake samples. The reverse pattern was shown by *Diatoma elongatum, Fragilaria waucheriae, Gomphonema parvulum, Nitschia palea* and *Synedra ulna*.

Seasonal patterns of abundance also varied with species and location. The four species with the highest tolerances were *C. meneghiniana, Nitzschia thermalis, N. filiformis* and *Navicula confervacea*. This last was formerly considered as a sub-tropical species but was common to the discharge canal and up to 5 km downstream.

One of the most detailed studies of benthic micro-algae in heated discharge areas was that in the Missouri River at the Cooper and Fort Calhoun power stations (Farrell and Tesar 1982) where the work extended over 6 years before and after the stations began operating. At the Cooper station populations decreased in the discharge area when temperatures exceeded 30°C though they recovered as temperatures fell later in the season. The diatom *Gomphonema intricatum* declined in relative abundance in the heated water in summer but *Navicula tripunctata* var. *schizonemoides* was relatively more abundant than in the control areas. Of the green algae (Chlorophyta), *Cladophora* and *Stigeoclonium* species were dominant at all sampling stations and *Chaetophora* sp. declined in the warmest water. Among the Cyanophyta, *Lyngba* sp. dominated the artificial substrates in the discharge area during July–September. Blue-green algae were most abundant throughout the summer when water temperatures ranged from 32 to 35°C. As temperatures fell the diatoms replaced the blue-greens. Of these diatoms, however, *Navicula luzonensis* was the dominant species in the discharge canal in summer.

At the Fort Calhoun station the total abundance of periphytic algae in the discharge area was 30–40% lower in summer than at the control sampling station. The reverse occurred in autumn. Again, as at the Cooper station, *N. tripunctata* var. *schizonemoides* was more abundant in the heated areas.

Optimal temperatures for various species of algae were calculated using least squares regression analysis of the log-densities of cells colonising glass slides in relation to water temperature. Figure 6.7 shows the temperatures at which the maximum population densities (P_{max}) occurred for sixteen species. For most species the densities declined at over 30°C. Optima for various blue-green algae were between 24 and 26°C though this did not seem to reflect their total temperature tolerance ranges.

These data should be regarded with some caution as factors such as flow, current velocity, turbidity and water chemistry, all of which could influence colonisation, were not included in the analysis.

In the heated effluent from the Hale generating station (Squires *et al.* 1979), *Cocconeis placentula* dominated the benthic diatom flora throughout the year but was common only in summer at the control site. *Achnanthes minutissima* was the only other species abundant in the heated water. During a winter period when the power station was not operating, the diatom flora in the previous plume area became diverse but this was soon replaced by the 'effluent' flora after the power station began operating again. Discriminant analysis separated the control and effluent flora as a function of temperature alone. The total area over which the changes in the benthic flora were measurable in the Provo river below the Hale effluent was about 135 m downstream by 20 m across the width of the river, that is about 0·27 ha in total. Although the main subject of this volume is thermal discharges from manufacturing processes, the thermal effects of impound-

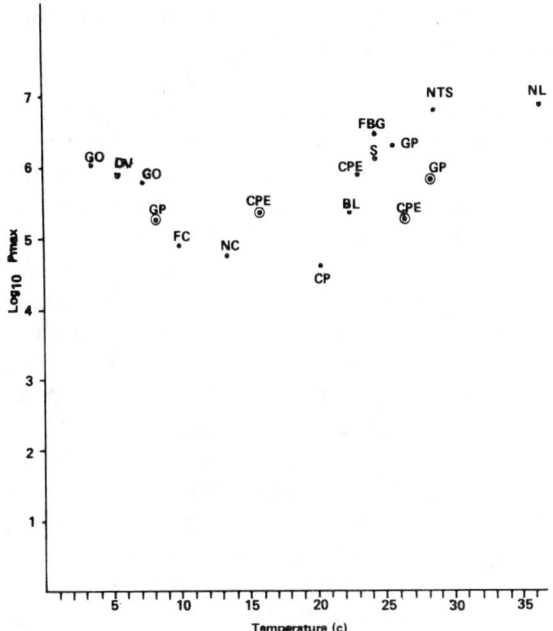

FIG. 6.7 Relationship of P_{max} for 16 species of algae near the thermal discharges from two power stations on the Missouri River (from Farrel and Tesar, 1982). GO = *Gomphonema olivaceum*, GP = *G. parvulum*, DV = *Diatoma vulgare*, FC = *Fragilaria construens*, NC = *Navicula cryptocephala*, NTS = *N. tripunctata* var. *schizonemoides*, NL = *N. luzonensis*, CPE = *Cocconeis placentula* var. *euglypta*, CP − *Cocconeis pediculus*, BL = *Biddulphia laevis*, S = *Stigeoclonium* sp., FBG − Filamentous blue–green algae.

ments on benthic algae downstream of dams is worth noting (Ward 1974; see Ward and Stanford 1979a).

(ii) Lakes and reservoirs
The studies of benthic algae in lentic waters near thermal discharges have shown similarly variable results to those in rivers. Fleming (1970) found that the total biomass of benthic algae measured as chlorophyll 'a' and 'b', was greater in the discharge of the Trawsfynnydd nuclear power station than in the intake area. The increase was entirely due to the dominance of blue-green algae.

At two power stations on Lake Wabamun the mean standing crop of epipelic algae in the discharge area was almost twice that in the control area mainly because of increased densities in autumn and winter (Hickman 1974; Hickman and Klarer 1974, 1975).

Here, the main cause of the increased densities was regarded as the increased light in winter where the heated effluent had removed the usual ice cover (Hickman 1974). As at other sites blue-green algae were abundant in the heated areas but not elsewhere. Of the diatom flora, *Navicula cuspidata* was abundant in the discharge canal at the Wabamun power station but not in colder waters. Some 13 indigenous species were absent from or less abundant in the rest of the lake. *N. cuspidata* was also considered as favoured by temperatures above 30°C in Polish waters (Stangenberg and Pawlaczyk 1960). *Cryptomonas erosa* was absent from both intake and discharge canals, apparently because the faster water currents restricted the settlement of this obligate lentic species.

Epiphytic species showed high standing crops in the effluents because of the massive standing crops of macrophytes (Hickman and Klarer 1975) (see p. 195). Densities of epiphytes were higher on the rush *Scirpus validus* in the heated waters than in unheated waters during spring and autumn but not in summer. Fewer species were found in the heated waters and there was a clear species succession at both stations. Three species of epiphytes began to grow earlier in the heated areas than in the cold waters. Blue-green algae did not dominate the epiphyte flora in the heated waters.

Primary productivity of the epiphyton showed the same trends as the standing crops with the largest differences between heated and unheated sites in May and October. When the heated discharge ceased following the redesign of the cooling system (Hickman 1982), the species favoured by the discharge conditions gradually declined in the previous plume area, while those species which had been excluded recolonised over a period of 2–3 years.

(iii) Estuaries and coastal waters

At Oyster Creek power station on Barnegat Bay in New Jersey, Hein and Koppen (1979) used styrofoam balls as artificial substrates to compare the benthic micro-algae of intake and heated discharge areas. The habitat comparisons were made difficult by the differences in salinity in addition to heat and chlorination. The mortality of cells entrained through the station would also limit the number available for colonisation. In the discharge canal, temperatures ranged from 2·8–34·4°C and the daily means exceeded 29°C for 87 days between June and October. Salinities differed by over 50% on two occasions.

Only 43% of the taxa recorded were common to intake and discharge canals. The greatest differences in the diatom assemblages as a whole occurred in August and October when salinities were similar in both habitats and temperature differences were 20°C and 9°C respectively. There was little real evidence that temperature alone caused the differences even though the authors were convinced that this was the major factor.

Kalin (1970) concluded that benthic diatoms showed lower community diversity and biomass in the effluent plume at the Northport power station than in control areas. The total area of effect was estimated as 1·5 km².

Lackey (1974) reported that green and blue-green algae colonised slides in the effluent at Turkey Point power station at temperatures up to 40°C.

At the Martigues-Ponteau power station on the Mediterranean, the population densities, biomass and community structure of the benthic algae differed with distance from the outfall (Verlaque 1977; Verlaque *et al.* 1981), with the major adverse effects at the nearest point.

6.5 MACRO-ALGAE

6.5.1 Fresh Waters
Cladophora sp. is reported as prolific in the outfall areas of at least three power stations (Langford 1974; Squires *et al.* 1979; Effer and Bryce 1975) and also where a hypolimnial discharge from a reservoir warmed a river in winter (Ward 1974). In this last example and in the area of the Provo River heated by a thermal discharge (Squires *et al.* 1979), *Cladophora* replaced the cold water species *Hydrurus foetidus* during the spring, autumn and winter. In summer both species were replaced by other algae. The upper limiting temperature for *H. foetidus* was estimated as 7·5°C and the upper limit for *Cladophora* was found by other workers to be around 23–25°C (Wong *et al.* 1978). In the discharge area of the Pickering power station on Lake Ontario, Effer and Bryce (1975) noted that *Cladophora* grew best in 'areas exposed to good currents and a temperature range of 15–25°C'.

6.5.2 Saline Waters
There has been a general concensus that the normal flora of algae and sea-grasses off the mouth of the discharge canal of the Turkey Point power stations in Florida was destroyed and replaced by a mat of blue-green algae (Roessler 1971; Lackey 1974; Thorhaug *et al.* 1974). The area affected was reported by some workers as about 122 ha (300 acres) (see Lackey 1974). However, others reported that the area affected was 12–20 ha with a less affected area of 8–10 ha outside this (Thorhaug 1980). The cause of the

disappearance of the normal flora was, supposedly, temperatures exceeding 33°C and this agreed with studies in other sub-tropical waters (Kolehmainen *et al.* 1975; Blake *et al.* 1976; Thorhaug 1980). However, Lackey (1974) noted that rocky substrata in the actual cooling canal began to be colonised with red and green algae similar to those apparently destroyed in the affected area off the canal mouth. Thus, although Thorhaug and her coworkers reported that the area inside the +5°C isotherm was denuded of the normal algae, temperature alone may not have been responsible for the change.

The intertidal species *Batophora oestidii* and *Acetabularia crenulata* survived in the heated areas though *Halimeda* and *Penicillus* spp. declined where temperatures reached 35–36°C. Factors other than temperature were also causing changes in the algal flora in the heated discharge at the Guyanilla Bay power station (Kolehmainen *et al.* 1975). Table 6.3 shows the number of species recorded to be reduced by only one in the heated areas and the lowest number of species occurred at the sampling station with the second lowest temperature regime to the control. The greatest average rise was over 8°C. The blue-green algae *Ulothrix, Lyngba, Oscillatoria, Schizothrix* and *Microcoelus* were most tolerant of high temperatures, occurring at up to 40°C. *Enteromorpha* was the most tolerant green algae, being found at temperatures up to 39°C. Blue-green algae also dominated the intertidal and sub-tidal zones at the Kahe Point power station in Hawaii, where the effluent was discharged across a sandy beach (Thorhaug 1980).

Intertidal algae normally tolerate extremes of temperature and some desiccation and would, therefore, be expected to survive in heated effluents.

TABLE 6.3

Numbers of species and biomass of algae recorded at given water temperatures and sampling sites near the thermal discharge at Guyanilla Bay, Puerto Rico (from Kolehmainen *et al.* 1974)

	Sampling site							
(Mean)	Control	1	2	3	4	5	6	7
T(°C)	25·5	26·0	26·4	27·2	28·9	31·5	33·0	34·5
No. spp.	7	4	5	8	7	5	5	5
T(°C)	30·0	31·1	31·8	33·2	34·5	35·5	36·7	37·0
No. spp.	6	3	5	8	7	4	5	5
Biomass	1·6	—	21·5	23·4	90·3	0·3	0·5	1·1

However, Arndt (1968) found that both *Ascophyllum* sp. and *Fucus* sp. were eliminated from a rocky shore where water temperatures reached between 27 and 30°C. They were replaced by *Enteromorpha intestinalis*. Coughlan (1969) also found *Enteromorpha* and *Cladophora* colonising the thermal discharge area at Pembroke power station. At both sites, current velocities were high in the affected area and chlorination was used. Filamentous algae also prospered in the outfall area of Louisla nuclear power station in Finland (Ilus and Keskitalo 1987).

At the Maine Yankee power station a series of studies showed changes in the intertidal algae in the discharge canal and immediate vicinity after about 2 years of power station operation (Vadas *et al.* 1976a; Fig. 6.8). At this time the discharge flowed across mudflats into Montsweag Bay and temperatures fluctuated from 7 to 15°C above natural ambients. In the first 2 years, four species of red algae new to the bay were recorded from the edge of the discharge canal extending the number for the bay to 38 species.

After 2 years the coverage of *Ascophyllum nodosum* and *Fucus vesiculosus* began to decline at one of the main sampling sites in the path of the effluent

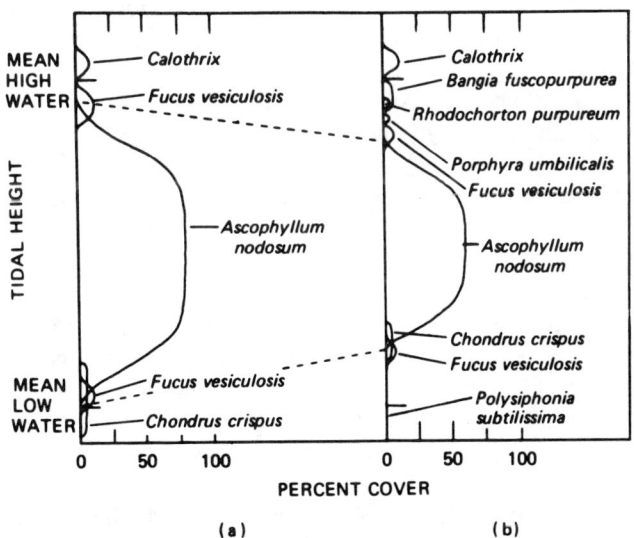

FIG. 6.8 Graphic illustration of the effects of a thermal discharge on benthic marine algae at Foxbird Island, Maine. (a) Before thermal discharge. (b) After 2-year exposure to thermal discharge. Dashed lines represent compression of vertical distribution of *A. nodosum* (after Vadas *et al.* 1976a).

where temperatures were about 5–7°C above natural ambients. Although the plants grew faster and showed earlier development of reproductive bodies the population still declined. The conclusion was that high temperatures were the main cause though other major changes had also occurred in the bay as a result of the removal of a causeway and increased freshwater flows.

Subsequently the shore discharge was replaced by a submerged diffuser outfall (see Chapter 2). Water temperatures were thus reduced in the direct path of the previous discharge and both *A. nodosum* and *F. vesiculosus* recovered at the affected sites. Productivity and reproduction reverted to pre-discharge levels. Whether the cause of the original changes were temperature, biocides or scour (or any combination) the new outfall design was effective in restoring the status quo.

The Pacific kelps *Macrocystis* spp. are basically cold water species. North (1969) predicted that kelp beds around thermal discharges on the California coast would be badly affected by the higher temperatures. Adams (1969) found, however, that at the Diablo Canyon station only about 0·7 ha were noticeably damaged. The cause was heat, biocide or possibly copper concentrations which were higher than usual in the effluent. Ernst (1970) also found that the kelp *Laminaria agardhii* was distributed only in the cooler water under a thermal plume at the Northport power station on Long Island Sound. The substrate was a jetty which was in the path of the plume. The cold control areas contained the most diverse algal flora and there was a gradual reduction in diversity toward the effluent outfall.

Thermal discharges appear to cause changes in the vertical zonation patterns at some sites. Thus at the Maine Yankee station the zone occupied by *Ascophyllum nodosum* was reduced in the discharge (Fig. 6.8) and sub-littoral species were found extended into the littoral zones which were permanently inundated by the effluent (Vadas *et al.* 1976a, b, 1978). Straughan (1980a) also noted the extension of species of algae normally found in the middle intertidal zones to rocks in the lower intertidal zones affected by thermal discharges. The changes occurred within 100 m of the outfalls. The most likely cause of these changes was the altered pattern of inundation caused by the effluent.

In temperature gradients caused by power station discharges on the west coast of the USA and Mexico, Devinney (1980) found that large phaeophytes were eliminated where temperatures exceeded 7°C above ambient. At 10°C above ambient there was a very restricted flora which was ephemeral.

6.6 ANGIOSPERMS

6.6.1 Fresh Waters

Hynes (1960) concluded that the exotic angiosperm *Vallisneria spiralis*, usually found in heated aquaria, has colonised heated reaches of rivers and canals in the UK because of increased winter temperatures and the species has been found in similar situations in France and the USA (Khalanski 1975; Massengill 1976a).

Many indigenous macrophytes are also commonly found in heated waters. For example *Elodea* spp. and *Potomageton* spp. have been recorded in thermal discharges at British and North American power stations (Trembley 1960; Langford 1972; Haag and Gorham 1977; Nichols *et al.* 1980), where temperatures in these habitats exceeded 30°C. In the River Severn at Ironbridge, *Ranunculus fluitans* was abundant in the reach below Ironbridge power station where water temperatures regularly exceeded 30°C in summer over many years (Langford 1971a, 1975).

In the discharge canal of the Peterborough power station there was a diverse indigenous macrophyte flora which included *Nuphar lutea*, *Myriophyllum* spp., *Potomageton* spp., *Typha latifolia*, *Iris pseudacorus* and *Phragmites communis*. In this canal the effluent temperatures were at one time 6–12°C above the natural ambients all year round, exceeding 30°C for periods of up to several weeks in warm summers (Cragg-Hine 1968, 1971).

There was evidence of replacement of macrophyte species in the thermal plume of the Wabamun power station. The natural flora in the shallow areas of the lake was dominated by *Myriophyllum exalbescens* and *Chara globularis*. Soon after the power station began operating, the community at the mouth of the discharge canal became dominated by *Potomageton pectinatus*. This species has also been found to be abundant in the heated and polluted reaches of the River Trent in the UK (Langford 1972). Some 4–5 years after its introduction to the lake, *Elodea canadensis* replaced *P. pectinatus* as the dominant species and also replaced *C. globularis* in the discharge area of another power station on the lake (Haag and Gorham 1977).

The standing crop of *E. canadensis* in the heated parts of Lake Wabamun was greater than at control sites, notably where winter temperatures were around 20°C, some 16–18°C above natural ambients. Growth and flowering were also advanced in spring, though light was the major factor controlling synchrony. There was no winter dormancy in *Elodea* though this was not the case for the other plants in the discharge.

Two periods of growth were reported with an advance of 2–3 months in the early part of the year. Although higher temperatures clearly encouraged growth the increased penetration of light in winter caused by the removal of the ice cover in the warm plume was also a vital factor. Thus when lower levels of insolation and increased turbidity in 1975 led to lower annual productivity of *Elodea* than in 1974, the reduced winter canopy allowed a more diverse flora to develop.

Vallisneria sp. dominated the angiosperm community in the heated discharge at the CP Crane power station (Nichols *et al.* 1980). In deeper water a stratified community was present. In the most heated creeks there was a diverse flora but the total density was lower than in control areas. *Vallisneria* showed advanced seasonal growth in the warm water.

The highest temperatures at which angiosperms have been recorded in man-made thermal waters are around the Savannah River plant. Here *Typha latifolia* survived in water exceeding 35°C, but *T. domingensis* would not grow where water temperature exceeded 30°C all year round (Gibbons *et al.* 1975). Neither species flowered in the warmest water which could exceed 45°C (see p. 47). Even at 50°C, however, some of the enzyme systems, namely the malate dehydrogenases in the *Typha* spp. were found to be stable. In the heated areas some of the semi-aquatic plants such as the swamp primrose, *Ludwigia leptocarpa* and swamp loosestrife, *Ammania coccinea* flowered earlier and produced more fruits and seeds.

The hottest effluents clearly had adverse effects on the swamp cypresses *Taxodium distichum* and gum *Nyssa aquatica*. In 1951, before the reactors began operating, there was a healthy forest in the 3020 ha of swamp. Trees died over 67% of this area once the reactors began operating. Some 227 ha of trees were killed in the area nearest the discharge where water temperatures were 25°C above natural ambients. No submerged plants grew in this area. Although the tree mortalities were greatest in the hottest water, it was considered that some mortalities in the cooler waters were caused by continuous inundation of the roots together with scour and bank erosion (Sharitz *et al.* 1974). Where the thermal discharges ceased, tree recovery was slow though the community clearly began to revert to the original state. As a contrast, there was some increase in the production of two shoreline plants in these waters as a result of prolonged growing and reproduction seasons in the warmer areas (Christy and Sharitz 1975). Both *Ludwigia leptocarpa* and *Ammania coccinea* showed greatest densities and standing crops in water at 35–40°C and lowest at normal ambient temperatures. Flowering and fruiting were 4–6 weeks in advance of their normal season at the optimal temperatures.

6.6.2 Saline Waters

(i) Sea marshes

Along the marshes of the Patuxent River estuary, Anderson (1969) found that *Potomageton perfoliatus* was replacing the usual estuarine sea-grass, *Ruppia maritima*, where thermal discharges warmed the creeks. Experiments showed that *P. perfoliatus* continued to respire at temperatures up to 35°C.

Results also showed that *Spartina alterniflora* plants growing in a heated effluent doubled their length where temperatures exceeded 35°C. These results did not agree with those of later work (Vadas *et al.* 1976b) at the Maine Yankee nuclear power station on Montsweag Bay. Here, after some initial effects on the growth of new shoots, the population of *Spartina* in the plume collapsed totally, including the rhizome system. Although water temperatures exceeded 30°C at times in the effluent and δTs were up to 15°C, there were also changes in turbidity, salinity and inundation regimes which could have exacerbated the effects on the *Spartina* beds.

Following the installation of a new diffuser outfall (see p. 194) and other physical changes in the bay, the moribund *Spartina* beds recovered to some extent in the previously heated area but declined in the control areas because of the altered salinity regime (Keser *et al.* 1978).

Similar studies of a *Spartina* marsh heated by the effluent from the Crystal River power station in Florida showed no evidence of effects. Here the effluent flooded over the marsh as the tide flooded. There was thus no scour or notable turbidity effects. In the heated areas the density of stems and net production of foliage over 3 months were higher than in the control areas though the mean weight of stems was 15% lower and decomposition loss greater (Young 1974). Night-time respiration rates were up to twice those of the control beds and the biological and chemical processes were accelerated in the marsh. Temperatures reached 37°C in summer.

Clearly, the results of the studies on the three systems produced somewhat conflicting conclusions. Temperature *per se* was not the only factor and it is evident that the other physical properties of thermal effluents are important to the effects on these marsh habitats. Biocides are not mentioned in the reports, though they were used at all the power stations and probably had significant effects.

(ii) Sea-grass systems

The most comprehensive studies of macrophyte communities in relation to thermal discharges have been those of the sea grasses in the shallow bays around the Turkey Point power station complex on Biscayne Bay and Card Sound in Florida (Lackey 1974; Thorhaug 1974, 1979).

The sea-grass ecosystem of the shallow bays of the Greater Caribbean is highly productive and forms the major habitat for many invertebrates and fishes. The plants are also the substrate for many species of epiphytes (McNae 1968; Brock 1975; Thorhaug and Roessler 1977).

Biscayne Bay and Card Sound, on the Atlantic coast of Florida, about 15 km south of Miami, are shallow coastal areas, fringed by mangrove swamps and with weak tidal flushing regimes. The depth of water in Biscayne Bay varies from about 1 to 3 m at mean low tide with very shallow bed gradients. Natural water temperatures range from 10 to $31.5°C$ with a winter mean of $17°C$ and a summer mean of $30°C$. The oil-fired Turkey Point power stations began operating in the mid-1960s discharging about $35 m^3 s^{-1}$ of heated water via a short discharge canal into Biscayne Bay. Discharge temperatures averaged about $5°C$ above ambients though the maxima could be $6–7°C$ above (Thorhaug *et al.* 1974). The thermal discharge plumes were relatively stable because of the low flushing rates in the bay. Also, because of the shallow waters, the warm water impinged directly on to the substrates with little immediate stratification. The potential for maximum adverse effects was clear. Roessler (1971) quotes areas of 10–12 ha heated to $4–5°C$ above ambient, 60 ha at $3–4°C$ above, 120 ha at $2–3°C$ above, 250 ha at $1–2°C$ above and over 400 ha at $0.5–1°C$ above. The discharge changed the normal current flows in the bay but strong winds could cause it to move in the opposite direction to its norm at times.

Apart from the temperature rise, the effluent also contained concentrations of copper up to $72 \mu g$ litre^{-1}, iron up to $300 \mu g$ litre^{-1} and other chemicals from industrial discharges nearby. The cooling water was also chlorinated using the standard 8-hourly doses of up to $1.0 mg$ litre^{-1} for periods of 30–60 min (Nugent 1970; Langford 1983a). The effluent plume from this first power station carried higher concentrations of suspended solids than the normal bay water, and the direct flow of the discharge would almost certainly scour the soft sediments from the mouth of the discharge canal. Sediments in the plume areas showed higher than normal concentrations of nickel, copper and vanadium (Thorhaug *et al.* 1974). Salinity was highly variable but dissolved oxygen concentrations were almost always at saturation levels. Clearly, therefore, the composition and physical effects of the discharge were complex irrespective of the temperature. When the thermal discharge first began, the bed of *Thalassia* disappeared completely from an area of 9.3 ha off the mouth of the canal within the boundary of the +5°C isotherm (Thorhaug *et al.* 1974). Figure 6.9 shows the data for the production of *Thalassia* in relation to the thermal plume over 3 years. Within the +3 to +4°C isotherm the cover generally declined by 50%. Roessler (1971) noted algae disappearing over an area of 10–12 ha in total.

Diplanthera colonised the denuded areas in winter, dying as temperatures rose. The maximum temperatures in the area of 120 ha of affected *Thalassia* reached 35°C. The denuded area was colonised by blue-green algae in summer.

Thorhaug (1980) in summarising the studies at Turkey Point concluded that the critical isotherm of the onset of adverse effects on the sea-grasses was at about +1·5°C. Within the +1°C isotherm there was some increase in the annual production of *Thalassia* caused by stimulated winter growth.

Although the effects on sea-grasses are relatable to the temperature changes in the bay *and* all the major publications imply that these were the major stresses, it is remarkable that there are few attempts to assess the effects of the other factors such as scour, biocides, turbidity and other chemicals (Nugent 1970). Clearly, in such a complex habitat the separation of component effects is difficult from field studies, but apart from limited mention their effects are apparently not considered quantitatively. The absence of chlorination data is particularly significant. Lackey's findings (Lackey 1974) that 'the sea-grasses *Thalassia*, *Diplanthera* and *Syringodinium* have persisted during all 4 years (of his studies) as healthy dense beds in areas receiving the plant effluent' are not consistent with temperature effects though he rarely recorded readings of over 35°C.

He also notes that the 'bare spot', that is the denuded area at the mouth of the effluent canal, was there before the power station began operating and was recolonised by *Diplanthera*. It is possible that the digging of the canal could have displaced sediments which blanketed the original beds.

Although the work is long completed and other events have overtaken the area, the conclusions are vital to the real understanding of heat alone as a significant pollutant in such habitats.

The original canal was closed in 1973 and a new canal system dug to accommodate a nuclear power station complex. This cooling canal displaced over 1500 ha of mangrove swamp and discharged to Card Sound to the south of Biscayne Bay. Before the oil-fired station closed and the nuclear station opened the discharge canal was modified to a 9·3 km long channel to Card Sound. The digging disturbed sediments which smothered an area of 2–3 ha of *Thalassia* under a 10 cm layer as the canal opened and the effluent began discharging. Subsequently, the areas of sea-grasses

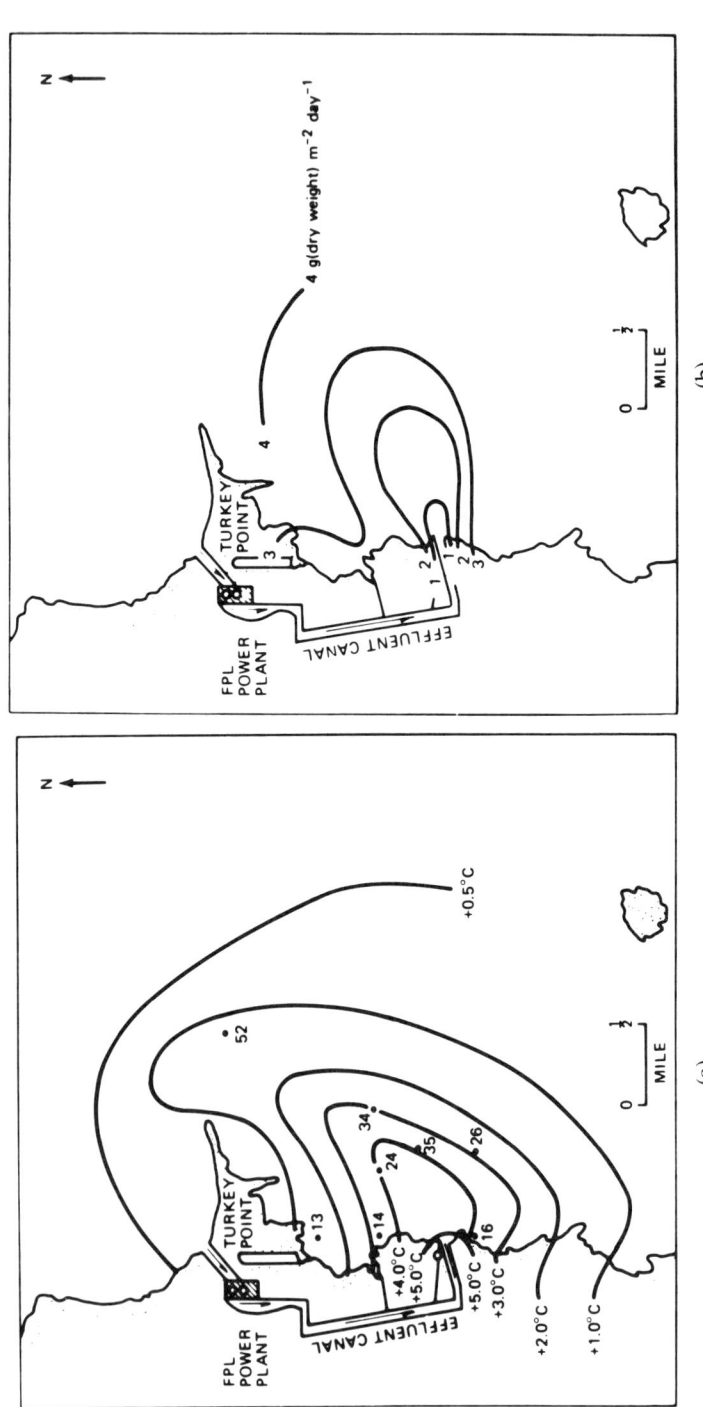

Fig. 6.9 The production of sea-grasses (*Thalassia testudinatum*) in the thermal plume from Turkey Point power station, Florida (after Thorhaug, 1974). (a) Mean temperature elevation above bay ambient in 1970 for the Turkey Point area with stations 16, 26, 35, 24, 14, 13 and 52. (b) Turkey Point production of *T. testudinatum* blade material for 1970.

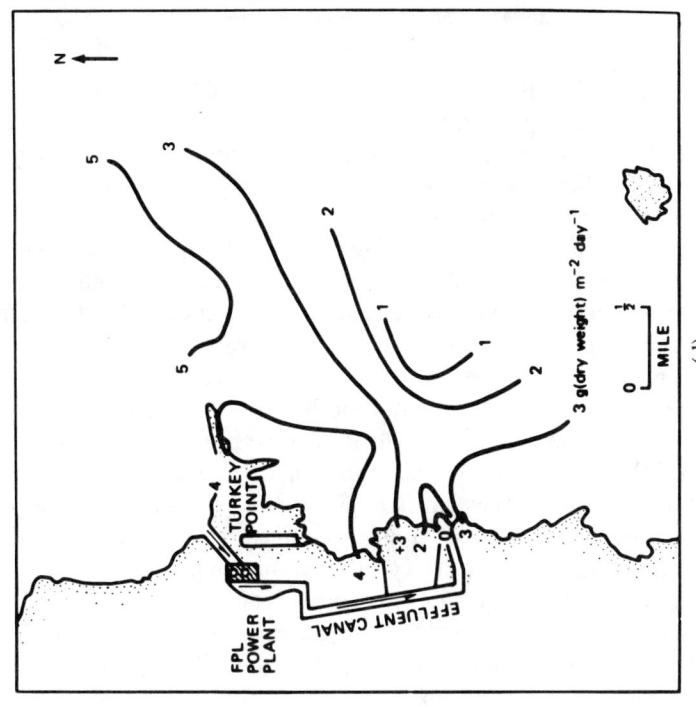

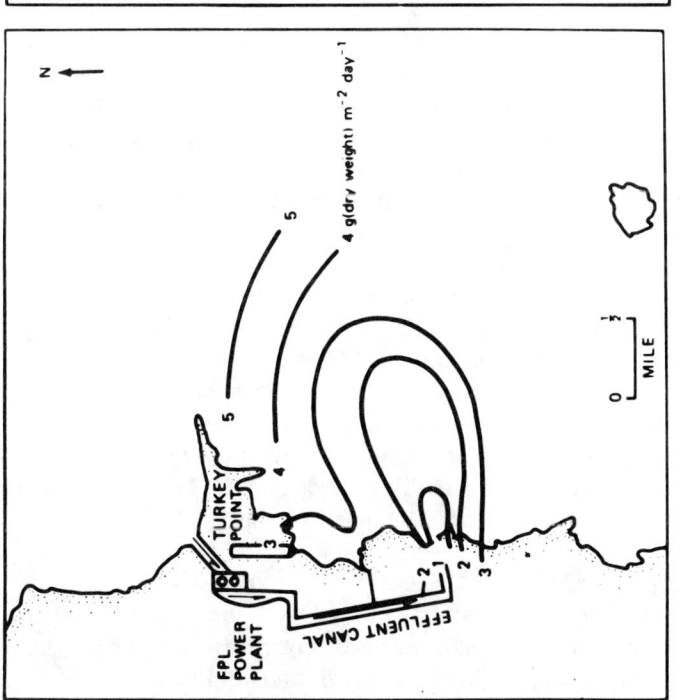

Fig. 6.9—*contd.* (c) Turkey Point production of *T. testudinatum* blade material for 1971, and (d) for 1972.

affected in Biscayne Bay have partially recovered and there is little ecological damage reported from Card Sound relatable to the new discharges. However, a large area of mangrove swamp has been destroyed to accommodate the new cooling system though the significance of this to the total swamp area is apparently small.

Areas of Biscayne Bay denuded of sea-grasses by various other human activities have been recolonised by transplanting schemes (Thorhaug 1987). Some 114 ha of barren sea-bed is intended for restoration and by 1987 some 47 ha were again productive and colonised by a diverse fauna. The value of such sea-grass beds is assessed as 187×10^3 dollars annually.

In Tampa Bay on the Gulf coast of Florida, some 81 ha of *Thalassia* were destroyed in a thermal discharge where temperatures reached $40.7°C$ in summer (Blake *et al.* 1976). No mention is made of biocides in this work though there was almost certain to be antifouling measures taken at the power station. Transplanted stands of *Thalassia* failed to survive temperatures in the discharge exceeding $31°C$ though they clearly survived temperatures higher than this in the control areas.

To add further to the conflicting data, transplanted stands of *Thalassia* showed no signs of dying in the effluent at the Guyanilla Bay power station in Puerto Rico at temperatures up to $35°C$ over a 9 week exposure. However, production was reduced at the higher temperatures. The natural beds of *Thalassia* in the effluent plume here were sparser than in unheated areas (Kolehmainen *et al.* 1975) but the authors of this paper stated that 'it was impossible to determine whether low biomass of turtle-grass was caused by elevated temperatures or strong currents and changes in the coarseness of sediments'. At several of the power stations mentioned by Thorhaug (1980) similar physical and chemical complexities are notable. For example at Cutler Ridge, north of Turkey Point, 35 ha of sea-grass were removed in the effluent plume but this effluent contained 6 times more sediment than the normal sea water. At Key West, where areas of *Thalassia* were destroyed, the effluent contained $1766\,\mu$g litre^{-1} of copper.

From the evidence it seems reasonable to conclude that the physical effects of scour, sediments, salinity, biocides and other chemicals are more responsible for the destruction of sea-grass beds in some thermal discharges than water temperature, unless this exceeds about $37°C$ for long periods. Even so the other factors can dominate (Nugent 1970).

(iii) Mangrove swamps
The mangrove root systems fringing the edges of sub-tropical and tropical shores provide a major habitat for algae, invertebrates and fish. The root system is complex and highly productive (McNae 1968).

In the heated effluent at Guyanilla Bay, the mangroves continued to reproduce when water temperatures around their roots' natural ambients reached 35–41°C (Kolehmainen *et al.* 1974, 1975). The trees nearest the outfall died eventually as a result of scour and sediment changes. As the power station was enlarged and the effluent volume increased new mangrove shoots began to die in the hottest water. Detailed studies (Banus and Kolehmainen 1976), showed that the seedlings of the red mangrove (*Rhizophora mangle* L.) near the outfall were 'visibly stressed' and were significantly smaller than those from the unheated areas. Seedlings developed negative buoyancy and initial roots at a faster rate in the heated waters but survival was poor and none were found rooted in the hottest waters. Critical temperature for *R. mangle* seedlings was reported as 37–38°C.

Some synergistic effects may have been present as high concentrations of heavy metals were found in these mangroves from the bay (Thorhaug 1980).

It is clear from the data that the discharge of effluents from power stations has affected the communities of angiosperms in the areas of plumes. It is not clear, however, which of the component influences is the stronger at most sites. Effluent temperature, even at the extremes, seems to exert a weaker effect than water currents, turbidity and possibly chemical substances such as metals or biocides. From the results discussed there is no unequivocal case for reducing effluent temperatures alone for all discharges in any region to protect angiosperms or other plants. The evidence indicates that alleviation of effects would come best from reducing scour and biocides and through more careful construction of canals or outfalls.

CHAPTER 7

Rotifera and Micro-crustacea

7.1 TOLERANCE AND DISTRIBUTION

This chapter deals mainly, though not exclusively, with those animals which live most of their lives in the water column as part of the 'zooplankton'. They are most abundant in ponds, lakes, reservoirs, slow and impounded rivers, estuaries and the sea. Their lateral transport is dependent on water currents and they are thus particularly vulnerable to entrainment in water currents at intakes and to displacement by effluents.

The 'micro-crustacea' comprise the Copepoda, Cladocera and Ostracoda, all of which have a wide distribution. These and the Rotifera include planktonic, epibenthic and epiphytic species. Some adult and larval insects together with the eggs and larvae of fish also become part of the planktonic communities but these are discussed in the relevant chapters (see Chapters 8 and 9).

Rotifera are not recorded from geothermal waters at temperatures much higher than those which occur in unheated waters. Thermophilic genera, for example *Hexartha* and *Ptygura* (Anderson and Lenat 1978), occur at temperatures over 35°C in naturally warm lakes.

Among the Ostracoda, *Heterocypris balneria* has been recorded in thermal springs at temperatures of 40–50°C (see Brock 1975), and these are regarded as having the highest thermal tolerance of any aquatic invertebrates (Castenholz and Wickstrom 1975). The thermal tolerances of the most common species of planktonic crustaceans are well known (see Coutant and Talmage 1977; Schubel *et al.* 1978; Talmage and Coutant 1980). As examples, some species of fresh-water Cladocera can tolerate short exposures to at least 30°C and estuarine species of copepods have $LT_{50,30}$s of 35°C (Strickland 1969; Goss and Bunting 1976; Schubel *et al.*

1978). Gonzalez (1974) showed that *Acartia tonsa*, acclimated to 27°C, could survive for 4 h at 37°C. This work also suggested that *A. tonsa* demonstrated a temperature tolerance which was determined genetically and linked to geographical distribution.

In many waters zooplankton populations migrate vertically on a daily or seasonal basis and thus there may be concentrations of these animals at different depths and different temperatures at different times of day (Hutchinson 1957; Sverdrup *et al.* 1963; McLaren 1963; Kinne 1970a). The relative locations of intakes and outfalls in different waters (see Chapter 2) can thus be vital in determining the extent of entrainment or thermal discharge effects. Gehrs (1974) showed in experiments that zooplankton species would avoid heated surface waters by moving into colder deeper water.

7.2 METHODOLOGY

Experimental studies have employed both orthodox laboratory scale methodology and large scale simulations (Poje *et al.* 1981).

In the field, the basic methods for sampling zooplankton in intake and discharge areas are similar to those for sampling phytoplankton (see Chapter 6), that is mostly using nets, pumps or other volumetric samplers (Lawler, Matusky and Skelly, Engineers 1979). One of the major problems with net samplers is the rate of mortality caused by impingement and abrasure on the net surfaces, which is in turn related to the water velocity in the net. Mortalities also occur in pump samplers as a result of passage through the pump impeller. Specially devised low-velocity samplers, with the net placed before the impeller, have been successful in reducing sampling mortalities (e.g. Utting and Millican 1977; Coughlan and Fleming 1978).

The criteria used to assess the effects of entrainment and effluents include abundance and living-to-dead ratios based on 'motility' or 'viability', though both of the latter are difficult to assess without long observations following the sample collection. For example, it is difficult to know whether a 'twitching' individual at the point of collection will recover or die in the longer term (Reeve and Cosper 1970; Koops 1972, 1975; Davies and Jensen 1974; Davies *et al.* 1976).

One of the more successful methods of assessing mortality has been through the use of vital strains (Dreesel *et al.* 1972; Fleming and Coughlan 1978), though again errors can arise from subjective judgements on the survival of individuals which only partially absorb the stain. Few field studies have attempted to assess the precision of results or the proper sampling effort required for statistical analysis (see Lawler, Matusky and Skelly, Engineers 1979). For example it was shown that to detect a 20% change in abundance between a control area and a thermal discharge plume in coastal waters at least 20 replicate samples are necessary at each station on each sampling occasion (Carpenter *et al.* 1974a). Detection of a 10% and a 5% change would require 75 and 300 samples respectively. No sampling programme has reached the second level of sampling effort and it is therefore arguable that none is capable, so far, of detecting changes of 10% or less.

7.3 SITE STUDIES

7.3.1 Effects within Cooling-Water Systems

Most of the site studies of the effects of cooling-water systems on zooplankton have been related to the effects of entrainment and passage through direct systems. Experimental studies have thus concentrated on short-term exposures to high temperatures, chlorine or combinations of both (see Schubel and Marcy 1978; Lawler, Matusky and Skelly, Engineers 1979; Hall *et al.* 1981; see Jolley *et al.* 1987b, 1980, 1983a,b; Yost and Uziel 1981; see also Chapter 4).

(i) Recirculating systems

It has been suggested that mortalities caused by indirect systems were likely to be greater than in direct systems because of chemical contaminants and prolonged exposure to high temperatures (Schubel and Marcy 1978; Majewski and Milller 1979). These predictions are difficult to understand as the available evidence already showed that many species of animals and plants survive in cooling-tower systems (Markowski 1962; Langford 1971b; Ross and Whitehouse 1973). Micro-crustacea were found to be abundant in the cooling-tower ponds and at the outfalls of British power stations in the 1950s and 1960s (Markowski 1962). The Ostracod *Cypriodopsis vidua* occurred 'in masses' at Leicester power station and a diverse zooplankton fauna existed in the tower ponds. Copepods were clearly reproducing in these habitats irrespective of the heat and intermittent chlorination. Milner (1984) subsequently also found copepods, cladocerans and ostracods to be abundant on artificial substrate samplers placed in the chlorinated waters of the cooling tower systems at Ratcliffe-on-Soar and Ironbridge power stations despite the fact that the water quality differed markedly at the two sites.

At most of the sites studied cooling-pond temperatures rarely exceeded 32°C though chlorine residuals (TRC) could reach 0·8 mg litre^{-1} for short periods (see p. 88).

(ii) Once-through (direct-cooled) systems
(a) Fresh waters. Markowski (1959) used plankton nets with 74 meshes cm^{-1} for qualitative samples and pumps for quantitative samples of animals from the intake currents and outfalls at 12 power stations. At two river cooled stations using once-through cooling there were no qualitative differences in the zooplankton samples between the sampling sites. The reservations about Markowski's data are, however, that at some sites he trailed nets in discharge currents and animals from the receiving water may have become entrained in the effluent flow and thus not have passed through the cooling system.

The abundance of zooplankton was found to be reduced following passage through only one of five cooling-water systems in the USA (Lawler, Matusky and Skelly, Engineers 1979). At this one site reductions in abundance occurred only during chlorination but there is no clear explanation of the cause. At the KORI nuclear power station, the density of the zooplankton was also reduced at the outfall after entrainment (Huh 1980). As with phytoplankton there are no obvious reasons for reductions in abundance after entrainment (see p. 166). Small zooplankton are unlikely to be disintegrated *en masse* during passage and the simplest explanation is the integration of discontinuously distributed animals by the mixing processes in the system. At one site the abundance was greater at the outfall than at the intake and this was also a result of mixing in the system.

The effects of entrainment on the motility of zooplankton are more readily detectable and understandable. In Fig. 7.1 the data from studies of entrainment of zooplankton are summarised using the mean reductions in the proportions of organisms retaining motility as indications of effects. At most sites a loss of motility occurred in less than 20% of the animals entrained, irrespective of heat, chlorination or mechanical stresses. However, the actual range of effects varied with site and season. The maximum effect at a river-cooled site was a loss of motility among 90% of animals. At lake-cooled sites the maximum in any sample was 100%. The greatest effects were usually in the hottest period of summer during chlorination. Outfall temperatures exceeding 30°C were associated with increased mortalities at several sites.

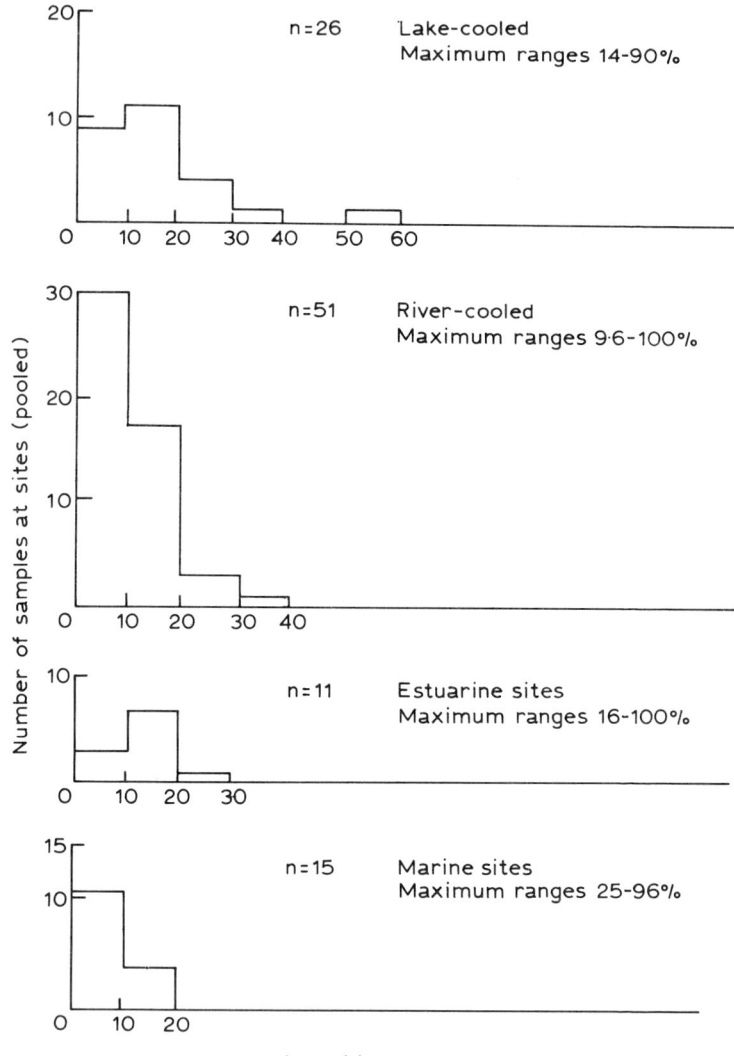

%o loss of motility among zooplankton

FIG. 7.1   Frequency distribution of mean losses of motility among zooplankton entrained at power stations on various waters (see text for references).

Chlorine dosed at 0·25–1·0 mg litre^{-1} reduced the motility ratio by 40% at the Indian River power station. At levels exceeding 1·0 mg litre^{-1} survival was reduced by 85–100% (Davies and Jensen 1974).

There was a direct inverse relationship between temperature and mortality at the James River power station (Davies and Jensen 1974) based on motility ratios (i.e. corrected % motility $= 204\cdot8 - 6\cdot3T_a - 1\cdot19\Delta T$, where T_a is the natural water temperature and ΔT is the temperature rise at outfall).

At the Peach Bottom nuclear power station on the Susquehanna River entrained Cladocera suffered 90–100% mortality during summer and the average for all zooplankton over the period from June to September was 30%.

At the Quad-Cities site on the Mississippi the total effect of entrainment in July was a 94·7% mortality. Here there was a direct linear relationship between the mortality rate and the size of the animal over a length range of

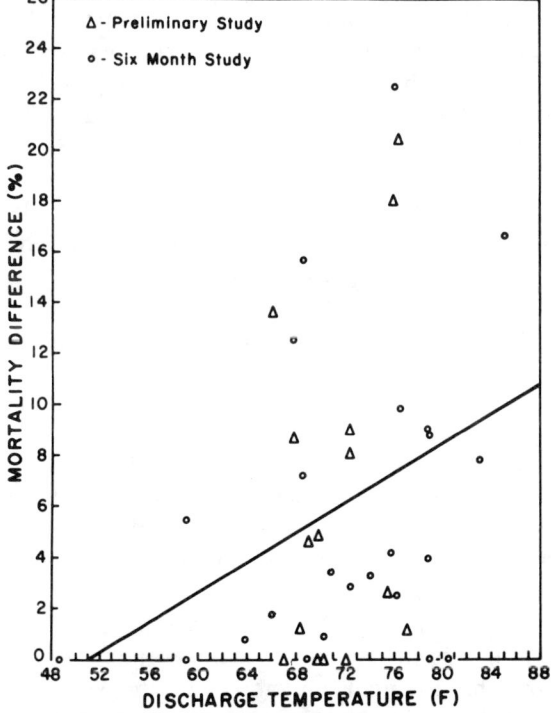

FIG. 7.2 A linear regression analysis of mortality differences and discharge temperatures is depicted for data collected during the preliminary and six-month study periods (from Icanberry and Adams 1974).

0·4–1·35 mm (see Lawler, Matusky and Skelly, Engineers 1979) (see also p. 275). The relative mortalities of the various taxonomic groups were different and highly variable from site to site. In most of the studies at river sites the assessment of mortality rates was made immediately after entrainment and few longer-term observations were made. Donze (1978), however, incubated the zooplankton collected from the outfall of Harculo power station on the River Issjel, in holding tanks for 10–14 days at temperatures of 4°C above natural ambients. Rotifers increased in abundance in the heated tanks, though a parasitic protozoan and a fungus, stimulated by the high densities of animals in the cultures, caused some mortalities.

Cladocera showed the greatest entrainment mortalities. *Bosmina* spp. densities in the holding tanks declined initially after a 13·5°C shock but recovered to comparable densities to the controls after 5 days.

The data from 26 surveys at 16 power station sites using lake water for cooling showed mean potential mortalities of entrained zooplankton mainly less than 20% (Fig. 7.1). At some sites intake mortalities exceeded outfall mortalities probably as a result of net abrasion. The highest mortality rate was 100%, and, as with the river sites, different taxa showed different mortality rates (see Lawler, Matusky and Skelly, Engineers 1979).

In one of the more comprehensive surveys (Evans *et al.* 1978), samples were taken from the intake and discharge forebays of the Donald C. Cook power station on Lake Michigan over a period of 23 months. The average instantaneous mortality rate for all zooplankton was 11·8% at the intake and 12·0% at the discharge and these were not significantly different ($p < 0.05$).

Among the various taxa the mortalities averaged from nil to 44·3% in different months but only *Diaptomus* spp. and *Eurytemora affinis* copepodites showed statistically significant entrainment mortalities over the whole period. The precision of the estimates was such that mortalities of less than 12% between intake and outfall were not statistically significant. Higher mortalities were found at other power stations but statistical tests were not always carried out.

At the Ginna and Nine Mile Point power stations increases in mortality after holding ranged from 0 to 7% but these were not significantly different ($p < 0.05$) from the instantaneous mortality rates (Storr 1974). de Nie (1982) found very low mortalities (2–3%) for most small Cladocera entrained at the Bergum power station in Holland, but the larger species showed 25% mortalities, mainly because of mechanical damage.

(b) Estuaries and coastal waters. Figure 7.1 shows that the mean mortality rates after entrainment at these sites, expressed as motility ratios, have mostly been less than 30%, though maximum rates of 100% have been recorded at several sites under extreme conditions of temperature and chlorination (Carpenter *et al.* 1974b,c; Lawler, Matusky and Skelly, Engineers 1979). The data from studies at coastal and estuarine sites show similar variability to those from the freshwater sites. At the Indian Point power station, the reductions in motility caused by entrainment were greater in the summer than in winter. At this site the abundance of zooplankton was greater at the outfall than at the intake during the 1971 and 1972 surveys but the reverse applied in the 1974 surveys. In 1975 there were no differences between the two (see Lawler, Matusky and Skelly, Engineers 1979). The indications were that mixing in the system and varying distribution patterns in the intake water were the most significant factors.

Using the loss of motility as an indicator of mortality, most of the studies at estuarine power stations have shown that mortalities exceeding 40% can occur at operating temperatures and normal chlorination rates though there is again a high variability with site and season. Coughlan and Davis (1983) using vital staining techniques found mortality rates of up to 58% at power stations in the UK. There was an increase in mortality rates from around 5% immediately after sampling to the higher levels on prolonged exposure in the outfall water. Such delayed effects were not evident after exposure of zooplankton, collected after transit through the CP Crane power station, to temperatures only 1–2°C above natural ambients (Davies *et al.* 1976).

At the Morgantown power station on the Potomac estuary intake samples were found to contain large percentages of dead planktonic animals (Heinle *et al.* 1974), and it was concluded that there was recirculation of effluent water to the intake during slack water at high and low tides. The sensitivity of the estimates of entrainment mortality was thus much reduced. The plankton was highly stratified at the intake and the skimmer wall which reached to a depth of 10 m and excluded the densest surface populations. Heavy cropping of these organisms also occurred at the intake by large numbers of fish, mostly Atlantic Menhaden, *Brevoortia tyrannus* which congregated there. The copepod *Acartia tonsa* showed the highest mortalities during chlorination but the abundance of living barnacle larvae (*Balanus* spp.) actually increased in the effluent as a result of the spawning of the adults colonising the culverts.

Where cooling ponds or long discharge canals are interposed between the condenser outlets and the receiving water, the effects of entrainment on the behaviour of both live and dead animals can reduce the numbers of individuals reaching the receiving water. At the Millstone power station, for example, some 70% of the copepods entrained and entering the cooling pond did not leave the pond (Carpenter *et al.* 1974c). The effluent water had a residence time of about 9 h in the pond and the δT was about 13°C. Vital stain techniques showed that the mortalities of copepods were less than 15% at the condenser outlets but the losses between the intake and the cooling pond outfall to the open water were between 67% and 94% depending upon temperature and chlorination. Subsequent experiments showed that many animals were not killed during entrainment but as they entered the isohaline, isothermal cooling pond they sank rapidly to the bottom. The sinking rate of entrained copepods was 2·5 times faster than that of control animals. The population in the cooling pond at Millstone was 16 times more dense at 25 m than at 2 m depth. However, 60% of these animals were dead at 25 m compared with 10% at 2 m. Clearly mortality was delayed and the rates exacerbated by retention at high temperatures in the pond. The total annual loss of copepods from Long Island Sound was calculated as 0·1–0·2% of the standing crop.

7.3.2 Effects on Receiving Waters
(i) Fresh waters
In one of the earliest studies of zooplankton in heated waters, Churchill and Wojtalik (1969) found less dense populations in the discharge area than at the intake of a power station on the Green River in Kentucky but further downstream the density was greater than at the intake. At the Gorgas power station on the Black Warrior River in Alabama, the total abundance of zooplankton was reduced by 48–78% between the intake and discharge areas. Here stratification of the animals at the intake and subsequent mixing in the system were regarded as the causes of the change (see Lawler, Matusky and Skelly, Engineers 1979). In the oligotrophic Trawsfynnydd Lake, used as the cooling pond for a nuclear power station, the dominant planktonic crustaceans were the copepods *Cyclops abyssorum* and other *Cyclops* species. Whitehouse (1971) reported that there was no evidence of differences in the species composition or abundance of the zooplankton between the cold and warm water areas.

Several studies have shown marked differences in the zooplankton of heated and unheated lake waters (e.g. Polivannaya and Sergeyeva 1971a,b; Kititsina 1973; Brauer *et al.* 1974; Goryanjova 1975; Kostylev and Yesipova 1982). In this last study, on Lake Monona near the thermal discharge from the Blount Street power station, the high zooplankton densities were associated with high water temperatures when the power station was operating but there was more sample-to-sample variation than at the intake. Here, and at the Kursk power station in the USSR, the increased densities were apparently caused by a redistribution of animals from the deep water offshore intake to the shallow littoral zones via the onshore outfall. Living plankton passed through the Blount Street cooling-water system except when temperatures reached 39–40°C at the outfall.

The densities of zooplankton in the heated water declined below those of the unheated waters at the three sites in summer. At the Kursk site the authors concluded that at temperatures above 26°C zooplankton development was inhibited. However, the data are poorly presented and show that when water temperatures in the heated areas were at 30–35°C, some 10–12°C above natural ambients, the zooplankton were more abundant in the heated water. At temperatures of 26–30°C in the heated water and 18–22°C in the cold water the reverse was true. The mean biomass was about 23% greater in the colder waters than in the heated waters, and the maximum exceeded that in the warm water by almost 95%. Species composition differed in that *Daphnia* spp. and *Cyclops* spp. dominated the warm water zooplankton while *Daphnia* spp. and Rotifera dominated the unheated water populations.

In the Bergumermeer, the zooplankton was less abundant near the outfall of the power station than at the intake but this was a very localised change and did not apply to the lake as a whole (de Nie 1982). Microcosm studies showed effects on both survival and population structure of zooplankton after heating (Donze 1978).

In Belews Lake, a 600 ha cooling lake (Anderson and Lenat 1978), two warm water stenothermal Rotifera increased in total density in the warm areas and a third showed increases in densities. Six taxa declined in the lake but the main decline was in the unheated waters. Discharge temperatures reached 39°C in summer.

Several of the differences between the zooplankton communities of the heated Lake Sangchris and the control, Lake Shelbyville in the USA, were described as related to the temperature differences between the lakes, irrespective of the other hydrographic differences (Larimore and McNurney 1979; Larimore *et al.* 1979a,b).

Species of Rotifera and micro-crustaceans showed longer periods of growth or were present for longer periods of the year in Lake Sangchris. The rotifer *Brachionus caudatus* occurred 2 months earlier in the warmer water of Lake Sangchris than in Lake Shelbyville and *Brachionus angularis* reproduced throughout the year instead of only over its usual 7 months. In contrast, less tolerant species occurred and reproduced during a shorter period of the year in the heated water than in the cold lake.

Temperature was not the limiting factor for some of the species of Cladocera and Copepoda in the lakes. *Bosmina longirostris*, for example, was found only in winter and spring in the cooler lake but throughout the year in the warm lake. The reason was that predatory pressure by the common cladoceran *Leptodora kindtii*, which was itself intolerant of the warm effluent and thus less abundant, eliminated predation in summer. This is an excellent example of an indirect or secondary biotic effect of a thermal discharge (see p. 105).

There was less seasonal variability in the major community parameters of the zooplankton in the warm lake than in the colder one. The standing crop and production were greater in Lake Sangchris than in Shelbyville in autumn but lower in the summer of 1976. Other factors influencing the differences between the zooplankton were reported as inflow, flushing rates, macrophyte densities, fish densities and the effects of entrainment. It seems clear, however, that temperature was implicated in the seasonal differences and in the changes in growth and the reproductive cycles.

The zooplankton populations of the Conowingo Pond on the Susquehanna River in the USA are subjected to three man-made stresses, namely:

—highly variable flushing rates because the 'pond' is a river impoundment,
—entrainment and heat because the pond is the cooling water source for the Peach Bottom nuclear power station, and
—drawdown and pumping effects because the pond is the lower reservoir of the Muddy Run pumped storage power station.

It is hardly surprising therefore that attempts to quantify the effects of the heated effluent on the zooplankton of this complex system were somewhat equivocal (Mathur *et al.* 1980). There was no change in the community or populations which could be clearly related to the thermal discharge. The main variable affecting the population density was river flow and there was a significant inverse correlation between the two.

(ii) Estuaries and coastal waters
The dynamic nature of plankton populations in the mobile waters of estuaries and nearshore make the quantification of effects more difficult than in the calmer waters of large rivers and lakes.

In one of the earliest studies, Raymont and Carrie (1964) found a marked increase in the abundance of zooplankton in the plume area of the Marchwood power station on Southampton Water in the south of England. The main cause was the early production of barnacle nauplii, particularly of *Elminius modestus*, from the large colonies in the cooling-water culverts and on structures near the outfall.

Studies at ten power stations in the USA showed no consistent patterns of change in the heated plumes (see Lawler, Matusky and Skelly, Engineers 1979). At two sites, namely Bowline Point and Calvert Cliffs, seasonal increases in the abundance of zooplankton were also a result of large numbers of barnacle nauplii from the colonies in the cooling systems. Other species were, however, less abundant in the plume and the appearance of fragments in samples indicated that some animals were killed by mechanical forces in the cooling systems.

At the Roseton power station on the Hudson estuary, cluster analysis distinguished two separate communities of plankton in the intake and plume areas respectively. The cause of the difference was the occurrence of benthic invertebrates in the intake samples entrained in the water currents. Chlorination was associated with the lower abundance of planktonic crustacea in the effluent plumes at several sites, though integration of patchy distribution was probably the main cause (see p. 166).

In contrast to other studies, that at the Morgantown power station on the Potomac estuary found that the copepodites of several species of planktonic crustacea were less abundant in the discharge area than at the intake (Heinle *et al.* 1974). In the discharge canal and the estuary near the Fort Meyers power station in Florida, there was a negative correlation between the abundance of zooplankton, temperature and dissolved oxygen concentration. However, the reason for the reductions in abundance were considered as caused by dead animals sinking to the bottom and thus not being caught by surface and mid-water nets (see p. 205).

At four truly coastal sites the same pattern of variability was found. At the Millstone plant the differences in zooplankton abundance between plume areas and unheated water were not statistically significant ($p < 0.05$), though the reason was concluded as 'too few samples' (see p. 206). At the Honolulu station in Hawaii, *Sagitta* spp. and *Lucifer* spp. were more abundant ($p < 0.001$) at the outfall and offshore sampling stations than at the intake. Cluster analysis of the plankton and icthyoplankton showed three different assemblages caused by the transport of littoral species from the nearshore intake to the offshore outfall area. In contrast, at the Cutler power station in Florida and the South Coast power station in Puerto Rico, the zooplankton was less abundant in the discharge area than at the intakes. At the Cutler site four surveys showed reductions of 20–65% but only one of the surveys was made while the station was operating. These effects were regarded as the result of 'long-term' influences of the effluents though the precise mechanisms through which the changes occurred were not specified.

In the heated waters near the Dunkerque power station, Brylinski (1981) found close correlation between temperature and the rate of development of the life-history stages of the calanoid copepod, *Temora longicornis*. Several stages were advanced seasonally in the heated water. The population densities of the benthic ostracod, *Haplocytheridea setipunctata* varied in direct relation to temperature in a thermal discharge in Tampa Bay, Florida (Stiles and Blake 1976) during the period between March and June, but in the warmer months the populations became depleted in comparison to the controls. At 35°C no living animals were present.

Bamber (1987a) found that the American benthic ostracod (*Sarsiella zostericola*) introduced to the Medway estuary in the UK did not survive in the most thermally stressed reach of the discharge canal of Kingsnorth power station (Bamber and Spencer 1984). In the less stressed reaches the individuals were larger and more fecund than those in the indigenous North American populations. Effects on shell physiology were believed to be the cause of the species not living in the stressed reach.

Interestingly, Bamber (1987a) considered that the ostracod was not an 'exotic' species specifically associated with thermal discharge (p. 316).

7.4 EFFECTS OF THE DIFFERENT STRESSES

7.4.1 Temperature
Markowski (1959) sampled animals passing through power station cooling systems and found very little effect of effluent temperatures up to 32°C and δT's of 17–22·5°C on entrained zooplankton. Also Markowski (1960) and Milner (1984) found large populations of zooplankton in cooling tower ponds where temperatures reached 30°C consistently during the year.

At the Donald C. Cook power station on Lake Michigan the mortalities of the dominant zooplankton species were significant over a wide temperature range. However, the mortalities did not increase as seasonal temperatures rose in summer. At the Fort Calhoun station, in contrast, Rogers (1978) concluded that 'thermal stress' was the principal factor causing zooplankton mortalities. Here maximum temperatures reached 35–37·1°C at the outfall but the mortality rates did not exceed 10–11% above those at the intake. Subsequently, however, the final conclusions of the work at both Fort Calhoun and the Cooper power stations were that temperature was *not* a major cause of zooplankton mortalities (Repsys and Rogers 1982), because of the short exposure times, mostly less than 2 min. The opposite was true for the mortalities of zooplankton at the Connecticut Yankee power station where the organisms were exposed to temperatures of 31–39°C for 50–100 min while passing through the discharge canal. This prolonged exposure resulted in 100% mortalities (Massengill 1976b).

At both the James River and Californian coastal power station sites (Davies and Jensen 1974; Icanberry and Adams 1974) regression analyses showed significant relationships between temperature and zooplankton mortalities. At the latter sites, the relationship was barely significant and showed huge variability at both higher and lower temperatures (Fig. 7.2). At the Lake Norman site the greatest mortalities occurred in winter, at the lowest natural temperatures, when the δT was doubled to 20·4°C because of a reduction in pumping rates (Davies and Jensen 1974). Effluent temperature was 34°C, higher than natural summer values. Figure 7.3 shows the mean weekly mortality rates of zooplankton entrained at the Wyman Power station outfall in relation to the temperature of the discharge. Mortalities exceeded 75% at temperatures over 32°C but ranged between 12% and 68% at temperatures of 20–30°C. Davis and Coughlan (1978) found negligible mortalities of marine zooplankton at effluent temperatures of 24–27°C. Gaudy (1981), however, attributed up to 16% of the mortality of zooplankton to temperature alone at the Martigues–Ponteau power station. Here, temperatures could exceed 30°C. Carlson (1974) used plastic swimming pools to expose natural mixed communities of freshwater plankton to different temperature increases, namely 4·8, 7·4, 13·5 and 20·8°C over natural ambients. Mean temperatures over a 35 day study were 25, 27, 31 and 42°C in the heated tanks and 23°C in the controls. In the hottest pool all the crustaceans died within 6 days. At 31°C densities were less than 5% of those in the other pools. At 27°C densities were also slightly reduced compared with the controls. The most common Cladoceran, *Scapholeberis kingii*, showed a clear trend of population instability with higher temperatures. This type of result had previously been shown for *Daphnia magna*. Following the microcosm experiments, Carlson assessed the temperature tolerances of *S. kingii* from a natural pond, a cooling canal and a heated pond at the Savannah River plant. Median survival times

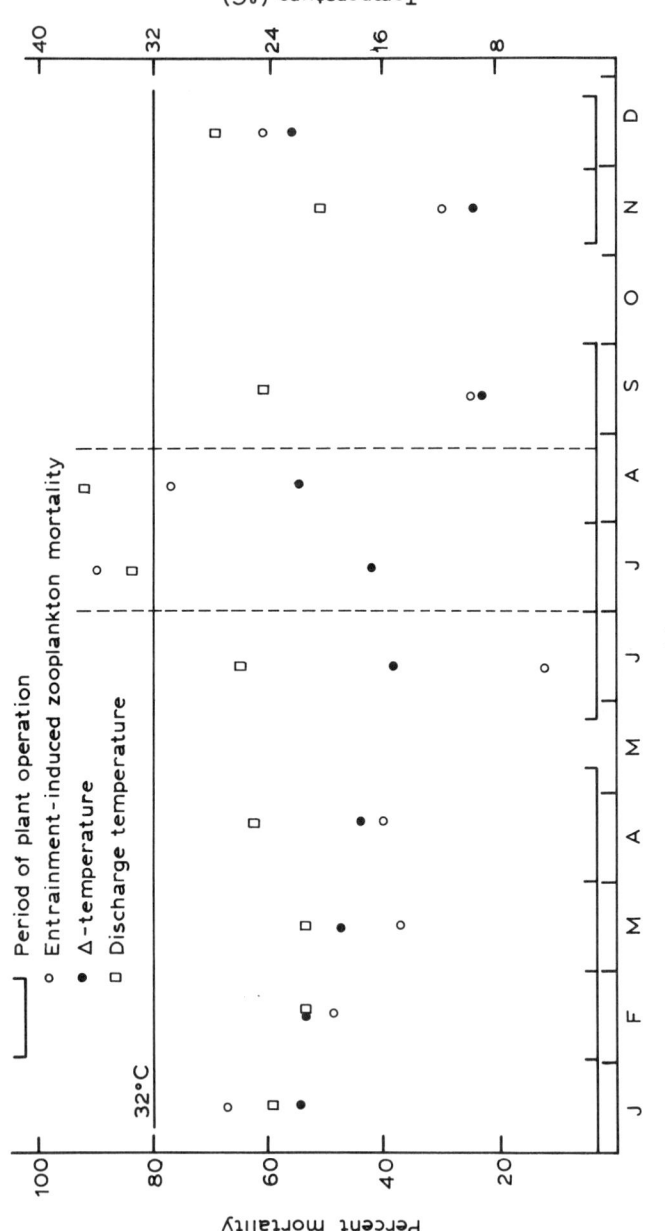

FIG. 7.3   The relationship of zooplankton mortality, discharge temperature, and temperature, at Wyman Station Unit No. 3. Data points are means of weekly determinations (after Lawler, Matusky and Skelly—Engineers 1979).

were from 80 h at 33°C to 12 h at 37°C. The two populations showed little difference in their tolerances except that the heated water population appeared to be more resistant to temperature shock than the other population.

The microcosm approach for predicting the effects of stress on communities has been used in several studies with varying success but the technique has limitations for thermal discharge studies. For example receiving waters are rarely uniformly heated, δTs are rarely consistent and there are considerable and irregular fluctuations in temperature in open waters as we have seen in earlier chapters (see Chapter 2).

Two other techniques have been used to evaluate thermal effects on plankton, namely large-scale physical models, simulating condenser tubes through which the animals could be passed (Ginn et al. 1978), and captive communities in small 'cages' which were floated along a discharge canal to study long-term exposure (Alden et al. 1976). In this last study, samples of zooplankton from the intake and outfall areas of a Florida power station were held in fine mesh cages and floated along the 1·6 km long discharge canal. This passage took 2 h and temperatures ranged in the various experiments from about 20 to 37°C. It is assumed that no chlorine was used in the system as the authors do not mention it. Immediate mortality rates of the four most common copepods at the outfall were 0–20% until the threshold temperature, above which the mortality rate increased exponentially. Below the threshold, salinity buffered the thermal shock. The critical entrainment temperatures ranged from 31 to 35°C depending upon the species. The large species, *Labidocera* sp., showed high immediate mortalities at all temperatures and was probably killed by mechanical stresses in passage through pumps and screens. Exposure in the discharge canal reduced the critical temperatures for all the species by 2–3°C and lowered salinity and longer exposure also reduced temperature tolerance. Figure 7.4 shows clear differences between the separate and combined stresses of heat and mechanical shock at a Lake Michigan power station. Here temperatures up to 32°C caused two- to sevenfold increases in mortality.

7.4.2 Chlorine and other Biocides

As with the phytoplankton (see p. 173), chlorination apparently causes higher mortalities of zooplankton than temperature in both fresh and saline waters at least up to around 32–33°C, though the combined stresses are most harmful (see Chapter 10). In site studies, mortalities were found to be significantly higher following entrainment where chlorine was used

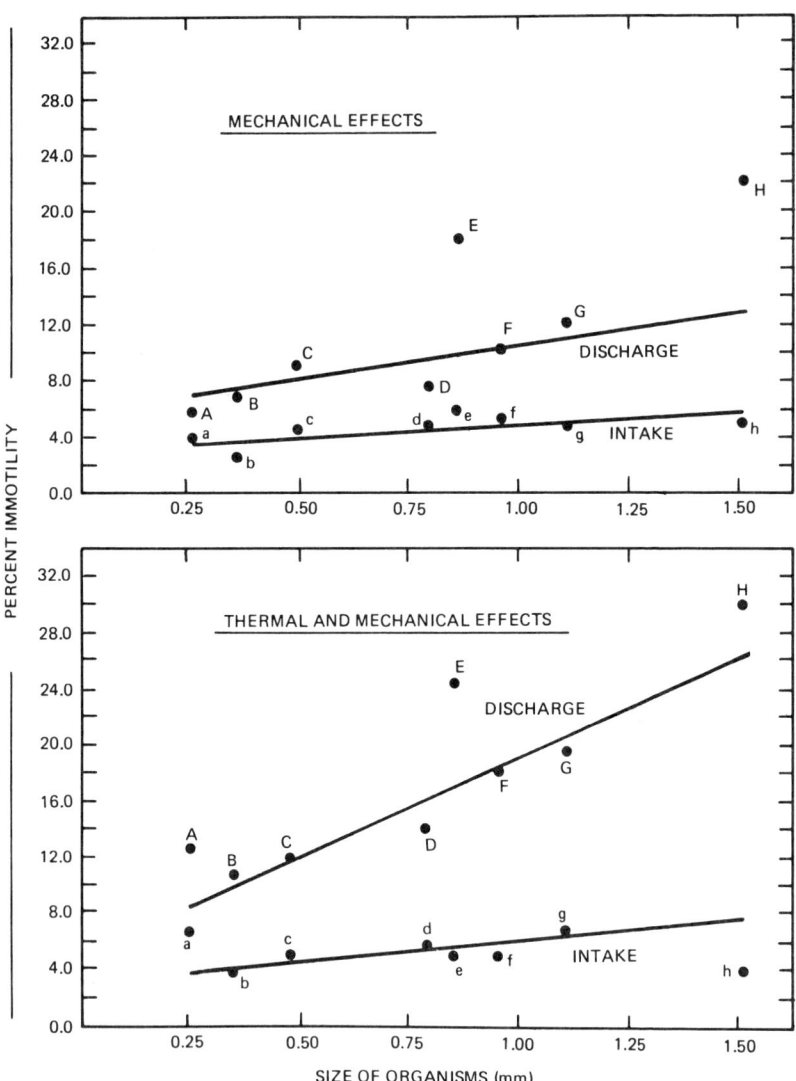

FIG. 7.4   Binomial regression analysis of condenser passage effects with and
without heat exchanged across the condensers (from Wetzel 1974).

(McLean 1973; Heinle *et al* 1974; Lauer *et al.* 1974; Heinle 1976; Gaudy 1981). Coughlan and Davis (1983) found mortality rates ranging from 5% to over 30% at dosed TRCs of 0–2·7 mg litre^{-1} at coastal sites, but mortalities were almost negligible when no chlorine was present (Fig. 7.5). The regression values for the relationship between chlorine concentrations and zooplankton mortality are shown in Table 7.1.

In a simulated plume experiment, a δT of 5·1°C and a TRC concentration of 0·44 mg litre^{-1} caused significant mortalities of zooplankton but the δT alone did not (Lanza *et al.* 1975). There was an inverse correlation between TRC concentrations causing mortalities and exposure time in experiments carried out with marine zooplankton (Gentile *et al.* 1976). This work

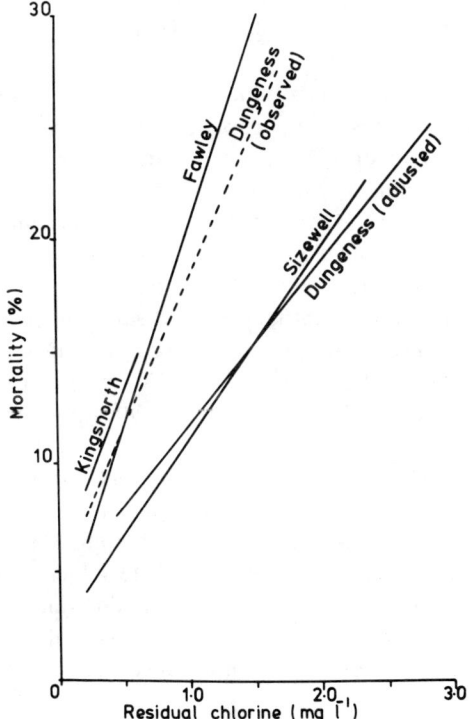

FIG. 7.5   The effect of 'residual chlorine' on the mortality rate of adult calanoid copepods within 1 h of entrainment. The observed chlorine concentrations at Dungeness were adjusted to a 6 min value from decay rate data. The maximum 1·5 mg/litre observed translates to 2·73 mg/litre on the adjusted line (from Coughlan and Davis 1983).

TABLE 7.1
Effect of chlorine concentration on zooplankton[a] (from
Coughlan and Davis 1983)

Site	b	a
Fawley	17·59	2·81
Kingsnorth	15·21	5·88
Sizewell	8·81	2·46
Dungeness	12·92	4·98
Dungeness[b]	8·39	4·25

[a] From the relationship: % Mortality $= b(R.Cl) + a$

  where R.Cl = the concentration of residual chlorine
  a and b = regression coefficients.

[b] Denotes the adjusted data for Dungeness.

indicated that exposure to less than 1 mg litre^{-1} for 5 min or less would cause little mortality. Fresh-water copepods show a wide tolerance range to free chlorine with temperature having a marked effect on the tolerance of some species (see Schubel *et al.* 1978).

Under the conditions experienced during intermittent chlorine dosing (see Chapter 3), animals would be subjected to a high initial dose and rapid subsequent decay. In experiments it has been shown that the 96 h TL_{60} was about 1·0 mg litre^{-1} for the initial dose and subsequent decay at 20°C. In comparison, the concentration tolerated in continuous sustained exposure was less than 0·05 mg litre^{-1} (Morgan and Carpenter 1978).

Food was found to be a limiting factor affecting the survival of the cladoceran *Daphnia pulex* after exposure to the combined stresses of temperature and chlorine (Buikema *et al.* 1978). If food was not provided, stressed animals did not survive the 24-h incubation period. The main stress in these experiments was temperature. The experiments also included a rare component. The animals were fed on algae which had also suffered simulated entrainment stresses. Growth, fecundity and the production of live young all decreased in those animals fed stressed algae, though as far as can be seen from the published data, no unstressed animals were used in control experiments.

7.4.3 Mechanical Effects
It is predictable that the larger organisms, particularly those with long appendages, will be more susceptible to damage from collisions, abrasions, turbulence and shear forces (Schubel and Marcy 1978), inside cooling-

water systems. For example, at the Crystal River power station in Florida, larger plankton, notably *Labidocera* spp., showed consistent mortality rates at all temperatures after entrainment whereas the smaller species did not. Other workers found similar effects with a range of species (Heinle 1969; Edsall and Yocum 1972; Bunting 1974; Bradford and Burns 1977). The largest percentage of mortality attributed to the effects of mechanical stresses was at the Millstone power station though there is no explanation. At both this site and the Marsden 'A' site in New Zealand dead organisms were most abundant on the bottom of the receiving water some distance from the outfall.

A generalised model relating size to mortality rate as a linear function is expressed as:

$$\text{Arc-Sine\% Mortality} = 0{\cdot}4053 \, (\text{Size}) + 0{\cdot}07571$$

where size is usually measured as overall length. The empirical data upon which this is based include measurements of both crustaceans and larval fish (see Schubel *et al.* 1978).

Separate studies at the Zion power station on Lake Michigan also indicated size-related mortalities though the data for separate taxa showed 'considerable variability in the size–mortality relationship' (see Lawler, Matusky and Skelly, Engineers 1979). Different species showed different susceptibilities but this is only to be expected with the vast range of physical conditions at different sites. Figure 7.4 shows that mechanical shock was a minor cause of mortality at the Kewaunee plant. At the KORI power station in Korea, mechanical shocks killed 37·6% of the entrained zooplankton, and thermal stresses about 11% (Huh 1980). The mortality rates of any one species can vary with site. For example at four Californian power stations the mortality rates of *Acartia tonsa* were 0, 9·09, 38·2 and 58·9% as a result of mechanical stresses (Icanberry and Adams 1974). At the Martigue-Ponteau site 28% of the zooplankton mortality was attributed to mechanical shock (Gaudy 1981).

7.5 POPULATION EFFECTS ON ZOOPLANKTON

The numbers of zooplankton killed by cooling-water systems have been translated into quantitative population effects at only a few sites. Table 7.2 shows some of the estimated cropping rates. None were considered as significant to the populations in the vicinity of the sites. The highest rate

TABLE 7.2

Percentage cropping of zooplankton populations by cooling-water systems at US power stations

Power station	Habitat	Effect
Cooper Nuclear	Impounded river	0·0–0·89% loss of population passing intake
Fort Calhoun	Tidal river	0·1–0·4% loss
Haddam Neck	Tidal river	1·2–4·9% loss (mean 4·2%)
Havana	River	0·5–1·5% loss (mean 0·7%)
Hennepin	River	2·5% loss
North Omaha	River	0·1–0·12% loss
Quad cities	River	mean 0·8% loss (max 4%)
Wyman	Tidal river	1·8% (mean) loss
Millstone	Ocean sound	0·1–0·2% loss in Eastern Long Island Sound

See text for references.

was 4·2% estimated for the Connecticut River near the Haddam Neck nuclear power station.

In Chapter 5, we discussed the predicted 'Cushing' effect on phytoplankton whereby even a slight reduction of the peak production would have a reflected effect through the food chain. No empirical data exist to support the hypothesis for phyto- or zooplankton and the long-term effects of additional cropping, whether of 0·1% or 20% of the stock is unknown, though mathematical modelling suggests that entrainment effects on populations are likely to be small at cropping rates up to 15% in some locations (Horwitz 1981). The dynamic nature of plankton populations in space and time and their rates of reproduction make even the definition of a population difficult. Defining the population limits in time and space will also affect the predicted effects of any cropping rate probably by several orders of magnitude (Lawler, Matusky and Skelly, Engineers 1979; Horwitz 1981; Murarka *et al.* 1981; Summers and Polgar 1981).

CHAPTER 8

Macro-invertebrates

8.1 TOLERANCE AND DISTRIBUTION

Macro-invertebrate communities have been used as indicators of pollution or stress in water bodies for many years and are perhaps most commonly used for the biological monitoring of pollution (Hynes 1960; Hawkes 1962; Perkins 1974; Hellawell 1986; Bilyard 1987).

Cummins (1972) stated that 'Operationally, the distinction between macro- and micro-invertebrates is based on the size of fully mature specimens, 3–5 mm long at minimum'. This definition applies adequately here though there will be exceptions.

The upper temperature limits for macro-invertebrates in natural waters are 45–50°C (Table 4.1). The most tolerant group are the Coleoptera and Chironomidae found in hot springs (see Brock 1975). Reviews of temperature tolerance and effects are given in a large number of publications (e.g. Macan 1963; Clarke 1967; Hynes 1970; Kinne 1970a; Coutant and Talmage 1977; Talmage and Coutant 1980; Cravens et al. 1983). Natural distributions of the various groups in relation to temperature are discussed in relevant sections later in this chapter. The effects of current velocities, wave action and substrate composition on the composition of invertebrate communities are also well known (e.g. Macan 1963; Hynes 1970; Newell 1970; Perkins 1974). There are also many data on the effects of contaminants such as chlorine, heavy metals, radioactivity and combinations of stresses. Specific references will be given in the appropriate section.

8.2 METHODOLOGY

The methods of sampling fresh-water invertebrates, analysing and presenting the data have been reviewed most recently by Elliott (1971),

Elliott and Tullett (1978), Hellawell (1986). Benthic sampling of marine fauna has been reviewed by Holme and McIntyre (1971). The two most commonly used approaches are:

—Direct sampling by nets, dredges, core-samplers or specially designed samplers which remove a portion of the substratum or weed beds. The animals are then sorted from the material for further analysis.

—Indirect sampling using artificial substrate samplers which are colonised by species over a set period and then removed for analysis.

The basic principles apply to fresh-water, estuarine and marine habitats (Southwood 1966). In intertidal zones or on rocky substrates direct observations are often made using quadrats, transects and, where necessary diving techniques. More recently observational techniques have included the use of film or videotape cameras either remotely operated, from submersibles or hand-held by divers.

Many habitats call for specially adapted sampling apparatus but again the basic principles are common.

8.3 SITE STUDIES

8.3.1 Invertebrates within Cooling Systems
The strong currents created by large intakes can entrain macro-invertebrates which are already in the water column and carry them into and through cooling-water systems. Some species can colonise surfaces within the system, resulting in fouling communities and consequent operational problems (e.g. Holmes 1970b; see Acker *et al.* 1972; Mangum *et al.* 1973; Whitehouse *et al.* 1984) (see also Chapter 3). Entrainment mortalities of drifting invertebrates in rivers, could, if on a large scale, affect the recolonisation of downstream reaches (Williams and Hynes 1976).

(i) Direct (once-through) cooling systems
Methods for sampling entrained macro-invertebrates are similar to those for zooplankton (see p. 205). Markowski's studies during the 1950s (Markowski 1959, 1960) which appeared to show that macro-invertebrates survived entrainment through power station cooling-water systems are now not regarded as unequivocal evidence. He used nets and pump samplers to sample in the discharge currents near to outfalls but it is possible that some living invertebrates were entrained in these effluent currents from the receiving water. These would not have passed through

the system (see also Chapter 7). However, at some sites where cooling towers were in use invertebrates colonising the ponds would have easily survived the passage from the pond outlet to the discharge (Markowski 1960, 1962; Langford 1971b; Milner 1984).

The studies at the Fort Calhoun site on the Missouri were much more detailed than those of Markowski but sampling mortalities, and the location of sampling nets could also have influenced the data considerably here (Table 8.1) (Carter 1978; Carter *et al.* 1982; Schlesinger *et al.* 1982). For some species mortalities were greater at the intake than at the outfall (Carter 1978). The average mortality rate of drifting invertebrates entrained in the cooling system was assessed as 7·7% over the period of the study with the mortalities for different taxa varying from 0·9 to 19·8%. Mechanical stresses accounted for less than half of the mortalities which increased markedly with higher temperatures. Assessed as a proportionate effect on the total drift fauna of the river the average mortality caused by the power station was about 0·2% of the standing crop ranging with year from 0·1 to 0·6% (Table 8.2). The maximum mortalities occurred in summer when discharge temperatures could exceed 35°C with maxima of 36–37°C. At the Quad-Cities power station on the Mississippi a δT of 16·2°C was reported to cause an increase of 24·3% on the mortality of entrained macro-invertebrates (see Carter 1978).

TABLE 8.1
Differential mortalities of drifting macro-invertebrates between the intake and discharge of a power station cooling system

Years	Taxon	Differential mortality (% range)	Total (all years)
1973–77	Ephemeroptera	2·6–8·6	6·7
	(*Caenis* spp.)	2·0–15·8	6·9
	Hydropsychidae	3·1–8·1	4·3
	(*Hydropsyche orris*)	3·3–19·5	6·9
	(*Potamvia flavia*)	2·7–4·7	—[a]
	Other Trichoptera	0·9–8·3	5·3
	Chironomidae	4·9–10·1	8·8
	Other Diptera	2·5–19·5	8·9[a]
	All other species	6·2–17·2	9·6

Figures for Fort Calhoun, Missouri River, USA (from Carter *et al.* 1982).
[a] Mortalities at intake sampler exceeded those at outfall at times.

TABLE 8.2
Effects of entrainment mortalities on total invertebrate drift—Fort
Calhoun power station, Missouri River, USA (after Carter *et al.* 1982)

Year	Average river flow $(m^3 s^{-1})$	% River flow through system	Average differential mortality (%)	Effect on total drift mortality (% increase)
1973	729	3·1	7·4	0·2
1974	866	2·6	7·0	0·2
1975	1 259	1·8	8·6	0·2
1976	989	2·3	6·9	0·2
1977	704	2·9	4·4	0·1

King and Mancini (1976) found that the mortality rates of most
invertebrates entrained through the cooling-water system of the Wabash
power station differed little from those caught at the intake. At both the
Wabash and Fort Calhoun power stations larvae of the caddis-fly
Potamyia flava colonised in the outfall areas, having apparently survived
entrainment and transit through the system. Mortalities at the intakes were
greater than at the outfall because of the effects of net abrasion in the faster
intake currents.

A specially modified pump sampler was used to collect animals entrained
through four power stations on the Hudson River (Cannon *et al.* 1978). The
survival rates at the intakes were mostly above 90% for the four most
abundant species. The mortality rates of the amphipod *Gammarus daiberi*
differed markedly at the Roseton and Bowline power station outfalls even
though the discharge temperatures were similar. At both sites the
mortalities increased consistently on retention.

In contrast the survival of the mysid shrimp *Neomysis americana* did not
increase during retention for 4 days, though discharge temperatures over
32°C were associated with mortality rates of 20–30% at the outfall.
Chaoborus punctipennis and the amphipod *Monoculodes edwardsi* survived
entrainment at all temperatures and subsequently, retention for 24 h.

In these, and the Fort Calhoun studies, the time taken to pass through the
cooling system was between 1 and 7 min. Longer exposures increase
mortality rates. For example, passage time for entrained animals ranged
from 6 to 33 min at Indian Point power station, depending upon the
number of generators operating and hence cooling water flow (Lauer *et al.*
1974). The survival rates of the three most abundant macro-invertebrates

TABLE 8.3

Mean mortality rates of three macro-invertebrate species entrained in a cooling-water system. Indian Point Nuclear Power Station, Hudson River, USA

Taxon	*Outfall temp (°C) (range)*	*Mortality rate*		*% With chlorination (outfall only)*
		Intake (% range)	*Outfall (% range)*	
Gammarus sp.	20–21·7 (no heat)	2·6–3·8	1·7–1·3	8·1–10·4
	28·2–31·3 ΔT (4·4–7·2)	2·6–3·6	3·4–5·7	8·9–10·8
	32·2–33·2 ΔT (7·1–8·3)	1·0–2·3	2·6–5·9	12·4–23·5
Monoculodes edwardsi	16·4–21·7 ΔT (5·6–10·6)	4·1–6·6	7·7–11·7	26·8–38·7
	28·2–29·8 ΔT (4·4–8·0)	3·2–6·3	4·3–5·2	(no data)
	31·1–33·3 ΔT (5·5–7·9)	1·9–5·0	5·9–8·3	(no data)
Neomysis americana	19·7–26·1		16·9–28·9	42·1–52·5
	ΔT (5·6–7·2)	12·9–16·3		
	28·2–31·1 ΔT (3·9–8·0)	10·4–13·8	13·3–23·6	35·7–54·2
	32·2–33·3 ΔT (7·3–7·9)	5·4	32·2–45·1[a]	29·6–44·1[a]

[a] NB the data in the original paper contain 95% confidence limits (see Lauer *et al.* 1974).

were related differently to temperature, chlorination and the times of exposure (Table 8.3). Chlorination reduced survival rates at most temperatures.

Simulations based on temperature under-estimated the effects of entrainment mainly because of the combined effects of temperature and mechanical stresses in the real systems which were not encountered in the experimental apparatus. There is circumstantial evidence from the colonisation of thermal discharge culverts and canals by marine invertebrates that many species survive entrainment through the cooling-water systems either as adults or in the larval stages (Markowski 1962; Icanberry and Adams 1974) though tests with the prawn *Palaemonetes varians* by Markowski (1962) found most specimens killed by mechanical damage in passage. Similar effects were found in other studies (e.g.

Massengill 1976b; Marcy *et al.* 1978). The individuals which survive are usually the smaller crustaceans with a relatively tough integument. In some tidal channels or culverts colonisation could occur from animals carried up by tidal influxes at times when the cooling system is not operating (e.g. Cory and Nauman 1969).

(ii) Indirect (recirculating) systems
Thriving populations of fresh-water invertebrates have been observed in the cooling-tower ponds of several British power stations (Markowski 1959; Langford 1971b; Ross and Whitehouse 1973; Milner 1984) at high temperatures and chlorine dosing up to 1.0 mg litre^{-1}. Milner (1984) in a comprehensive study of cooling-tower ponds at Ratcliffe-on-Soar and Ironbridge power stations in the English Midlands found over 30 taxa in the cooling-tower ponds. The diversity of the faunas collected on artificial substrate samplers was generally lower in the tower ponds (Fig. 8.1) than in the respective rivers, probably as a result of lower habitat diversity rather than water conditions in the tower ponds. It was significant that diversity was reduced in the cooling system before heat and chlorine were applied. Many of the taxa recorded in the rivers were found in recirculating cooling systems at the power stations (Milner 1984), though there were clearly several which were favoured by the conditions, notably the molluscs *Potomapyrgus jenkinsi* and *Physa acuta*, flatworms and the hydroid *Plumatella* sp.

Cooling towers are rarely used at marine sites and there are no published data on the communities colonising them. Clearly some species of barnacles and molluscs can colonise outlet culverts if the conditions are not too adverse (see p. 71).

(iii) Fouling and heat treatment
The concrete and metal surfaces together with the crevices and spaces in cooling-water culverts and pipework provide suitable places for sessile invertebrates to colonise (Crisp 1965; Whitehouse *et al.* 1984; CEGB Research 1990). Surfaces in discharge canals and thermal plume areas also become colonised readily. In several locations settlement and growth are advanced in these heated waters (Nauman and Cory 1969; Young and Frame 1976). Once the surfaces have been colonised by bacteria, fungi and algae, animals can find adequate food in such places. Most of the fouling of cooling-water systems occurs in the culverts carrying unheated water and is thus outside the scope of this volume (Acker *et al.* 1972; Mangum *et al.* 1973; see Langford 1983a; Whitehouse *et al.* 1984). At a number of power

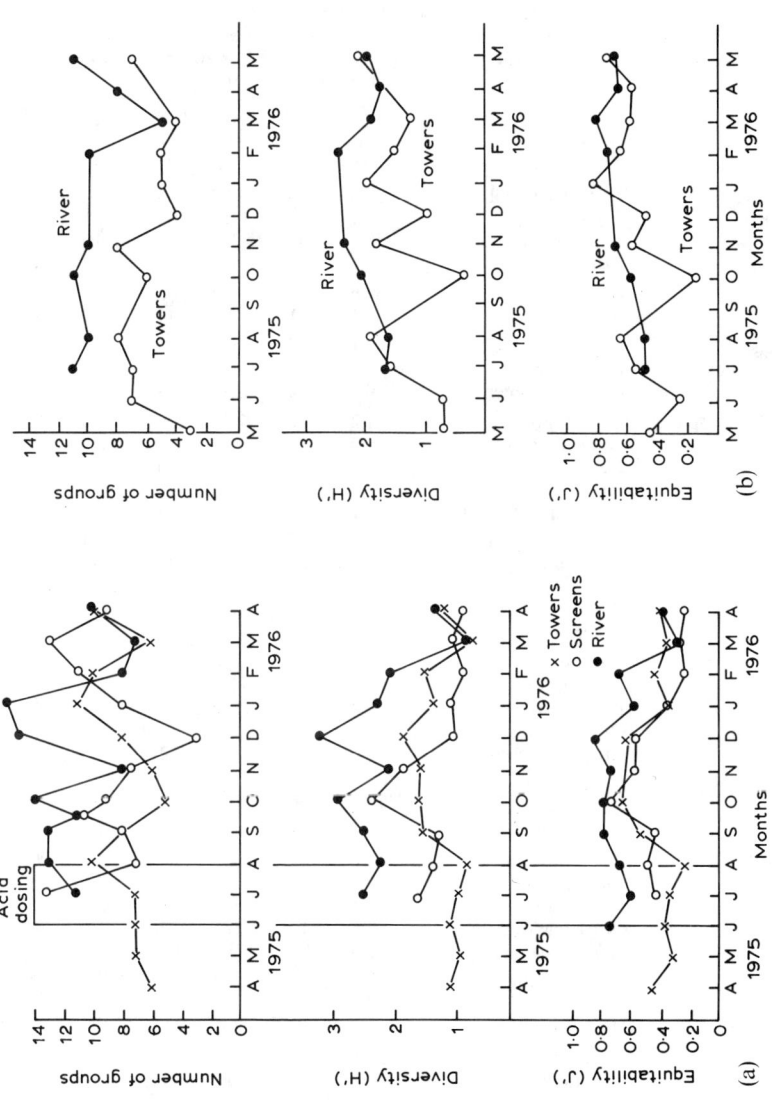

Fig. 8.1 Numbers of groups, diversity and equitability indices for communities colonising artificial substrates in (a) the Ratcliffe-on-Soar cooling water system; (b) the Ironbridge 'B' cooling system.

TABLE 8.4
Total numbers of animals and percentage occurrence of taxa on monthly sampling visits to sampling stations upstream and downstream of Ironbridge power station, River Severn, UK, 1965–67

Group	Species/genus/family	Upstream (1·0 km)		Downstream (0·5 km)		Downstream (1·6 km)	
		%C[a]	%F[b]	%C	%F	%C	%F
Flatworms	All species	0·4	67	2·9	77	0·5	82
Worms	Oligochaeta (all)	22·9	100	14·9	100	18·3	100
Leeches	Erpobdella sp.	0·4	83	0·5	94	1·1	100
	Glossiphonia sp.	0·3	94	0·2	77	0·6	82
	Other spp.	0·1	56	<0·1	39	<0·1	41
Crustacea	Gammarus sp.	32·6	100	34·1	100	20·5	100
	Asellus sp.	0·1	61	1·3	94	0·9	100
Insects Plecoptera	Taeniopteryx sp.	0·1	44	0·2	44	0·1	47
	Leuctra spp.	<0·1	44	<0·1	33	<0·1	24
	Other Plecoptera	<0·1	44	<0·1	33	<0·1	24
Ephemeroptera	Caenis spp.	1·2	94	3·7	100	2·6	94
	Ephemerella sp.	1·9	50	3·1	67	2·9	47
	Heptagenia sp.	1·7	100	2·1	100	1·9	100
	Baetis spp.	2·4	56	1·8	61	1·4	59
	Other Ephemeroptera	<0·1	100	<0·1	33	2·1	100
Trichoptera	Hydropsyche spp.	2·6	100	1·6	94	4·3	100
	Psychomyia sp.	2·2	89	3·2	94	1·8	100
	Leptoceridae	0·3	78	0·2	61	0·1	65
	Other caseless spp.	<0·1	33	70·1	22	<0·1	44
	Other cased spp.	<0·1	44	70·1	44	0·1	35
Coleptera	Elmis sp.	1·3	100	1·0	83	1·9	100
	Oulimnius sp.	0·9	94	0·9	83	0·8	88
	Limnius sp.	0·4	89	0·4	78	1·0	100
	Other Elmids	—	0	<0·1	6	<0·1	12
	Other Coleoptera	<0·1	6	<0·1	12	<0·1	6
Diptera	Simulium spp.	(11 915)	100	(2 762)	94	(5 385)	94
	Chironomidae	19·4	100	14·8	100	22	100
	Other Diptera	0·1	39	<0·1	12	<0·1	24
Molleisca	Ancylastrum sp.	3·0	100	4·3	89	3·5	100
	Bithynia sp.	0·8	94	1·9	83	3·3	100
	Viviparus sp.	<0·1	10	0·2	28	<0·1	24
	Other Gastropods	<0·1	10	<0·1	12	<0·1	6
	Sphaerium spp.	2·6	94	3·8	89	5·9	100
	Anodonta sp.	<0·1	44	<0·1	12	<0·1	24
	Other bivalves	<0·1	44	0·2	56	0·2	41
Arachnids	Hydracarina	0·3	61	0·6	29	0·7	94
	Hydra spp.	<0·1	22	0·6	22	0·3	29
	Nematoda	0·1	33	0·1	17	0·2	19
Totals	including Simulium spp.	59 650		33 181		37 179	
Totals	excluding Simulium spp.	47 735 (18)[c]		30 419 (19)		31 794 (17)	

[a] %C = % composition by number. [b] % F = % frequency of occurrence. [c] (18) = number of collection dates.
Both columns underlined = common and abundant; % F underlined = common.

stations, however, heated water recirculated from the outfall is used to try to control fouling organisms and the tolerance of these species to heat is thus of prime importance to the effectiveness of the process.

In the 1920s water temperatures raised to 42°C were used to kill mussels (*Mytilus edulis*) which had colonised culverts at a Scottish power station (Ritchie 1927). The procedure was to reverse the cooling-water flow for several hours each month, but because of the extra heating required to make it fully effective, the process was costly and eventually discontinued. In tropical waters where natural ambients are already high the system may be more efficient.

Power stations on the west coast of the USA use heat-treatment systems particularly where there are constraints on the use of chlorine or other antifouling chemicals (Stock and Strachan 1977). As an example the San Onofre Generating Station (SONGS), recirculates about two thirds of its cooling water back through the condensers to absorb additional heat. The reversed flow reaches 52°C and is then mixed with colder water to maintain 40–42°C at the intake for about 2–6 h every 5 or 6 weeks. The effect on the temperature of the discharge plume is significant but of short duration.

The efficacy of this method is clearly related to the thermal tolerance of the fouling species. Fox and Corcoran (1957) found that the effective temperature killing for *Mytilus californianus* was similar to that for *M. edulis*, that is over 37°C, but Gonzalez and Yevich (1976) found that *M. edulis* did not survive temperatures above 27°C in a thermal discharge canal. In this later study there was no mention of chlorination but it is possible that this was a significant factor determining mortality. Factors other than temperature were regarded as responsible for the reduced growth of mussels in another thermal plume (Kastendiek *et al.* 1981).

Stock and Strachan (1977) noted that at Californian power stations, most of the local sessile taxa were found growing in the cooling systems between heat treatments. Figure 8.2 shows the effective heat doses for the three most troublesome species at three Californian power stations. Clearly there is variable resistance to heat by different species and this variation is one factor which hampers the fully efficient use of heat as an anti-fouling method. In Britain, several of the species which colonise cooling tower systems (see p. 230) have caused operational problems including gastropod molluscs, *Gammarus* sp. and the bryozoan *Plumatella* sp. (Langford 1983a; Milner 1984).

8.3.2 Effects in Receiving Waters
(i) Communities and species composition
(a) Rivers. One of the earliest and still most convincing demonstrations of

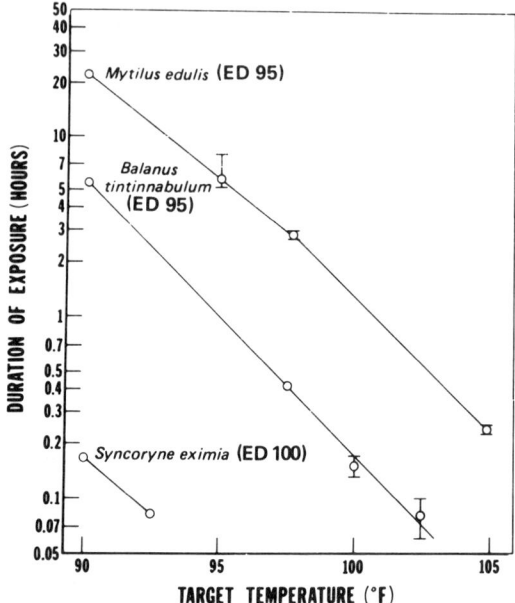

FIG. 8.2   Comparison of thermal tolerances and confidence limits between the bay mussel (*Mytilus edulis*), a barnacle (*Balanus tintinnabulum*) and a hydroid (*Syncoryne eximia*). From Stock and Strachan (1977).

Target temperature (°F)	ED95 (Hr)		
	Bay mussel	Barnacle	Hydroid
90	22·3	5·6	0·2
97·5	2·7	0·4	—[a]
105	0·2	—[a]	—[a]

[a] Immediate mortality at target temperature.

the effects of a thermal discharge on a macro-invertebrate community was at the Martins Creek power station on the Delaware River (Coutant 1962). The study was part of the comprehensive work on the river ecosystem (Tremblay 1965). The heated water spread across the riffle areas but the plume kept mainly to one half of the channel. There was no thermal stratification near the outfall and the plume impinged on the river bed for some distance downstream. A transect across the river included both heated and unheated water. Discharge temperatures reached 42°C and the effluent was chlorinated.

Using direct sampling methods, Coutant (1962) showed that the species composition and abundance of the macro-invertebrate faunas were similar in the winter in both heated and unheated areas despite the 10°C temperature rise but as temperatures rose Ephemeroptera, Trichoptera and Coleoptera were eliminated immediately downstream of the outfall. As temperatures exceeded 37°C in July and August the mean number of species found immediately downstream declined from 25 to 5. The mean density of individuals declined from 1763 to 186 m^{-2}, and the biomass declined from 10·4 g m^{-2} to 0·9 g m^{-2}. At the time of the highest temperatures there were 40 taxa recorded upstream and 4 downstream of the outfall. The critical temperature range was about 32–35°C near the outfall though at temperatures of up to 35°C a mile downstream the fauna was not affected markedly. This suggests that chlorine may have exacerbated the effects near the outfall. Recolonisation of the denuded areas began as soon as temperatures began to fall in the autumn and after about 2 months the fauna was almost as diverse and abundant as in early spring.

Very few other site studies have ever shown effects as clear as this and one of the major advantages of the site was that the control and heated stations were similar in most respects other than the presence of the thermal discharge. At the Haddam Neck on the Connecticut River, the long-term studies which began before the Connecticut Yankee power station began operating first showed the effects of the scour as the water pumps started up prior to any heat discharge (Massengill 1976a). The densities of benthic organisms fell from 617 m^{-2} to zero over 2 years, though during a temporary shutdown they rose rapidly to over 1000 m^{-2}. This site, in contrast to Martins Creek, was extremely complex in that habitats both upstream and downstream of the outfall received heated water because of tidal movements. The stations heated most, by 10°C or more, were also those at which the substrates had changed most.

The fauna in the discharge canal was affected by both high current velocities and temperatures. Over 37°C only the worm *Limnodrilus hoffmeisteri* survived from the winter fauna. At 40°C the numbers of *L. hoffmeisteri* were reduced from 900 to 30 individuals m^{-2}. The autumn and winter benthos was denser than in summer but a large immigration of benthic feeding fish was believed to have suppressed the peak densities (Massengill 1976a). Dipteran larvae were observed passing alive through the power station and apparently augmented the invertebrate fauna of the discharge canal.

Using artificial substrate samplers, Proffitt (1969) and Bender and

Profitt (1974) showed that only the amphipod *Hyalella* sp. and two species of Ephemeroptera colonised at temperatures over 34°C in a heated effluent. At 40°C no animals were collected. The chironomid *Glyptotendipes lobiferus* was the only species occurring consistently in the discharge canal. In the cooler seasons three genera of caddis flies were more abundant in plume samples than upstream. The reverse was true for the mayflies *Stenonema, Baetis, Caenis* and *Tricorythodes*. The total area of effect as estimated by biological changes in the White River was estimated as 4·5 ha.

The effects of location on colonisation were clearly important here though not easily evaluated. Poor colonisation would be expected where samplers relied at least partly on animals which had passed through the cooling system as many of these would be killed or damaged in transit. The use of artificial substrate samplers in discharge canals may not, therefore, give a true representation of the tolerance of all the local species to conditions in the discharge water.

Churchill and Wojtalik (1969) found significant though 'not severe' changes in the invertebrate fauna of the Green River in Kentucky below the thermal discharge from the Paradise power station. The installation of a cooling tower reduced the temperature of the thermal plume and subsequent surveys found that the fauna was reinstated.

There are inconsistencies in the published data when the effects of temperature alone are considered. For example, in contrast to the studies described above, Wurtz and Dolan (1960) found few of the common species absent from a heated effluent in the Schylkill River even where temperatures reached 38–40°C for short periods. However, in this river the species were mainly those tolerant of other forms of pollution. Changes in the macro-invertebrate fauna in the river downstream of the Glen Lyn power station were not related to temperature alone (Cairns *et al.* 1970). At all sampling stations Trichoptera and Ephemeroptera were relatively abundant though Mollusca declined in the heated reaches. There was some increase in Chironomidae but the changes were generally not significant except for the molluscs. Even up to 3·2 km downstream of the effluent molluscs were still relatively scarce. Clearly some other influence than temperature was present though the published account does not identify it. The maximum water temperatures reached about 30°C at this site.

In a large-scale experimental study, using artificial channels heated by a power station effluent, the maximum standing crop and the smallest number of taxa were found on samplers in the warmest channel (Alston *et al.* 1978). The temperature differences between controls and the warmest channel varied from 3 to 10°C. The highest temperature was 36·7°C in

August. Oligochaeta and Chironomidae were dominant, numerically, in all channels.

In the Sheep River (Alberta, Canada), the most marked changes in the invertebrate community occurred where a thermal and a chlorinated sewage discharge were combined (Osborne and Davies 1987). The dominant effect was that of the sewage effluent. At the one site where only the thermal discharge was present, however, an increased abundance of Oligochaeta was associated with mats of blue-green algae (see Chapter 6, p. 188).

A more specific study of the insect communities in the heated streams and swamps around the Savannah River nuclear plant (Howell and Gentry 1974; Gentry *et al.* 1975), found larvae of 7 genera of damselflies and dragonflies (Odonata) in the heated waters, compared with 23 species found in the unheated waters. No species could tolerate the areas which exceeded 40°C. The most tolerant species, all belonging to the family *Libellulinae*, were well adapted to the silted areas near the reactor effluents and could clearly tolerate changes of 15–20°C when the reactors shut down and restarted. Current velocities in the discharge areas, siltation and the loss of vegetation were also important to the distribution and survival of species.

The insect fauna of heated streams was dominated by *Chironomidae* and *Corixidae*. More species of insects were recorded in the control and previously heated streams than in the heated streams. Plecoptera were relatively more than twice as abundant as *Chironomidae* in the controls. The diversity of the insect fauna was higher in the controls than in the other streams.

In the UK very few long-term community studies have been carried out in relation to thermal discharges in fresh waters and most data are from short-term surveys (Langford 1971a,b, 1975; Langford and Howells 1976).

In the River Trent (Whitehouse and Aston, 1964; Langford 1971a, 1974; Langford and Aston 1972), the major changes in the invertebrate fauna in the 1960s occurred downstream of the entry of the grossly polluted River Tame. The changes in communities downstream of the thermal discharges were mainly a result of chlorine, substrate changes or aeration. Water temperatures rarely exceeded 30°C in most places. The decreases and increases in the numbers of species around the power station outfalls bore little relation to the temperature rises in the river.

It is predictable that the species found in polluted rivers like the Trent or the Schylkill should be tolerant of other stresses such as heat. The benthic and epibenthic communities of the Trent, for example, were dominated at this time by a 'pollution fauna' (Hellawell 1986) comprising

Limnodrilus hoffmeisteri, Tubifex tubifex, Asellus aquaticus, Erpobdella octoculata and *Chironomidae* for many kilometres of its length.

In contrast, the invertebrate fauna of the River Severn around the Ironbridge power stations was very diverse (Langford 1967, 1971a,b), consisting mainly of species generally regarded as intolerant of pollution (Hynes 1960; Hellawell 1986). Detailed studies of the invertebrates in the reaches upstream and downstream of the thermal discharges were made over almost 10 years (Langford 1967, 1970, 1971a,b, 1972; 1975; Langford and Daffern 1975; see Langford 1983a). As at Martins Creek (see p. 234) the effluent water was well mixed over the riffle areas and there was no vertical or horizontal stratification at the downstream sampling stations. Maximum water temperatures reached 30°C up to 2 km downstream of the outfalls and almost 33°C immediately downstream on occasions.

It is surprising, in view of the species composition of this clean-water fauna, that there were no consistent community effects which can be attributed to the differences in water temperature (Fig. 8.3). No species (taxon) found upstream including Plecoptera, Ephemeroptera or Trichoptera (Table 8.4), known to be intolerant of many pollution stresses (Hynes 1960; Hellawell 1978), was consistently absent from the downstream reaches even in the warmest periods, and there were no significant differences. The absence of marked community changes in this intolerant fauna is even more remarkable when it is considered that the Ironbridge 'A' power station had been operating for more than 30 years when these studies began, mostly on a much more consistent basis all year round. Intermittent chlorination was used but there was no evidence of effects on the river community. The newer power station on this site, Ironbridge 'B', used cooling towers, and produced only small temperature changes downstream of the outfall.

Fey (1977) found that the Plecoptera were virtually eliminated downstream of the Elverslingen power station on the River Lenne in Germany where maximum temperatures reached only 25°C. Experiments showed that *Dinocras cephalotes* (Plecoptera) had a 24-h LT_{50} of 25–33°C depending upon acclimation, while *Epeorus sylvicola* (Ephemeroptera) had a 24-h LT_{50} of 25–29°C. The former was absent from the thermal plume, the latter present. *Hydropsyche pellucidula* (Trichoptera) dominated numerically at all sampling stations but was least abundant in the outfall area. Chironomidae and *Erpobdella octoculata* were more abundant downstream than upstream.

In slower, deeper rivers in eastern England, several invertebrate species found in unheated reaches were consistently absent from the discharge

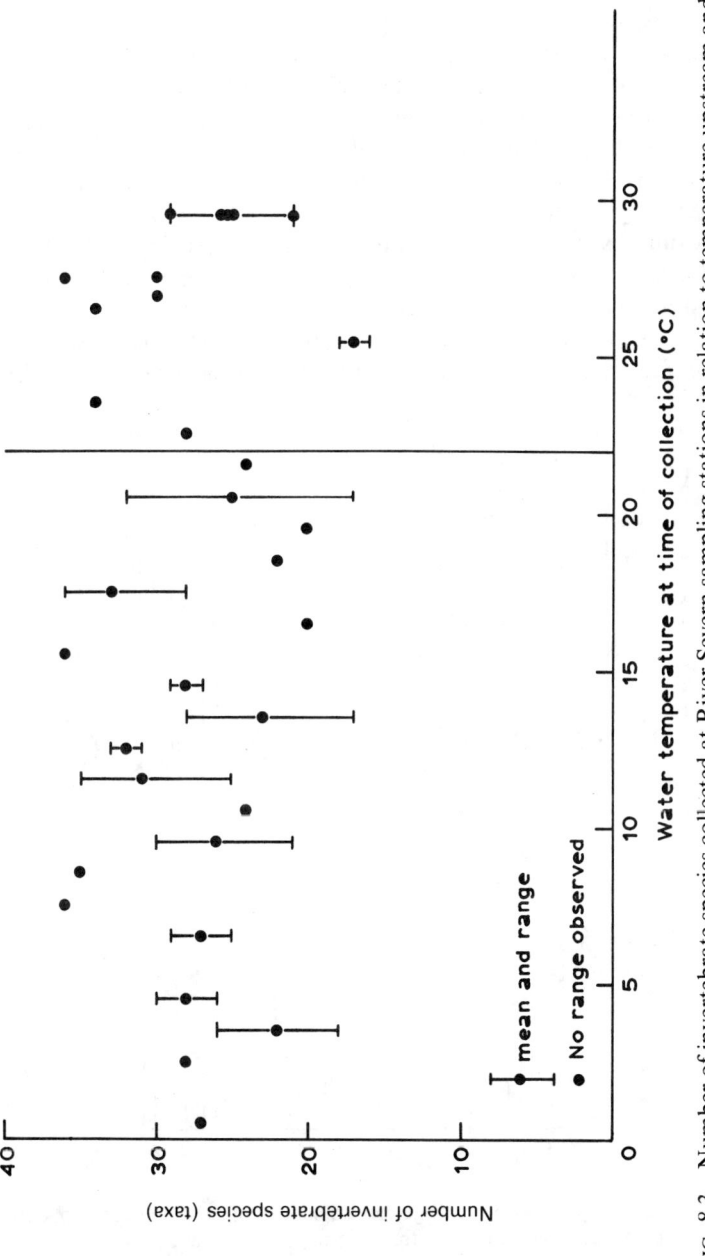

FIG. 8.3   Number of invertebrate species collected at River Severn sampling stations in relation to temperature upstream and downstream of Ironbridge 'A' power station, 1965–1975.

areas of power stations (Mann 1965; Langford 1971b). The changes in the faunas were, however, probably more related to the effects of scour than heat. Flatworms (Turbellaria), leeches and some mollusc species were apparently most affected by the discharges. The lamellibranch *Pisidium* spp. and the gastropod *Bithynia tentaculata* disappeared from the discharge area of the Gentilly power station on the St Lawrence River where temperatures reached 30°C (Effer and Bryce 1975), but both were common below outfalls in British rivers at temperatures exceeding 30°C (Langford 1971b). Scour caused by the effluent currents was the most likely cause of the disappearance. Moller (1978b) also considered scour to be a major factor affecting benthic communities near thermal discharge outfalls.

The fact that temperature shocks can trigger invertebrate drift is

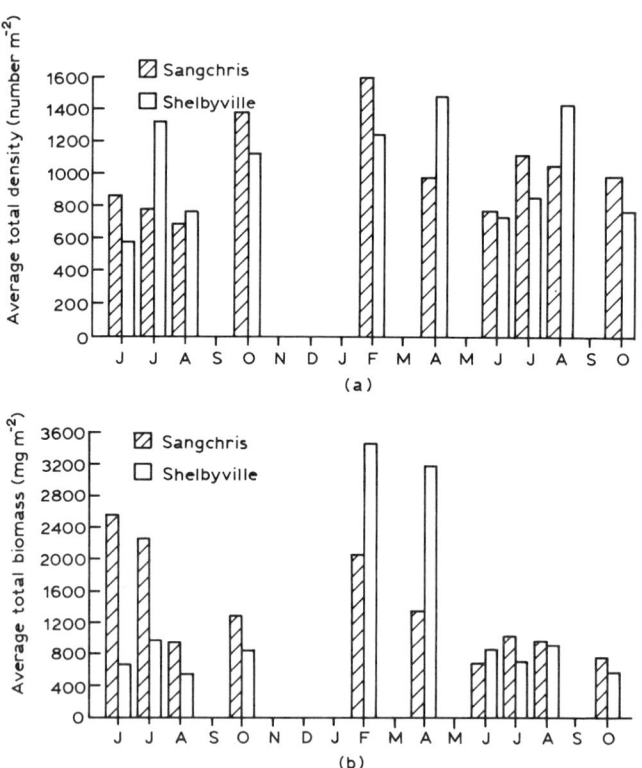

FIG. 8.4   (a) Comparison of average total densities of benthic organisms and (b) comparison of average total benthic biomass at Lake Sangchris and Lake Shelbyville June 1976–October 1977 (from Larimore *et al.* 1979b).

recorded (Wojtalik and Waters 1970) but no direct studies of the effects of a power station operating sporadically have been published. Invertebrates have, however, been recorded drifting in discharge canals and plume areas, though these may have originated from the outfall, after entrainment (Durrett 1972; Durrett and Pearson 1975).

The effects of downstream temperature changes in rivers caused by impoundments are not easy to distinguish from other effects such as water quality, flow variations and plankton overspill (Lemkuhl 1972; Fraley 1979; see Ward and Stanford 1979a,b). Fraley (1979) however, considered that temperature rises averaging 3–4°C stimulated productivity, allowed warm-water species to dominate and stimulated early emergence of some insects below a shallow reservoir on the Madison River. The more constant thermal regime also caused decreased diversity and a failure of emergence in the cold-water stonefly, *Pteronarcys californica*.

(b) Lakes, reservoirs and cooling ponds. Relatively few of the lakes used as sources for cooling water are truly natural. Most are either impounded rivers or lakes which have been artificially deepened or extended by damming the outlets. In the dryer parts of the world, cooling-water reservoirs and ponds have been created solely for the purpose (Harmsworth 1974; Yost and Talmage 1981). Rothwell (1971) in her work on Trawsfynydd Lake found a sparse benthos in the warm lagoon and the effluent channel near the outfall, but the highest diversity and abundance of benthos were also found in the warm channel a short distance further away from the outfall where water temperatures were little different from the nearer sampling points. Rothwell concluded that factors other than temperature were restricting the fauna, mainly scour, heavy metal concentrations in the sediments and the nature of the substrate which was fairly sterile peat.

In Lake Shelbyville and the heated Lake Sangchris, the predominant macro-invertebrates were *Chironomidae*, Oligochaeta and *Chaoborus punctipennis* (Diptera), and the average densities of organisms were 1015 and 1049 m^{-2} respectively. In the heated Lake Sangchris 76 taxa were recorded in comparison to the 88 taxa in the unheated Lake Shelbyville (Larimore *et al.* 1979b). The difference was accounted for by the occurrence of a shallow, flowing inlet stream in the unheated lake which contained species not present in the heated lake which had no such habitat. The density of animals and the biomass were greater in Lake Sangchris in 7 out of 10 months though the reverse applied for biomass in February, March and April (Fig. 8.4).

There were also differences in the species composition in the lakes. For example, the mayfly *Hexagenia limbata*, the Asiatic Clam, *Corbicula manilensis* and *Chironomus* spp. were relatively more abundant in the heated lake. The clam was most abundant in the effluent canal of the power station where the fast water currents exposed the coarser substrates which the animals prefer (Cherry *et al.* 1980; Graney *et al.* 1980). Oligochaeta increased in abundance in the warmer waters and the exotic species *Branchiura sowerbyi* was most prolific in the effluent canal. Most of the differences in the species compositions of the benthos in the two lakes were related to factors other than temperature or vegetation, mainly water currents, substrates and the formation of the thermocline.

In the open waters of Lake Erie, the benthos was not affected at all by the thermal discharge from the Monroe Power Station (Cole and Kelly 1978), because of stratification and variable plume behaviour. Here, however, the authors could demonstrate changes in the fauna caused by the effluent in the small creek to which it was discharged prior to entering the lake. *Branchiura sowerbyi* was recorded before the power station began operating but, surprisingly, did not increase in abundance in the heated waters. The population changes of some 24 littoral macro-invertebrate taxa were monitored around the thermal discharge into Belews Lake (Lenat 1978), before and after the discharge began. Some 150 taxa were recorded over the 3 years of study. Macrophytes did not appear in the lake until the discharge began but once present were more abundant in the cooler waters than in the discharge. Maximum water temperatures were 34·8°C and 36·7°C in the 2 years after the station began operating. Control maxima were 27·8 and 29·8°C respectively.

Ten of the 24 taxa showed changes in population densities in the discharge areas. Among those that declined, the chironomid *Polypedilum digitifer* showed significant changes between both year and location, but the amphipod *Hyalella azteca* declined at both heated and unheated sampling stations. However, the decline was total at the most heated station and this was predictable considering that the ultimate lethal temperature of *H. azteca* is 32°C (Sprague 1963). The worms *Nais communis* and *Limnodrilus hoffmeisteri* increased in densities at both heated and unheated stations once the discharge began but declined at the latter in the second year of operation.

The mayfly *Hexagenia munda* also increased in density markedly at both heated and unheated sampling stations, but much more rapidly at the former. The increased densities of the worms *Dero* sp. and *Nais* spp. and the dipteran *Palpomyia* sp. were associated with the blooms of blue-green algae in the discharge. The seasonal and spatial variations in the densities of the chironomids *Eukiefferiella* sp. and *Endochironomus nigricans*, and the caddis-fly *Cyrnellus marginalis* were associated with the appearance and distribution of macrophytes. In Baldwin Lake, a cooling reservoir in Illinois (Parkin and Stahl 1981), no fauna existed where temperatures exceeded 35°C and all genera of *Chironomidae* were specific to specific substrates where temperatures did not exceed 32°C.

The studies on the macro-invertebrate faunas of 14 cooling water reservoirs in the USA were of variable value (Becker *et al.* 1979a,b). Very few were comprehensive enough to reach sound conclusions and the results were often conflicting. The factors mostly responsible for any apparent differences in the abundance of invertebrates or species composition of the community between heated and unheated areas were regarded as:

—water currents,
—the location of the sampling stations and poor controls,
—the presence or absence of macrophytes,
—the presence or absence of predators,
—poor sampling or statistical design.

In the more detailed studies at power stations on Lake Wabamun (Gallup *et al.* 1975), there were demonstrably different communities in the heated and unheated areas although standing crops were similar. The decline of the total standing crop and of the population of the snail *Physa gyrina* at sampling sites was associated with the harvesting of the dense macrophyte growths. Rasmussen (1982) in later studies found that some heated areas had higher standing crops and that tubificid worms were abnormally abundant and chironomids very sparse.

The population of the lamellibranch *Anodonta* sp. in the thermal discharge canal of the Sundance power station on the lake was unaffected by the sudden fall in temperature from 21·8°C to 4·9°C over a 24-h period in February 1973 when the station ceased operating, though this was a considerable temperature shock.

As in the USA, the communities of heated Eastern European lakes showed changes in the relative abundance of species near the discharges though the species recorded were similar in warm and cold areas (Kititsina 1973). The total biomass of the benthos increased or decreased depending upon the temperature rise at any site. Thus for example, in the Konin Lakes the benthic biomass in the most heated areas was only about half of that in natural lakes. Where summer temperatures exceeded 33°C at one site the biomass was reportedly reduced by 90% compared with natural lakes. However, this sampling site was close to the 'spillway' (outfall) and was most likely affected by scour. Clearly the temperatures are not excessive in comparison with some recorded in other waters. Near the outfall of the Konakova district power station, the invertebrate fauna associated with the pondweed *Potomageton perfoliatus* was more abundant in spring and autumn than in unheated waters but less abundant in summer (Wroblewsteigo 1977; Zgarheva and Mordukhai-Boltovskoy 1980).

Oden (1979) reported a comprehensive study of the benthic meiofauna of Par Pond (see Chapter 2). The fauna was dominated by Nematoda, Rotifera, Ostracoda, Cladocera and Hydracarina comprising together over 90% of the numbers present. Sediments were similar at all sampling sites. The lower density and biomass of meiofauna at the heated sampling sites was concluded to be as a result of the higher water temperatures though the mechanism of change was not clear. Oden suggested that the temperature differences caused differences in the reproduction of species suppressing some and enhancing others. At the highest temperatures, however, species may have been eliminated by directly lethal temperatures.

The communities of Rotifera in Par Pond were analysed in more detail than the other meiofaunal groups. Some 372 species were identified from all the habitats. About 28% of the species were common to all sites, 29% were only found at the unheated site and about 8% only at the heated site. The communities showed low similarities as indicated by the Czekanowski coefficient (C_z) (see Hellawell 1986) but this was also evident between replicate samples at the same site. The variability between samples was such that no significant differences could be demonstrated between sites. The assemblages in the unheated area were distinguishable into summer and winter groups but those in the heated water showed no seasonal distinctions. Diversity and evenness indices (H' and J') (see Hellawell 1986) showed no significant differences between sites.

As in many other such studies no conclusive explanation is given for the differences found in the communities in the heated water. Differences in reproduction are suggested and it seems likely that continuous reproduction of some species could explain the absence of seasonal fluctuations in abundance at the heated sites.

There are few examples of replacement faunas in heated lakes though exotic species such as the oligochaete worm *Branchiura sowerbyi* and the gastropod snail *Physa acuta* thrive in heated lentic waters.

(c) Estuaries and coastal waters. The changes in the macro-invertebrate

fauna of Queens Dock, Swansea found by Naylor (1959) were mainly a result of exotic species replacing the indigenous species in the area affected by a thermal discharge. In particular the exotic barnacle *Balanus amphitrite* replaced the indigenous species *B. crenatus* and the common immigrant *Elminius modestus*. As the power station operated less frequently the indigenous species re-invaded and the exotics disappeared (Naylor 1965a,b). Tunicata were abundant near the outfall even when the power station was operating fully but *Ascidiella aspersa*, which was eliminated by the high summer temperatures, reappeared as the natural conditions were reinstated.

Markowski (1959, 1960, 1962) used a dredge and concrete slab artificial substrates to investigate the macro-invertebrate communities in the thermal discharges of six British power stations. Few of the species recorded at any of the six power stations were consistently absent from the heated areas but the prawn *Palaemonetes varians* was noted as having a 'predilection for the warm water of the outfall' at one site.

The results from the long-term studies at the Bradwell nuclear power station on the Blackwater Estuary (Barnes and Coughlan 1970, 1972; Hawes *et al.* 1975; Coughlan 1978) have been interpreted somewhat differently by different authors. Objectors to the building of the power station originally predicted that the chlorinated thermal discharge would cause dramatically adverse changes in the macro-invertebrate fauna, including the destruction of the oyster (*Ostrea edulis*) stocks. Coughlan (1978) reporting on a survey of the benthos carried out 13 years after the power station began operating, concluded that there was evidence of the 'continued existence of 109 benthic invertebrate species of which the oyster, *Ostrea edulis* was the fifth most widely distributed species'. Hawes and his co-authors (Hawes *et al.* 1975) noted that the greatest loss of recorded species, mostly those of the littoral and sub-littoral zones, was a result of the severe winter of 1962/63 when natural water temperatures were at $-2.5°C$ for 58 consecutive days (Crisp 1964). By 1970 most of the species lost or declined in numbers had returned or recovered their losses (Coughlan 1978). Bamber and Henderson (1981), in their review of the Bradwell data noted that the loss of species could have been obscured by a change in the method of sampling in a 1965 survey. In fact they concluded that very few of the data were comparable from survey to survey because of changes in sampling method or sieve mesh sizes. Without the knowledge of these changes it could be concluded that the greatest loss of species from the estuary coincided with the opening of the power station and not with the severe winter. It is clear from the work, however, that the drastic effects on the fauna predicted by the objectors to the power station have not occurred in the estuary.

The discharge stratified within a short distance of the offshore outfall. In the close vicinity of the outfall there were minor changes in the benthic and epibenthic fauna which were related to the discharge. As a general trend the number of species and abundance increased with distance from the outfall though, at the extremes, the control sampling point produced the same number of species but a less dense fauna than the warmest point. Similarity analysis clearly separated the control and outfall communities (Bamber and Henderson 1981) and only two species, the exotic crab *Pilumnus hirtellus* and the dog whelk *Buccinum undatum* were common to both areas.

Barnes and Coughlan (1970) considered that changes in substrate composition caused by scour were the main cause of the faunal differences near the outfall, though the higher temperatures might be expected to favour the exotic species. The major problem with the Bradwell studies is that the data were not quantified and thus have limits to their use either as a monitoring record or for predictive purposes.

A more quantitative approach showed that there were no effects on the composition of the subtidal communities of the sandy substrates heated by the thermal discharge from the Hunterston nuclear power station in Scotland (Barnett and Hardy 1969; Barnett 1972). The dominant species was the bivalve mollusc, *Tellina tennuis* and although the population densities fluctuated over several years the fluctuations were not related to the effluent.

Crema and Pagliai (1980) concluded that the effluent from the Prombino power station in Italy did not cause changes in the sub-littoral communities near the outfall though they did not sample closer than 150 m. They found no evidence of community disturbance and the diversity and abundance of the fauna was related to distance offshore and depth of sampling. Among the abundant polychaetes near the outfalls was *Heteromastus filiformis* which, as we will see, is common to several saline waters.

Two separate, detailed studies of the benthos of thermal discharge channels concluded that temperature fluctuations rather than maxima or prolonged raised temperatures were responsible for elimination of species and the creation of stressed communities.

At Kingsnorth power station on the Medway estuary in the UK, the discharge canal is a dredged muddy creek for much of its length. The benthos of this channel was a dynamically stable community within the normal limits of estuarine faunas, but Bamber and Spencer (1984) concluded that in the reaches of the channel which were affected by temperature fluctuations of up to 10°C tidally, the community was a classical stress community. About 50% of the species which might have been present were missing and the diversity was generally low.

The rapid temperature changes were reported to have eliminated many of the species even though the maximum temperatures rarely exceeded 30°C. The changes were limited to a relatively short reach of the canal.

The resident stable fauna at the most temperature-stressed sampling station comprised mainly the oligochaetes *Tubificiodes benedeni* and *T. amplivasatus* and the polychaetes *Caulierella zetlandica*. *Nereis diversicolor* and *Heteromastus filiformis* were also common.

The sampling station nearest the outfall, whilst experiencing the highest maximum temperatures but less diurnal fluctuation, was dominated by an epifaunal community and species characteristic of coarser substrates. These substrates were a result of scour by the discharge. Despite the consistently high water temperatures the number of species recorded near the outfall was similar to that at the control station (Bamber 1985).

Most of the molluscs recorded in the Medway estuary were absent from the other stressed areas of the channel where temperature fluctuations were greatest, with the exception of *Macoma balthica*. Also all the species resident in the warmest areas were littoral species and thus adapted to temperature fluctuations (see also Walters 1977). Locan and Maurer (1975) found very similar effects in the discharge canal at the Indian River power station. Here the chlorinated discharge was about 10°C warmer than natural ambients and maximum temperatures reached 36·5°C. At the Gladstone power station in Australia, current velocity and scour were the major factors affecting the fauna of the discharge canal (Saenger *et al.* 1982). The infaunal species of mud were replaced by epifaunal species as the mud was removed by the effluent currents. The number of species, densities, diversity and evenness indices were all lower at the warmest sampling station than at the control. The benthos was dominated by *Nereis succinea* and *Heteromastus filiformis*. These species were also dominant during the summer in the discharge areas of the Calvert Cliffs nuclear power station (Loi and Wilson 1979), though other species became dominant in winter, and sediments rather than temperature were the important factor. Reductions in the diversity of the benthos were also noted in summer at sampling stations within 400 m of the outfall of the Yorktown power station on the York River, Virginia (Warinner and Brehmer 1966). Only about 20% of the winter species were recorded in August. Again, although this was a sublittoral zone rather than a canal or creek, tidal temperature fluctuations were up to 10°C. Here, however, some changes also occurred in the summer at the control sites and these were regarded as caused by the natural decomposition of mud and detritus. Interestingly, *N. succinea* and *H. filiformis* were again the most tolerant species. One marked change near the Yorktown outfall was the absence of eel-grass (*Zostera*) which occurred at the control site. This may have been eliminated by the effluent through scour, temperature or chlorine but its absence was most likely the cause of the reduced diversity and abundance of the benthos and epibenthos.

(d) Sub-tropical and tropical waters. Among the longest-running and most comprehensive series of studies of marine benthos in relation to thermal discharges are those in the sub-tropical waters along the coast of Florida in the USA (e.g. see Thorhaug 1980). These and studies in tropical waters (e.g. Kolehmainen *et al.* 1975; Zeiman and Ferguson-Wood 1975; Thorhaug 1980) are important because the organisms in such waters are reportedly living close to their thermal death points at natural maximum temperatures. Thus small temperature increments could be lethal (Saenger *et al.* 1980a,b).

The long-term studies at Turkey Point in Florida (see p. 191), showed that there was some relationship between the numbers of species and the numbers of animals collected by otter trawl, and the isotherms produced by the effluent (Table 8.5). Regression analysis showed that the relationships were not significant for all groups. Predictive models indicated that most species and most individuals would be caught where average temperatures were close to 26°C. Some 50% of species would be excluded at 33°C and 75% at over 33°C. Analysis of variance using the mean density of animals showed that few individuals of the common species were present where the temperature rise was between 4 and 5°C. At a rise of 2–3°C catches were low in summer but higher in winter and spring when water temperatures ranged between 12 and 18°C but overall within the +2°C isotherm areas no significant effects of the effluent were demonstrated (Thorhaug *et al.* 1974, 1978; Thorhaug 1979). The area of maximum effect on the macro-invertebrate fauna was about 60 ha or 0·0023% of the Biscayne Bay area. The diversity and density of the epibenthos were probably more related to the density of the vegetation than directly to temperature in Biscayne Bay. Other factors, including chlorine, heavy metals, scour or sedimentation may also have been significant (Nugent 1970; see Langford 1983a). Few species would be expected to survive water temperatures exceeding 35°C for long periods (Thorhaug *et al.* 1978). Roessler (1971) quoted lethal temperatures of around 33–34°C for the common invertebrates and plants in Biscayne Bay.

TABLE 8.5

Catch of algae and animals per drag of an otter trawl in the vicinity of the thermal discharges at Turkey Point, Florida, USA
(after Roessler 1971)

Station	Ambient temp. (°C)	Effort	Weed (kg)	Fish	Molluscs	Crustaceans	Sponges	Coelenterates	Echinoderms	Misc.	Total
N1	+0·30	128	4·6	1·8	20·5	40·6	1·7	0·16	0·00	0·00	64·7
N2	+0·34	128	3·2	4·8	43·9	14·1	0·1	0·04	0·01	0·01	62·9
N3	+0·27	128	6·4	4·4	129·5	51·3	0·7	0·02	0·07	0·00	186·0
N4	-0·16	54	3·1	2·6	89·1	81·9	2·1	0·06	7·48	0·00	183·3
N5	+0·07	54	8·9	23·1	361·2	47·1	0·0	0·02	0·00	0·00	431·5
NE1	+0·33	128	3·4	1·9	32·1	39·2	1·4	0·02	0·52	0·00	75·1
NE2	+0·23	128	5·0	2·1	28·7	60·2	1·7	0·03	1·03	0·00	93·8
NE3	+0·01	128	4·0	2·3	38·2	65·3	0·7	0·08	4·98	0·00	111·5
NE4	-0·13	54	1·0	1·3	8·9	12·6	2·1	0·43	6·33	0·01	31·7
NE5	-0·32	54	0·1	0·8	1·9	3·6	0·9	0·22	1·61	0·00	9·1
SE1	+3·25	128	1·4	0·9	28·4	11·4	0·2	0·02	0·02	0·00	40·8
SE2	+0·86	128	3·2	1·6	54·9	62·1	1·5	0·02	7·03	0·02	127·1
SE3	+0·34	128	2·7	2·2	40·1	61·4	1·9	0·00	3·73	0·00	109·2
SE4	-0·02	54	1·0	1·5	16·6	6·8	1·0	0·24	1·39	0·00	27·6
SE5	+0·25	54	0·1	2·4	1·6	1·0	0·5	0·56	6·57	0·00	12·7
S1	+3·42	128	0·7	0·6	5·7	7·0	0·1	0·00	0·04	0·00	13·4
S2	+1·99	128	3·5	1·6	75·4	36·6	0·4	0·00	2·22	0·00	116·1
S3	+0·86	128	1·7	0·9	23·2	29·6	0·7	0·02	5·78	0·00	60·3
S4	+0·75	52	0·5	1·2	10·7	4·5	7·6	0·04	8·23	0·00	32·3
S5	+0·52	54	0·1	1·4	2·6	1·3	4·4	1·65	6·24	0·00	17·7
A	+1·39	114	2·3	1·3	41·4	14·8	1·0	0·00	0·05	0·00	58·6
B	+0·84	86	2·4	1·1	16·6	17·8	0·3	0·01	0·21	0·02	36·1
C	+0·08	86	1·5	1·6	17·0	24·2	0·9	0·05	5·29	0·00	49·1
D	20·45[a]	86	0·8	0·9	15·5	8·1	20·3	0·17	0·70	0·03	45·8
E	+0·28	86	1·7	1·0	24·4	15·1	0·6	0·03	5·27	0·01	46·5
F	+1·81	86	7·0	1·9	146·6	78·9	6·6	0·05	0·03	0·00	234·1
G	+4·20	86	0·0	0·2	0·5	1·7	0·0	0·00	0·00	0·00	2·5
H	+1·39	86	3·5	1·7	45·2	24·0	0·7	0·08	0·23	0·00	71·8

[a] Station D was considered the temperature control station.

The blue crab (*Callinectes sapidus* was more common in the heated effluent canal than in colder waters except during July and August when temperatures exceeded 35°C. The opening of two nuclear power stations on Biscayne Bay increased the effluent volume by 7·5 times that of the earlier power stations. The predicted area of maximum biological effect would therefore have been about 450 ha (<0·02% of the bay area) using simple linear extrapolation. However, a 9 km long canal was dug through the mangrove swamps to take the effluent to Card Sound. The canal construction probably destroyed some 1500 ha of mangroves (Nugent 1970). The temperature rise from the mouth of the canal to the open water of Card Sound was 2°C. The diversity and abundance of the benthos and epibenthos of Card Sound were not affected by this temperature rise (Thorhaug *et al.* 1974, 1978; Thorhaug 1980). However, the opening of the canal eliminated the vegetation over about 2–3 ha of Card Sound mainly because of the discharge of suspended solids as the canal was opened.

The benthos and epibenthos of the area of Biscayne Bay affected by the original effluent recovered in both diversity and abundance once the older power stations closed, though in areas adversely affected by other of man's activities, recovery did not occur until the sea-grasses were restored by deliberate transplanting (Thorhaug 1987). At another Florida power station on Tampa Bay, it was concluded that some 35 of about 104 species of macro-invertebrates were excluded from the warmest areas, though 9 species were introduced (Blake *et al.* 1976). Here as at Turkey Point the macro-vegetation was eliminated or reduced near the outfall. The elimination of invertebrates was not as marked when only sandy substrates were compared, indicating that the removal of the vegetation from the mud substrates was a primary factor affecting the fauna. The polychaete *Prionspio heterobranchia texana* and the bivalve *Mysella planulata* were more abundant in the heated areas than in the controls (see also Thorhaug *et al.* 1978). The total abundance of benthic animals did not differ significantly ($p < 0·05$) between all intake and discharge sampling points. No mention of chlorine or other anti-foulants is made in the published papers, though some anti-fouling procedure was probably used at all sites.

At the Guyanilla Bay power station in Puerto Rico, water temperatures reached 40°C in the effluent during 1972. Very few species of macro-invertebrates survived here (Table 8.6), (Kolehmainen *et al.*, 1974, 1975). Only 10 species were recorded at temperatures over 37°C close to the outfall, though when temperatures were around 34°C some 27 species occurred. The number of species at the control site was 87. Water temperatures exceeding 35°C apparently eliminated most of the indigenous

TABLE 8.6

Maximum temperatures at which invertebrate species were found
in Guyanilla Bay, Puerto Rico (from Kolehmainen *et al.* 1974)

Temperature (°C)	Number of species found	Cumulative number of species
40	1	1
39	2	3
38	8	11
37	2	13
36	17	30
35	12	42
34	17	59
33	8	67

species though they recolonised temporarily at seasonally lower temperatures, particularly the ascidians, polychaete worms and crabs.

From Fig. 8.5 it is clear that factors other than temperature limited the occurrence of species in the heated areas. For example the temperature rise at station 1, furthest from the outfall, was less than at other locations but the number of species was lowest here probably as a result of scour caused by the water currents which were strongest here and at station 7. A rise of 5°C was considered critical, but it is clear that rises of 2 and 4·5°C were also associated with low numbers of species because of other factors.

No details of the concentrations of contaminants are given for the Guyanilla site though it was noted that 'free chlorine disappears quite fast... in the discharge canal' (Kolehmainen *et al.* 1974).

The reduction in the abundance of Nematoda recorded in the effluent cove was reputedly caused by high concentrations of heavy metals and petrochemical residues in the substrates rather than heat. There were clear differences in the communities of Foraminifera between heated and unheated areas but again substrate differences were the major cause. The complete absence of any macro-benthos in the mud and sand substrata in the heated cove was remarkable even though temperatures were high. Some temporary recolonisation should have occurred in the colder seasons as in the mangrove root areas, unless other factors were involved.

The lower biomass of macro-invertebrates in the heated areas of mangrove roots (Table 8.7), was caused mainly by scour. In the closest other sampling location, where temperatures differed by only 0·1–0·2°C and no scouring occurred, the biomass was much greater. The tree oyster *Isognomon alatus* and the barnacle *Balanus amphitrite* var. *pallidus*

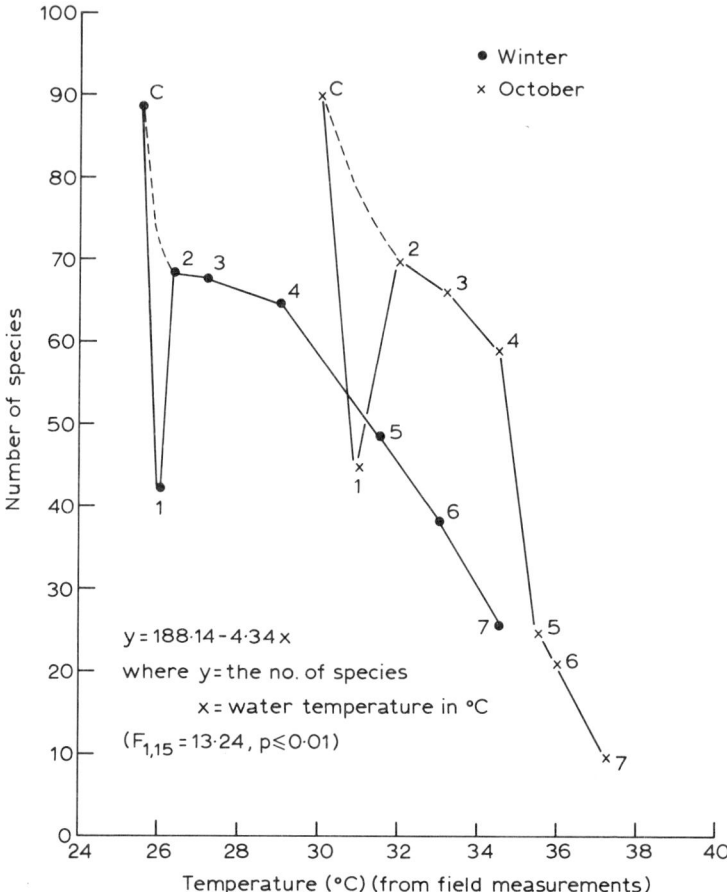

FIG. 8.5 Guyanilla Bay numbers of invertebrate species recorded at various temperatures, taken from two separate surveys. C = control sampling site; 1–7 = effluent sampling sites (after Kolehmainen *et al.* 1974).

dominated the biomass in the heated cove, molluscs at the control stations, ascidians and cirripedes at other locations. There was no relationship between biomass and water temperature, but strong water currents were clearly a major factor. When a second larger power station was opened at Guyanilla Bay, the tree oysters and barnacles again increased in biomass.

The heated cove near the outfall where the biological effects were most marked, covered an area of about 4 ha of the total 140 ha of the bay itself.

At the KORI nuclear power station in Korea, the thermal discharge did

TABLE 8.7

Biomass of organisms on mangrove roots at eight sampling stations near the thermal discharge at Guyanilla Bay, Puerto Rico, October 1971 (from Koleh-mainen *et al.* 1975)

	Biomass (*wet weight*) g/root							
	Control	1	2	3	4	5	6	7
Algae	1·6		21·5	23·4	90·3	0·3	0·5	1·1
Porifera		55·6	4·6		14·7		5·1	
Coelenterata	0·1							
Annelida	29·2	7·0	27·2	15·1	34·4		0·5	0·2
Mollusca	98·1	154·2	30·3	70·3	13·9	63·1	622·6	
Cirripedia	50·9	33·4	11·8	37·7	7·5	50·5	22·2	13·1
Other crustacea	6·9	20·6	2·9	9·4	7·0		0·1	
Bryozoa	5·1	5·0	40·3	0·5	0·7	0·1	0·1	
Echinodermata			0·3	1·0	0·5			
Chordata Ascidiacea	48·0	28·3	40·7	26·6	36·7	0·1		
Total biomas (wet weight) g/root	240·0	304·1	179·6	184·0	205·7	114·1	651·1	14·4
One standard deviation	71·3	120·1	43·5	41·6	54·9	32·4	103·6	2·81
Biomass (wet weight) g/cm of root	3·87	6·76	2·64	2·70	3·74	2·54	7·98	0·41
One standard deviation	1·15	2·67	0·64	0·61	1·00	0·72	1·27	0·08
Temperature (°C)	30·0	31·0	32·0	33·3	34·5	35·6	36·0	37·2

not cause changes in the intertidal or subtidal faunas though it was apparent that data from very near the outfall were not published (Huh 1980; Soon Kil Yi 1987). Water temperatures reached 33–36°C in the effluent.

At the Kahe power station in Hawaii the hermatypic corals were killed within an area of 0·7 ha from the outfall where water temperatures rose by 4–5°C above natural ambients in summer. Within the area heated by 2–4°C the zooxanths of the corals were bleached in summer but regained their colour in winter. The effect suggests the influence of low concentrations of chlorine rather than temperature alone. In a study covering the periods before and after the opening of the Gladstone power station in Australia the number of taxa in the macro-benthos of the discharge canal declined from 255 to 225, but the percentage composition of the major species remained fairly constant (Saenger *et al.* 1982). Changes in the sediments were believed to be the cause of the changes and it was concluded that temperatures of 30–33°C, and an average rise of 8·2°C would not result in dramatic reductions in the macro-fauna where scour did not occur.

Thorhaug (1980), with her considerable experience of field and experimental research on tropical flora and fauna, concluded that temperature increases of as little as 1–2°C were damaging to tropical ecosystems. Kolehmainen *et al.* (1975) considered that a δT of 5°C was critical at Guyanilla Bay. However, whilst thermal discharges have no doubt caused changes in tropical coastal ecosystems, water temperature was only one of the possible factors and no critical temperature rise (δT) could be predicted objectively from the field evidence (Langford 1983c). Water currents, scour or biocides were clearly more important at some sites as we have already seen for plants (Chapter 6). All the field and experimental data suggest, however, that water temperatures above 35°C inhibit many species and that no species, permanently inundated by a thermal discharge would tolerate temperatures over 40°C persisting for more than very short periods.

(e) Rock and other hard substrates. In most of the studies outlined above, the substrates were soft and thus relatively easy to sample quantitatively, except where scour exposed hard clay or rock. Studies of rocky substrates exposed to heated effluents are scarce. Quantitative sampling in the sublittoral zone is difficult and is usually best carried out by divers using transect or quadrat techniques.

Straughan (1980a,b) found minimal changes in the faunas of rocky breakwaters and rock substrates at the King Harbour and Morro Bay power stations in Southern California and at a power station in Hawaii (Straughan and Straughan 1972). The affected areas were within 300–500 m of the outfalls. Maximum effluent temperatures were between 7 and 10°C above natural ambients.

In contrast, the chlorinated thermal discharge from the Martigues-Ponteau power station near Marseilles caused a distinct reduction in the number of species on rocks near the outfall compared to nearby unheated rocks (Arnaud *et al.* 1981). The temperature rise was 7°C on average. Once outside the immediate area of the outfall there was no gradient of numbers of species or animals which could be related to temperature. However, there were large areas in the path of the effluent over which blackened tubes of Serpulidae occurred. The worms had apparently died.

All 11 species of Hydroida recorded in the locality were found in the path of the effluent away from the outfall. The tropical species *Corydendruian parasiticum* was found in the warmest area. *Sertularella ellisi* var. *mediterranea* was generally the most common species and was particularly abundant near the outfall. Mollusca were also least abundant in the discharge basin. Chlorine residuals rather than temperature caused the changes in the fauna and the mortalities of Serpulidae. At the Kahe Point power station in Hawaii, chlorine also seemed to be the most damaging factor for the hermatypic corals in the heated plume (Jokiel and Coles 1974).

Moller (1978b) found that the macro-benthos was reduced in diversity and abundance over an area of about 1 ha near the discharge of a German power station. Maximum temperatures in the effluent were 3°C and chlorine concentrations 0·5–1·0 mg litre^{-1}. The maximum numbers of mussels *Mytilus edulis*, and barnacles, *Balanus crenatus*, occurred in the higher current velocities near both intake and outfall. The abundance of most macro-invertebrate species was greater in the intake and effluent zones than at the control stations. Surprisingly, numbers declined rapidly over 12 days after the power station closed down in August. The reasons were probably the decline of filter feeders because of the loss of currents on which they depended, and heavy predation by the crabs, *Carcinus maenas*, and eels, *Anguilla anguilla*, which invaded both intake and outfall areas in very large numbers as soon as the cooling water ceased to flow.

Mytilus edulis was found growing in large numbers during autumn and winter in the discharge canal of a Massachusetts power station but were killed in summer. Mortalities began as temperatures exceeded 27°C. During July–September maxima exceeded 30°C for long periods (Gonzalez and Yevich 1976).

McCain (1975) used fouling panels to demonstrate the differences in colonisation by epifaunal species between heated and unheated areas at the Waiau power station. The critical temperature was about 32°C.

On the rocky shore near Wylfa nuclear power station in north Wales, the population of the barnacle *Balanus balanoides* decreased near the thermal discharge outfall. Bamber (1990) considered this an effect of mean temperature though he did not say whether the effluent was chlorinated.

(ii) Growth, life histories and reproduction
(a) Annelida. The biology of some aquatic worms (Oligochaeta) and at least one species of leech (Hirudinea) was affected by the thermal discharges from British power stations. In a series of carefully executed field and

experimental studies Aston (1968a) showed that the exotic oligochaete, *Branchiura sowerbyi*, produced cocoons over a longer period at higher temperatures than at ambient temperatures. Growth was also faster in the heated waters of the River Thames near Earley power station than in the colder waters of the River Avon though somatic growth was eventually inhibited in the warm water population because of increased gonadal growth and higher cocoon production. The optimal temperature for cocoon production in both populations was 25°C. As we have seen, *B. sowerbyi* is found in thermal discharges though populations also occur in natural temperate waters (see p. 268). The species was also found in the experimental streams heated by a power station effluent (Alston *et al.* 1978). The populations showed seasonally advanced sexual maturity and were very abundant.

In subsequent studies, Aston (1973) found that the proportions of sexually mature specimens of the indigenous oligochaetes *Limnodrilus hoffmeisteri* and *Tubifex tubifex* were higher in the populations downstream of a thermal discharge than upstream (Fig. 8.6). The temperature rise ranged from about 7 to 12·5°C over a year. In experiments *L. hoffmeisteri* increased egg production in relation to increasing water temperature up to 25°C and eggs were still being produced at almost 30°C. In contrast *T. tubifex* showed a lower and consistent egg production over a temperature range of 15–25°C. *L. hoffmeisteri* was clearly favoured by the higher temperatures in the effluent, which reached 27·5°C during the year of the field study. In other parts of the world *L. hoffmeisteri* is also affected by thermal discharges. In the heated areas of the Ivankovskoe Reservoir in the USSR, the species bred 3 weeks earlier than in the unheated waters and there were two generations each year instead of one. Also the production of this species was found to be 1·5 times greater in another heated reservoir (see Kititsina 1973) than in colder waters.

Euilyodrilus hammoniensis also showed increased production by a factor of 1·2 in the same location. Lenat (1978) found that *Nais communis* declined in numbers in the area near a thermal discharge outfall but increased where algae were prolific (see p. 188).

In the River Trent, the leech *Erpobdella octoculata* was found to show changes in growth, ultimate size and life cycles in reaches variously affected by organic pollution and the thermal discharges from four power stations (Aston and Brown 1975). Figure 8.7 shows the ranges of water temperatures at sampling stations in relation to the average weights of adult leeches. The largest leeches occurred at the most polluted location which was not heated by a thermal discharge. The smallest specimens were

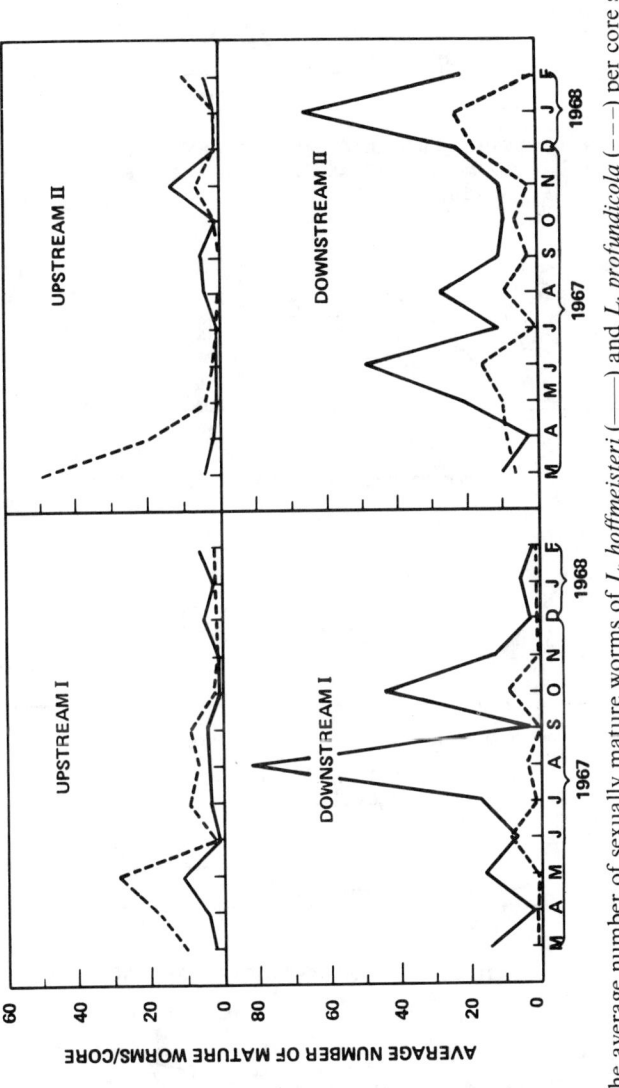

FIG. 8.6   The average number of sexually mature worms of *L. hoffmeisteri* (———) and *L. profundicola* (– – –) per core sample taken from the River Trent upstream and downstream from Drakelow Power Station (from Aston 1973).

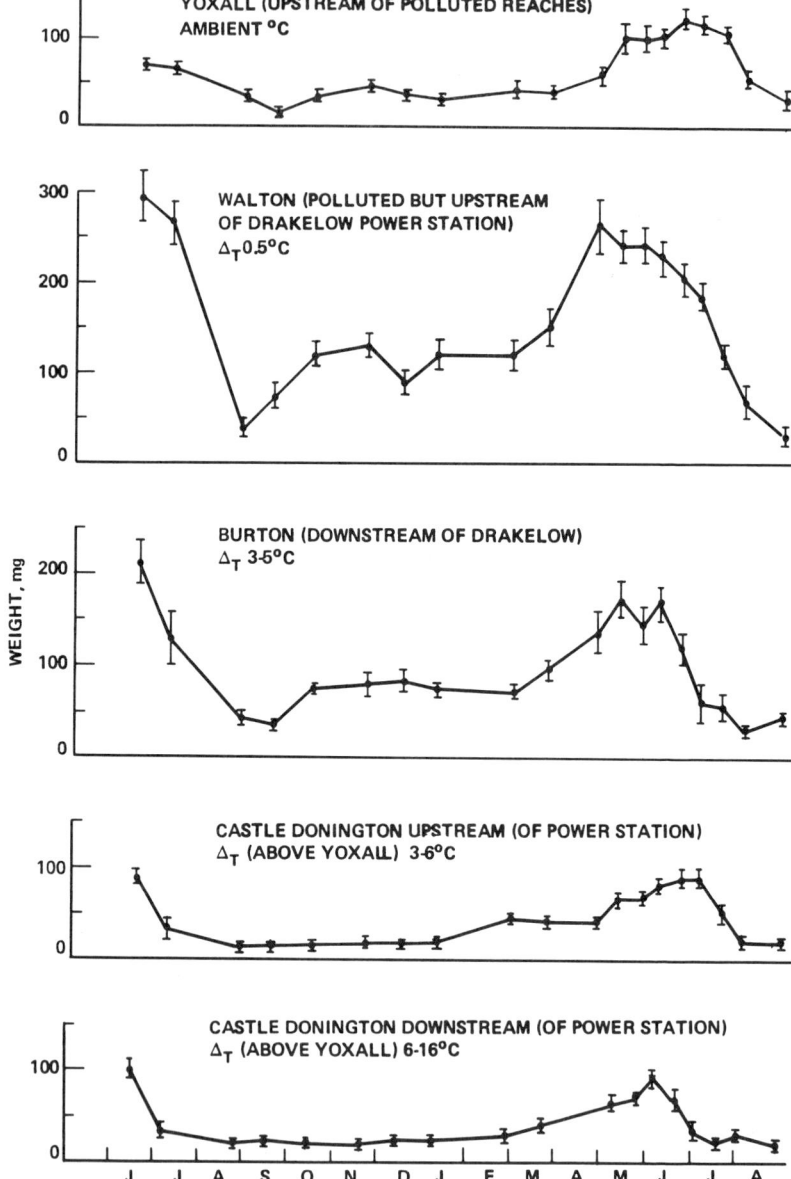

Fig. 8.7   Seasonal variation in the mean weights of leeches collected from the
River Trent (from Aston and Brown 1973).

found at the second warmest location. Newly hatched leeches occurred 2–4 weeks earlier in the heated reaches than in the colder waters. Cocoon production was also earlier. Sexually mature specimens appeared earlier than expected at both heated and the most polluted location. The lowest densities were at the least polluted but warmest location.

The differences in weight and the timing of the life histories were not related simply to increased temperatures. The faster growth and higher cocoon production in the most polluted reach were a result of the increased abundance of oligochaetes on which the leeches fed. The evidence from this study indicates that *E. octoculata* completed its life cycle in 1 year in the most polluted and in the warmest reaches compared with 2 years in the colder and cleaner reaches.

Mulstay (1971) noted that sexually mature and gravid individuals of the polychaete worm *Scoloplos* sp. were present in the heated effluent at the coastal Northport power station in April whereas the usual reproductive period is in June or July. *Polydore ligni* was also reported as spawning earlier than usual.

(b) Crustacea. Two studies in Britain showed that the life cycle of the isopod *Asellus aquaticus* was shortened in thermal discharges. The species had a 1 year cycle in the heated area of the River Thames at Earley power station compared with 2 years in other reaches of the river. Studies on the River Trent (Aston and Milner 1980) found that the life cycle could be completed in less than 1 year in the heated reaches. There was continuous recruitment through the summer at both heated and unheated sampling locations. Juveniles, less than 3 mm long were present in all except a few winter samples. Ovigerous females were present over a longer period in the heated waters. Spring juveniles grew faster in the warmer water and matured in autumn instead of overwintering as immature adults. Late summer juveniles did overwinter to become mature the following spring. As with the leeches (*E. octoculata*), the largest individuals of *A. aquaticus* occurred in the most polluted but not the most heated location. The population densities also followed the same pattern.

The amphipods *Pontogammarus crassus* and *P. rubustoides* also completed their life cycles in up to 2 months less than normal in the heated waters of two reservoirs in the USSR (see Kititsina 1973). The biomass of the fresh-water shrimps increased up to five-fold in these waters and adults were reported as 1·2 times larger than in colder waters. The average number of young produced by each female increased from 5 or 6 to 8 or 10 in the heated waters of the Khurakhova reservoir (see Kititsina 1973).

In contrast, the average size of the mysid shrimps, *Mesomysis kowalevskyi*, decreased in the heated waters of Lake Liman (Kititsina 1973). Irrespective of the smaller size of these mature adults the growth rate to maturity was faster and the number of generations increased from 4 to 5, or 5 to 6, though the significance of this increase is difficult to assess. *L. benedeni* also reproduced all year round in the heated water instead of in summer at natural temperatures.

Bamber (1990) has recently reviewed the life-histories and biology of some marine crustaceans in thermal discharges. He considered that problems with survival were associated with upper incipient lethal temperatures of around 33°C.

The burrowing amphipod *Urothoe brevicornis* began to be gravid about a month early in the thermal discharge from Hunterston power station in Scotland (Barnett 1971). The females in the unheated waters carried eggs longer than in warm water in direct contrast to the fresh-water crustaceans discussed previously (see p. 259). Overwintering adults were larger in the discharge area than in the control areas though the rates of increase in length were very similar in both habitats. The final difference in mean length of adults was about 20%. The bottom-living harpacticoid *Asellopsid intermedia*, also bred about 2 months earlier than normal in the Hunterston effluent. The reproduction and survival of *Corophium ascherusicum* was enhanced by the conditions in the effluent at Oyster Creek nuclear power station such that it replaced the indigenous *C. tuberculataum* which was dominant at the intake.

Another isopod, *Cyathura carinata*, grew throughout the winter and bred earlier than other temperate populations in the thermal discharge canal at the Kingsnorth power station (Bamber 1984b, 1990). The conclusion was that the population was showing the growth rate and life history characteristics of natural southern populations in a northern temperature location.

Extensions of breeding periods in thermal discharges are also reported for *Corophium ascherusicum* and the wood-boring isopod *Limnoria tripunctata* in a heated dock (Naylor 1965a). *Limnoria* also colonised new timber more rapidly in heated waters and showed longer reproductive periods causing particular problems with wooden structures.

The immigrant barnacles, *Balanus amphitrite* and *Elminius modestus* showed enhanced breeding in thermal discharges in Britain (Crisp and Molesworth 1951; Pannell *et al.* 1962). In the heated dock at Swansea (see p. 269) the indigenous barnacle *Balanus crenatus* did not breed at temperatures above 20°C but the exotic species *B. amphitrite* did, with the result that the latter replaced the former. As the power station ceased to operate the situation was reversed (Naylor 1965a,b). Similarly at the Oyster Creek nuclear power station, *Balanus eburneus* showed an extended breeding season and optimal growth conditions in winter instead of in summer. At the Martigues-Ponteau site on the Mediterranean, adults were larger at maturity in the unheated than in the heated waters.

Breeding was halted in the shore crab *Carcinus maenas* in the heated docks at Roosecote and Swansea (Markowski 1962; Naylor 1965a,b). *Carcinus* also moulted during the winter which was abnormal, particularly if females are carrying eggs. As the temperatures declined at Swansea, breeding *Carcinus* re-appeared.

Naylor (1965a) states that 'there is no experimental evidence to support the view that raised winter temperatures might prevent breeding in *C. maenas*'. The fact that this did occur in the heated docks may therefore be a result of other influences, for example chlorine. Although no mention is made of biocides in the published work it is known that both cooling-water systems were chlorinated (Coughlan Personal communication). It is also known that chlorine can inhibit the reproduction of at least one species of amphipod, albeit from freshwaters (Arthur and Eaton 1971).

Blue crabs, *Callinectes sapidus*, showed seasonal patterns of migration into and out of the heated effluent at a Texas power station (Gallaway and Strawn 1975). Avoidance temperatures were estimated as between 38 and 40°C in summer, though individuals made short-term incursions into the plume at night to feed. Similar migrational patterns were made by white shrimp, *Penaeus setiferus* (Williams and Strawn 1980).

(c) Insecta. It has been demonstrated clearly in experiments that aquatic insects with aerial adults will emerge in advance of their normal season when exposed to higher temperatures in the laboratory (Gaufin and Hern 1971; Nebeker 1971a,b; Elliott 1972, 1978). At the higher temperatures there is some evidence that emergence is suppressed. Nordlie and Arthur (1981), for example, maintained an experimental channel at 10°C above the natural ambient river temperatures and found that emergence of several species began 1–4 weeks earlier than in the control channel, but few insects emerged at summer maxima in the heated channel. In experiments on thermal tolerance and shock (Sherberger *et al.* 1977), mayflies (*Isonychia* nr. *Sadleri*) continued to emerge throughout until the temperature rise reached 14°C and the exposure temperature 34°C. With a temperature rise of 8°C and 11°C emergence rates did not differ from the controls. The hatching time of if females are carrying eggs. As the temperatures declined at Swansea

insect eggs is also clearly related to water temperature (e.g. Humpesch and Elliott 1980), and could therefore be advanced in heated waters.

It was predicted early on in the thermal pollution debate that increased water temperatures in fresh waters near thermal discharges would affect the life histories of emergent insects such that adults would emerge early into lethally cold air and thus populations would be eliminated (Bregman 1969). However, very few field studies have been carried out to try to validate the predictions and the relatively few data which exist are somewhat equivocal. In the heated waters of Trawsfynnydd lake (p. 212) the emergence of three species of Trichoptera, *Oecetis lacustris*, *Cyrnus flavidus* and *Phryganea striata* was advanced by up to 3 weeks. Chironomids also showed similar advances in their emergence periods in the lake and in the heated effluent from the Konakova power station in the USSR (Kititsina 1973). Another caddis fly *Hydropsyche pellucidula* began to emerge 4 weeks earlier than normal in the heated effluent from the Elverslingen power station on the River Lenne in Germany (Fey 1977), though the main emergence period was the same as in the colder reaches. Larvae continued to grow in the warm water during winter in contrast to the cold water population. Growth accelerated in February and pupation was advanced in contrast to the cold water population. Fey concluded that the pupae were particularly sensitive to sudden changes in temperature which stimulated early emergence. Coutant (1967) also concluded that a temperature rise of only 1°C caused a significant 2 week advance in the emergence of Hydropsychiidae downstream of the Hanford reactor cooling-water outfalls.

In the River Severn in the UK a series of studies on the insect fauna showed very few effects of the heated effluent from the Ironbridge power stations on life histories (Langford 1971a, 1975; Langford and Daffern 1975). The stonefly, *Taeniopteryx nebulosa*, which emerges in the winter and hatches from the eggs in early spring, showed some inhibition of hatching in the warmer water. In contrast the nymphs of the mayfly *Ephemerella ignita* appeared earlier downstream of the outfalls than upstream by a week or so but the main hatching and emergence periods of all species of Ephemeroptera and Trichoptera were similar in the heated and unheated reaches.

Most of the common species showed prolonged hatching and multi-voltine patterns which obscured any effects of the heated water in mid-summer. Emergence patterns were not disrupted by water temperatures up to 28°C though increased river flows caused reductions in emergence. It was concluded that the absence of clear effects was a result of the unstable temperature regime and the complexity and flexibility of the life histories of the species. Tenessen and Miller (1978) found that nymphs of the mayfly *Hexagenia bilineata* exposed to a thermal plume in winter emerged at the same time as those at the control site, mainly because these latter grew very rapidly in spring to reach the emergence stage.

The influence of water temperature on insect emergence is not fully clear. Most recent studies associate the periodicity of emergence with cumulative effects of temperature, though species vary in their responses (e.g. Macan 1958b; Elliot 1972, 1978). Macan and Maudsley (1966) showed a correlation between the onset of emergence and the first date on which 11°C occurred. Other authors showed that day length and temperature have complex and interrelated effects (see Beck 1968; Hynes 1970; Danilievskii 1985). It seems clear that in a stable thermal plume some individuals will emerge early, even into cold air, and emergence may be suppressed at times, but because of the variation in stages of development among individuals in a normal species population many will also emerge during the normal period.

(d) Mollusca. There were indications that both the fresh-water lamelli-branch *Anodonta anatina*, and the estuarine gastropod *Hydrobia ulvae*, showed seasonal advances in their life histories in the thermal discharges at Earley (Negus 1966) and Kingsnorth (Walters 1977) power stations, respectively. Juvenile gastropods, particularly *Potomapyrgus jenkinsi*, were also recorded all year round in cooling-tower ponds, indicating a prolonged reproductive period (Markowski 1959; Milner 1984).

Physa virgata, a common gastropod in the USA, has three generations each year in the southern states compared with two in the north. McMahon (1975) found that the species tolerated temperatures of 39·5°C in a heated Texas lake but that the spawning of the first generation was delayed by up to 1 month. Later generations showed little change from normal for the region. In the second year of the study, the spring spawners also produced significantly fewer eggs in each egg mass laid in the heated waters.

Mattice (1976) showed that growth, survivorship, reproductive success and production of the gastropod *Limnea obtusa* were directly related to temperature up to the optimum after which all declined in the classic pattern (see Chapter 4). Temperatures over 30°C inhibited the growth of the fresh-water mussel *Dreissenia polymorpha* (see Kititsina 1973), but the species was found to colonise, grow and reproduce faster in power station cooling ponds in the USSR than in natural waters. In the USA *D.*

polymorpha has now become a major fouling species in fresh waters (Nuttall 1990).

Several studies showed that the thermal discharge from the Glen Lyn power station had marked effects on the population of the Asiatic clam. *Corbicula fluminea*. During the very cold winter clams survived well in the plume but not in the unheated river. In general, size classes were more diverse in the plume than in the cold water over 2 years and densities were also higher (Graney *et al.* 1980). In experiments and artificial channel studies the clam was found to be resistant to intermittent chlorination and heavy metals but accumulated many elements (Cherry *et al.* 1980). In tidal waters, the wood boring mollusc *Teredo navalis* increased its feeding and breeding rates in the heated waters of Swansea dock such that wooden structures were replaced by concrete, and wooden working boats were replaced by metal punts (Bell 1949). At sites in the USA, teredinids were found to extend their geographic and habitat range in heated waters (Hoagland and Turner 1980).

The American hard-shell clam, *Mercenaria mercenaria*, exposed to the thermal discharge of Poole power station in southern England, showed early onset of growth, a 100% increase in instantaneous growth rate and a prolonged growing season when compared with cold water animals (Ansell 1963). Mature individuals reached spawning condition earlier but the spawning season was curtailed in the warm water. The conclusion was that the species was near the northern limits of its range but could be cultured in warm water. It is significant that a commercial clam fishery developed in Southampton Water in the vicinity of the heated plume from the Marchwood power station (see p. 270). At a nearby site, oysters, *Ostrea edulis*, were apparently unaffected by the heated plume from the Fawley power station outfall and a major fishery developed in adjacent waters (Key and Davidson 1981).

In studies of another marine lamellibranch, *Tellina tennuis* (Nair and Saraswarthy 1971), living in the thermal discharge area of the Hunterston nuclear power station, young individuals were collected in autumn instead of the following spring as was usual in control populations (Barnett 1971). Specific growth rates increased to produce a 40–50% difference in size between warm and cold populations by the second winter of life.

The effects on the prosobranch, *Nassarius reticulatus*, were much more obvious. Spawning was advanced by up to three months (Fig. 8.8), beginning in January and February instead of April or May (Barnett 1972). The peak was in early May instead of July or August and the total spawning period extended to 6–7 months instead of 4 or 5. Fecundity varied over the

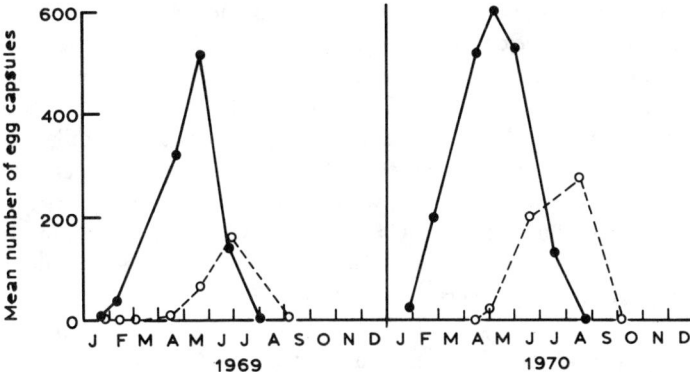

FIG. 8.8 The mean number of egg capsules of *Nassarius reticulatus* in the vicinity of a thermal discharge from Hunterston nuclear power station, Scotland; —●— heated, −−○−− unheated (from Barnett 1972).

spawning period in the warm water from about 260 to 200 eggs per capsule but this did not occur in the control populations. Subsequent experiments showed clearly that the hatching time of *Nassarius* eggs was directly related to water temperature. There was no evidence of population decline in the warm water though Barnett (1971) postulated that the early release of veliger larvae and lack of food could affect the populations over a long period.

The American oyster *Crassostrea virginica* grew better in the thermal discharge canal at a power station in the winter than in unheated areas (Tinsman and Maurer 1974), though in summer they showed lower condition factors and higher mortality rates in the warm water. Studies in other discharges showed variable effects on spawning at different sites ranging from early development of gonads and extension of the spawning period (see Tinsman *et al.* 1976) to no noticeable changes in either (Roosenburg 1969).

Factors other than temperature caused reductions in the growth rates of mussels, *Mytilus edulis*, at the outfall of the San Onofre nuclear power station (Kastendiek *et al.* 1981). The animals showed a significant decrease in growth rate when both heated and unheated water was discharged but the cause was not identified. In contrast the reproductive cycles of *M. edulis* and *M. californianus* were not affected by a heated effluent (Hines, 1979).

Transplanted bay scallops, *Argopecten irradians concentricus*, suffered higher than normal mortalities in heated effluents at Florida coastal power stations (Studt and Blake 1976). Suspended solids and chlorine may have been more damaging than temperature alone. Reproduction, gamete formation and tissue composition were all altered in the warmest areas. In simulation experiments on three species of molluscs it was found that *Kelletia kelletia*, a warm water carnivore grew faster in warm water than in the control tanks. *Roperia poulsoni*, a species with a similar natural distribution, grew little in either and *Astrea undosa*, a grazer on algae, actually lost weight in the warm water but not in the controls (Ford *et al.* 1978). Actual effluent from a California power station was used for the warm tanks but no mention was made of chlorination which could have, if applied, had marked effects.

(e) Other invertebrates. It is clear that species found in cooling-tower ponds, notably species of Hydracarina, the bryozoan *Plumatella* spp., the leech *Erpobdella octoculata* and flatworms (Turbellaria) (see p. 230) were reproducing for longer periods than normal as young were collected in the ponds during most months of the year (Markowski 1959; Milner 1984). Similarly, in the heated waters of the Konin lakes in Poland, several species of Hydracarina showed an increase from 1 to 2 or even 3 generations each year. Nine species of Turbellaria were classified as thermophilous, some showing increased reproduction (Wroblewsteigo 1977: Biesiadka *et al.* 1978). Kititsina (1973) also reported that fresh-water bryozoans developed rapidly in cooling ponds causing fouling problems. In the early studies on heated docks. Naylor (1965a) found that the two ascidians, *Ascidiella aspersa* and *Ciona intestinals* reproduced twice each year instead of once as in colder waters. There was also some temporal separation of the species in the warmest waters such that *Ascidiella* was most abundant during winter and spring while *Ciona* was most abundant in summer and autumn. The maximum size at maturity of *Ascidiella* was less than in normal populations. Naylor suggested that differing temperature tolerances were responsible for the seasonal separation but it is also possible that intermittent chlorination (see p. 261) in summer inhibited the settlement of larval *Ascidiella*. The sea anemone, *Arthropleura elegantissima* spawned up to 2 months earlier than usual in the outfall area of the Morro bay power station in California. Transplanted individuals acclimatised rapidly and also showed advanced spawning (Jennison 1978).

In simulation studies using effluent water from a California power station (Ford *et al.* 1978), two species of echinoid echinoderms, *Strongylocentrus purpuratus* and *S. franciscanus* showed significantly lower survival rates in heated tanks than in the controls. After 14 weeks some 20–60% survived, compared with 95–96% in the controls. In the intermediate temperature ranges no effects were noted. In contrast, the survival of two asteroid species, *Piscester ochraceus* and *P. giganteus* was unaffected in the heated tanks.

All four species grew more slowly in the warmest tanks and the mean size of *S. franciscanus* actually decreased over the experimental period as a result of shrinkage by some individuals. Larger individuals lost their spines and could not erect the remainder. Similar phenomena were reported in the effluent plume of the station. The mean weights of *P. ochraceus* also declined in the warmest water.

(iii) Behaviour and thermoregulation
There is ample experimental evidence that temperature and chlorine can stimulate avoidance responses in both fresh-water and marine invertebrates (e.g. Capuzzo and Reynolds 1980; Crawshaw *et al.* 1980; Hall 1980; Hillman 1980; Williams and Strawn 1980).

Larvae of Odonata acclimatised to heated waters showed a distinct increase in their preferred temperature in experimental situations (Gentry *et al.* 1975). However, there was evidence here that a genetic adaptation was occurring rather than a simple behavioural response. In other experiments the nymphs of the mayfly *Isonychia* nr. *sadleri* took longer to orientate to currents after exposure to increased temperatures than control animals (Sherberger *et al.* 1977).

Sudden or regular increases in water temperature can be associated with increases in the abundance of drifting animals in natural streams (Muller 1963; Pearson and Franklin 1968; Wojtalik and Waters 1970). Clearly there could be increases in drift caused by the avoidance behaviour of invertebrates if conditions become adverse. Also, epibenthic organisms narcotised by heat or chlorination would, predictably, lose their natural rheotropic behaviour and drift.

Among marine invertebrates, the lobster *Homarus americanus* shows fairly clear temperature preferences and increased walking rates as temperatures rise (Reynolds and Casterlin 1979; Capuzzo and Reynolds 1980).

The records of behavioural responses to thermal discharges *in situ* for invertebrates are rare (see Hocutt *et al.* 1980). One of the most interesting is that of the tube-building amphipod *Corophium insidiosum* which was observed in free-swimming swarms in the discharge canal at the Narrangassett Bay power station (Gonzalez and Yevich 1976). The animals had clearly left their tubes to avoid adverse water conditions and experiments showed that this occurred when water temperatures reached

27–28°C though adults were still swimming actively at 34°C. This is perhaps the clearest record of a mass behavioural response in the field. At the P. H. Robinson power station on Galveston Bay, Texas, there was evidence that blue crabs, *Callinectes sapidus*, and shrimps, *Penaeus aztecus* and *P. setiferus*, avoided the highest temperatures in the discharge canal (see Williams and Strawn 1980).

At temperatures below 35°C, the crabs were abundant in the effluent, but at about 38°C they emigrated, though a few individuals would enter the plume at night for short periods to feed. Salinity and biocide concentrations affect the temperature tolerance of this species to a great extent (see Hall 1980).

The shrimp populations in the effluent declined as temperatures rose above 18°C but increases occurred at 35°C though the reason for this is not known. There was an accumulation of shrimps in a cool bypass canal when the thermal discharge temperatures rose above 35°C, which was attributed to avoidance behaviour. Similar phenomena were reported for the two shrimp species at the Cedar Bayou power station in Texas (see Williams and Strawn 1980). However, it was concluded that current velocities were the major factors inhibiting the colonisation of the Cedar Bayou effluent by shrimps rather than water temperature, especially when the power station was operating at full output. At the Guyanilla Bay power station (see p. 250), the crab *Pachygrapsus transversus* avoided the worst conditions by climbing out of the water on the mangrove roots (Kolehmainen *et al.* 1975).

(iv) The occurrence of exotic species
The occurrence of exotic invertebrates such as the Asian worm, *Branchiura sowerbyi*, the clam, *Corbicula fluminea* and the mediterranean snail *Physa acuta*, in the vicinity of thermal discharges has already been noted (see pp. 242 and 244). *Branchiura sowerbyi* is common in warm water ponds in many European botanic gardens (Aston 1968a) and was recorded in the thermal discharges from Earley and Coventry power stations in the UK in the 1950s and 1960s (Mann 1965; Aston 1968a). Aston also found the species in two unheated locations. The occurrence and success of the species in heated waters is reviewed by Brinkhurst and Cook (1974).

The records of *B. sowerbyi* from the USA suggest that the species is also present in unheated waters though this is not always indicated. In Lake Sangchris, the species comprised about 0·5% of the benthos by weight (Larimore *et al.* 1979b).

B. sowerbyi has a very restricted distribution even in regions with a large number of thermal discharges. For example, of the 20 power station effluents sampled in Britain (Langford 1971b) the species was recorded at only two sites. Clearly the species does not spread rapidly. In contrast, *Physa acuta* was found throughout the rivers of the English Midlands and in the cooling tower ponds of 12 power stations (Langford 1971b, 1974; Milner 1984). *P. acuta* has only been recorded by the author from rivers, canals or ponds which have received heated water although the species has been reported from a canal which apparently did not receive such effluents.

The replacement of indigenous species by exotic species in heated fresh waters is not widely reported, but it was clear that in the heated reaches of the Rivers Trent and Witham, the indigenous snail *Physa fontinalis* was rarely found where *P. acuta* was common and abundant. *P. acuta* also appeared in the cooling tower ponds at the new Ironbridge 'B' power station on the River Severn within 12 months of its opening (Langford 1983a). The species was not previously recorded from the river over a 5 year study, even though the river was heated by an older power station. Ironbridge was a thriving industrial community in the 18th and 19th centuries and the species may have been surviving in disused engine ponds. The site is also some 30 km from the River Trent where *P. acuta* flourished. The high thermal tolerance of the Physidae is also indicated by the studies of Coutant (1962) and Dahlberg and Conyers (1974) who found *Physa* spp. to be among the last to be eliminated from thermal discharge areas.

The disappearance of molluscs from the thermal discharge areas at the Gentilly power station was reportedly caused by higher temperatures (Vaillancourt and Couture 1975) but as the maxima reached only 29°C and the species recorded have been found at higher temperatures, other factors may have been more influential.

The largest number of immigrant warm-water marine species in the UK were recorded during the 1950s and 1960s in the heated waters of docks at Swansea and Shoreham (Crisp and Molesworth 1951; Naylor 1965b; Raymont 1976) and few new records have been made since then (Bamber 1987, 1990; Barnett and Hardy 1984). The heated docks clearly provided the optimum criteria for the survival of exotic immigrants, namely regular shipping and higher water temperatures. In the Queen's Dock at Swansea and in Shoreham harbour the exotic barnacle *Balanus amphitrite* var. *denticulata* and the wood-boring isopod *Limnoria tripunctata* replaced the indigenous species. Raymont (1976) also considered that the damaging outbreaks of *Teredo navalis* and *Lyrodus* spp. were both stimulated directly by the heated effluents, though in more open waters such problems have not arisen (Coughlan 1977a,b).

L. tripunctata has a higher breeding threshold temperature (14°C) compared with the more widespread species *L. lignorum* and *L. quadripunctata* (10°C), though all three species overlap in their ranges (see

Naylor 1965b). In a power station effluent in New Jersey two species of sub-tropical *Teredo* (*T. bartschi, T. furcifera*), not reported previously, developed breeding populations alongside the native species (*Bankia gouldi* and *T. navalis*) (Masnik 1979).

Several species of warm water immigrants have colonised natural temperate waters including the ascidian *Styela clavata* and the copepod *Acartia tonsa*, though Raymont (1976) suggested that the species were initially favoured by heated effluents. Bamber (1990) considered that the only two 'exotic' crustacean species really associated with thermal discharges in the UK were the crab *Brachynotus sexdentatus* and the barnacle *Balanus amphitrite*.

One of the classic cases of a successful warm water immigrant adapting to temperate waters is that of the American hard-shell clam *Mercenaria mercenaria* in Southampton Water (Ansell 1963; Ansell *et al.* 1964; Mitchell 1974). The clam was supposedly introduced in the late 19th or early 20th century and thrived so well that it became the basis of a successful commercial fishery. The thermal discharge from the Marchwood power station clearly enhanced the reproduction and survival of the species. Densities of up to 160 clams m^{-2} were recorded near the outfall whereas in other countries 3 clams m^{-2} is regarded as a reasonable density. In the thermal discharges at Marchwood and at Poole power stations the growing seasons were 173 and 204 days respectively instead of about 123 days in unheated waters (Ansell 1963; Ansell *et al.* 1964; see Raymont 1976). The percentage daily increments for young clams were up to 54% in warm waters and 14% in colder waters. In Southampton Water the clams spread over a wide area away from the thermal discharge and were not influenced directly by the warm water, though recruitment was clearly enhanced by larvae carried from the outfall area. In its natural range *M. mercenaria* spawns at temperatures over 24°C but the threshold in the UK is around 18°C.

The closure of the Marchwood power station in 1983 has apparently been followed by a decline in the clam fishery, probably as a result of poor recruitment and overfishing. The clams have not spread to the outfall area of the Fawley power station some 10 km distant mainly because the hard sand and shell debris in this area do not form a suitable substrate.

The closure of several coastal power stations has also led to the decline of other warm water immigrants (Naylor 1965a; Raymont 1976; Bullimore *et al.* 1978), though of the immigrant species in the heated dock at Swansea (see p. 269) only the polyzoan *Bugula neritina* disappeared completely when the thermal discharge ceased.

CHAPTER 9

Vertebrates

9.1 FISH

9.1.1 Tolerance and Distribution

Most of the data discussed here concern fish. Relevant studies of other aquatic poikilotherms are mostly of turtles and alligators, in heated waters in North America. The studies of aquatic mammals are scarce (Colson 1974; Reynolds and Wilcox 1985).

Of the major taxonomic groups found in water, fish are generally the least tolerant of high water temperatures (Brock 1975) (see Table 4.1). The maximum natural temperature at which fish populations are recorded is 37°C in the hot springs at Hammam Meskoutine in Algeria (see Brock 1975). There are, however, as we will see, records of fish making excursions of short duration into water heated to 40°C. The most eurythermal species are those of temperate fresh waters and intertidal pools, where annual temperature ranges can exceed 30°C. In intertidal pools even daily ranges can exceed 20°C (see Chapter 2).

Brock (1975) concluded that 'no fish should ever be able to maintain a population' in an ecosystem heated above 38°C. He stated that 'Even where this temperature is exceeded for only a few days or weeks this should be enough to eliminate the population'. This may well apply where the whole water body is at that temperature but the effects of heating only a portion of a water body on animals that are both highly mobile and sensitive to adverse conditions are much less predictable. Not all naturally thermal waters contain fish. The small number of species which are present belong mainly to the family Cyprinidae (Brock 1975). Surprisingly the hot springs of Yellowstone do not contain thermophilous species though the genera *Cyprinodon* and *Crenichtus* are found in the Californian hot springs on the other side of the Rocky Mountains. Mason (1939) found that *Barbus* sp.

could survive for long periods in pools which reached 37°C and some individuals would migrate temporarily to pools at 38°C but none were ever found at higher temperatures. Experiments clearly distinguished two or three 'thermal' populations of *Barbus*, based on their lethal temperatures. Studies in the naturally heated reaches of the Yellowstone River have indicated that genetic adaptations have occurred among the resident trout (*S. trutta*). As we will see, genetic adaptations can also occur in artificially heated environments (see p. 314).

There have been many experimental studies on the temperature relationships of fish and many of these are reviewed by Fry (1967), De Sylva (1969), Raney and Menzel (1969), Brett (1970), see Garside (1970), Alabaster and Lloyd (1980), Stauffer (1980), Menasveta (1981), Talmage and Opresko

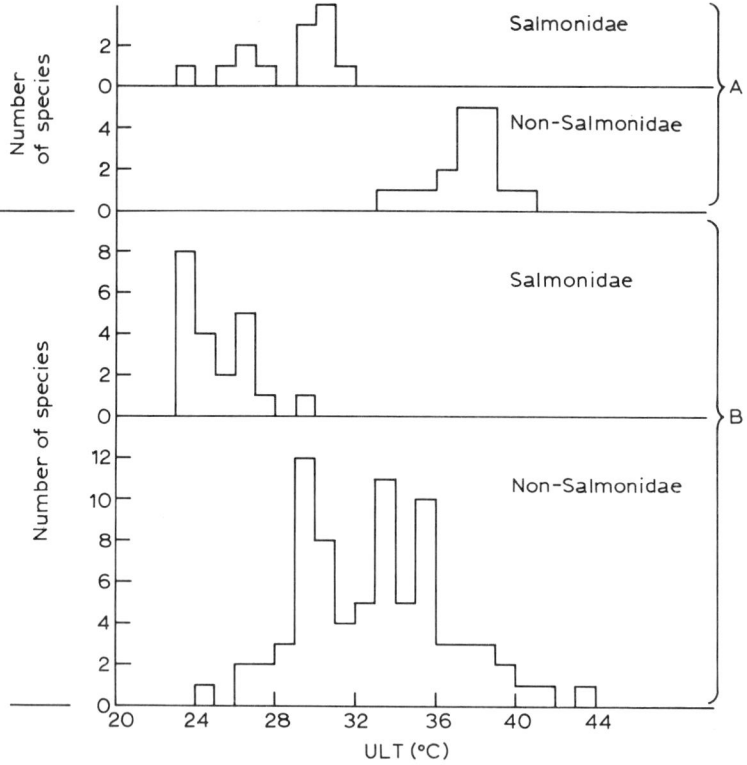

FIG. 9.1   Frequency distributions of upper lethal temperatures (ULTs) for fish. A, after Alabaster and Lloyd (1980) (mostly European species); B, After Houston (1982) (mostly N. American species).

1981), Houston (1982), Langford (1983a) among others. The ultimate lethal temperatures (ULT) for various major groups are between 28 and 33°C (Fig. 9.1), though clearly some species can survive short exposures of up to 40°C (see p. 290). Among the most tolerant species outside naturally thermal waters are the killifish (*Fundulus heteroclitus*), the common carp (*Cyprinus carpio*), the goldfish (*Carassius auratus*) and the fathead minnow (*Pimephales promelas*). De Sylva (1969) noted that research workers have a propensity for testing the tolerance of such species 'all of which appear to be virtually immortal'. Later work has included species with wide ranges of tolerance. The temperature tolerances of many species can be found in the various publications listed above. As we have seen the tolerance of any individual can be markedly affected by many internal and external factors (Chapter 4). The tolerance of fish to contaminants such as heavy metals and chlorine and radioactivity are well known (e.g. Klein 1962; Jones 1964; see EPA 1976; Thurston *et al.* 1979; Alabaster and Lloyd 1980.

The toxicity of chlorine and its residuals has been well studied (e.g. EPA, 1976; Schubel and Marcy 1978; see Thurston *et al.* 1979; Morgan 1980; Houston 1982). In field studies of chlorinated sewage effluents no fish were present where chlorine residuals exceeded 0.37 mg litre^{-1} and the species-diversity index was zero at 0.25 mg litre^{-1} (Tsai 1975). At concentrations of around 0.10 mg litre^{-1} the diversity index was less than half the value of that in the control area. Near chlorinated sewage effluents, residuals of 1.0 mg litre^{-1} caused fish mortalities (Bellanca and Bailey 1977) and, as we will see later, p. 283, chlorine is implicated in mortalities near thermal discharges.

9.1.2 Methodology

As with the other groups of organisms, the methods of field sampling, data analysis and presentation generally follow the standard procedures used by biologists over many years (see Southwood 1966; Ricker 1975). Methods have been adapted to suit unusual habitats such as intakes and outfall culverts or channels as we will see.

Small fish and the juvenile stages of larger fish, like other organisms in the water column, are susceptible to entrainment and transit through cooling-water systems. The size of fish passing through any system will depend upon the smallest aperture in the screens. For example, Turnpenny (1981) showed that the mesh size on the drum-screens of a power station with apertures of 9 mm square, would allow most fish under 40 mm long to pass through, though this also depended on the shape and the projections on the body of the fish. The size of fish captured on screens also depends on the

Ecological Effects of Thermal Discharges

effects of temperature on escape swimming speeds (Turnpenny 1988)
(Fig. 9.2). Almost all the studies of entrained fish larvae and juveniles
(ichthyoplankton) have been made at direct-cooled power stations (e.g.
Kerr 1953; Knutson *et al.* 1976; Kelso and Leslie 1979). In many cases nets

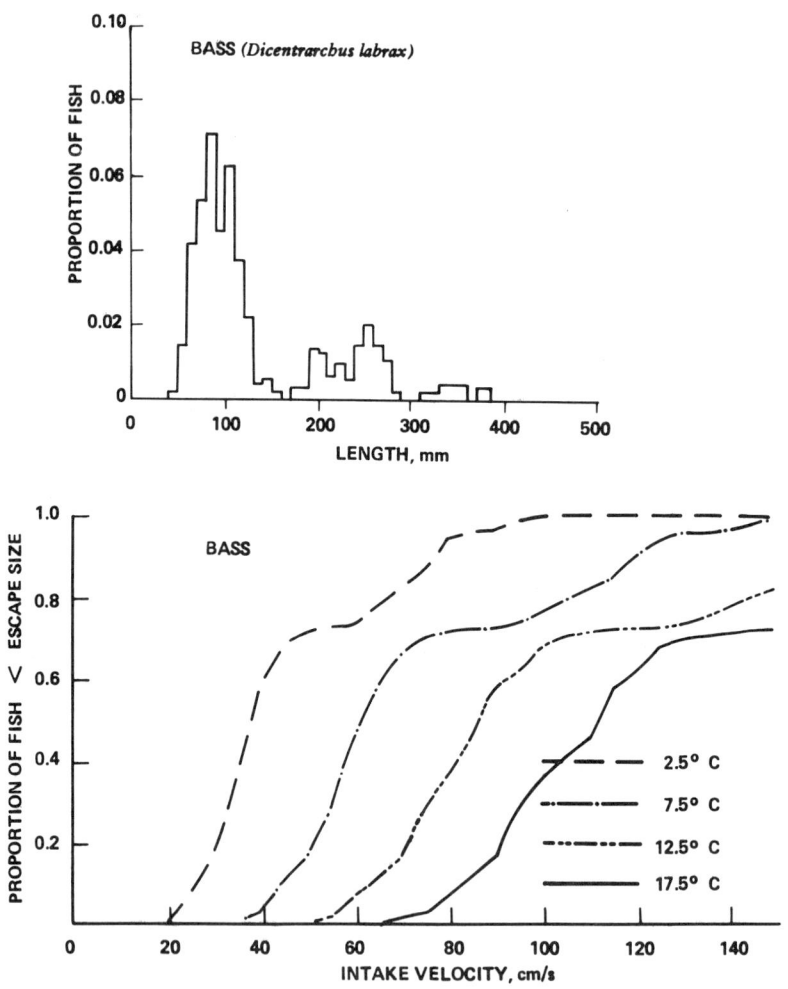

FIG. 9.2   Curves depicting the proportion of fish too small to escape from water
intakes in relation to water velocity at different water temperatures. The upper
histogram shows the fish length distribution on which the curves have been
calculated (from Turnpenny 1988).

of some description were used to sample at intakes and outfalls though the net mortalities were high at both locations in many studies. In the most recent studies, specially designed pumps or catching systems were used to try to minimise sampling mortalities (e.g. Leitheiser *et al.* 1978; Ney and Schumacher 1978). Bowles and Merriner (1978) have reviewed the methodology.

In receiving waters, the methods have included most types of fixed nets or traps and classical methods such as gill-netting, trawling or seining.

Experimental systems using large-scale simulations of condenser tubes have been used for some predictive studies though the data are of limited application to real cooling water systems (Kedl and Coutant 1976).

9.1.3 Fish within Cooling-Water Systems
(i) Direct cooling systems
(a) Fresh waters. Using a diaphragm pump, Jude (1976) showed a high day to day variability in the number of fish eggs and larvae entrained at the Donald C. Cook power station on Lake Michigan. The assumption was that the eggs and larvae would all be killed in passing through the system.

At Fort Calhoun and the Cooper Nuclear power stations on the Missouri River (King 1978; Hergenrader *et al.* 1982), net sampling in the high current velocities of the intake caused 80–100% sampling mortalities in the early studies. After modifications and corrections for sampling mortalities, true entrainment mortalities were assessed at 47–100% for all species. Both thermal and mechanical factors were involved. Higher survival occurred when discharge temperatures were at 36·0°C than at 29·0°C (Table 9.1). Larger larvae of the freshwater drum (*Aplidonotus grunniens*) survived better than smaller larvae. Suckers (catastomatids) showed high survival rates as larvae, usually over 50%. Using an average mortality rate of 14·3% it was estimated that about 3·2% of the ichthyoplankton which drifted past the Fort Calhoun power station each day was killed. The natural mortality over the larval period was estimated as about 40% in the reach as a whole.

Wrenn (1976b) could not estimate the effect of the Colbert power station on entrained larvae as most of the species studied spawned in the discharge canal where the post-entrainment samples were taken. Teleki (1976), on the other hand, estimated that an entrainment rate of 2·9% of the local larval population each hour near the Nanticoke power station on Lake Erie, the total larval stock of the 4·3 km² study area could be entrained in 35 h. At mortality rates of 96–99% this represented 43 t of a single year class lost to the fishery or about 12% of the annual catch of a local fisherman.

TABLE 9.1

Survival rates of larval fish passing through the condenser cooling system at Fort Calhoun Station on different dates (from Hergenrader *et al.* 1982)

Date	Sample size (No.)		Observed intake mortality (%)	Adjusted intake mortality (%)	Survival rate (%)	Discharge temperature (°C)	ΔT (°C)
	Intake	Discharge					
2 July 1975	594	195	90·7	22·6	9·3	36·0	9·5
9 July 1975	284	27	95·8	40·5	31·1	36·0	9·4
21 June 1976	364	355	91·8	70·3	9·5	27·8	7·7
28 June 1976	284	138	96·8	73·2	10·8	30·0	8·5
6 July 1976	364	134	97·5	70·5	15·2	31·1	8·9
1 June 1977	98	24	53·1	29·6	17·8	32·4	11·0
6 June 1977	312	60	63·8	34·9	7·7	34·0	10·7
15 June 1977	1 264	159	77·3	31·2	2·7	32·2	10·9
23 June 1977	265	26	47·5	19·6	52·6	33·4	10·8
30 June 1977	277	31	36·1	17·7	23·5	34·3	9·9
20 June 1978	55	124	96·4	69·1	15·7	28·5	6·0
27 June 1978	72	65	95·8	23·6	0·0	29·0	5·0
5 July 1978	412	850	78·9	19·9	1·0	34·0	7·0
12 July 1978	412	68	69·4	23·3	3·8	32·0	9·0

In this study some 49·5% of the larval mortality was caused by mechanical shock and 34·5% by heat. The mortality rate of smelt (*Osmerus mordax*) was inversely related to size ($r = -0·943$) at a time when no heat was applied in the condensers. At the Palisade power station over 50% of the mortality was caused by heat alone (Edsall and Yocum 1972). King (1978) also showed a clear temperature effect on fish eggs and fry entrained at the Wabash River power station. The threshold temperature was approximately 31°C. Some 3–3·5% of the drifting larvae were killed in June and July. Gammon (1976) later adjusted these values to 7·5 and 8·9% though he used densities at intake and outfall for comparison without investigating the innate patchiness and stratification of the organisms (see p. 166).

Hadderingh (1978) concluded that up to 5% of the larvae of coarse fishes could be entrained from Bergum Lake into the local power station but survival rates here ranged from 90 to 100% (see also van Densen and Hadderingh 1982).

(b) Tidal and saline waters. Marcy (1976a,b), in his extensive and thorough studies, found that 77% of the egg production of the 16 major fish species in the Lower Connecticut River occurred upstream of the Connecticut Yankee Power Station at Haddam Neck. Alewives (*Alosa pseudoharengus*) and blueback herring (*Alosa aestivalis*) comprised over 97% of the eggs and larvae. Only 0·2% occurred in the coves near the station, and thus effects of entrainment upon eggs was predicted to be minimal. Some 61% of the larvae in the reach were, however, within a short distance of the intakes. About 4% of the ichthyoplankton passing through the river reach was entrained by the power station and mortalities were from 92 to 100% between intake and the distant end of the 1·8 km long discharge canal. Exposure to high temperatures was maximised in summer. Marcy (1976b) estimated that about 72–87% of the total mortality was due to mechanical effects, larger fish being most affected. In other studies mortalities at saline water sites have ranged from 27 to 100% (e.g. Marcy and Jacobson 1976; Schubel and Marcy 1978; Marcy *et al.* 1980; Yost and Uziel 1981).

No fish larvae or post-larvae survived 100% after entrainment and subsequent exposure in the Connecticut Yankee discharge canal at temperatures exceeding 28°C. At temperatures lower than 28°C, 34·5% of entrained larvae survived to the outfall before passage through the canal. At outfall temperatures in the 29–33·5°C range, no larvae survived even to the outfall.

In the Hudson River, the mortalities of striped bass (*Morone labrax*) were

5–40% after entrainment and passage through a power station cooling system, and there were no significant differences with and without heat (Lauer *et al.* 1974). Sampling mortalities were, however, high for both this species and the bay anchovy (*Anchoa mitchilli*). Chlorination of up to 0.11 mg litre^{-1} increased the mortalities. Data from later studies (Cannon *et al.* 1978), showed that the mortality rate of the bass larvae was primarily related to discharge temperature. At discharge temperatures exceeding $33°C$ few larvae of any of the major species survived passage through the systems of four power stations. Latent mortalities occurred when larvae were maintained in the discharge water, cooling naturally in containers for 4 days. The mortality rates varied with species. Figure 9.3a shows a clearer relationship between the mortality rate of *Morone americana* and temperature at two Californian power stations. There was a linear increase in mortality from 10 to 100% over a range of $31–38°C$ (Stevens and Finlayson 1978). Below $31°C$ no mortalities occurred which were attributable to heat though about 13% mortality was caused by mechanical damage. The LT_{50} calculated from the data was $33.9°C$ which agreed well with the $31–33°C$ estimated from experiments (see Schubel *et al.* 1978). Analysing the mortality data in relation to δT showed that there was a linear increase from δTs of 8 to $15°C$ (Fig. 9.3).

The comprehensive studies at the Millstone Point site on Long Island Sound showed highly variable mortality rates for 17 species ranging from 0 to 60% depending upon season and species (Nawrocki 1977). All of the mortalities were attributable to mechanical stresses.

In the UK, few fish larvae were found to survive entrainment at an estuarine site (Dempsey 1983). Clupeid larvae were particularly vulnerable though total effects on the population were regarded as minimal.

(*c*) *Significance of entrainment mortalities.* There is little doubt that large numbers of fish eggs and larvae are entrained into and through cooling-water systems and suffer mortalities as a result of the thermal, mechanical and chemical shocks within the systems.

At the same time there are mortalities to juvenile and adult fish caused by impingement on the intake screens (see Van Winkle 1977; Langford 1983a for references), and although these mortalities have not been the subject of this volume, the assessment of the total effect of a cooling-water system on a fish population must consider all these factors in addition to the effects of discharge mortalities.

The cropping rates caused by entrainment of larvae and eggs ranges from about 2.5 to 20% of the localised or drifting stock in most waters for which

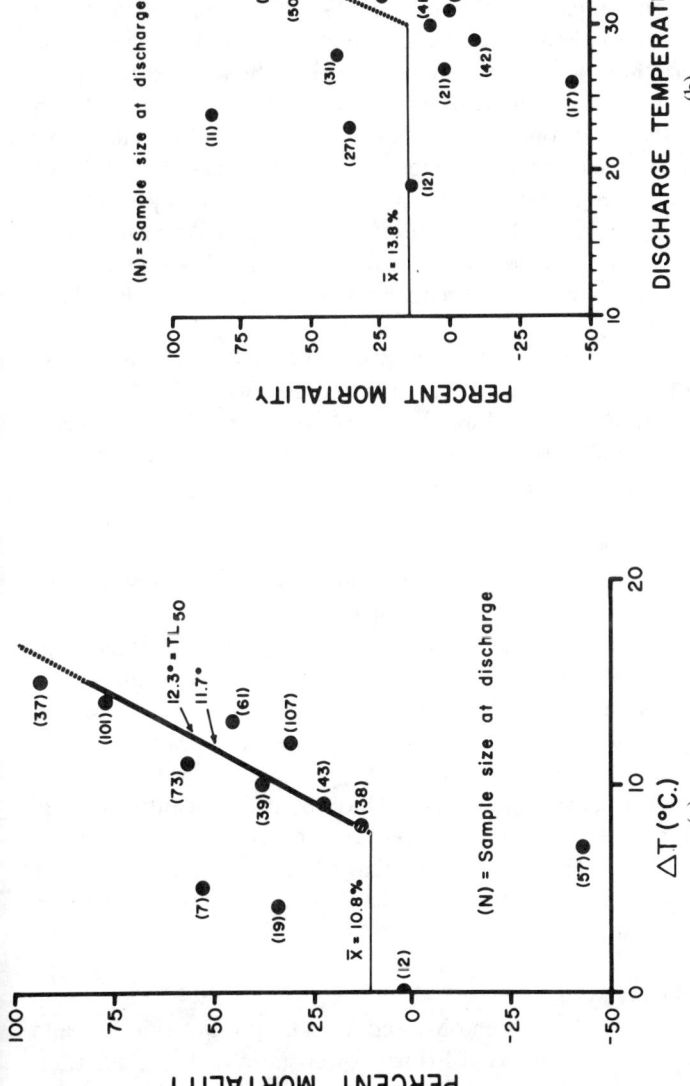

FIG. 9.3   The mortality rates of young striped bass in relation to ΔT and discharge temperature after entrainment through a cooling water system (from Icanberry and Adams 1974). (a) Regression line for estimating TL_{50} temperature was fitted to points between 8 and 15°C. Estimate of mortality owing to factors other than heat (10·8%) is an arithmetic mean of mortality estimates at ΔTs between 0 and 7°C. (b) Regression line was fitted to points between 32 and 37°C. Estimate of mortality owing to factors other than heat (13·8%) is an arithmetic mean of mortality estimates at temperatures between 19 and 31°C.

such values have been estimated. In most cases the implications for the adult fish or catchable fish populations are not estimated though this has been the subject of many mathematical models (see Van Winkle 1977; Yost and Uziel 1981; Mitsch *et al.* 1981; Langford 1983a for references).

In general the total effects of impingement and entrainment at any one site are low when compared with the cropping rates of commercial fishing on adult stocks (Turnpenny *et al.* 1983). Perhaps the highest cropping rates for larvae were those quoted by Navrocki (1977) who showed that up to 20% of the local stock of larvae of winter flounders (*P. americanus*) were entrained by the Millstone power station. However, using catch and entrainment data, the calculated annual mortality to the adult population of this species in Long Island Sound was 0·01 and this was predicted to cause a reduction of 9% in the population over 35 years. At a rate of 0·12 the population could decline to zero over a very long period of time, provided that no compensation mechanisms or immigration occurred (see Van Winkle 1977).

Using the catches recorded at Sizewell power station on the Suffolk coast, it was shown that the cropping effect of the cooling water intakes on the North Sea populations of several species was equivalent to a single, inefficient trawler or about 1000–100 000 times less than the normal commercial fishing operations in the area. No data were available for larval or egg entrainment but the area is not known as a high density spawning location (Turnpenny *et al.* 1983).

A long-term study of the sand-smelt (*Atherina boyeri*), a small inshore fish particularly susceptible to entrainment and impingement at the Fawley power station in southern England, has shown that, although large numbers of fish of all age groups are killed each year by impingement there has been no measurable effect on the population structure over more than 10 years (Henderson *et al.* 1984a).

It seems clear, therefore, that there are no cases as yet where the population of a fish species can be said to have been significantly depleted by cooling-water systems, either by impingement, entrainment or thermal discharge mortalities. This is not to say that there are no locations where such effects might occur. Very few studies have been carried out in restricted nursery areas of estuaries.

(ii) Indirect cooling systems
Various species of fish have been observed in cooling-water channels and ponds in cooling-tower systems of British power stations. The predictions that all organisms entrained into such systems would be killed (Schubel *et*

al. 1978; Majewski and Miller 1979) is clearly not valid (see p. 206). Species recorded as abundant and surviving for long periods in cooling-tower ponds are listed and include roach (*Rutilus rutilus*), carp (*Cyprinus carpio*), eels (*Anguilla anguilla*), chub (*Leuciscus leuciscus*) and barbel (*Barbus barbus*) (see Markowski 1960; 1962; Langford 1974; Milner 1984; Milner personal communication). In several of these cooling-water systems the fish had survived temperatures exceeding 30°C, regular chlorination with residuals exceeding 0·8 mg litre^{-1} (see p. 88), acid treatments to reduce chemical scaling (see p. 231), other water treatments and concentration factors of up to 1·5 times. At the Spondon site it was remarkable that many coarse fish were found on draining the tower ponds and channels following the legal judgement that the power station cooling water was at least partly responsible for the destruction of the fishery in that reach of the Derwent (All England Law Reports 1952).

9.1.4 Effects in Receiving Waters
(i) Thermal discharge mortalities.
Surprisingly, in view of the vast amounts of literature dealing with thermal discharges, very few large-scale mortalities have occurred which can be unequivocally related to high temperature in any effluent.

In natural waters fish mortalities occur as a result of exteremly high and extremely low temperatures and the data are reviewed by Brett (1970) and Barnett and Hardy (1984). In lentic fresh waters, particularly shallow ponds, the phenomena known as 'summer kill' and 'winter kill' are related to the seasonal temperatures but the cause of the mortalities is usually low oxygen concentrations rather than temperature (Barica 1975). Mortalities of salmonids occur even in temperate rivers during low flows and hot weather (e.g. Brooker *et al.* 1977; Edwards and Brooker 1982).

Records of mortalities caused by high temperatures in the sea are rare (Brett 1970; De Sylva 1969) but Brett (1970) records nine references of fish mortalities caused by extreme cold in various latitudes. Most mortalities occurred in shallow waters though Woodhead (1964) reported mortalities among trawled fish around the British coast when water temperatures fell to 0·6°C.

In sub-tropical waters, fish mortalities occur fairly frequently during unusual cold spells (Bohnsack 1983). Schwartz (1964) suggested that natural mortalities may be more frequent than is reported in oceanic habitats but because of their remoteness they are not observed.

In the UK, fish mortalities in thermal discharges have been very small. Even in the classic 'Pride of Derby' legal action (see p. 9), in which a thermal discharge was implicated in the destruction of a river fishery, no large-scale mortalities were recorded. The small mortalities which were recorded (see Langford 1983a) were associated with low oxygen concentrations rather than water temperatures of 30°C. Small numbers of cyprinids, usually less than 100, have been reported killed in extreme conditions in two rivers at temperatures exceeding 33°C (Alabaster 1962; see Langford 1983a). No mortalities were observed in either of the thermal discharge canals at Peterborough or Kingsnorth power stations, the former fresh water, the latter estuarine, where water temperatures exceeded 30°C regularly each summer (Cragg-Hine 1971; Langford 1983b, 1987). There was evidence of summer emigration from both of these channels but this did not appear to be related simply to temperature.

In the USA, larger-scale mortalities have been reported from the vicinity of power station cooling-water outfalls, though the phenomena are not common. Of over 400 stations operating with direct cooling in 1968, only four were associated with fish mortalities. Of these one mortality was probably caused by a release of boiler-cleaning fluid into the discharge canal, the other three were related to water temperature (USAEC 1971). The number of fish killed by all forms of effluent in 42 states was estimated as 15 236 000 of which less than 0·1% were probably killed by high water temperatures near power stations.

Since 1968 there have been other mortalities attributed to high temperatures in thermal discharges (Talmage and Opresko 1981; Langford 1983a; Barnett and Hardy 1984). Of sixty species on which observations were reviewed by Talmage and Opresko (1981), mortalities of only three species were associated with thermal discharges. Gulf menhaden (*Brevoortia patronus*) and sea catfish (*Arius felis*) died when water temperatures reached 39°C in the effluent canal at the P. H. Robinson power station on Galveston Bay (Gallaway and Strawn 1974), and caged American shad (*Alosa sapidissima*) died when drifted down the effluent canal at the Connecticut Yankee nuclear power station as part of a research project. Water temperatures exceeded 32°C and the fish died in 4–6 min (Marcy 1976a).

The temperature rise associated with the mortalities of menhaden (*Alosa* sp.) at Northport nuclear power station, was 15°C above ambient and the ultimate temperatures of 37–38°C were predictably lethal. At a Massachusetts power station, chlorine was believed to be the lethal factor rather than heat. In Par Pond the mortalities of blueback herring and bluegills were caused by the fish being unable to escape lethal conditions after swimming into the discharge to feed or being chased in by predators (see p. 290).

At several sites, a rapid fall in water temperature was the cause of fish mortalities. At the Oyster Creek site, for example, a shutdown at the nuclear power station caused the temperature to fall by 7°C over 24 h, reportedly killing the resident fish which had acclimatised to the thermal discharge (USAEC 1971; IAEA 1972). Further analysis of this mortality suggests, however, that temperature was not the only cause of death. A fall of 7°C over 24 h is not very rapid and should have been tolerable even to the acclimatised fish. Also, power station staff reported that the station has been shut down for maintenance five times in previous winters but no fish were killed. Chlorine or some other toxic substance was the most likely cause.

Young (1974) reviewed other cold-shock mortalities. Clearly the Atlantic menhaden *Brevoortia tyrannus* is susceptible to such occurrences though the species is known to suffer huge natural mortalities after spawning (see Reintjes and Pached 1966; Langford 1983a). At the Lake Wabamun power station in Canada, where the water temperature in the discharge canal fell from 21·8°C to 4·9°C in 30 min, some 158 000 spottail shiners (*Notropis hudsonis*) and 250 pike (*Esox lucius*) died (Houston 1982). Irrespective of these records, Coutant (1977) concluded that cold-shock mortalities associated with thermal discharges were relatively rare. In none of the reported cases of mortalities in thermal discharges has there been evidence that a portion of a population large enough to endanger it was killed. Barber (1872) suggested that most fish kills at power stations are not recorded though, since that time, numerous reports of mortalities at intakes have been recorded (see Langford 1983a) and it seems unlikely that discharge mortalities would not be similarly reported.

Some species of fish are clearly extremely eurythermal. For example, populations of the mosquito fish, *Gambusia affinis* survived well in temperatures of 12–19°C in unheated waters and 28–40°C in heated waters near the Savannah River nuclear plant (Bennett and Goodyear 1978).

(ii) Abundance and large-scale movements in relation to thermal discharges
As we have seen, the physical and chemical effects of a thermal discharge are complex and provide several kinds of stimuli for organisms in the receiving water. The research on fish around thermal discharges reflects these complex multiple stimuli but there are basic patterns of abundance and movements which have been identified and associated with the effluents.

Almost every in-situ study shows a similar seasonal fluctuation in the abundance of fish, that is, increased abundance in spring, autumn and winter and decreased abundance in summer. The simple conclusion is that the fluctuations are caused by behavioural thermo-regulation (see Chapter 4), that is active temperature selection. Analysis of the data suggest, however, that other factors may be involved, including:

—the abundance of food organisms at an outfall,
—the effect of temperature on catchability through enhanced activity,
—natural migrations,
—specific differences in temperature preferences through which one species may become excessively abundant at the expense of others,
—the effects of water currents on the natural movements,
—the effects of chemical changes, for example avoidance of chlorine or other contaminants, or the avoidance of low oxygen concentrations or areas of gas supersaturation (see Chapter 4),
—or any combination of these factors.

There appeared to be clear temperature-related mass movements of fish in and out of the thermal discharge canal at the Connecticut Yankee power station on the Connecticut River (Marcy 1976a). There was an increase in abundance during October which correlated with a decline in abundance in the unheated areas. Seasonal peaks occurred in spring and in winter months, when water temperatures were in the 15–24°C range in the canal. In the unheated areas, peak catches were common in July and August. Marcy concluded that each of the 23 most common species left the canal in late spring and early summer and there were few fish left at temperatures over 35°C. However, even a fall in temperature of 1°C allowed fish to re-enter the canal (Fig. 9.4). Five species (Table 9.2) were collected from the canal at temperatures around 40°C though it seems unlikely that individuals would survive for long at such temperatures.

In the River Trent in the UK, it was suggested (Alabaster and Downing 1966; Alabaster 1969) that the increased catches of roach (*Rutilus rutilus*) and gudgeon (*Gobio gobio*) in the heated reaches confirmed a temperature preference of 22–24°C estimated from experimental work. Sadler (1980), in a later study of the same reaches of the river, found that population densities, the number of species and diversity of the community were all, on average, greater in the heated than unheated reaches. There was some correlation between roach densities and water temperature, possibly as a result of the effluent delaying an upstream migration, but the general conclusion was that the seasonal fluctuations in abundance of most species were a result of natural mass movements. Cragg-Hine's (1971) work on the coarse fishes of the River Nene at Peterborough in the UK showed similar

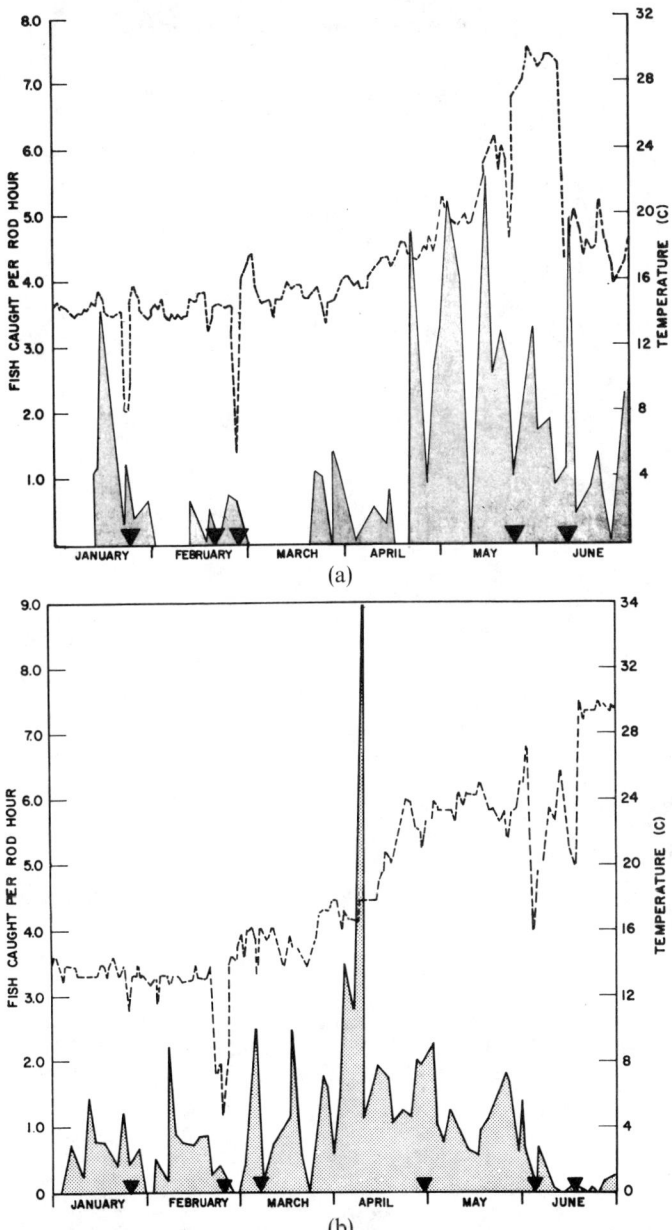

FIG. 9.4 Mean daily water temperatures (dotted line) and fish caught per rod-hour of effort (solid line) in the Connecticut Yankee heated discharge canal: (a) January through June 1972; (b) January through June 1973. Triangles indicate dates of major drops in power output (from Marcy 1976a).

TABLE 9.2

Minimum, maximum, and mean temperatures for fish species (ranked from most to least tolerant) and their seasonal occurrence in the CYAP discharge canal, 1968–1972 (after Marcy 1976a)

Species	Temperature (°C)			Occurrence											
	Minimum	Maximum	Mean	Winter months						Warmer months					
				N	D	J	F	M	A	M	J	J	A	S	O
Pumpkinseed	11·9	40·0	28·4	×	×		×	×		×	×	×	×	×	×
White perch	5·7	40·0	27·5	×	×	×	×	×		×	×	×	×	×	×
Golden shiner	6·7	40·0	24·0	×	×	×	×	×		×	×	×	×	×	×
White catfish	4·5	40·0	20·2	×	×	×	×	×	×	×	×	×	×	×	×
Brown bullhead	5·1	40·0	19·6	×	×	×	×	×	×	×	×	×	×		×
Redbreast sunfish	10·1	39·2	27·0				×						×	×	
Carp	5·1	39·2	21·6	×	×	×	×	×	×	×	×	×	×	×	×
Spottail shiner	5·1	39·2	20·1	×	×	×	×	×	×	×	×	×	×	×	×
Mummichog	14·7	37·0	22·8						×	×	×	×			
Bluegill		35·0	35·0								×				
American eel	11·9	35·0	20·5			×		×	×	×	×				
Hogchoker	32·0	32·5	32·3								×				×
Blueback herring	6·7	32·5	20·7						×	×		×			
Alewife	5·7	31·0	15·8						×	×		×			
Johnny darter	20·1	28·8	24·5												
Largemouth bass	14·8	28·7	21·3	×	×				×	×		×			
Channel catfish	5·1	28·5	15·2	×	×	×	×	×	×			×		×	
Black crappie	16·5	27·3	21·3	×						×	×				
Northern pike	12·9	22·0	16·0		×	×	×	×			×				
American shad		22·0	22·0								×				
Common shiner		20·0	20·0								×				
White sucker	10·0	19·9	14·4	×	×	×	×	×	×						
Rainbow smelt		5·7	5·7						×						

patterns to those in the Trent but the populations were denser being 1·5–6 fish m^{-2} compared with 0·2–0·6 fish m^{-2} found by Sadler (1980).

Brown (1979), in his study of under-yearling fish in the Nene reported aggregations in the thermal discharge from Peterborough power station beginning in September or October and ending in April as the small fish dispersed. The fluctuations in catches at the Montereau power station on the River Seine showed clear seasonal patterns (Leynaud and Allardi 1975) but without control catches it was not possible to assess the reasons.

In most of these studies the increased abundance became evident in autumn and continued through to April. The pattern was in most cases the reverse of that found in the unheated waters where catches declined in autumn and remained low in winter. The actual temperatures in the heated habitats differed from year to year and location to location and it seems evident that the emigration into the thermal discharge areas was at least initiated by natural mass movement rather than temperature *per se*.

British coarse (rough) fishes are not generally regarded as migratory but several studies have shown clear seasonal patterns of natural aggregation and movement among various species (see Langford *et al.* 1979). Where these fishes come into contact with a thermal discharge during their seasonal movements, the higher water temperatures stimulate them to remain active in the water column and thus potentially more catchable by both passive and active fishing methods in the colder seasons. Marcy (1976a), for example, noted that there was a natural overwintering migration of brown bullheads (*Ictalurus nebulosus*) in coves and connecting ponds along the Connecticut River during late autumn and this matched with the aggregation of the species in the thermal discharge canal at the Connecticut Yankee nuclear power station. In cold water the fish normally burrow into sediments or remain inactive on the bottom, but in the warm water they remained active in the water column.

The influence of the natural emigration pattern on the summer population was not clear. There was no apparent threshold temperature effect because fish did not return as temperatures fell below that which ostensibly caused them to leave the canal (Fig. 9.4). Surprisingly a sharp fall in water temperature during spring caused a decline in catches in the canal. The cause was thought to be the sudden cold shock which induced lethargy in the fish. The water currents then carried the torpid fish out of the effluent canal.

Fish catches in thermal discharges reflected temperature preferences at other sites (White *et al.* 1977; Wilkonska and Zuromska 1977; see Stauffer and Edinger 1980) though the data were variable and equivocal. At the J.

M. Stuart power station, for example (Yoder and Gammon 1976), seasonal ranges of temperature preferences were defined from the temperatures at which species were caught but most species were, in any case, most commonly caught in the effluent canal and the highest catches occurred in winter. At least two species, the skipjack herring (*Alosa chrysocloris*), and the largemouth bass (*Micropterus salmoides*) followed their preyfishes into *their* preferred habitat.

In the New River, Virginia (Stauffer *et al.* 1974, 1975a,b,c; see Stauffer 1980), the abundance of fish in heated reaches generally reflected preference and avoidance temperatures estimated from experiments, though the field data again showed wide variability. The distribution and abundance of the spottail shiner, (*Notropis spilopterus*), were clearly not affected by water temperature even up to 35°C. Here again, predatory behaviour may have been more significant than temperature preferences in determining the distribution of species. High temperatures clearly excluded some of the 64 species of fish in the heated streams around the Savannah River plant but impoundment also eliminated species. As streams cooled fish reinvaded (McFarlane 1976).

In lakes and reservoirs receiving thermal discharges, similar patterns of community structure, abundance and distribution occur as in rivers. Langford and Sherwood (1970) showed a complementary pattern of catches in fixed gill-nets for brown trout (*Salmo trutta*) and perch (*Perca fluviatilis*) on the heated and unheated waters of Trawsfynydd Lake in Wales. Catches of perch were higher than those of trout in the heated lagoon and the reverse was true for the cold waters. More trout were caught in the warm areas in winter than in summer, when water temperatures could exceed 28°C. Trout catches could not be related directly to water temperature but those of perch were significantly so (see also Caneet *et al.* 1981). A similar complementary pattern was shown for salmonids and non-salmonids caught by anglers at Point Beach power station outfall on Lake Michigan (Romberg *et al.* 1974).

In the Missouri River impoundments, different species showed different patterns of seasonal response (Hesse *et al.* 1982a). Carp (*C. carpio*), and gizzard shad (*D. cepedianum*), showed complementary spring and autumn migrations into the heated waters. In North Lake, Texas, none of the 30 species of fish showed a distribution or abundance pattern which could be related to a heated effluent which reached 42·2°C (McNeely and Pearson 1974). Here the discharge was strongly stratified and most species could avoid adverse conditions by moving to deeper waters. There was some evidence that carp (*Cyprinus carpio*) and white crappie (*Pomoxis annularis*) aggregated in the warm water in spring and early summer.

The effects of food availability and water currents have been reported as major factors attracting some lake fishes to thermal discharge outfalls. For example, the aggregation of largemouth bass (*M. salmoides*) was associated strongly with the abundance of gizzard shad (*Dorosoma cepedianum*) in the heated waters of Lake Sangchris (Larimore *et al.* 1979a). In the unheated Lake Shelbyville no such seasonal aggregations were observed.

In contrast, the migrations of the white bass (*Morone chrysops*) into the thermal discharge canal in Lake Sangchris during spring were regarded as a response to the water currents prior to spawning. In lakes with inflow streams the fish normally migrate upstream to spawn but the discharge canal was the only inflow to Lake Sangchris.

In a Tennessee River reservoir, sauger (*Stizostedion canadense*) fed on the threadfin shad (*Dorosoma petenense*) which were so densely packed into the discharge canal of the Johnsonville power station that outboard motors could not operate (Dryer and Benson 1957). The shad normally migrated into shallow waters in winter where many died. In the discharge canal they remained actively feeding throughout the winter. Clearly, water temperature, migrations and food supply were acting together on different species in this habitat. In other locations the factors influencing the distribution and abundance of fish seem equally complex. For example, at the Ginna power station on Lake Ontario, white perch (*Roccus americanus*) congregated near the outfall when the power station was operating but moved away when only cold water was pumped through, indicating that temperature rather than water currents or the presence of entrained food organisms attracted the fish at this site.

Temperature was also regarded as the main attraction to the spottail shiners (*Notropis hudsonis*) in the heated discharge of a South Carolina nuclear power station (Prince and Mengel 1981). However, at the Nine Mile Point station, yellow perch (*Perca flavescens*) did not congregate in the warm effluent but congregated near the outfall when the effluent ceased. The authors (Storr and Schlenker 1974), concluded that temperatures of 25°C were too high for the fish but once the effluent ceased water flow and food supply in the area attracted them. Food organisms, particularly *Gammarus* spp. were 2–3 times more abundant in the plume area compared with unheated waters. Forage fish, notably *Notropis hudsonis and Alosa pseudoharengus* also congregated in the area and these three species comprised 70–100% of the food of the perches.

Here, the perches did not remain during winter as was observed with other species in discharge canals. Fish tagged near the outfalls were found to migrate, as normal, to spawning grounds 32 km from the outfall. Similar dispersal was found for other species. Spigarelli (1975) also found that tagged salmonids migrated up to 480 km from their tagging point which was the thermal discharge canal at the Point Beach power station. This type of temporary residence has been shown for other species in other locations and it seems clear that shoals of any species in a thermal discharge, or perhaps any other aggregation point, may be dynamic in composition with individuals moving in and out continuously (e.g. Neill and Magnusson 1974; Marcy 1976a; Langford *et al.* 1979; Larimore *et al.* 1979a; Talmage and Opresko 1981).

Using acoustic techniques, Spigarelli (1975) showed that the densities of alewives (*A. pseudoharengus*) and smelt (*Osmerus|mordax*), were not affected by thermal discharges in Lake Michigan either during operation or shut-down periods. Aggregations which took place before and during spawning were most likely a response to water currents. Later studies (Kelso and Minns 1975; Minns *et al.* 1978; Spigarelli and Thommes 1979; Spigarelli *et al.* 1982) confirmed that these currents and the topography of the lake bed were the most important influences on fish distribution. This strong rheotropic response was believed to be the cause of fish swimming into lethally high water temperatures at thermal discharges in other parts of the Great Lakes (Effer and Bryce 1975). However, the stimulus of food organisms present in discharges may also cause such behaviour.

A series of observations in Par Pond has shown the effects of food and other factors on the behaviour of fish in a thermal discharge. In this habitat, high densities of plankton, mainly Cladocera, Copepoda and *Chaoborus* occurred at different times of year (Janssen and Giesey 1984). Very few zooplankton survived in the effluent pond at temperatures exceeding 36°C. Blueback herring (*Alosa aestivalis*) were most abundant on days when zooplankton densities were high and were found to be feeding almost exclusively on these planktonic animals. Large shoals of *A. aestivalis* remained near the thermal discharge outfall for most of the winter. Largemouth bass (*Micropterus salmoides*) were observed chasing the herring and feeding on them. The bass lived mainly in cooler waters and made short term feeding movements into the effluent or alternatively 'ambushed' the herring as they migrated into the warmer water. Both herring and bluegills (*Lepomis macrochirus*) were sometimes killed by the higher effluent temperatures even though they were free to escape. Schools of these fish were observed swimming into the effluent only to die immediately on entering the plume. The bass fed on the dying and dead herrings and bluegills as they were discharged from the cooling-water pond to the main receiving water. Bass were caught from water at temperatures up to 46°C but their lower body temperatures indicated that they were making excursions of very short duration into the warm water to pick up prey (see also Bennett 1971). Clearly the feeding stimulus in these species was much stronger than the temperature avoidance stimulus.

Studies in estuarine and marine habitats have shown similar patterns of distribution and abundance around thermal discharges (see Allen *et al.* 1970; Grimes and Mountain 1971; Young and Gibson 1973; Hillman *et al.* 1977; see Hocutt *et al.* 1980).

Neuman (1977, 1979a,b,c, 1982), in summarising data from studies around Swedish power stations, classified the preferences of species as 'warm' or 'cold' based on the size of catch at various temperatures over a range from 0 to 30°C. The locations covered a range of salinities from fresh to two-thirds sea water. Most fresh-water species were more abundant at higher temperatures than the marine species. Catches of roach (*R. rutilus*) and perch (*P. fluviatilis*) were higher in the warmer water than in colder water, but ruffe (*Acerina cernua*) showed no pattern. At the more saline sites, only corkwing wrasse (*Crenilabrops melops*), flounders (*Platichthys flesus*) and eels (*Anguilla anguilla*) were common and abundant in heated areas.

Intermittent operation of an estuarine power station gave opportunities for comparing effects of warm and cold outflows in these studies (Neuman 1979c). There was no clear trend. Cold-water outflows were associated with both increased and decreased catches of various species when compared with control areas. A similar pattern occurred in the heated waters. There was strong evidence that water currents reduced the catches of burbot, *Lota lota*. The data suggested that at all sites natural movements and migrations initially brought fish into the sampling sites. The catches in directional traps showed that the movements of perch, *Perca fluviatilis* and roach, *Rutilus rutilus* in one direction were closely correlated with those in the other direction in certain seasons rather than with water conditions (Neuman 1979c). The effects of natural migrations were also apparent from the author's research at the estuarine power station at Kingsnorth in southern England (Langford 1987). Here young bass (*Dicentrarchus labrax*) began to enter the thermal discharge canal in September. Peak catches occurred mainly in winter. A similar pattern of catches occurred on the power station intake screens in unheated waters. There was a pattern of serial migrations in and out of the discharge canal involving both mass movements and individual migrations. Water temperature was clearly not the initial stimulus bringing the fish into the canal though they remained active and feeding once in the warm water.

The decline in numbers in summer was also not related simply to temperature. Bass and other species occurred in the canal over a temperature range of 4–32°C (Langford 1983b). Here as in some freshwater locations, food was present throughout the winter, particularly organisms passing through the power station cooling-water system. Bass and several mullet species (*Chelon labrosus*, *Liza ramada* and *Liza auratus*), are well known from warm water outfalls (Wheeler 1969; Langford 1983a,b). Gallaway and Strawn (1974) also noted 'tremendous numbers of mullet' at the mouth of a thermal discharge in Galveston Bay, Texas.

The abundance of cod (*Gadus morhua*) at the outfall of a German power station (Moller 1978b) declined when the warm water was not discharging, but the effects of food and water currents, in the absence of heat, were not discernible.

Salt-marsh creeks are common nursery areas for inshore fish, but three studies of creeks in Florida showed that small fish were less abundant where thermal discharges entered, particularly in summer (Nugent 1970; Carr and Giesel 1975; Homer 1976). Over 12 months, the density of the fish fauna of heated creeks near the Crystal River power station remained much more constant than in control creeks. Massive increases in biomass during September and November were caused by catches of large mullet. These comprised over 30% of the total annual catch in the control creek and over 60% in the discharge creek. Water temperatures reached 36°C in summer in the discharge.

In other semi-tropical and in tropical waters the data also show fluctuations in abundance in thermal discharges. Around the Turkey Point power station, Nugent (1970) found fewer fish in the heated areas than in the controls. Exclusion temperatures for some of the inshore species were as low as 26–27°C (Roessler *et al.* 1975; Thorhaug 1980). At the Guyanilla bay power station only 28 species of fish were found in the outfall area compared with 103 species in the open waters of the bay, and 254 species in unpolluted bays nearby Kolehmainen *et al.* 1975; Thorhaug 1980).

Chlorination was apparently responsible for the emigration of fish from the thermal discharge plume at the Tanguisson Point station in Hawaii. It is also possible that anti-fouling procedures and contaminants cause avoidance and depletion of the local fish community at other sites but there are no published quantified observations. In a study which did not involve a thermal discharge, it was clearly demonstrated that chlorination of wastewaters was the single most important factor restricting the fish community of streams polluted by sewage (Paller *et al.* 1983). The role of chlorine in thermal discharges and its effects on fish communities has not been established from the field studies so far (see p. 316).

(iii) Movements of individual fish in relation to thermal discharges
The use of 'active' tagging methods, involving radio or ultrasonic transmitters (see Stasko and Pincock 1977), particularly those with thermistors measuring the internal or external temperatures, has helped to clarify the behaviour of fish in relation to thermal discharges. In some studies thermo-luminescent tags which fade at a rate related to temperature exposure have been used to estimate cumulative or maximum residence times in thermal plumes (Spigarelli 1975). Also, passive tags of many types have been used to assess migrations of fish from thermal discharge areas and effects on homing behaviour.

(a) Freshwater species. From passive tagging studies, Storr and Schlenker (1974) concluded that thermal discharges were 'not a strong enough attraction' to deter individual yellow perch (*P. flavescens*), from their normal annual migrations to spawning grounds in Lake Ontario. Fish travelled distances ranging from 1·6 to 59 km from the plume area. The authors also concluded that both yellow and white perch (*Roccus americanus*) were attracted to a thermal discharge under various conditions though as we have seen (p. 289) food was the major attractant.

In similar studies (Romberg *et al.* 1974; Spigarelli 1975) 14% of tagged salmonids were recovered from their capture and release area in the thermal plume from the Point Beach power station, but over 64% of the 253 tag returns showed that the fish had migrated over distances up to 484 km. Several tagged fish were found in thermal discharges other than that into which they were originally released. Thermo-luminescent tags on 22 fish indicated lower than expected residence times in heated water with most fish spending less than 10% of the time at higher than ambient temperatures. Tag returns may also have reflected the favoured locations of the anglers who made the returns.

Most of the active tagging studies have indicated that residence times of fish in thermal plumes were relatively short, and that natural movements and migrations of potadromous species are unlikely to be disrupted in the longer term by the presence of a warm water plume. Kelso (1974, 1976) found that 8 out of 10 brown bullheads (*Ictalurus nebulosus*) tracked, had left a thermal discharge area within 24 h of release and similar responses occurred in other species.

There were indications of altered behaviour patterns when fish were in the plume. They turned more frequently and swam more slowly over the bed than fish in control locations. However, this also applied when only cold water outflows were present and Kelso concluded that water currents initiated the responses. Other authors (e.g. Langford *et al.* 1979; Janssen and Giesey 1984), have suggested that the more erratic behaviour was a result of feeding excursions into the outfall area. The exposure of different individual white perch (*R. americanus*) and white suckers (*Catostomus* sp.) to the highest discharge temperatures ranged from a few minutes to 9 h (Kelso 1976).

Largemouth bass (*M. salmoides*) tracked with radio or ultrasonic tags have generally been observed to be more active and mobile in waters receiving thermal discharges than in unheated waters. In both Lake Sangchris and Par Pond, fish caught and released in the discharge areas had greater annual ranges of movement than those in cold water, mainly because of their immigrations from the plume in the warmest periods of the year. Displaced fish roved over wider areas then those replaced at their point of capture (Clugston 1973; Larimore *et al.* 1979b). In Lake Sangchris the fish were more active in winter in the warm than in the cold waters as might be expected. In both locations, the tracked fish did not select the warmest water in winter but those caught from the warm areas stayed mainly in temperatures 2–4°C higher than those from the cold areas.

The longer distance movements of *M. salmoides* in Par Pond were exceptional and directly related to size (Quinn *et al.* 1978). Fish moved infrequently between warm and cold areas. The tracking of two *M. salmoides* in rivers near thermal discharges (Wrenn 1976a; Moss *et al.* 1978), showed that the fish moved through temperature gradients of 6–7°C within minutes. Selected temperatures were mostly lower than the maximum available though one fish was recorded at 34–35°C on one occasion, the maximum temperature available at the time. Coutant (1975) found that *M. salmoides* tracked around the Bull Run power station on the Tennessee River tended to favour small home territories in the discharge canal, selecting a rock or stump as a location rather than a preferred temperature. The fish left the canal when the power station closed down temporarily even when unheated water was being discharged, but returned when it began operating again. Several fish made short excursions into the main river even when the warm water was discharging. In a natural lake fish fitted with transmitters selected temperatures of around 27°C rather than higher temperatures at the surface (Coutant 1975). In the heated waters of Par Pond, tagged largemouth bass *M. salmoides* retreated to cool waters when the reactor effluents caused temperatures above 50°C, but as the receiving waters cooled the fish reinvaded (Block *et al.* 1984). They generally avoided temperatures above 31°C. Bluegills, *Lepomis macrochirus*, stayed in water at 32–37°C.

Different species obviously respond differently to the discharges. Flathead catfish (*Pylodictis olivaris*) moved in and out of a discharge canal apparently irrespective of temperature while one individual walleye (*Stizostedion vitreum*) was unusual in that it stayed in a thermal discharge canal for over 30 days moving less than 50 m per day over a temperature range of 19·2–27·4°C (Wrenn 1976a; Moss *et al.* 1978).

In British rivers coarse fish, mainly bream (*Abramis brama*) and pike (*Esox lucius*) showed similar short residence times in thermal discharge plumes to those fish in the American studies (Langford 1974; Langford *et al.* 1979). Sonic tracking studies also indicated that the bream congregated in winter shoals away from the discharges, but that the composition of the shoals were dynamic with some individuals migrating up to 8 km from the shoal and then returning within 1–6 days. High river flows displaced fish over distances up to 6 km though this was usually followed by compensatory upstream movement. Natural winter migrations also occur in cyprinids in other waters (e.g. Johnson and Hasler 1977).

There was little evidence of temperature selection in any of the field studies using active tagging methods which could be closely related to temperature preferences demonstrated in the laboratory. For example individual yellow perch (*P. flavescens*) travelled over the whole temperature range available in a Mississippi reservoir including a thermal discharge plume (Ross and Siniff 1982). Very few fish selected the highest temperature of 18°C though the preference temperature was estimated as 24·2°C in tanks and 19–21°C under natural conditions. However, acclimation may be important, as McCauley (1977b) found that after acclimation at 5°C the species selected a range of 12–14°C. The mean temperature selected by fish at all locations was 5·4°C in a range of 0–18°C. In the discharge the mean temperature selected was 6·3°C. There was great individual variation as in other studies, and most fish divided their time between cold ambient water and the discharge plume.

Body temperatures of fish caught in a thermal plume were at times lower than discharge temperatures indicating that the fish had only recently entered the plume and probably had short residence times (Spigarelli *et al.* 1974).

(b) Diadromous fish. Several authors have suggested that the migrations of diadromous fish such as salmonids or shads could be inhibited or deflected by the water currents and higher temperatures created by thermal discharges (Naylor 1965b; Stewart 1968; Hawkes 1969; Nakatani 1969; Bush *et al.* 1974; see Leggett 1976; Tongiorgi *et al.* 1986). There are few

quantitative data. There was no indication, from the catch records, of a dramatic decline in salmon (*S. salar*) or trout (*S. trutta*) populations in the River Severn (UK) between 1932 and 1975, during which time Ironbridge 'A' power station was operating and heating the full width and depth of the river by up to 8°C for at least 2 km downstream of the outfalls (Langford 1970, 1971a, 1974). Temperatures reached 30°C for some distance and exceeded 32°C near the outfalls in the warmest summers. High river flows decreased the temperatures (Langford 1970). The main salmon migrations probably occurred outside the periods of highest temperature though whether the seasonal migrations were ever delayed is not known.

In the Columbia River migrating Chinook Salmon (*Onchorhynchus tschawytscha*) and Steelhead trout (*Salmo gairdneri*), tracked using ultrasonic transmitters, were found to follow a similar upstream route along the river bank opposite from the cooling water outfalls of the Hanford reactors (Becker 1973). This preference was most apparent during the highest water temperatures. It appeared that the fish were deliberately avoiding the outfalls (Coutant 1969b; Becker 1973). However, Nakatani (1969), indicated that the preference for the same side of the river occurred in unheated reaches.

The conclusions from the tracking of 650 fish were that:

—The spawning run was unaffected by either on-shore or mid-river thermal discharges.
—Where fish encountered warmer water their migration was unaffected.
—Where necessary, salmon were able to avoid adverse temperatures and continue their migration.

Clearly in this large river the discharges occupied only a small proportion of the width and depth of the river (see Chapter 2).

In similar studies on American shad (*Alosa sapidissima*) in the Connecticut River tracked fish were found to show little response to the thermal discharge from the Connecticut Yankee nuclear power station (Leggett 1976). The observations showed that for upstream migrants:

—The natural transition from salt to fresh water caused some meandering.
—There was a clear chosen route for migration.
—There were no marked deviations from this route once the power station began operating, though some shad made minor exploratory excursions toward the mouth of the thermal discharge canal.
—Seaward migrants were undisturbed by the discharge.

It seems clear, therefore, that the migration of diadromous fish in rivers will not be upset, provided that:

—The plume stratifies vertically or horizontally with sufficient space for fish to avoid adverse temperatures, or
—where the whole width and depth of the river is heated, the maximum water temperatures do not exceed the lower avoidance temperatures of the species and do not coincide with the main migration period.

(c) Species in saline waters. Nyman (1975) tracked individual eels (*Anguilla anguilla*), trout (*Salmo trutta*) and ide (*Leuciscus idus*) around thermal discharges on the Swedish coast. The indications were that the eels were activated by the warm water as a power station began operating after a period of shutdown. Tagged eels left the shelter of rocks at the edge of a bay and moved into the discharge and remained there as temperatures rose to the maximum of 24°C. Trout entered the warm water in winter but not in summer. Ide swam into water at temperatures up to 24°C but emigrated from the area in autumn irrespective of temperature.

(d) The causes of aggregations in thermal plume. In summary, it is apparent that many authors have observed mass aggregations of fish, around or within thermal discharge plumes in both fresh and saline waters. There is also evidence that the population densities in such areas fluctuate within a fairly predictable pattern which is sometimes the reverse and sometimes the same as that shown by fish in unheated habitats. Individuals show widely varying responses *in situ*, ranging from immigration in response to the warm water to apparent avoidance of adverse conditions. In most studies, individuals spent varying times in the discharge plume and shoals were dynamic in composition.

The explanation of active temperature selection although demonstrable in the laboratory is clearly too simplistic to explain the phenomena recorded in field studies around thermal discharges. All the factors which comprise the total physical and chemical characteristics of any thermal discharge can be significant in influencing the mass movements or individual behaviour patterns in relation to a thermal discharge. As a general rule, the sequence of any movements is probably as follows:

—Natural movements or migrations bring fish into contact with the thermal discharge in autumn.
—The higher temperatures cause the fish to remain active, feeding and

catchable in winter, while fish in cold natural waters have become torpid and do not feed.

—Once the fish are feeding, the food originating from active prey in the warm water together with items passing through the cooling system form a strong attractant to the fish and many remain near the outfall. There are, however, individuals which migrate away and may or may not return.

—As water temperatures rise naturally in spring, the fish in the cold water become active and catchable again, while the fish in the thermal discharge emigrate from the area in their normal seasonal migration pattern. If the temperatures are very high, fish may take avoiding action, though they may make short-term feeding excursions into the plume. Warm and cold water populations thus mix seasonally.

—The pattern repeats itself each year, but only a proportion of the previous year's 'effluent' fish may return to the discharge plume or canal.

(iv) Angling in thermal discharge canals

Thermal discharge canals are noted for good fishing in various parts of the world, particularly in winter. Landry and Strawn (1973) noted that between 50% and 90% of the angler visits in a reach of the Patuxent River were to the thermal discharge canal of the Chalk Point power station. Several studies have used rod catches to try to assess the population fluctuations in such canals. There are two main sources of error in the method, first that angling is highly selective for species and size (Dryer and Benson 1957) and second the fish may be feeding more actively than those in colder water but are not necessarily more abundant.

In the study by Dryer and Benson (1957) it was evident that catfish were abundant and frequently caught in the thermal discharge canal at the Johnsonville steam plant. However, while net surveys found that sauger (*Stizostedion canadense*) were common in the canal very few were caught by anglers. Marcy (1976a) concluded that there was no direct relationship between rod catches and water temperature between 4·8 and 30°C in the Connecticut Yankee discharge canal. However sudden falls in temperature in winter and spring were associated with declining catches. The catches also declined at the highest summer temperatures (Fig. 9.4), but increased again as soon as temperatures fell by 1°C. The winter–spring fishery in the discharge canal was 2–10 times more productive in rod catches than in unheated reaches of the river.

In Lake Sangchris fishing pressure was greatest near the thermal discharge outfall (Larimore *et al.* 1979a). Only channel catfish (*Ictalurus punctatus*) and carp (*Cyprinus carpio*) were consistently more abundant in rod catches from the heated water than from unheated waters. White bass (*Morone chrysops*) were more abundant in winter and spring but not in summer. As we have already seen (p. 290) water currents rather than temperature were regarded as the reason for the aggregations of several species in the discharge. At Point Beach on Lake Michigan, catches of lake trout (*Salvelinus namaycush*) were zero at temperatures over 21°C in the thermal discharge. Rainbow trout (*S. gairdneri*) were not caught in water at over 25·5°C, but non-salmonids were prevalent at temperatures over 22°C (Romberg *et al.* 1974). In another study, at the P. H. Robinson plant in Texas, angling pressure was greatest during November to April and catches were greatest during October, January and February. There was no direct correlation between catch and either discharge volume or temperature.

A detailed study of the River Trent (UK), using angler catch records, showed that there was no obvious relationship between water temperatures and catch in either heated or unheated reaches (Whiting *et al.* 1976). Interestingly, however, several record cards included remarks that catches were low because the thermal discharge from a relevant power station was not operating.

Surprisingly, none of the studies using angler catches has found a consistent significant correlation between catch and water temperature, though, as with studies using other catch methods, there were consistent seasonal patterns (Elser 1965; Moore and Frisbie 1972; Landry and Strawn 1973; Moore *et al.* 1973; Gallaway and Strawn 1975; Whiting *et al.* 1976; Cane *et al.* 1981). The declines in catches in summer in most situations were associated with water temperatures exceeding 32–33°C. The relationship between angling visits and water temperature follows the same pattern as that of the rod catches in most places and it seems likely that this factor is also the greatest influence on the catch data.

(v) Food and feeding in thermal discharges
Fish clearly remain active in heated waters at times of year when they might normally be torpid. It is, therefore, predictable that if food is available, they will feed at times when they may be normally fasting. The available food will be other organisms which are kept active by the warm water or, alternatively, organisms entrained at the cooling-water intake and discharged at the outfall, dead or alive. Many fish are opportunistic feeders and will take advantage of the most readily available food organisms present (e.g. Langford, 1963, 1966).

(a) Fresh waters. The dietary composition and feeding rates of fish in naturally heated waters were investigated by Naiman (1975) and Kaeding and Kaya (1978). In those reaches of the Firehole River, Montana, heated by hot springs and geysers, brown trout (*S. trutta*), fed mainly on Diptera, the gastopod snail, *Physa heterostropha*, Ephemeroptera and plant material. Emerging insects were also taken all year round.

In the unheated reaches the diet was dominated by Trichoptera, with some Plecoptera and Ephemeroptera and a few emerging insects in summer. The differences in diet clearly reflected the differences in the fauna and flora of the river caused by the hot springs and geysers, but the trout adapted readily. It was also evident that the feeding rates of the fish were, predictably, higher in the warmer waters (see Chapter 4). Some food selectivity was shown by perch (*Perca fluviatilis*) in the heated waters of the cooling-tower ponds at the Grove Road power station in London (Dandy 1964a,b). They fed mainly on sticklebacks (*Gasterosteus aculeatus*) which were abundant in the ponds. Perch also contained Cladocera at times. None fed on Ostracoda even though these were extremely abundant. The sticklebacks fed on algae, crustacea and insects. Sticklebacks also formed the main food item of large trout (*S. trutta*) in the heated, brackish-water dock at Roosecote power station (Dandy 1964b), leading the author to conclude that the species could be a successful food for consumable fish cultured in heated waters.

In these studies there were no controls for comparison, but in other studies where comparisons have been made, differences in diet have been noted between heated and unheated waters. For example, Cragg-Hine (1969b) showed that species of coarse (rough) fish fed throughout the winter in a thermal discharge instead of fasting as normal. The diets were less diverse than in the colder waters and there was evidence that the fish took more vegetable material in the discharge canal. The canal was shallow with a mud substrate and was much richer in plants than the unheated river which had a less stable, clay-based substrate. Fish continued to feed through temperatures of 27–28°C which is some 3°C below the lethal temperature of the most common species present (see Alabaster and Lloyd 1980).

More obvious changes in diet in heated waters were shown by Rachyunas (1973) and Polivannaya (1974). The former found that roach (*R. rutilus*) in the cooling ponds of a Lithuanian power station fed mainly on the most abundant invertebrate, the zebra mussel *Dreissenia polymorpha*, with small amounts of algae, macrophytes and insects. In the unheated waters *Dreissena* was much less common, and the plants, *Elodea*,

Potomageton, Ceratophyllum and *Chara* were predominant in the diet. A similar phenomenon was reported by the author for the roach and bream (*A. brama*) of an unheated lowland river in the UK where the flora and invertebrate fauna were sparse (Langford 1963, 1966). Here, the diet of all fish over 15 cm long was almost exclusively *D. polymorpha*, the only abundant invertebrate. These results indicate that fish can be opportunist feeders and that adverse conditions, leading to a restricted invertebrate fauna, do not necessarily cause difficulties for fish.

At the Kurakhova power station, perch (*P. fluviatilis*) fed on different species of zooplankton in the summer in heated and unheated waters. The diets here reflected the different plankton faunas. The diet of ruffe (*Acerina cernua*) in the heated waters included more gammarids than in the unheated waters. In all the artificially heated waters fish fed through the winter.

Rachyunas (1973) also showed that rudd (*Scardinius eryophthalmus*) fed on plant material in both heated and unheated waters but in the former the dominant species were *Potomageton lucens* and *P. perfoliatus*, in the latter *Ceratophyllum, Elodea and Myriophyllum*. The silver bream (*Blicca bjoernka*) fed mainly on the freshwater mussel (*D. polymorpha*) with some macrophyte material and chironomids in the heated waters, but in the unheated waters they fed on insects (of various groups), gastropods and oligochaetes. Tagging studies indicated that both roach and silver bream spawned in cooler waters but fed consistently in the heated waters. This was clearly not simply a case of availability but active selection of food in the heated zones.

The same phenomenon was observed in Par Pond (see p. 290) where the largemouth bass (*M. salmoides*) left cooler waters to feed on the blueback herring (*A. aestivalis*), which in turn were entering heated ponds almost at lethal temperatures, to feed on dense populations of zooplankton.

Channel catfish (*Ictalurus punctatus*) increased their proportionate intake of fish both in the thermal discharge canal at the Connecticut Yankee power station (Merriman and Thorpe 1976) and, in summer, in the heated areas of North Lake Texas (McNeely and Pearson 1974). Discharge temperatures at the latter site reached 42°C in summer, above the lethal level for the species and its prey. The authors concluded that the catfish were foraging in the cooler water beneath the stratified plume. Juvenile salmonids had a similar diet in both heated and unheated reaches of the Columbia River near the Hanford reactors (Becker, 1969).

Near the outfalls of the Nine Mile Point station on Lake Ontario, the usually varied diet of the yellow and white perch (*P. flavescens, R. americanus*) was restricted to three taxa, viz. gammarids, alewives and spottail shiners (*N. hudsonis*) all of which were unusually abundant in the area (Storr and Schlenker 1974).

The diel patterns of feeding of bluegills (*L. macrochirus*) were different in the cold and heated waters of two Texas reservoirs (Sarker 1977). In the heated water the fish filled their stomachs by 0900 hours but in the latter full stomachs were mainly recorded at 1400–1600 hours.

(b) Saline tidal waters. Data on the diets and feeding of fish around marine and estuarine thermal discharges are scarce. At Chalk Point, the polychaete worm *Nereis succinea* was more common in the diet of white perch (*R. americanus*) taken from the effluent canal than in unheated waters. The reverse was true for the isopod *Cyathura polita* (Moore *et al.* 1975).

At the Kingsnorth power station, Langford (1987) found that the diets of bass in heated and unheated waters mainly reflected the effects of substrate on the invertebrate fauna. Thus polychaetes were more common as food in both heated and unheated zones with soft mud substrates, while decapod crustaceans dominated in the area with the harder substrates. Markowski's early studies (Markowski 1959) also showed some effects of different substrates on fish diets. In a heated dock where the substrate contained fine sediments, the diet of the flounder (*P. flesus*) was mainly the polychaete *Nereis diversicolor* and the gastropod *Potomapyrgus jenkinsi*. In the open waters where there was more sand in the substrate, the amphipod *Corophium volutator* and the lamellibranch *Macoma balthica* were the main food items.

It is clear that even in places where the normal dietary items are available in unheated waters, fish will actively select to enter thermal discharges to take the different food items which have congregrated in the plume or discharge canal. In the author's experience fish will also feed on organisms which have passed through cooling-water systems and enter the receiving water alive or dead (Langford 1987). Thus at Kingsnorth power station, it was evident from the appearance of freshly eaten sprats (*Sprattus sprattus*) and shrimps (*Crangon crangon*) in the stomachs of bass (*D. labrax*) that they had been taken more or less immediately on discharge from the power station outfall.

The fact that organisms are killed during entrainment does not therefore mean that they are lost to the energy flow through the ecosystem. In fact the supply of such food items to fish remaining active in thermal discharge canals and ponds allows them to survive the winter instead of dying in the extreme cold of natural waters, especially if fat deposition is enhanced in the autumn and size-selective mortality is a major factor regulating the population (see Baranava 1980; Henderson *et al.* 1988).

(vi) Growth and condition in thermal discharges
The growth of fish can be controlled by temperature, provided that other conditions are constant and favourable (see Chapter 4). However, the instability of thermal discharge plumes and the variable residence times of fish do not encourage the simple prediction of the effects of the plume on growth. The site studies carried out so far have mainly been concerned with 'population growth' (Ricker 1975), and few have used growth measurements on marked individuals. In most cases the duration of exposure to the elevated temperatures has not been known.

(a) Fresh waters. In the naturally heated streams of Yellowstone, brown trout (*S. trutta*) spawned and hatched 4–5 months earlier than normal and showed two growth periods rather than one. Consequently they were much longer at a specific age than cold water populations (Kaeding and Kaya 1978). There was evidence, however, of the retardation of growth at the highest summer temperatures. In this river and in the River Severn (Langford 1983a) trout were caught in summer at temperatures exceeding their incipient lethal temperatures, suggesting active feeding.

There are highly variable results from field studies. In the Connecticut Yankee discharge canal some resident individual catfish (*I. nebulosus*, *I. catus*) lost the equivalent of 1–1·5 years' growth over the winter, while some individuals clearly gained weight (Marcy 1976a). The effect on the population in the reach of the river was small because only 2·5% and 0·5% of the species, respectively, returned to the canal for a second year and there was thus a considerable mixing of canal and river populations. This has also been shown for other species and locations (Brown 1973, 1979). The catfish made up their weight loss during the summer, feeding in the cooler waters of the main Connecticut River.

The reasons for the loss in weight in winter are not clear as both species fed throughout the winter. Merriman and Thorpe (1976) concluded that the causes were:

—higher metabolic rates in the warmer water,
—higher expenditure of energy used while keeping station in the fast water of the discharge currents,
—poor food availability because of overcrowding and competition in the effluent canal.

Poor food supply supposedly accounted for the poor growth of carp (*Cyprinus carpio*) in Lake Sangchris (Larimore *et al.* 1979a, b), and largemouth bass (*M. salmoides*) in Par Pond (Bennett and Gibbons 1974; Gibbons *et al.* 1978). The poor condition factors of all the species in North Lake, Texas (McNeely and Pearson 1974), and the bluegills (*Lepomis macrochirus*) in Par Pond (Graham 1974), were no doubt caused by higher metabolism and poor food but no data on these are presented in the publications. In studies on Lake Michigan fish (Spigarelli and Smith 1976), there was no clear pattern of increased or decreased growth of fish tagged with thermoluminescent tags. Some trout (*S. trutta*), exposed to a heated plume, grew faster than some control fish but the growth rates were not significantly faster than the best controls at natural ambient temperatures.

Similar individual variability was shown by *M. salmoides* in the heated areas of Par Pond (Janssen and Giesey 1984). There were two groups with differing growth rates which supposedly reflected their residence times in the heated water. Those with higher condition factors had large ranges of movement and short residence times while those with lower condition factors were more or less resident in or near the heated pond. This second group were longer for their age than the first group but much thinner. The conclusions were that long-term residence in the heated effluent caused increased metabolism in summer which was not matched by food intake. Hence fat reserves were used up and condition factors decreased.

In Par Pond and adjacent waters, the fat content of mosquito fish, *G. affinis*, was not correlated with water temperature though there were significant differences between locations (Falke and Smith 1974). Roach, *Rutilus rutilus*, in a heated effluent at a power station in the USSR, showed increased amounts of fat in the body compared with those in unheated waters (Baranava 1980).

Accelerated growth has been demonstrated for several species in heated effluents in various locations. Salmonids reared in the warm water from the Hanford reactors were larger than those reared in unheated waters (Nakatani 1969). In the River Fiddich and its tributaries juvenile salmon (*Salmo salar*) grew more rapidly in reaches heated by a distillery cooling-water discharge than in unheated reaches (Morrison 1989). Enhanced production of invertebrate food was considered to be the main reason. The salmon also migrated as 2 + year old smolts instead of 3 + year old. Brown (1973, 1979), found increases of over 30% in the lengths of bleak (*Alburnus alburnus*), common bream (*Abramis brama*) and chub (*Leuciscus cephalus*) caused by their residence in a thermal discharge canal, compared with unheated waters. Cragg-Hine (1971) concluded that the growth season of these species could be extended by up to 2 months in the same canal. As in other locations the increases in growth would be obscured in the total river population as the cold and warm water groups mixed in early summer. In several other locations in Europe the growth of young coarse (rough) fish was also enhanced by heated effluents (see Alabaster and Lloyd 1980; Karas and Neuman 1981; Neuman 1982). In most studies, winter growth occurred.

Enhanced growth rates have also been reported for bluegills (*Lepomis macrochirus*) (Serns and Strawn 1975) in Texas reservoirs, largemouth bass (*M. salmoides*) from other US reservoirs (Larimore *et al.* 1979a,b; Perry and Tranquilli 1984), breams and other species in Eastern European lakes (see Alabaster and Lloyd 1980). At the Konakovo power station, bream (*A. brama orientalis*) grew faster in the heated waters until temperatures exceeded 28°C when the rate slowed considerably (Sappo 1976). Similarly, growth rates of bream over 2 years old slowed down from an enhanced rate in the heated Konin Lakes (Marciak 1977) though the reasons were not clear. In several locations the larger size attained by fish within given times was related to a longer growing season rather than an increased growth rate *per se*.

The abundance and size of prey species was regarded as the major reason for enhanced growth of three species of sport fishes in a heated South-Carolina reservoir, rather than the effects of the discharge (Barwick and Lorenzen 1984).

Spot (*Leistomus zanthurus*) trapped in the discharge canal at a Virginia power station (White *et al.* 1976), apparently survived their predicted upper avoidance and lethal temperatures and intermittent chlorine residuals of up to 0.2 mg litre^{-1} and yet were longer at ages 1, 2 and 3 than fish in cooler waters. Clearly the rapid temperature fluctuations and possibly areas of cooler water may have allowed the fish to survive (see Chapter 2).

In the Forsmark test basin (Sandstrom 1985), perch (*Perca fluviatilis*) increased their growth rates such that at 1 year old they were up to 2.5 times heavier than before the thermal discharge began (Karas and Neuman 1981; Neuman 1982). In contrast, fish at 2, 3 and 4 years old grew more slowly than in similar unheated waters. Temperatures in the heated water exceeded the optimum growth temperatures for perch (see Talmage and Coutant 1980) in summer.

(b) Saline tidal waters. Langford (1987) found that the bass (*Dicentrarchus labrax*) population in a thermal discharge canal grew faster in length than in the unheated waters of the Medway estuary and other British waters.

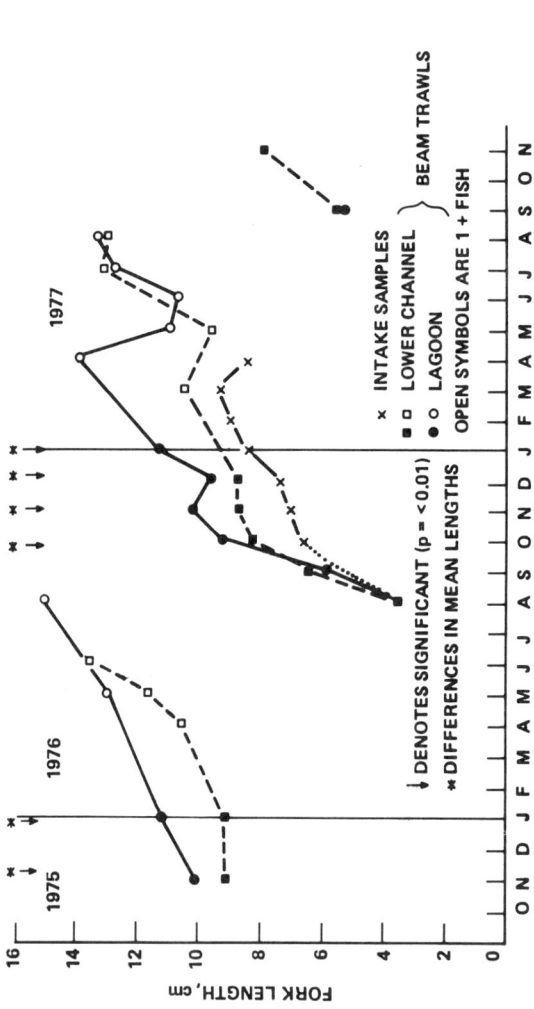

FIG. 9.5   The seasonal growth of 0+ bass, *Dicentrachus labrax* in heated and unheated habitats in an English estuary. Intake = unheated; lower channel = most fluctuation; lagoon = most heated (from Langford 1987).

Post-larval fish entered the discharge canal in late summer and early autumn and then grew faster in the warmer water (Fig. 9.5), feeding on invertebrates from the canal itself and those passing through the cooling-water system. At the end of the first year's growth the bass were larger than in other waters in the UK. The effects on the total population of the estuary were probably small because of the same kind of seasonal mixing of warm and cold water populations which occurred in freshwater fish populations (see Langford *et al.* 1979; see p. 305). Scale anomalies among the bass indicated that fish entered the canal at various ages from under 1 year to about 4 years old and residence times varied from a few weeks to 2 years or more. The apparent slowing of growth after 2 or 3 years (Fig. 9.6) was an artefact caused by this variable residence, serial migrations of fish of various ages and variable growth rates.

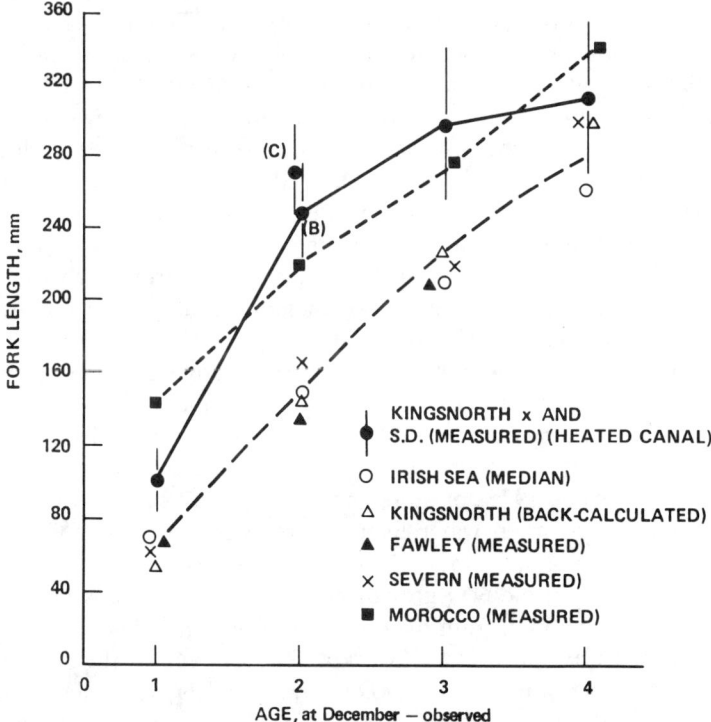

FIG. 9.6   Growth of bass (*D. labrax*) in a thermal discharge canal in comparison to other locations (from Langford 1987). NB, back calculated lengths are smaller because they probably relate to cold water growth.

In another study near an offshore outfall, the growth rate of the black goby (*Gobius niger*) was slightly faster than in other places in Europe (Vesey and Langford 1985), though the influence of the thermal plume was not clear as it was strongly stratified and unlikely to come in contact with the bottom-living fish. Increased availability of food in the vicinity of the outfall originating from entrained organisms was the probable cause of the enhanced growth.

(vii) Ageing and annulus formation

To compare growth rates, the ages of fish must be known. The markings usually known as 'checks' or 'annuli', caused by the different seasonal rates of growth of scales, operculae, vertebrae or some other skeletal structure are used for age determination in most species (see Ricker 1975). The growth of these structures is related to the growth of the fish and checks are usually formed during periods of slow or no growth. Thus if a fish grows all year round, or if the seasonal growth patterns are altered, check formation can be upset and age determination becomes difficult.

In the naturally heated waters of the Firehole River, Yellowstone, both brown and rainbow trout (*S. trutta* and *S. gairdneri*) showed two distinct growth checks in the warmest water. One was associated with the cessation of growth in the coldest part of the winter, the other with the cessation of growth caused by temperatures exceeding the upper limit for growth in summer. The main growth seasons were therefore in spring, early summer, autumn and early winter instead of in mid-summer as in cold-water populations (Kaeding and Kaya 1978).

Anomalies in the growth of scales have been recorded in several studies of fish around thermal discharges (see Perry and Tranquilli 1984). In heated fresh waters, the largemouth bass (*M. salmoides*) produced indistinct winter growth annuli (see Larimore *et al.* 1979a,b), in some cases no growth annuli (Siler and Clugston 1975), or more annuli than expected because of extra annuli laid down when water temperatures were too high for growth (Galloway and Kilambi 1988).

Continuous or prolonged growth also resulted in the absence of the normal seasonal annuli from the otoliths of flounders (*Platichthys flesus*) in the heated Cavendish Dock (Markowski 1966) and in the absence of checks from the scales of bass (*Dicentrarchus labrax*) in the thermal discharge canal at Kingsnorth (Langford 1987). In this latter habitat, annuli appeared in some fish up to 3–4 months earlier than in the unheated waters. This was caused by fish entering the canal after they had stopped growing in the cold water, and beginning to grow again in the warm water.

Because of the difficulties in ageing such fish, it is clear that to measure the effects of thermal discharges on growth the use of fish of known ages or individually identifiable fish is advisable.

(viii) Spawning and reproduction
Brett (1970) noted that the temperature range over which salmon spawn naturally is about 25–33% of the total tolerance range for adult fish. Houston (1982) showed that the temperature ranges for spawning tend to be higher for those with higher tolerance ranges. Several species of fish spawn in natural waters at temperatures within 3–5°C of their critical thermal maximum. Alabaster and Lloyd (1980) compiled the data on the temperature ranges for spawning and temperatures for embryonic development of many European fishes. Both Brett (1970) and Cragg-Hine (1971) concluded that photoperiod was the primary influence on the onset of spawning though temperature was an important secondary stimulus. Other factors are also significant, for example water currents, salinities or food availability.

Hokanson *et al.* (1977) found that for some percids, higher than normal winter temperatures could inhibit gonad development and hence spawning. In the Firehole River, brown trout (*S. trutta*) did not spawn successfully in the warmest waters. Their gonads did not mature and eventually degenerated (Kaya 1977). Recruitment was consequently poor. Rainbow trout (*S. gairdneri*), on the other hand reproduced successfully but avoided the hottest periods by a shift in the spawning season for spring to autumn. Thus egg hatching and the development of fry occurred when temperatures were optimal. This shift had clearly evolved in a relatively short time as the species was not introduced to the river until after 1889 (Brock 1975).

In the River Witham (UK), the thermal discharge from Lincoln power station heated the whole of the river reach between two locks and fish had little chance to avoid the higher temperatures. Thus Bray (1971) concluded that fish spawned up to 6 weeks in advance of their normal period. Similarly, roach (*R. rutilus*) showed advanced gonad development and early spawning (by 3 weeks), in the heated areas of the River Meuse downstream of the Tihange nuclear power station (Mattheeuws *et al.* 1981). Early spawning of roach (*R. rutilus*), perch (*Perca fluviatilis*) and bream (*Abramis brama*) occurred in heated lakes in Poland and in the Netherlands (Zawisza and Backiel 1972; van Densen and Hadderingh 1982). In Poland the advance was 3–8 weeks, in the Netherlands, at lower ΔTs, it was 1–2 weeks. In contrast, the thermal discharge canal at Peterborough was only a small cut-off from the river Nene and there was no evidence from the gonad development or the time of appearance of fry, that the fish population in this river spawned early (Cragg-Hine 1969a, 1971; Brown 1973, 1979).

There are observations of a number of species spawning actually within thermal discharge canals at higher than ambient temperatures and in apparently unstable temperature regimes. For example, fry hatched successfully from the nests of smallmouth bass (*Micropterus dolomieui*) in a power station thermal discharge canal when daily water temperatures were averaging 24°C and fluctuating widely between 14 and 28°C (Dunford 1977). Chinook salmon (*Onchorhynchus tschawytscha*) also spawned consistently in the heated Columbia River below the Hanford reactors. Between 1950 and 1967 the number of redds increased from 300 to 3300 irrespective of the thermal discharges (Nakatani 1969).

As we have seen, water currents attracted white bass to the thermal discharge in Lake Sangchris (p. 289) and the fish spawned in the heated water shedding their eggs and sperm in the strong currents of the effluent (Larimore *et al.* 1979a). In Par Pond, largemouth bass, *M. salmoides* did not show overall changes in the spawning period in the heated waters, though some individuals showed advanced gonadal development (Bennett and Gibbons 1975). In the same region, female mosquitofish (*G. affinis*) carried eyed embryos for much longer periods in heated waters than in unheated waters. The brood size varied inversely with temperature in the most heated areas (Ferens and Murphy 1974; Bennett and Goodyear 1978). Fish of various species were found to have spawned in the heated, chlorinated plumes at three Canadian power stations, though no data are available on the survival of eggs in the long term (Haymes 1980).

Data from field studies of saline water species are scarce. The main reason for the absence of mature bass from the Kingsnorth thermal discharge canal (Langford 1987) was that these fish migrate offshore and become solitary once attaining spawning size and this behavioural change is clearly necessary, irrespective of water temperature. Such pre-spawning behaviour and emigration from thermal discharge canals is the reason that many species do not spawn early in such places although gonad development may be advanced (Marcy 1976a; Langford 1987).

(ix) Fecundity and survival

The fecundity of various species in heated waters may be increased or reduced but the reasons are not always clear. In Lake Sangchris, genetic or growth rate differences were believed to be responsible for the decline in fecundity of largemouth bass (*M. salmoides*) in the warmest waters

(Larimore *et al.* 1979a,b). Females matured at 30 cm, which was attained in 3 years in the colder waters and 2 years in the heated waters. Older fish tended to be more fecund than the younger fish in both types of habitat.

There are no reported studies of the rates of hatching of eggs *in situ* in thermal discharges though, as we have seen (Chapter 4, p. 120) the process is closely related to temperature.

(x) Parasites in thermal discharges

There are several factors which might lead to the increased incidence of parasites or disease among fishes in the vicinity of thermal discharges, including:

—The winter aggregations could increase the probability of direct transmission of pathogens or parasites;
—enhanced growth, survival and reproduction of parasites or pathogens in warm water;
—increased stress among fish in heated areas could increase their vulnerability to infection or infestation;
—aggregations of birds in warm open water on otherwise frozen lakes or rivers could encourage the transmission of parasite stages from fish to their secondary bird hosts.

As we have seen in Chapter 4, there is evidence from experimental work that some parasites develop and become mature earlier at higher water temperatures. Field studies tend to support the experimental studies.

For example, in the heated waters of Lake Liden the destructive tapeworm (*Ligula intestinalis*) was reportedly more frequent than normal in the body cavities of cyprinids (see EIFAC 1968a; Alabaster and Lloyd 1980). However, this parasite is notoriously patchy in its distribution (e.g. Hoffman and Bauer 1971).

Detailed studies in the heated waters of Par Pond by several authors indicated that the presence or abundance of secondary hosts were the most important factors influencing the parasite load of fish in the heated waters (Eure and Esch 1974; Bowen 1976; Aho *et al.* 1976). For example, the most common parasite of largemouth bass (*M. salmoides*), the acanthocephalan *Neoechinorhyncus cylindratus*, was more abundant (numbers per unit body weight) in the heated waters than unheated waters during winter, spring and autumn. In summer the differences were not significant (Eure and Esch 1974).

The other groups of parasites did not show consistent patterns. The infestation rates of several species were related to the population fluctuations of the intermediate hosts such as the ostracod (*Physocypria pustulosa*) (Bowen 1976) and the bluegill (*Lepomis macrochirus*). As we have seen there is a close correlation between the thermal discharge and the occurrence of planktonic crustaceans and fishes in Par Pond (see p. 290).

The incidence of *N. cylindratus* in *M. salmoides* increased in autumn at which time the bass were feeding heavily on bluegills (Bennett and Gibbons 1972).

In the same habitat two trematode parasites *Ornithodiplostomum ptychocheilus* and *Diplostomum scheuringi* found in the mosquito fish *Gambusia affinis* showed contrasting patterns of occurrence in the heated waters. The former was more abundant in fish in summer and winter in the heated water and decreased in abundance with distance from the heated discharge. The reverse applied to *D. scheuringi*. Again the main reasons for the distribution and abundance of the parasites in the various habitats were related to the presence and abundance of the relevant intermediate hosts, in both cases gastropod snails and water birds (Aho *et al.* 1976).

In Lake Sangchris (see p. 289), the parasite distribution was not directly related to water temperature but to a combination of enhanced growth and reproduction of the trematodes in the warm bluegills (Larimore *et al.* 1979a). The infestation rates were, however, lower in the heated lake than in the unheated Lake Shelbyville.

Studies of parasites in fish from heated saline waters are scarce. Markowski (1966) listed the species found in fish from Cavendish Dock but comparisons with unheated waters were not made. Preliminary data from the Forsmark basin in Sweden indicate no increased incidence of parasites in the heated waters (Sandstrom 1985), though Thulin (1981), suggested that the increased water temperature in the discharge basin of the Barseback power station prolonged the time over which the young stages of *Lernaeocera branchialis* colonised the gill filaments of flounders (*P. flesus*).

(xi) Diseases and infections

Pathogenic bacteria, protozoa and fungi are well known from heated waters, as we have seen (Chapter 5), and it has been reported that the incidence and severity of viral and fungal infections of fish can be related to high water temperatures (Brett 1956; De Sylva 1969; Ross and Smith 1974). The incidence of columnaris disease in the Columbia River (Stroud and Douglas 1968) was clearly related to aggregations of fish at the bases of fish ladders rather than the thermal discharges from the Hanford reactors as was first believed (see Templeton and Coutant 1971).

The incidence of red-sore disease in Par Pond was not significantly increased by the heated water. The infection rate in fish was related to the abundance of the pathogen *Aeromonas* in the water column (see p. 150). Also, fish in poor condition were more susceptible than those in good condition (Esch and Hazen 1978). The poor condition could be related to direct temperature stress, higher metabolic rates and poor food in the warmest waters.

Eels (*A. anguilla*) showing the symptoms of the virus causing 'cauliflower disease' were more abundant in the heated waters of the discharge basin of the Barseback power station in Sweden, than in unheated waters (Thulin 1981).

(xii) Gas bubble disease
Although not a disease in the true sense of the word, gas bubble disease has been observed in fish caught from in or near thermal discharges (Demont and Miller 1971). It is, however, most associated with the supersaturation of nitrogen and its release below hydro-electricity dams (Rucker 1972; Fickeisen and Schneider 1976). In warm waters fish and other animals (see p. 357) suffer embolisms as gases become less soluble in their body fluids (e.g. Nebecker *et al.* 1979). Fish can avoid exposure to high levels of supersaturation (Gray *et al.* 1983) and the phenomenon rarely develops rapidly. Records of related mortalities are thus rare though the symptoms have been observed in a number of habitats.

The bulging eyes and embolisms characteristic of the disease were obvious in all the 13 species of fish caught from the discharge of the Marshall power station (Adair and Demont 1971; see Gallagher 1974). The mortality of *Pomoxis* spp. was not positively attributed to gas bubble disease though all the fish showed symptoms. Menhaden (*Brevoortia tyrannus*) were reported to have entered a thermal discharge at a nuclear power station on the east coast of the USA and subsequently died of gas bubble disease (Marcello and Fairbanks 1976). The species clearly did not avoid the conditions in the water and this also applies to other species (Gray *et al.* 1982, 1983). Carp (*Cyprinus carpio*) showed the symptoms unlike other species at the Waukegan and Zion power stations but as carp were the only species to reside in the discharges for long periods this was understandable.

(xiii) Morphological and meristic effects
It is well known that many environmental influences can cause morphological and meristic changes in fish, particularly during the early

stages of development (e.g. De Sylva 1969; Garside 1970; Boytsov 1974; Schubel and Marcy 1978; Alabaster and Lloyd 1980; Moodie 1985 for references). Temperature is only one of the many influences.

As we have seen in Chapter 4 (p. 120), thermal shock and high constant temperatures during hatching can cause deformations of the body and it is likely that such deformations are more common and obvious in cultured fish than in wild populations because they are not selected out by competition or predation at an early stage.

Different species showed different responses in the Ivan'kovskoye reservoir (Boytsov 1974). The morphological changes in roach (*R. rutilus*) and bream (*A. brama*) were however related to the current velocities of a thermal discharge rather than temperature, and supposedly increased the mobility of the species in the fast currents. The morphological adaptations of older bream were persistent and the suggestion was that these were related to a change from the normal benthic feeding to plankton feeding in the heated water. Sappo (1975) stressed that the bream population near the thermal discharge outfall were morphologically distinct from natural populations and the localised strain numbered some $1 \cdot 3 \times 10^6$ fish over an age range of 1–13 years. Natural lakes with different thermal regimes also produce bream with morphological differences.

Asymmetry in numbers of rays in paired fins, numbers of scales in the lateral line, scale radii, scale lengths and the size of otoliths has been observed in fish exposed to thermal and other stresses during development (see Garside 1970; Ames *et al.* 1979 for references). However, no asymmetry was observed in the fish from the heated waters of Par Pond though in an unheated pool having high concentrations of mercury and a low pH, unconnected with Par Pond, there was evidence of marked asymmetry among the fish (Ames *et al.* 1979).

In a recent study Bamber and Henderson (1985) have used meristic and morphological analyses to show that the supposed 'exotic' species of atherinid (*Atherina boyeri*) found in heated waters, in the UK is a form in a continuum, where the physical characteristics are related to lowered salinity and temperature. *A. boyeri* is, therefore, not distinguishable from *A. presbyter*, which was previously believed to be the indigenous British species and the authors concluded that *A. boyeri* is the correct name for all records for Britain.

(xiv) Genetic effects and adaptations
Many species show clear differences in temperature tolerance related to their geographic distribution and other ecological variables (McCauley

1958; Mitton and Koehn 1975; Hall *et al.* 1978; see Houston 1982), but this is not true for all species of either vertebrates or invertebrates (Hutchinson 1976) and may only apply at the level of sub-species (Houston 1982). Short-term genetic changes, mainly in allele frequencies, have been demonstrated in fish from artificially heated habitats.

In the Savannah River ponds, one variant of the Mdh allele (for Malake dehydrogenase) was more frequent in largemouth bass (*M. salmoides*) from the heated ponds than in cold or previously heated ponds (Yardley *et al.* 1974). This allele was believed to have some selective advantage in the heated water. The theory was supported by the finding of a similar increase in the same allele in the heated Lake Sangchris when compared with unheated waters (Larimore *et al.* 1979a).

In Par Pond and the other Savannah River ponds, Smith (loc. cit. Gibbons and Sharitz 1981) noted that the frequency of the Mdh allele returned gradually to the 'cold-water' level after the reactors closed down and the thermal discharge ceased. This work seems to demonstrate clear short-term genetic adaptation to changing conditions and could easily explain geographic differences in temperature tolerances.

There are reservations to the unequivocal acceptance that temperature was the sole cause of the short-term genetic changes. Firstly, the results show a high variability in allele frequencies among populations of *M. salmoides* from various habitats not affected by thermal discharges (Yardley *et al.* 1974; Larimore *et al.* 1979a). Secondly the data from the Savannah River did not show significant differences ($p < 0.05$) between fish from heated and unheated waters but there were much clearer differences between the lentic and lotic environments. Finally, none of the other species in Par Pond showed the same trends.

The genetic separation of the populations of the red shiner (*Notropis lutrensis*) in a regulated Texas reservoir was considered as caused by thermal alterations resulting from hypolimnial discharges at the dams. Tailwater fish were less tolerant to temperature than the reservoir fish and became genetically adapted after 40 years (King *et al.* 1985).

(xv) Exotic species in thermal discharges
Fish have been moved around the world so much by man, that a discussion of the meaning of truly 'exotic' species could be extremely long (e.g. Wheeler 1969; Wheeler and Maitland 1973). There are several situations, however, where the presence of a thermal discharge allows populations of species from basically warmer regions to maintain themselves in normally temperate waters (see Pullin and Lowe-McConnell 1982). The use of

thermal discharges for fish culture has led to introductions, probably followed by escapes and hence wild populations of non-indigenous species. Iles (1963) imported tilapias for culture in the cooling ponds of British power stations. In Belgium a population originating from heated ponds survives in a thermal discharge (Philipart and Ruwet 1982). In the USA populations survive precariously near such outfalls in colder regions.

A reported exotic atherinid, the sand smelt (*Atherina boyeri*) found near a British power station (Bowers and Naylor 1964; Palmer *et al.* 1979) was found to be more a 'phase' in a taxonomic continuum and not a separate species (Bamber and Henderson 1985). The main influences were salinity and location rather than temperature.

9.1.5 Effects of Chlorine Residuals in Thermal Discharges

As we have seen (p. 283), chlorine has been considered as a cause of fish mortalities in thermal discharges and the major stressor affecting communities in rivers affected by other wastewaters (p. 235). The tolerance limits for many species are well known (see Morgan and Carpenter 1978; Morgan 1980). Fish kills have been reported from areas affected by thermal

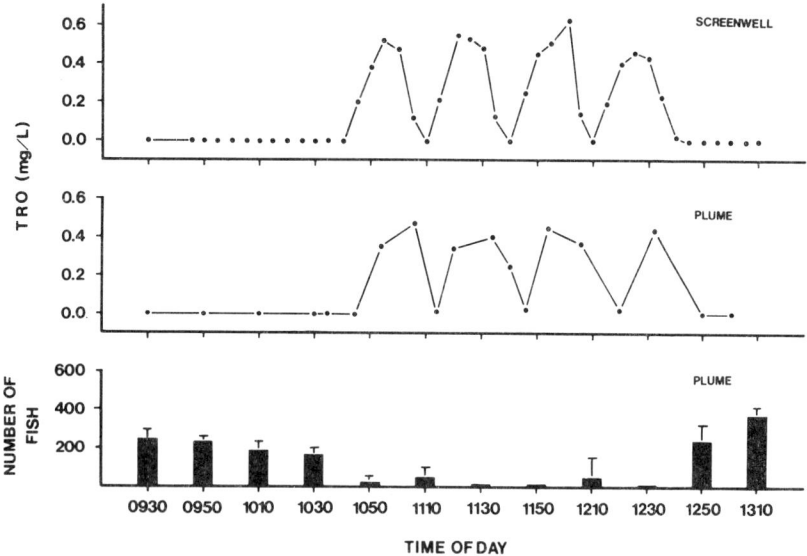

FIG. 9.7   Screenwell and plume total residual oxidant (TRO) concentrations and mean number of fish observed in the discharge plume of units 7 and 8 Redondo Beach Steam Generating Plant, King Harbor, California, during an experimental chlorination (24 June 1981). (From Hose *et al.* 1983c).

plumes in which residual chlorine was regarded as the lethal factor rather than temperature (see Morgan and Carpenter 1978). Residuals ranged from 0·01 to 3·05 mg litre^{-1}. At Indian Point, chlorine concentrations of 0·11 mg litre^{-1} caused increased mortalities among entrained bass (*Morone* spp.) (Lauer *et al.* 1974). Marcy (1976a) considered that chlorine concentrations of 0·1 mg litre^{-1} produced the same level of effect on entrained fish as mechanical effects at the Connecticut Yankee nuclear power station.

Chlorine residuals also produce sub-lethal effects in the field, namely avoidance behaviour (Cherry *et al.* 1977b,c,d, 1978; Cairns *et al.* 1981; see Morgan 1980). Few field studies have produced good quantitative data. The numbers of fish in the discharge plume at the Redondo Beach power station in California were markedly influenced by the chlorine residuals (Fig. 9.7). During the period of chlorination the numbers fell from around 200–400 to less than 20 or so (Hose *et al.* 1983c).

From studies on sonic tagged white bass (*M. chrysops*) in a thermal discharge canal (Grieve *et al.* 1978), it was concluded that the fish took avoidance action when the residual chlorine rose above 0·35 mg litre^{-1} (Fig. 9.8).

At the Kingsnorth power station (Langford 1983b), fish remained in the discharge canal throughout the year despite the chlorination schedules.

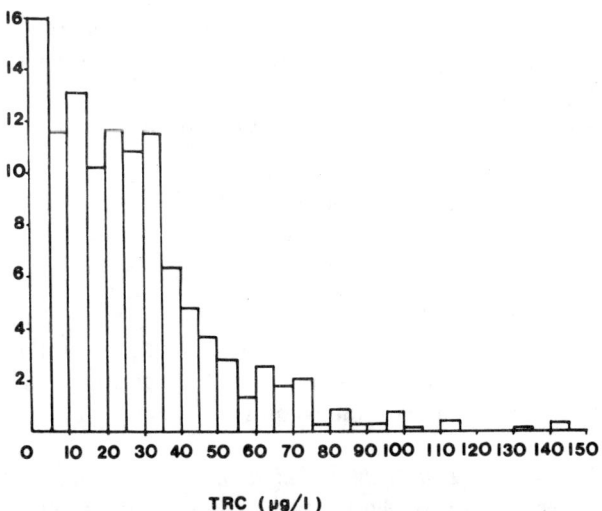

FIG. 9.8 Histogram representation of the total frequency of encounters of fish in an outfall channel in relation to total residual chlorine (from Grieve *et al.* 1978).

Spencer (1982) found concentrations up to 0.35 mg litre^{-1} but the decay was extremely rapid and from temperature studies it was evident that mixing in the canal was poor. Areas of low chlorine residuals were probably available into which fish could move to avoid adverse conditions.

9.2 OTHER VERTEBRATES

9.2.1 Amphibians and Reptiles
(i) Behaviour, growth and life histories
The behaviour, life histories and growth of both amphibians and reptiles can be affected by thermal discharges. The toad *Bufo terrestris* was found to spawn in March in both the heated and unheated ponds near the Savannah River reactors (Nelson 1974). In the hottest waters where temperatures ranged up to 38°C, the mortalities of eggs and embryos exceeded 95% and few larvae were observed. Most of the spawning and the highest survival occurred in the cool marginal seepages around the heated ponds or in the cooler ponds nearby. In the heated, sub-lethal waters embryos developed faster and metamorphosis occurred earlier and at a smaller size than in cooler waters.

Several studies of the yellow-bellied turtle (*Pseudemys scripta scripta*) have shown effects on growth (Fig. 9.9) and maturity in the heated ponds around the Savannah River reactors (see Gibbons and Sharitz 1981). Adult females grew faster than normal and attained a larger size at maturity at the higher temperatures.

Comparisons were made of turtle populations in:

—heated but otherwise unpolluted waters,
—unheated and unpolluted waters which had been heated in previous years,
—unheated and polluted waters (Christy *et al.* 1974).

Females again showed faster growth rates in the first two habitats. Growth continued after maturity in the heated waters, but not in the other two habitats. The higher growth rates in each of the habitats was believed to be related to a change from a generally omnivorous diet to a more carnivorous diet.

The American alligator (*Alligator mississippiensis*) is a major predator in these southern swamps of the USA, feeding on many groups of aquatic organisms and using the water courses for migrations and thermo-regulatory behaviour (Glassman and Bennett 1978). By shining lights at the

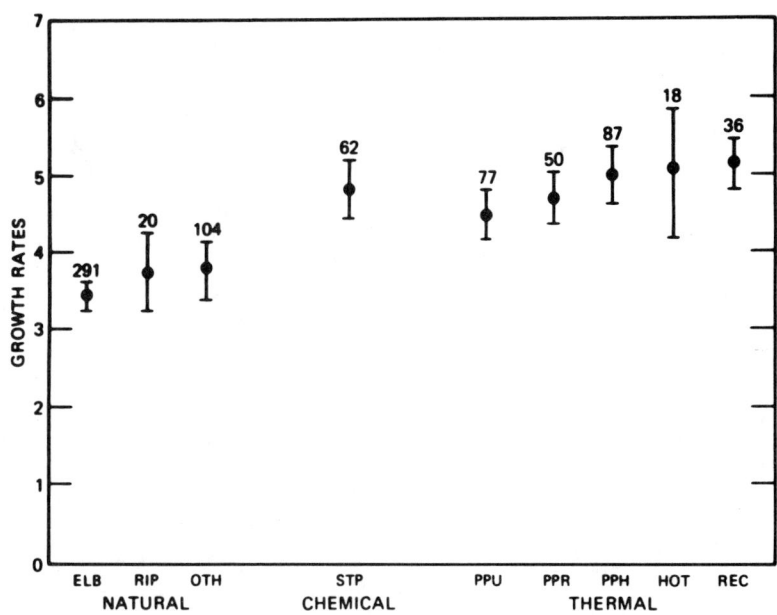

FIG. 9.9   Growth rates of turtles in areas differently affected by thermal effluents from reactors at the Savannah River Plant. Dots and bars indicate slope $(b) \pm$ two standard errors based on a linear regression analysis $(Y = a + bX)$ of each sample, where Y is the plastron length and X is the age. Only immature turtles that could be aged were used in the analysis, excepting a few thermal-area females that displayed continued rapid growth rate beyond the usual size at which maturity is reached. Natural sampling areas: ELB, Ellenton Bay; RIP, Risher Pond; and OTH, other natural areas. Chemical sampling area: STP, Steed Pond. Thermal sampling areas: PPU, Par Pond unheated areas; PPR, Par Pond recovery areas; PPH, Par Pond heated areas; HOT, Pond C and other areas receiving maximum thermal loading ($> 50°$C); and REC, recovery areas (Pond B and Steel Creek). From Christy *et al.* (1974).

eyes of the animals at night, Murphy and Brisbin (1974) were able to make head counts and plot the distribution of the species in relation to water temperature. As Fig. 9.10 shows, there was a winter aggregation of alligators in the heated water. More than 50% of the variation in the distribution was accounted for by water temperature. The winter temperatures equated well with the reported preference temperature for the species. In the warmer waters, alligators fed and remained active through the winter instead of becoming dormant as they do in natural waters. Remote sensing techniques, using radio-tagging, showed that alligators thermo-regulate efficiently by using both land and water as necessary

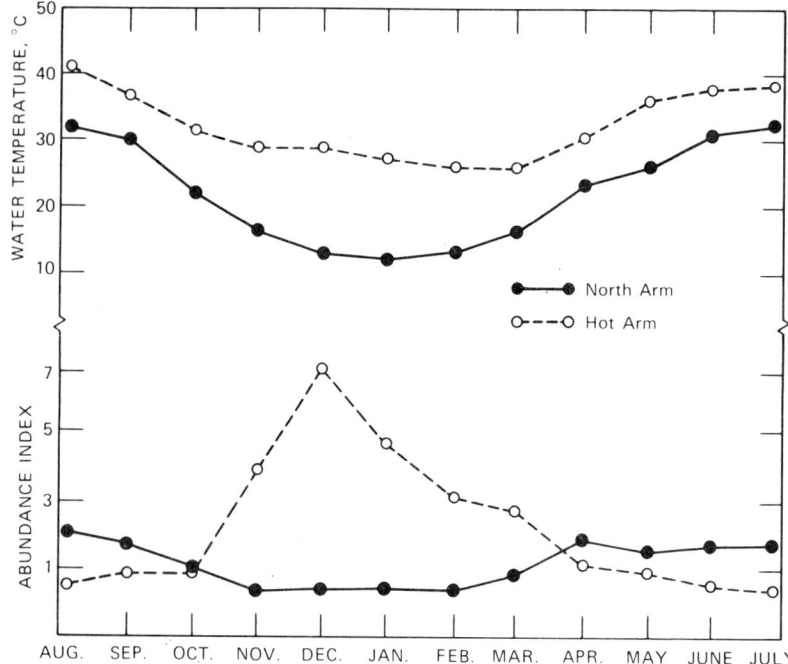

F‌IG. 9.10   Seasonal fluctuations in mean monthly surface-water temperatures and abundance indices of American alligators at the Hot Arm and the North Arm census stations of Par Pond. Means are based on determinations made three times a month between January 1968 and July 1972 at the Hot Arm and between March 1969 and July 1972 at the North Arm (from Murphy and Brisbin 1974).

(Spotila 1974). Using theoretical models, Porter and Tracy (1974) predicted that the life histories and ecology of both the Garter Snake (*Thamnophis sertalis*) and the frog (*Rana pipiens*) common in the Savannah River swamps would be affected by excessive heat but no empirical data are given to validate this.

(ii) Parasites and disease

The different groups of parasites found in turtles responded differently to the heated habitats (Bourque and Esch 1974). The acanthocephalans, *Neoechinorynchus* spp., were not significantly more frequent in heated than in unheated habitats, but the actual body burdens were greater by up to 15 times. In the river swamps, the incidence of the nematode *Camallus* sp. was directly related to the increased heat load but the reverse was true for Par

Pond. Trematode species showed more or less random incidence in both heated and unheated waters. However, the body burdens of both nematodes and trematodes were lowest in the warmest habitats. The intermediate hosts for the acanthocephalan *Neoechinorynchus* sp. are usually ostracods which are abundant in the Savannah River ponds. These, in turn, feed on the blooms of blue-green algae in the warm ponds.

Alligators exposed to water temperatures of 35°C caused by preventing the normal thermoregulatory behaviour, rendered them prone to infection by bacteria, particularly *Aeromonas hydrophila*, often resulting in death (Glassman and Bennett 1978). *A. hydrophila* is common in the heated waters of Par Pond (see p. 150).

9.2.2 Birds

The distribution of birds in relation to heated waters is of interest for two main reasons. Firstly, aggregations could lead to increased predation on fish or invertebrates, particularly where ice-cover is removed in winter. Secondly as intermediate hosts of parasites the birds could increase the incidence of such organisms in themselves and the relevant fish and invertebrates which act as the other hosts. Gulls and other pisciverous birds are commonly seen near cooling-water outfalls feeding on aggregations of fish or on small fish discharged after entrainment through the cooling system (Prentice 1969).

In the Savannah River cypress swamps, the bird communities were less diverse where heated effluents entered the streams and creeks. One of the major causes for the imbalance in the community structure was that trees which were killed by water temperatures up to 50°C (see p. 196) became attractive nesting sites for woodpeckers, crows and herons, which dominated the normally more evenly balanced community (Straney *et al.* 1974). The highest numbers of birds occurred in a creek which was recovering from thermal effects.

Similar studies on waterfowl on Par Pond again showed lower densities and diversities in the heated waters (Fig. 9.11). The lower diversities were attributable to the dominance of one or two species. There was clearly an increased density of waterfowl in the heated area during the ice and snow storms which occurred in December 1971 (Fig. 9.12) (Brisbin 1974).

The reasons for the different responses of species of waterfowl to the heated water was believed to be related to food rather than a direct temperature effect. Coots, for example, were feeding on the mats of blue-green algae which were prolific in the warmer waters. The grebe (*Podilymbus podiceps*) fed on the invertebrates and small fish. None of the

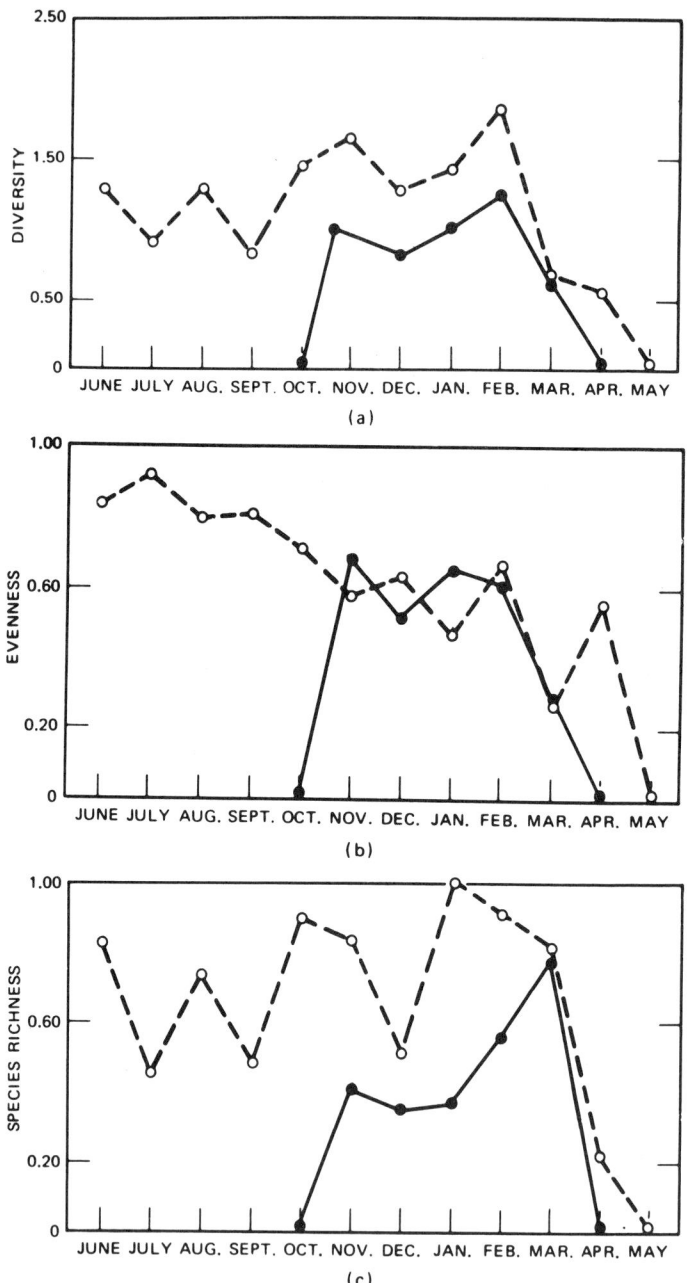

FIG. 9.11 Community indices for birds in heated (●——●) and unheated (○---○) areas of water (from Brisbin 1974).

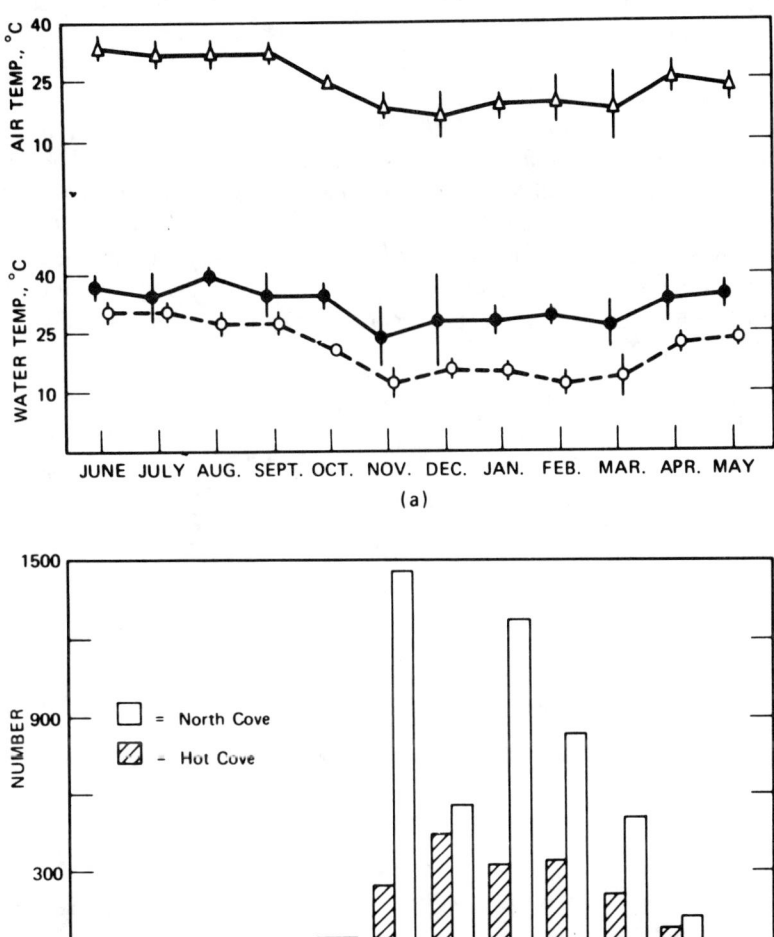

Fig. 9.12 Seasonal changes in (a) the average air and water temperatures and (b) the total abundance of all waterfowl species present. Vertical lines represent two standard errors above and below the average temperature. Δ, air temperature. ●, Hot Cove water. ○, North Cove water. Abundance is expressed as the total number of birds counted on three census visits to each study area during each month (from Brisbin 1974).

published studies have quantified the predicted effects of bird aggregations on predation or on the parasite infestations in aquatic invertebrates or vertebrates.

9.2.3 Mammals

The West Indian manatee (*Trichechtus manatus*) has been observed to congregate in thermal effluents in the south-eastern USA. There is reported to be a population of about 1000 individuals in the US. Aerial surveys of thermal discharges at five Florida power stations showed varying numbers of individuals in the plumes, ranging from 15 to 98 (Reynolds and Wilcox 1985). Regression analysis showed that water temperature and air temperature were very important, though not the only factors affecting distribution.

CHAPTER 10

Combined Effects, Uses of Waste Heat, Thermal Criteria and the Law

10.1 HEATED AND OTHER EFFLUENTS IN COMBINATION

It is well known that the toxicity of many contaminants increases with increasing temperature and that raised temperature can have synergistic effects on toxins (Jones 1964; Cairns and Messenger 1974; EPA 1976; Leland et al. 1977; Cairns et al. 1978; Thurston et al. 1979; Alabaster and Lloyd 1980) (see Chapter 4). This is not universally true, however, as a few substances such as phenols do not follow the same trend (e.g. Klein 1962; Jones 1964, 1978) and some of the data on heavy metal toxicity has shown that there can be differential effects with temperature and species (see Houston 1982). Using experimental data and models the potential effects of temperature on toxicity can now be predicted to a great extent for single toxins and mixtures (Alabaster and Lloyd 1980). The effects of heat on other processes such as the accumulation of heavy metals or radioactive isotopes in tissues is also clearly important though data from field studies are scarce. In fact, the number of *in situ* studies of most contaminants, other than chlorine, where the effects of heat can be assessed separately from other effluents are relatively few.

Langford (1983a) has discussed the effects of the major power station wastes on aquatic ecosystems of which some are associated or combined with thermal discharges. Without the complication of additional heat, these wastes, for example from ash-disposal, water-based flue-gas desulphuris-ation processes and fuel storage can produce both major disposal and ecological problems (e.g. Grimas and Ehlin 1975; Dvorak and Lewis 1978; Opresko and Hannon 1979; Chu and Olem 1980; Cherry *et al.* 1982a; Langford 1983a; Barnett and Hardy 1984; Duedall *et al.* 1985b; see also Chapter 3). In this section the emphasis is on the relevance of heat to the effects of such other effluents. As we have already noted early in this volume, most effluents from industrial processes are heated to some extent but in these cases heat is a secondary factor to the process wastes and again no published field studies have assessed the relative effects of heat and the chemicals in these discharges. In desalination plant effluents, the separate effects of high salinities, high temperatures and heavy metal concentrations are also difficult to separate in the field, though at some sites the undiluted effluent is clearly toxic and causes marked effects *in situ* (see Barnett and Hardy 1984).

10.1.1 Fuel-Waste Effluents and Thermal Discharges

Effluents from fuel storage on power stations, such as oil tanks or coal stocks, contain contaminants which may affect aquatic organisms (e.g. Nelson-Smith 1970; Witt 1971; Harrison 1977; Birge 1978; Langford 1983a) and the effects of heat on these may be synergistic (see Chapter 4).

The effluents from ash-settling ponds can cause marked adverse effects in aquatic ecosystems whether combined with heat or not, often eliminating the flora and fauna for large areas in fresh or tidal waters (Cairns *et al.* 1972; Bamber 1978; Forbes 1980; Forbes *et al.* 1981; Webster *et al.* 1981; Cherry *et al.* 1982a; Cairns and Cherry 1983; Langford 1983a; Bamber 1984a; see IAWPR 1982, and Duedall *et al.* 1985a; Iwanski and Chu 1985; Norton 1985). Generally, the most damaging component of the discharge, or solid waste disposal, is the physical effect of flocculation, deposition or accretion of the solid matter. The leaching of metals, pH and oxygen depletion are additional factors.

In the Wisconsin River, caged crayfish (*Oronectes propingus*) survived in an ash effluent for long periods but were found to accumulate most of the elements leached from the ash, notably chromium in the exoskeleton, selenium in the hepato-pancreas and muscle, zinc in all three and iron in the hepato-pancreas alone (Forbes *et al.* 1981).

10.1.2 Combined Effects of Heat and Ash disposal

In several locations ash settling effluents are discharged in close proximity to thermal discharges, enabling studies of the combined and sometimes the separate relative effects of each to be evaluated.

(i) Micro-organisms and plants

In the complex stream system around the Savannah River Project, 400D, Area, the greatest diversity and the greatest proportions of natural chromagenic bacteria were found in unpolluted control reaches (Guthrie *et al.* 1978). The sampling station with the highest water temperatures (24–44°C), but not the highest ash content, showed the most severe reductions in diversity and chromagens. At lower temperatures but higher ash concentrations the effects were not as severe. Irrespective of the low pH values (3·5–4·6) there were increases in the abundance of bacteria wherever ash was present, mainly caused by the increased nutrients leached from the ash. The bacteria accumulated selenium, zinc and chromium from these ash leachates (Guthrie *et al.* 1974a, b, 1978), which were passed through the food chain (Cherry *et al.* 1979a). These authors concluded that in this system the higher range of water temperatures limited the diversity and numbers of chromagens in the bacterial communities.

In contrast, the composition and diversity of the bacterial communities of the streams receiving ash effluents and thermal discharges from the Glen Lyn power station were mainly affected by the ash effluents (Larrick *et al.* 1981) and no additional effects were observed with the entry of a heated effluent to a stream receiving an ash effluent. Here water temperatures were always below 28°C. However, the bacterial community of the New River which received the main chlorinated cooling-water discharge, showed similar reductions in diversity and proportions of chromagens to that of the streams receiving ash effluent alone. Forbes (1980) also found that ash effluents retarded the microbial decomposition of leaves in streams.

The activity of bacterial communities (V_{max}) was markedly reduced by heat, chlorine and ash effluents. The physiological processes were, therefore, much more sensitive indicators of effects than community structure. The biomass, chlorophyll 'a' and carbon-fixation rates increased downstream of the heavy ash effluent at Glen Lyn (Cherry *et al.* 1979a). Temperature increases of up to 4°C and increased levels of nutrients caused the increases.

(ii) Invertebrates

The invertebrate communities were not markedly affected by the warm ash effluent at Glen Lyn until the last 6 months of operation, when Ephemeroptera, Plecoptera and Trichoptera began to decline. The beetle *Psephenus herricki* appeared to thrive in the worst affected reach (Cherry *et al.* 1979a,b). The invertebrate community began to recover 2 months after the ash effluent stopped discharging and recovery was complete after 10 months. Heavy rains flushed the accumulated ash deposits away and enhanced recovery.

In the Savannah River streams, ash effluents and thermal discharges together eliminated the invertebrate fauna from the worst affected sites (Cherry *et al.* 1979b). Temperatures reached 44·5°C. As temperature conditions became less severe the more temperature-tolerant species such as damselflies (Odonata) and chironomids reappeared first, though the ash content of the water column and sediments limited their densities. The absence of macrophytes in the areas of densest ash concentrations also inhibited the survival and colonisation of invertebrates. The major factors limiting the invertebrate fauna were concluded to be temperatures exceeding 35°C, and turbidity, suspended solids and pH in the ash effluent (Cherry *et al.* 1979a, b). There was some evidence of bioaccumulation in the invertebrates though this was most marked in the crayfish *Procambarus* sp. Considerable bioaccumulation of trace elements occurred in a stream receiving the ash effluent from a settling basin (Cherry *et al.* 1979a, b; Table 10.1). Accumulation factors exceeding 2000 were measured.

(iii) Fish
The only species of fish found in the warmest, ash-polluted streams around the Savannah River plant was the mosquito fish (*Gambusia affinis*) (Cherry *et al.* 1979b). Outside the warmest zones, pH was the main factor inhibiting the recolonisation of the swamp streams. *G. affinis* could not survive the extremes of up to 44·5°C even with its high thermal death point of about 39–40°C (Cherry *et al.* 1976). At the lowest temperatures recorded in the study areas, usually around 9°C, the fish were found to be torpid (Cherry *et al.* 1979a). After the installation of a new ash-settling pond, which reduced the solids in the effluent, temperature-tolerant invertebrates and *G. affinis* recolonised the streams where temperatures allowed. Some $1200 \, m^{-2}$ of swamp area were affected by the coal ash effluents. About 1–2·5 km of stream and river showed traceable biological effects of both the ash and thermal discharges.

At Glen Lyn power station, fish were more abundant downstream of the ash effluent (Cherry *et al.* 1979b) but the community was less diverse than in the control streams. The decreased diversity was caused mainly by a predominance of the stoneroller fish (*Campostoma anomalum*). As the settling pond became less efficient just prior to closure, the number of species downstream declined though the stoneroller became yet more abundant. In associated experiments, it was found that the avoidance reactions of fish to high temperatures and to chlorine were relatively easy to quantify, but the responses to ash effluents were less so (Cairns and Cherry 1983). The complexity of variables in the ash effluents, namely pH, suspended solids and temperature, and their interactions confused the interpretation of data.

Coutant and his colleagues (Coutant *et al.* 1978) considered that the effects of a coal-ash effluent at the Bull Run power station may have been incorrectly ascribed to the nearby thermal discharge. They found that the ash effluent was toxic to fish and that the flocculated ferric hydroxide complex $(Fe\,(OH_3)/FeOOH)$ discharged from the ash stream eliminated the invertebrate fauna. In the vicinity of the stream outfall, the ash effluent was clearly the major polluting influence. In the studies on bio-accumulation

TABLE 10.1

Concentrations of trace elements in abiotic and biotic components of an ash basin receiving stream (adapted from Cherry *et al.* 1979a)

Trace element	Concentration (*ppm*)				
	Abiotic water	Biotic			
		Benthos	Plants	Invertebrates	Fish
Aluminium	13·0	40 657·0	3 985·1	1 199·3	215·5
Iron	16·9	20 912·4	1 113·2	1 202·6	154·7
Potassium	6·1	8 149·2	1 803·6	2 666·2	1 946·2
Calcium	9·2	1 844·8	850·1	2 656·4	5 752·9
Magnesium	4·1	5 460·8	656·2	369·4	307·2
Titanium	0·9	2 388·5	109·4	71·5	15·1
Sodium	7·7	688·0	267·9	703·8	309·8
Chlorine	3·8	84·1	198·2	364·9	131·4
Barium	0·7	294·2	36·3	50·2	20·0
Strontium	0·3	236·0	60·3	48·4	36·3
Manganese	0·07	46·2	70·2	21·5	10·0
Cerium	0·2	129·7	9·7	4·3	1·6
Tin	0·1	85·0	18·0	20·7	3·4
Rubidium	0·4	51·6	8·2	29·0	8·5
Vanadium	0·04	63·9	4·7	4·4	0·6
Chromium	0·2	38·4	5·7	9·7	2·8
Zinc	0·4	6·4	5·0	14·9	11·8
Arsenic	0·06	19·7	4·2	2·1	0·5
Lanthanum	<0·01	20·3	1·4	1·4	0·1
Thorium	0·03	15·3	1·3	1·7	0·3
Bromine	0·1	1·2	3·0	10·1	2·9
Selenium	0·1	6·1	1·8	2·6	9·4
Cobalt	0·1	10·6	1·7	1·7	0·5
Iodine	0·1	4·6	1·3	3·4	0·4
Uranium	0·01	8·0	0·7	0·3	0·1
Cadmium	0·1	1·7	1·5	4·0	1·3
Caesium	<0·01	3·9	0·6	0·7	0·5
Antimony	0·07	1·0	0·8	2·1	0·7
Mercury	0·03	0·8	0·5	0·5	0·2

(Cherry *et al.* 1979a), fish generally showed lower factors than invertebrates and plants (Table 10.1), except for calcium and selenium.

10.1.3 Flue-Gas Washing and Desulphurisation Wastes

As we have seen in Chapter 3, the chemical effects of the heated flue-gas washing effluents on the River Thames (UK), in the 1950s and 1960s were very marked. The fauna of the river at that time was poor as a result of its overall polluted state and thus there were no data on the biological effects of the effluents. To date, field data are still scarce, even though the effects of the main components of flue-gas desulphurisation effluents, e.g. high temperature, low pH, solids and heavy metals are mostly well known separately (see Klein 1962; Jones 1964; Dvorak and Lewis 1978; Alabaster and Lloyd 1980). The different methods of removing sulphur will have different implications for aquatic habitats (see for example, McGuire 1977; Bettelheim *et al.* 1981; Moser 1981; Kyte *et al.* 1983; Kyte 1986), but it is likely that in some processes there may not be significant discharges of either solids or heavy metals if proper treatment processes are installed. Leachates from the disposal of FGD sludges are possible contaminants for groundwater or for watercourses in some situations (Burnett and Fedyko 1978; Woodyard and Sanning 1978). Effluents from coal cleaning and the chloride prescrubbers will also contain similar components (Kyte *et al.* 1983). The disposal of coal wastes as solid blocks forming artificial reefs in the sea (e.g. Woodhead *et al.* 1980) will produce leachates which may affect organisms in the close vicinity (see Duedall *et al.* 1985a), though colonisation of such blocks is usually rapid.

The potential biological effects of a proposed FGD effluent on a British estuary were assessed, using the scarce data available on some of the components of the discharge. The additional $3°C$ δT in the cooling-water discharge caused by the $45°C$ FGD effluent being diluted produced maximum temperatures of around $28°C$ which would have been tolerable for many estuarine species. The lowered pH of the cooling-water effluent, i.e. around 5.6 in the near field was regarded as being in the range to cause harmful physiological effects to marine or estuarine species exposed for long periods (Bamber 1987b) (Fig. 10.1). Knutzen (1981) also reviewed the data on pH and marine organisms and concluded that some effects may be noted on marine species exposed to reductions of 0·3 pH units.

10.1.4 Heavy Metals in Thermal Discharges

Those metals most commonly associated with large thermal discharges are listed in Becker and Thatcher (1973). The separation of temperature effects

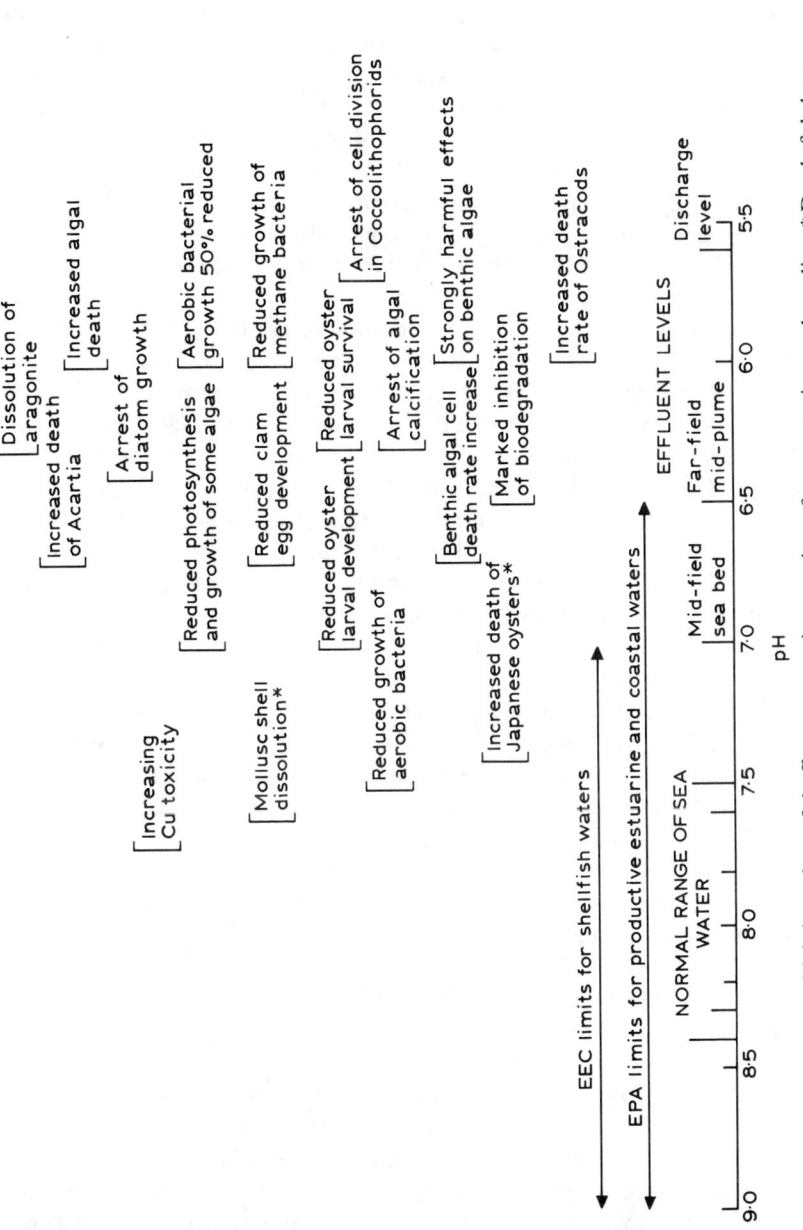

FIG. 10.1 pH values which have harmful effects on marine species—from experimental studies. * Doubtful data. Reproduced with permission of the CEGB.

from those of toxic metals is rarely possible in field studies, either around power stations or the industries such as metal finishing which produce heated toxic effluents (see Klein 1962; Jones 1978). Although Stratton and Lee (1975) and Birge (1978) showed that toxic concentrations of heavy metals could be associated with thermal discharges, neither described any effects in natural ecosystems *in situ*.

At the Fruitland power station cadmium and zinc concentrations in the cooling lake effluent exceeded the potentially toxic limits for sensitive species (Dreesen *et al.* 1977), but the highest concentrations of zinc occurred in the water entering the intakes from the cooling lake.

The concentrations of zinc and chromium compounds in some cooling-tower blowdown effluents were toxic to the algae (*Selanastrum capricornutum*) and to trout (*Salmo gairdneri*). Tests of the corrosion inhibitors Tetrahydo 1 and 4,2,H-oxanzine in the effluent from Oahu power station, Hawaii, also showed that the concentrations in the effluent were several orders of magnitude less than their LC_{50}s for mullet (*Chelon engeli*) and mosquito fish (*G. affinis*), at water temperatures up to 27·8°C (Becker and Thatcher 1973; see EPA 1976; Thurston *et al.* 1979).

The only mortality of aquatic animals directly attributable to heavy metals in a large cooling-water discharge was that of the abalone (*Haliotis* spp.) in the discharge area of the Diablo Canyon nuclear power station in California (Martin *et al.* 1977). Here concentrations of copper in the discharge were above the toxic concentrations for long periods.

The effects of desalination effluents were considered to be mostly a result of synergistic interactions between salinity and temperature (see Barnett and Hardy 1984). At two sites, the normal flora and fauna were practically absent from the path of the undiluted effluents, though some recovery occurred during periods when the plants were not working. At a third site, in Florida, some species of animals appeared to be attracted to the effluent though many of the indigenous species were unable to survive.

10.1.5 Bioaccumulation in Thermal Discharges

Experimental data indicate that the rates of bioaccumulation of substances such as radioactive isotopes, heavy metals and chlorinated compounds (e.g. PCBs) (see Chapter 4, and Houston 1982) may be enhanced by higher temperatures, but there are very few field data where accumulation has been compared at heated and non-heated sites (e.g. Murray 1979). In the studies on heated waters polluted by ash effluents (see p. 329) trace metals were accumulated by sediments, plants and animals (Table 10.1) but no comparisons of heated and unheated reaches were made. Selenium was found to be in greater concentrations in four fresh-water fish species in a cooling reservoir than in a water supply reservoir (Sager and Cofield 1984).

Roosenburg (1969) noted that oysters (*Crassostrea gigas*) turned green as a result of copper toxicity and accumulation near the outfall of a power station. The copper was believed to originate from the corrosion of condenser tubes in the cooling-water system.

Romeril and Davis (1976) compared the accumulation of metals in eels from power station cooling ponds and from unheated sites in the UK. They did not find any abnormal accumulation in those eels kept at the higher temperatures. The fastest growing eels in the warm water developed new tissues faster than they accumulated metals and thus had lower specific metal concentrations than the cold-water eels.

In Southampton Water in the UK, molluscs of various species were found to accumulate heavy metals near thermal discharges at other sites (Romeril 1971, 1972, 1974, 1975, 1976a, b, 1979; Davis 1977). However, the metals originated from the shipyards and docks around the Water and accumulation was greatest in these areas irrespective of the thermal discharges.

A comprehensive study of perch (*P. fluviatilis*) and roach (*R. rutilus*) in the heated effluent of a Swedish nuclear power station indicated that the perch showed significantly higher levels of organo-chlorine residues (DDTs and PCBs) in the warmer water than in the cold. Roach did not show the same effect (Edgren *et al.* 1981) (Fig. 10.2). The authors offered no explanation for the differences between the species but one possibility was differences in diet.

Experimental studies (Anderson 1983) have shown that other components of thermal discharges, for example chlorine, will increase the rate of some heavy metal accumulation in fish tissues when exposed to the combined stresses.

Most of the radio-nuclides released into the aquatic environment, under strict control and monitoring procedures, are diluted in thermal discharges from reprocessing plants or nuclear power stations (see IAEA 1966, 1969, 1971, 1972, 1974a, b, c, d, 1975a, b, c, 1976, 1979; see MAFF 1972 *et seq.*; Hetherington 1976; Eisenbud 1973; Streffer 1975; Hunt 1989) but comparisons with unheated sites are rarely shown.

Experiments on different species have shown different effects of temperature on accumulation of isotopes. For example, the effects of temperature were small on the uptake of ^{60}Co and ^{65}Zn in shrimps (*Crangon crangon*) (Van Weers 1975). Higher temperatures increased food

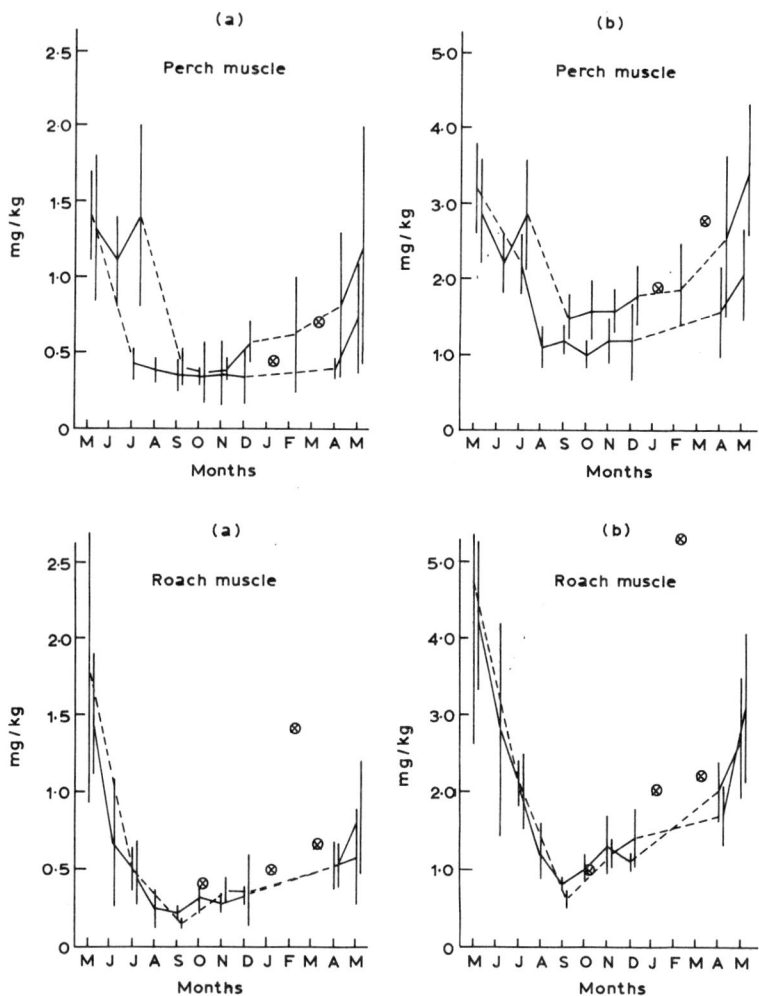

FIG. 10.2 Accumulation of PCBs in fish from heated and unheated waters. Monthly mean values and 95% confidence interval (vertical rules) for the (a) sDDt and (b) PCB levels in extractable fat from perch muscle and roach muscle (from Edgren *et al.* 1981). ■, Fish from intake area; ×, fish from cooling water recipient; ○, less than six specimens analysed; -----, lack of monthly mean values.

intake and moulting frequency which in turn increased the rate of turn-over of the isotopes.

In contrast, Fraizier and Ancellin (1975) considered that temperature was a major factor influencing the uptake of ^{59}Fe by mussels (*Mytilus edulis*) and the blenny (*Blennius pholis*). At the Humboldt bay power station in the USA, oysters (*Crassostrea gigas*) accumulated ^{65}Zn in the thermal discharge but the rate of uptake was related to the concentration of the isotope in the water rather than temperature (Salo and Leet 1968).

Observations in the vicinity of the thermal discharge from the Maine Yankee power station showed that *C. gigas* on experimental rafts grew faster and accumulated radioactive materials faster at the heated sites than at the unheated sites (Fig. 10.3).

In another field study, the uptake of iodine, caesium and cobalt isotopes were measured in bivalve molluscs exposed in a 1 km long thermal discharge canal (Patel *et al.* 1975). Natural ambient temperatures were 22–24°C and the thermal plume temperatures 27–35°C. The accumulation of ^{137}Cs, and ^{60}Co in the blood clam (*Anadara granosa*) were not affected by temperature along the course of the discharge canal. In comparison, the accumulation of ^{131}I was more temperature dependent, particularly after 30 days exposure. The effects of temperature on the other species and isotopes were not consistent (Table 10.2).

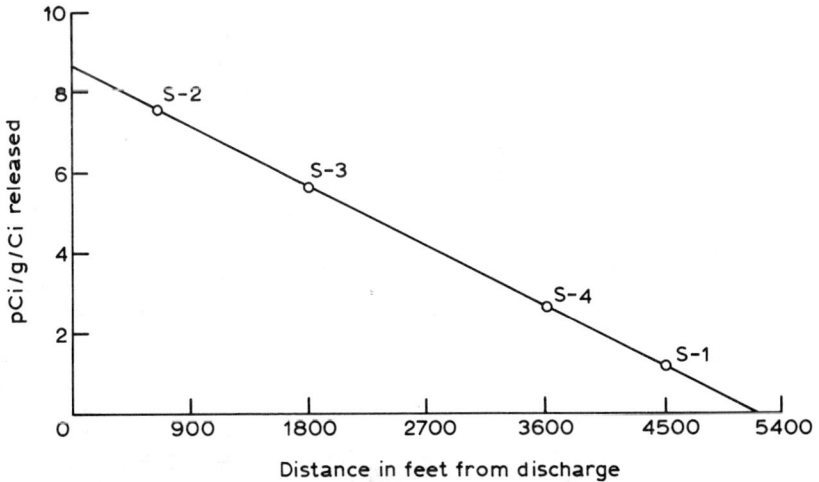

FIG. 10.3   Ratio of retention, *U*, versus distance from the discharge point measured along the most direct water route (from Salo and Leet 1968).

TABLE 10.2

Effects of a thermal discharge on accumulation of isotopes at a power station
(from Patel et al. 1975)

(a) Effect of temperature on the accumulation of ^{131}I, ^{137}Cs and $^{58,60}Co$ by blood clam Anadara granosa population, and ^{137}Cs in the discharge canal

Days after translocation	^{137}Cs in canal water ($pCi\ litre^{-1}$)	30–35°C				29–31°C				27–29°C			
		^{131}I	^{137}Cs	^{58}Co	^{60}Co	^{131}I	^{137}Cs	^{58}Co	^{60}Co	^{131}I	^{137}Cs	^{58}Co	^{60}Co
					(pCi/g dry tissue of A. granosa)								
5	30·0	21·2	9·8	3·9	13·5	8·9	8·7	2·8	11·3	9·9	9·2	3·7	10·9
10	31·7	13·5	6·3	2·4	18·3	12·4	5·5	2·6	22·3	9·2	5·8	2·4	17·7
15	23·4	9·5	5·7	2·8	14·5	—	—	—	—	6·8	3·7	1·7	16·3
19	—	—	—	—	—	—	—	—	—	10·9	4·4	1·8	19·9
30	30·4	53·7	14·3	4·9	19·2	45·0	15·0	3·5	17·3	12·7	9·8	8·2	23·6
43	71·5	—	—	—	—	36·6	26·0	8·9	98·1	36·0	13·6	5·6	37·7
46	67·6	18·1	24·5	4·4	28·0	—	—	—	—	—	—	—	—
64	61·0	—	—	—	—	—	—	—	—	25·5	20·8	7·3	111·4
94	28·6	—	—	—	—	—	—	—	—	15·7	9·1	5·7	151·2

(b) *Effect of temperature on the accumulation of ¹³¹I, ¹³⁷Cs, ⁵⁸,⁶⁰Co and ⁶⁵Zn in oyster* Crassostrea gryphoides *on exposure in the discharge canal*

Days after translocation	30–35°C					27–29°C				
	^{131}I	^{137}Cs	^{58}Co	^{60}Co	^{65}Zn	^{131}I	^{137}Cs	^{58}Co	^{60}Co	^{65}Zn
	pCi/g dry Crassostrea gryphoides tissue									
0[a]	0·4	1·0	—	1·8	—	0·4	1·0	—	1·8	—
5	23·4	6·4	4·5	58·0	18·6	24·7	4·6	7·3	44·8	—
15	11·2	4·8	3·5	24·3	4·3	—	—	—	—	—
38	71·5	17·5	12·0	43·3	13·0	31·7	15·7	9·0	33·6	6·0
59	—	—	—	—	—	10·5	28·7	8·0	59·6	7·0

[a] Radioactivity in population prior to transplant.

(c) *Effect of temperature on the accumulation of ¹³¹I, ¹³⁷Cs, ⁵⁸,⁶⁰Co in venerid clam* Katelysia opima *on exposure in the discharge canal*

Days after translocation	30–35°C				29–31°C				27–29°C			
	^{131}I	^{137}Cs	^{58}Co	^{60}Co	^{131}I	^{137}Cs	^{58}Co	^{60}Co	^{131}I	^{137}Cs	^{58}Co	^{60}Co
	(pCi/g dry K. opima tissues)											
0[a]	—	0·7	—	1·5	—	0·7	—	1·5	—	0·7	—	1·5
5	6·2	6·2	4·3	28·2	—	—	—	—	7·8	5·3	4·9	19·9
10	3·9	3·1	2·2	16·7	—	—	—	—	3·2	2·5	1·6	16·2
15	—	—	—	—	—	—	—	—	3·6	2·4	—	14·3
25	35·7	19·8	15·4	41·6	15·1	19·1	14·1	37·5	12·4	16·6	9·7	27·5
38	—	—	—	—	—	—	—	—	19·4	20·0	12·9	41·8
46	14·9	33·1	14·2	80·5	11·1	28·5	12·4	104·3	9·0	31·5	8·4	80·1

[a] Concentration in population prior to transplant.

10.2 ECOSYSTEM EFFECTS AND MODELS

It is clear from this book and from the other literature, that a thermal discharge can affect all the trophic levels of any ecosystem (e.g. Smith *et al.* 1974b; Cairns 1976; Cairns *et al.* 1979; Langford 1983a). Microcosm studies have demonstrated such effects under controlled conditions (e.g. Copeland *et al.* 1974). In most of the work reported from site studies, the results of the surveys or research programmes have been used more as monitoring data than for predictive purposes. To be useful for the future, the provision of some predictive models is necessary (Coutant 1971b).

There have been many attempts to create models to predict the effects of entrainment and intake mortalities on groups of organisms such as phytoplankton (e.g. Smith and Jensen 1974; Boreman *et al.* 1978), zooplankton (McNaught 1976; Polgar *et al.* 1976) and fish (Lawler *et al.* 1974; Englert *et al.* 1976; Freeman and Sharma 1977; Goodyear 1977a, b; McFadden 1977; see Van Winkle 1977; see Jensen 1978b; Ogawa 1979; Murarka *et al.* 1981; Swartzman *et al.* 1981; Summers and Polgar 1981; Turnpenny and Henderson 1981; Turnpenny *et al.* 1983, 1985; Henderson *et al.* 1984b). Models incorporating the effects of thermal discharges are less common, particularly models which attempt to include effects on the total ecosystem (e.g. Odum 1974).

Chung and Strawn (1978) used a simple stochastic model to try to predict the survival of organisms in a power plant cooling system but did not allow for non-thermal factors. They considered laboratory data unsatisfactory for operational predictions at sites.

Swartzman and co-authors (Swartzman *et al.* 1981) emphasised the state of ecosystem models in that 'it is recognised that simulation models ... present an ideal medium for exploring the possible consequences of various types of impact. On the other hand, no simulation models at present have proven effective at giving accurate quantitative predictions of the effects of impacts'. After reviewing a number of models, these authors concluded that although a mathematical model may provide a framework for a hypothesis, there is no substitute for 'on-site' investigations to provide the quantitative assessments. The situation has changed little with regard to total ecosystem models.

Ecosystem models were used to assess the effects of different combinations of power station cooling systems on Cayuga Lake, New York (Murarka *et al.* 1981). The basic components of the model are shown in Fig. 10.4. The overall assessment, using the model and 46 state variables, showed that there was a 'threshold' number of cooling systems involving a known volume of water, below which there would be little or no effect on the lake ecosystem. Above this threshold, the effects on some taxa would be greater than the linear increase which might be expected. In another modelling exercise, the large amounts of data collected from the studies of the heated Lake Sangchris and the unheated Lake Shelbyville (Larimore and McNurney 1979), many of which have been discussed already in various chapters of this volume, were used to try to validate predictions of the effects of heating on a lake ecosystem (Larimore and McNurney 1979; Larimore *et al.* 1979a, b; Tetra-Tech Inc. 1980; Murarka *et al.* 1981). The model was most useful in assessing the potential changes which would be caused by the interactions of trophic levels given the effects of temperature rises on one of these levels. Figure 10.5 for example, shows the predicted effect of changes in the success of spawning of gizzard shad (*D. cepedianum*) on zooplankton biomass, making the assumption that the spawning of the shad could be altered by temperature changes. The authors comment that the results of the modelling exercise 'must . . . be applied with judgement and experience'. This, of course, is seen to apply also to field data where long experience of sampling procedures and working on different ecosystems

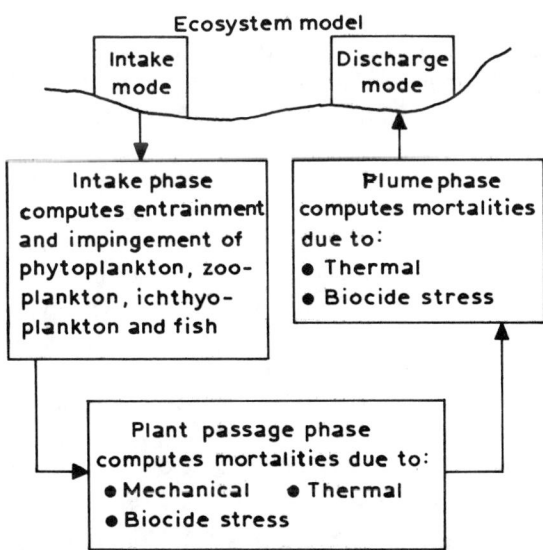

FIG. 10.4   Relationship of power plant simulation model to ecosystem model
(from Swartzman *et al.* 1981)

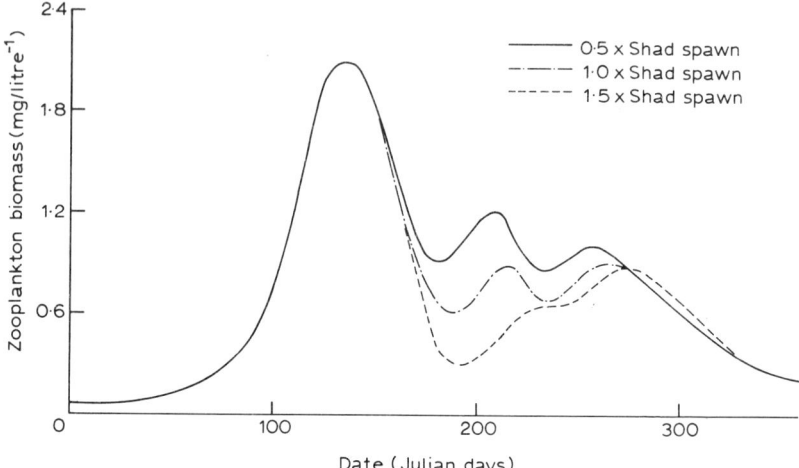

FIG. 10.5   Simulation results indicating the effect of changes in gizzard shad spawning success on zooplankton biomass (from Larimore and McNurney 1979).

must be an advantage to the interpretation of the data obtained. Data from 50 years of basic research on the ecology of the smallmouth bass (*M. dolomieui*) were used to construct a quantitative model as a base from which to predict the effects of the new Bai du Dore nuclear power station on the Lake Huron population (Shuter *et al.* 1985). Further data, collected over the 5 years after the station began operating were used to validate the model.

In this case the observed effects on the timing of spawning and growth rates were 'within the range defined by the forecasts'. Here, as in other locations, there were changes in the distribution of fish around the outfall which led to better angling catches (see p. 298). Such long runs of data are rare, but are vital to such modelling projects for accuracy.

One of the more interesting total impact models was that used to evaluate the need for cooling towers at the Crystal River power station in Florida (Kemp *et al.* 1977). Here the energy evaluation included total flow through from all relevant sources for the whole region, in relation to the ecological effects on the estuary and a cost of the towers. From their comprehensive calculations, the authors concluded that:

—The system of power plants and estuarine cooling, after an adaptation period, was economically and ecologically more competitive than cooling towers.

—The adapted ecosystems were somewhat different from unaffected ones, but were within the range of energy budgets, metabolism, diversity and productivity of other Gulf Coast estuaries.

In short, the building of cooling towers at this site was ecologically and economically unnecessary and this would no doubt apply to many other sites where the water supply is plentiful. Such models would seem to be vital where ecological constraints are proposed which are controversial.

There is clearly a long-term future for the predictive model in which the effects of physical and chemical changes in a habitat can be translated into effects on the respective ecosystem. However, it is unlikely that the mathematical model will easily be accepted as the ultimate evidence of the potential impact of a cooling-water system or any other disturbance. Pollution-control authorities or opposition groups will almost certainly expect follow-up field studies at specific sites, even if a model predicts minimum impacts. The improvement of models by detailed studies of ecosystems may, on the other hand, be a vital factor in helping planners with decisions and be usable in Environmental Impact Statements. Research data from long-term field studies are most cost-effective in the long term. In such studies the effects of different parameters can be evaluated from the differences between years. The possible manipulation of ecosystems using controlled changes would be extremely valuable as a source of these data.

The improvement of hydrological models and predictions is a major pre-requisite of ecological prediction (see Chapter 2).

10.3 THE RECOVERY AND USES OF DISCHARGED HEAT

The use of waste heat does not fit strictly into a volume on the ecological effects of discharges on natural systems, but the fact that any potential pollutant can produce benefits as well as problems is of practical importance to the future management and control of disposal. Also, positive methods for using waste heat can reduce the heat load to natural waters and help minimise effects. The potential uses of heat discharged from factories have been investigated for many years and they cover many fields including district heating, sewage treatment, irrigation, horticulture and aquaculture (e.g. Iles 1963; Shelbourne 1970; IAEA 1974d; Devik 1975, 1976; Hatle and Lampar 1975; Sylvester 1975b; Balligand *et al* 1976; Coutant 1976; Korneev 1976; Descamps 1977; Magnuson *et al.* 1977;

Robertson 1977; Grauby 1978; Muller-Fuega 1978; Kirk 1979; Lee and Sengupta 1977, 1982; Majewski and Miller 1979; Wilcox 1979; Howells *et al.* 1980; Tiews 1981; Langford 1983a; Barnett and Hardy 1984; Engel *et al.* 1985). Table 10.3 summarises some of the uses of waste heat in various parts of the world.

Buckland (1883) referred to an article in the magazine *Land and Water* (May 3, 1879, Number 693), in which an account is given of two goldfish breeding establishments in Germany and Austria, one of which uses the heated effluent from a hemp-spinning factory to enhance the growth and breeding of the fish. Some 120 ponds, covering over 3 ha were heated by steam pipes from the factory and the water was sometimes 'as high as 100°F' (39–40°C). Buckland commended the use of waste heat and steam to 'many men who have ponds connected with factories'. By the 1980s similar schemes had been tried at various types of factory discharging heat, including power stations, textile factories and whisky distilleries.

In fact, during the development of thermal power generation there have been many schemes to use the large amounts of heat discharged to the environment and more are planned for the future. There are, however, limitations to the use of the low grade heat discharged from power stations as effluents, mostly at temperatures lower than 40°C (IAEA 1974d). Also it is worth bearing in mind that:

—in general, waste heat is more usable in cold than warm climates;
—the higher the temperature of an effluent the more possibilities there are for its use;
    the uses of heat may be highly seasonal in temperate regions and may not match with the operating demands of a power station or other industry;
—there are no projects foreseen in which *all* of the waste, low-grade heat from the expanding power industry or other industries can be used for beneficial purposes;
—under the normal operating conditions it is not economic to run power station plant or industrial plant solely to maintain a system which is using the waste heat;
—plants designed to provide heat directly must be near the point of use;
—dual purpose, electricity and heat-providing plants tend to produce the electricity with lower efficiencies, though the total efficiency of fuel-use may be higher than in a normal power plant.

The uses of waste heat have been discussed at length since the 1960s (see Iles 1963; Aston 1988) (Table 10.3) but more recently the limitations to its use

TABLE 10.3
Energy centre heat applications[a] 2000–4000 MW(th)

Application	Particular use	Approximate quantity of product with heat
Central heating	Steam and hot water for residential, commercial and industrial heating	For a city of 500 000– 1 000 000 people
Central cooling	Evaporative cooling for residential and commercial needs	For a city of 500 000– 1 000 000 people
Manufacturing	Electricity and heat for (typical mix)	
	Evap. salt	2 775 t day^{-1}
	Petrochemical	60 000 barrels day^{-1}
	Arc process acetylene	220 t day^{-1}
	Polyvinyl chloride	500 t day^{-1}
	Sodium hydroxide	1 695 t day^{-1}
	Kraft paper	500 t day^{-1}
Desalination for municipal water	Waste water recycling	To 3×10^6 m^3 day^{-1}
	Brackish water distillation	To 3×10^6 m^3 day^{-1}
	Sea-water distillation	To 3×10^6 m^3 day^{-1}
Agriculture	Arid land irrigation with distilled water	To 3×10^6 m^3 day^{-1} (128 000 hectares)
	Arid land irrigation with condenser discharge water	To 5×10^6 m^3 day^{-1} (80 000 ha)
	Greenhouse heating and cooling	To 400 ha
	Poultry house heating and cooling	To 400 ha
Transportation	Stored steam for buses and trucks	54 kg water (condensate of exhaust vapour) per bus mile
	Ice-free shipping lanes	16–32 km ice-free water
Aquaculture	Warm water and sewage for culture:	
	Shell fish	Unknown
	Crustaceans	900 000 kg a^{-1} (400 ha)
	Fish	4 270 000 kg a^{-1} (860 ha)
	Algae	20×10^6 kg a^{-1}
Miscellaneous	Outdoor heating	To $3 \cdot 6 \times 10^6$ m^2 (410 ha)
	Snow melting	To $3 \cdot 6 \times 10^6$ m^2 (410 ha)

[a] The applications shown would each absorb the waste heat from plants in the range mentioned. Actual mix of applications would depend on the circumstances of each specific location.
From IAEA (1974d).

have been recognised better and schemes for the use of the heat have been assessed more prudently from an economic viewpoint. In recent years the use of nutrient-rich water from the depths of the sea, particularly for thermal energy schemes or warmed in shallow ponds has considerable potential for aquaculture (e.g. Othmer and Roels 1973).

10.3.1 Aquaculture
(i) Aquatic plants
Aston and Sadler (1981) concluded that the low grade heat from British power station effluents would provide water temperatures suitable for culturing the exotic water hyacinth *Eichornia crassipes*, which could be used to extract nutrients from sewage effluents as a final treatment before discharge. Winter light was the limiting factor and artificial light would be necessary for the system to be biologically viable. This would, in turn, render such a scheme uneconomic in comparison with the use of conventional sewage treatment. The culture of macrophytes in cooling waters is, however, being considered in a number of regions for use as animal feeds. One of the other systems by which plants could be cultured is where nutrient-rich water from lakes or the sea is pumped through cooling-water systems and after inoculating with suitable algae the system is used to provide the basis of a polyculture involving plankton, other invertebrates and fish (Fig. 10.6, Table 10.4) (Gundersen and Bienfang 1972; Beall 1973; Beall *et al.* 1977; Langford 1983a; Aston 1988). Aston (1988) summarised projects in EEC countries to 1987. Of 45 examples, 22 involved fish, 20 used horticultural products and 3 district heating.

(ii) Invertebrates
Crustaceans and molluscs have understandably been given most attention as culture organisms (IAEA 1974d; Hill 1977; PSEG 1977; Majewski and Miller 1979; Langford 1983a; see Barnett and Hardy 1984) (Table 10.5). In fresh waters, only the tropical fresh-water prawn (*Macrobrachium rosenbergii*) seems to have been useful (Guerra *et al.* 1976, 1979a) though experiments in the UK in the 1960s were not encouraging. Marine crustaceans are cultured in heated effluents in several parts of the world (Table 10.5). At an Italian power station, the prawn (*Panaeus kerathurus*) was reared from post-larvae to marketable size between September and April in the heated effluent water. A second crop was produced in the summer (Palmegiano and Saroglia 1981). Thus two crops were produced instead of the normal one per year. In Japan panaeid prawns are commonly cultured in such heated waters (see Yee 1972; Chiba 1981). In contrast to the

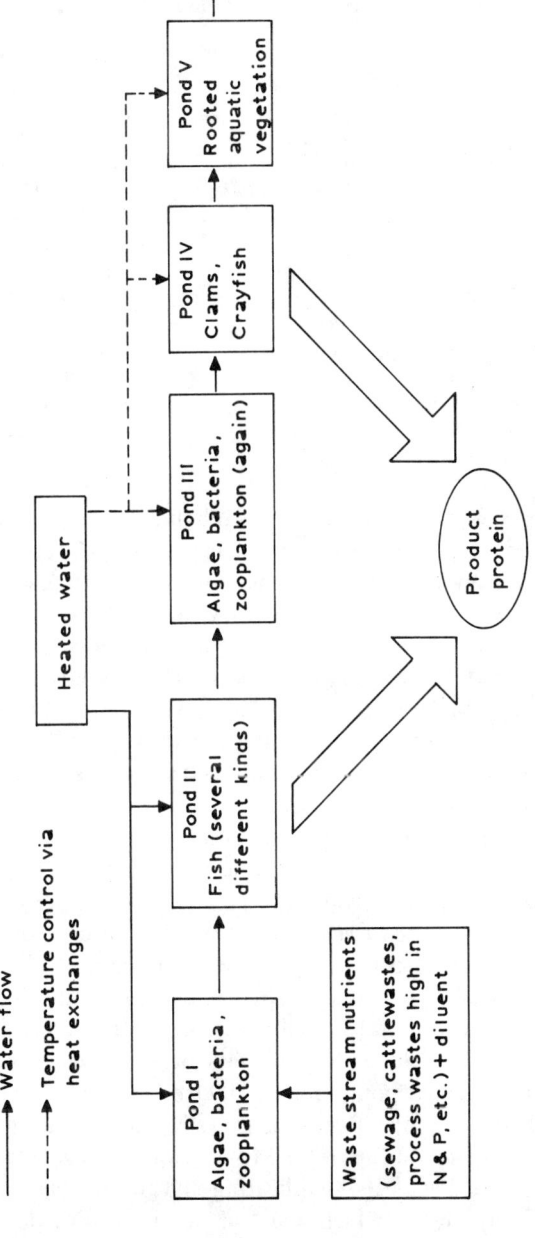

FIG. 10.6 Components of a polyculture system using waste heat.

TABLE 10.4
Two associations of species for waste heat aquaculture systems

Alternative	*Species*	*Primary feeding niche*
1. Carp association		
Grass carp	*Ctenopharyngodon idellus*	Large floating plants
Bighead carp	*Aristichtys nobilis*	Midwater zooplankton
Silver carp	*Hypophthalmichtys molitriy*	Phytoplankton in midwaters
Mud carp	*Cirrhinus molitorella*	Bottom feeder, faeces
Common carp	*Cyprinus carpio*	Bottom feeder
Black carp	*Nylopharyngdon piceus*	Molluscs
2. Tilapia association		
Nile tilapia	*Tilapia nilotica*	Omnivorous (esp. plants)
Java tilapia	*Tilapia mossambica*	Omnivorous (plankton)
Blue tilapia	*Tilapia aurea*	Plankton
Congo tilapia	*Tilapia rendalli* (*T. melanopleura*)	Rooted aquatic vegetation

positive aspects which many studies highlight, the growth of American lobsters (*Homarus americanus*) from the egg to 450 g over 3 years in the heated effluent from the Ecina power plant near San Diego, was considered too slow for commercial development (Aston 1976). Oysters of various species are considered as suitable candidates for warm water culture. At Northport, in the USA, a project was begun in the 1960s (Aston 1976). A two-phase system was used where the larval oysters were first reared from eggs. The larvae were then planted out on mesh trays in the power station discharge. After growing on for 6–12 weeks, the oysters were then planted out into Long Island Sound for their final growth. The warm water phase was claimed to be of considerable advantage.

At Californian sites and at the Maine Yankee nuclear power station heated effluents are apparently used successfully to produce fast growth in young oysters (*Crassostrea virginica*) and also for seeding other molluscs (Hess *et al.* 1978). In California, commercial use is made of a heated effluent and a natural upwelling of water rich in nutrients to stimulate good growth of phytoplankton as food for the molluscs. At the Tokai power station in Japan some 50 000 abalones and 100 000 prawns were expected to be produced from a heated pond area of $2000\,\mathrm{m}^{-2}$ (IAEA 1974c).

In contrast, the growth of mussels (*Mytilus edulis*) was slower in the heated waters near the outfall of the Maine Yankee power station than at the control site (Hess *et al.* 1979) indicating that culture was not feasible in that effluent. Organisms which may be used to provide protein for

TABLE 10.5

Organisms ranked in order of the number of projects in which they are under culture. FW = freshwater culture, M = marine culture, total numbers include commercial as well as R and D projects (from Aston 1988)

Species		No. of projects	
		Total	Commercial
Eels	FWM	9	6
Bass	M	8	2
Carp	FW	6	4
Bream	M	5	1
Turbot	M	4	1
Trout	FW	4	2
Tilapia	FW	2	1
Ornamental fish	FW	1	1
Shrimp	M	2	
Prawn	M	1	
Shell fish	M	1	
Sea trout	M	1	
Salmon	M	1	1
Silure (wels)	FW	1	
Roach	FW	1	1
Perch	FW	1	
Stevsson	FW	1	
Other FW species	FW	1	

processing have been considered for culture. One of the species favoured by British workers is the exotic oligochaete (*Branchiura sowerbyi*) (Aston and Milner 1981; Aston *et al.* 1982). These showed their highest reproductive rate in an activated sludge medium at temperatures of 21–29°C with a doubling rate of 1·6 weeks at 25°C (Aston *et al.* 1981). The sludge needed to be mixed with a coarser medium, such as sand, to give optimal results. The temperatures required were much higher than for temperate species (Aston 1984).

(iii) Fish

Some of the species listed by various authors as being, or having been, cultured in thermal discharge waters are listed by Aston (1976, 1988). As we have seen (p. 342) the idea of fish culture in warm water is at least 100 years old and it seems likely that well before this farmers or fishermen had realised that fish grew well in warm water and had utilised hot springs or waste heat from manufacturing processes.

In the UK, the first, well-documented efforts to grow fresh-water fish in power station effluent water were in the 1960s where three species were kept in cooling-water ponds and their growth monitored (Iles 1963). Some of the best work on the potential growth advantages of such effluent water has been done by Aston and his colleagues in the English Midlands (Aston 1976; Aston and Milner 1976; Aston *et al.* 1976, 1981; Sadler 1979) and by Descamps (1977) in France. The growth rates of eels (*Anguilla anguilla*) (Fig. 10.7) and carp (*Cyprinus carpio*) reared in power station cooling waters were found to be greatly enhanced (Aston *et al.* 1976). The fastest growth was found in the warmest water and marketable eels, for example, could be produced in $2-2\frac{1}{2}$ years instead of 10–14 years in the wild. As yet, however, commercial operation has not shown the profit advantages of using heated effluents for their culture, though in 1987, six of the European projects out of 9 producing eels were on a commercial basis.

Aston (1975) showed that a carp farm in East Germany was a profitable venture with a net profit of about 50% of the annual expenditure. Table 10.5 shows the species being cultured in Europe in 1987.

Many marine species have been used for culture in both temperate and sub-tropical regions.

In Japan the fish farms using heated effluents are considered as profitable ventures (see Barnett and Hardy 1984), mainly because of the ready market for the cultured species. In the USA, commercially viable farms were claimed at the Mason Power station in Maine and at the Lake Colorado City power station (Aston 1976; Beall *et al.* 1977). At the Hunterston nuclear power station in Scotland, the farms rearing flatfish and more recently fish such as sea breams and bass (*D. labrax*) have been in existence for well over 20 years (Nash 1968, 1969, 1970; Kerr 1976). In the earlier work, plaice (*Pleuronectes platessa*) and halibut (*Hippoglossus hippoglossus*) were found to grow on to marketable size about twice as quickly in the heated effluent as in cold water but, even so the costs of the maintenance, pumping, extra food and capital equipment rendered the culturing of these fish uneconomic.

Trials with turbot (*Scophthalmus maximus*) suggested economic advantages but these do not seem to have been fully realised. The most recent work utilises the heated effluents from power stations to enhance the growth of the early life-history stages before planting out into the sea. Barnett and Hardy (1984) noted that the use of heated brackish waters to grow both fresh-water and estuarine fish, for example eels (*A. anguilla* and

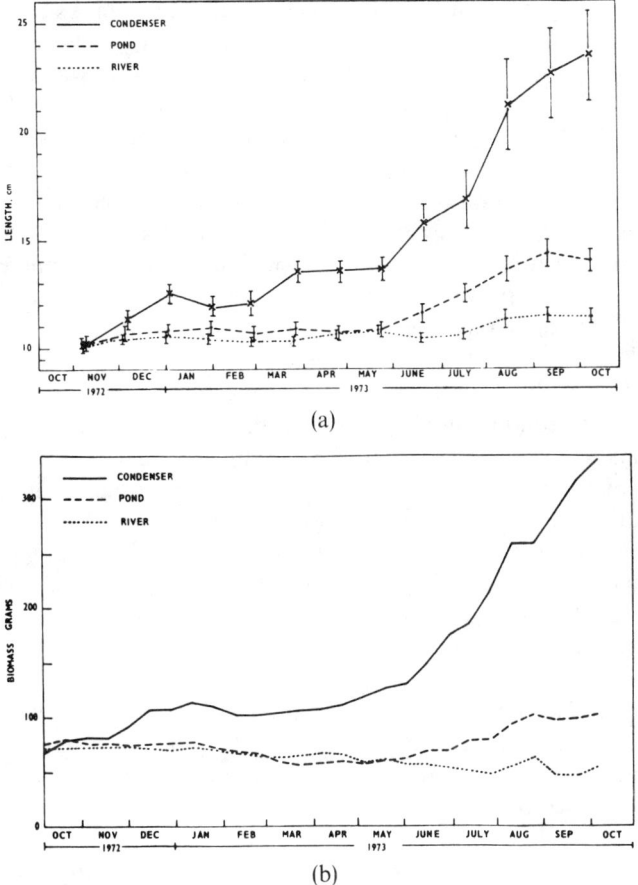

FIG. 10.7 (a) Mean lengths and standard errors of eels from condenser, cooling tower pond and river water from November 1972 to October 1973. (b) Total biomass of eels from condenser, cooling tower pond and river tanks from October 1972 to October 1973 (from Aston *et al.* 1976).

A. japonica) is successful in several countries because of the temperature tolerance of the fish and their high market value. The increasing use of additional heat to culture salmonids either for the table (Peterson and Seo 1977) or to speed their development for planting out (Bulleid 1974; Lawrie 1976; Saunders 1976; see Barnett and Hardy 1984) offers commercial and ecological advantages for the future and requires further development.

As we will see in the next section, even in temperate countries, heated effluents could be lethal for many species cultured in tanks or ponds in summer. Optimal use can be made of these tanks or ponds, however, by rotation of species with season or by tempering the effluent with cold water. For example, at the Mercer power station in New Jersey, *Macrobrachium* was grown at the higher summer temperatures and rainbow trout *S. gairdneri* in the winter (Eble *et al.* 1975). Combinations of salmonids and cyprinids could also be viable because of the different thermal tolerances (see Chapters 4 and 9). Cage culture in heated waters is another alternative which has been attempted in various places, but the added advantage of the heat is not yet proven commercially over cage culture at normal temperatures (e.g. Holmes *et al.* 1974; Hooper *et al.* 1978).

(iv) Problems with aquaculture in heated effluents
The direct and continuous use of a heated effluent for the culture of aquatic organisms is rarely possible. Whether the effluent originates from a power station or a manufacturing process, there are usually operational incompatibilities between the needs of the culture system and that of the industrial installation. For example, as we saw in Chapter 2, the maintenance schedules at a power station, the seasonal fluctuations in demand for electricity, differential fuel costs or sudden breakdowns of plant cause both predictable and unpredictable fluctuations in the volume and temperature of effluent produced.

The use of biocides, flocculants and all the other treatment chemicals in industrial systems must also be controlled, together with contaminants from other sources or the manufacturing process if an aquaculture system has to be protected.

(a) Temperature. The optimisation of the temperature for the best growth and food conversion rate is the major objective of any aquaculture system. Figure 10.8 shows the desirable range for several species of fish in relation to possible effluent temperatures from UK power stations. Using a cold-water tempering system, it was found that the daily growth rate of turbot (*S. maximus*) in a heated water farm could be increased by a factor of 2·7 (Jones *et al.* 1981), but there was a clear economic penalty from the costs of pumping cold water.

In a trial using the recirculating water at a Midland power station for growing carp (*C. carpio*), growth rates and food conversion were improved by tempering the condenser outlet water with cooled water from the cooling tower ponds such that temperatures did not exceed 28°C. Both mirror carp and common carp (Fig. 10.9) showed improved growth in the

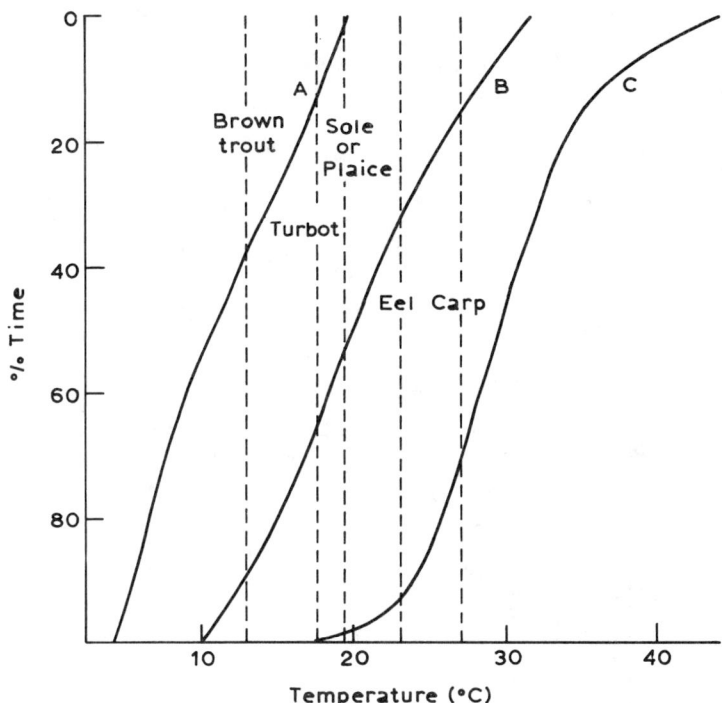

FIG. 10.8   Water-temperature distribution over a five year period in (A) a typical British River, (B) the condenser effluent from a coastal power station (Bradwell) and (C) the condenser effluent from an inland power station using cooling towers (Ratcliffe-on-Soar). Also shown with dashed lines are the temperature optima for growth in fish of commercial interest which have been reared at farms sited at power stations in the UK (from Turnpenny *et al.* 1985).

tempered water. The mean daily increase in weight for the former was from 2·6% in the condenser water to 3·2% in the tempered water. The respective figures for common carp were 2·8% and 3·3% (Aston *et al.* 1978). Food conversion ratios improved from an average of 3·7:1 to 3·0:1. Pumping costs for the temperature regulation would have been relatively high, but would have been reduced by 10% by surface cooling in large fish ponds.

In later work at this site, a commercial farm, on a pilot scale, grew cyprinids for stocking by floating cages in the cooling-tower ponds where temperatures were, on average, most advantageous for their growth (Aston personal communication). One important observation in this work was that, provided that temperatures were not lethal, the short-term

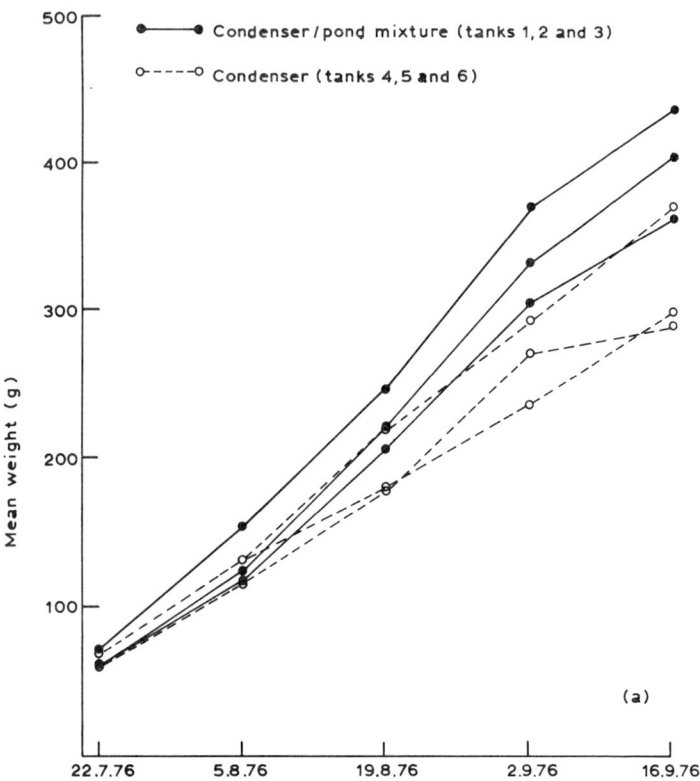

FIG. 10.9 Growth of (a) common carp and (b) mirror carp in condenser/pond water mixture (solid lines) and condenser water (dotted lines) at Ratcliffe-on-Soar power station. Each point gives the mean value for the weights of fish in each tank (from Aston *et al.* 1978).

fluctuations over a range of 3–10°C did not suppress growth advantages. There have been recorded mortalities of fish and crustaceans being cultured in heated effluents caused by lethally high temperatures (e.g. Strawn 1970; Guerra *et al.* 1976) and this will always be a problem for the less tolerant species. A bigger problem because of its unpredictability, is the sudden shutdown of power stations or other industries in winter. As we have seen (p. 283), this has caused fish mortalities in thermal discharge canals where acclimatised fish were suddenly exposed to a fall in temperature of up to 10–12°C. In tanks or floating cages there would be no chance to avoid or escape the adverse conditions. The stand-by equipment for this type of emergency at a large fish farm using heated effluent could be prohibitively expensive. The recommendation is, therefore, only to consider sites where

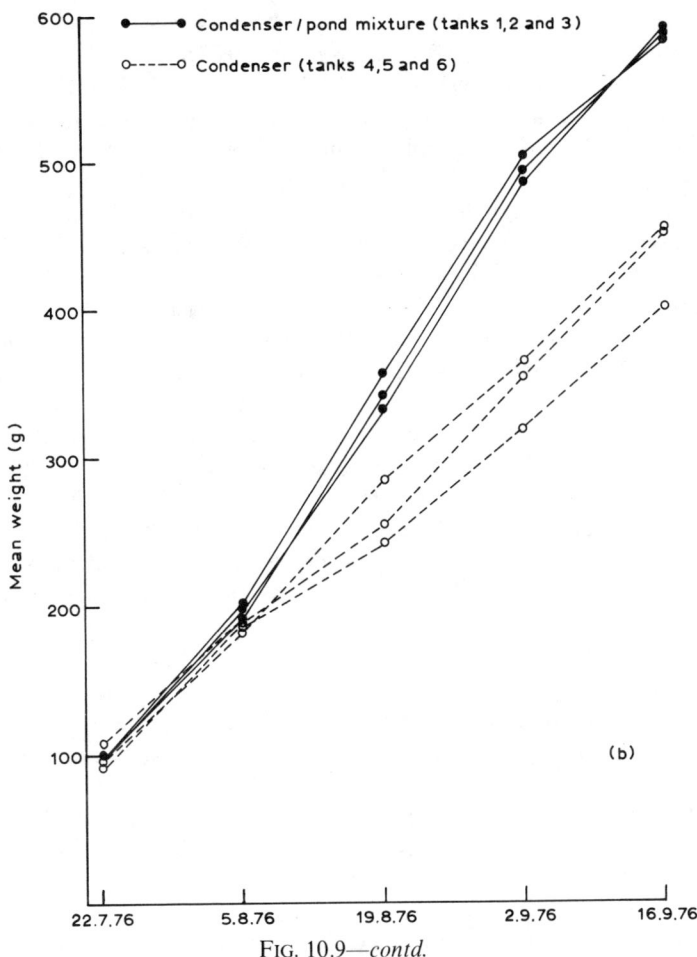

FIG. 10.9—*contd.*

there are multiple installations of plant producing heated effluents and thus the likelihood of all being out of use is reduced.

(b) Chlorine and other biocides. Chlorination was one of the earliest problems encountered by the marine fish farm experiments at Hunterston nuclear power station (Nash 1968; Langford 1983a; Barnett and Hardy 1984). Systems used to try to prevent mortalities have included chlorine alarms which either activate valves to shut off the water supply (see Aston 1976) or which notify staff to take some action (Page-Jones 1971). This last

author described the alarm at Hunterston which was set at a threshold of 0.3 mg litre^{-1} though it could detect as little as 0.03 mg litre^{-1}. Nash (1974) suggested that chlorine residuals of 0.02–0.1 mg litre^{-1} would have little effect in fish tanks if the exchange rate in the tanks was less than 5 volumes per day. As we have seen (Chapter 3) chlorine dosing at large power stations is usually aimed at producing concentrations of 0.2–0.5 mg litre^{-1} at the condenser inlets though decay and dilution reduce these concentrations at the outlets and in discharges. These levels are most likely to be toxic to larval stages of fish and crustaceans in culture systems and thus some dechlorination methods would be required (Saroglia and Scarano 1983). At fresh-water sites, intermittent dosing produces short-lived peaks which do not appear to have killed fish being kept in the cooling water (see Chapters 3 and 9).

Dechlorination of effluents has been attempted using various chemicals (see Langford 1983a). The most favoured have been thiosulphate compounds and sulphur dioxide. Seegert and Brooks (1978b) compared dechlorination methods, including activated carbon, ultra-violet light and sodium sulphite. They eventually recommended a method involving a combination of carbon and sodium sulphite which they considered as safe, effective and reliable.

Scott (1983) concluded that the dechlorination of chlorinated effluents by thiosulphate at a precise $1:1$ chlorine to thiosulphate ratio would decrease the degree of toxicity and physiological stress induced in marine invertebrates by the chlorine residuals. Excess thiosulphate could, however, be harmful.

It has been suggested that at fossil-fuelled power stations, flue gases which contain substantial amounts of sulphur dioxide (SO_2) could be used to dechlorinate effluent water at fish farms using the waste heat. Emberton and Turnpenny (1982) found that SO_2 bubbled into chlorinated sea water containing fish, produced better survival than in the untreated chlorinated water. The major problem was that the pH of the sea water was also reduced and could have been an additional lethal factor in the long term. The authors concluded, however, that, no chlorinated heated sea-water effluent from a British power station should be used in tank culture systems without dechlorination. Free living fish or mobile invertebrates would avoid adverse conditions (see Chapter 4).

Clearly, the safest systems to use for heated water aquaculture are those in which no chlorination or other biocide treatments are used (Aston 1976). At power stations, for example, those which use physical methods of condenser cleaning (Garey 1980) are preferable.

Sylvester (1975b) considered that chlorination could have adverse effects on the taste and marketability of cultured species, though there is no evidence to substantiate this. In some situations, however, there may be substances in the cooling-water source which could cause such problems. For example, the author and colleagues found that salmonids and bass *D. labrax* kept in the intake water at the Fawley power station in southern England became inedible after a few weeks because of the taint of hydrocarbons originating from the legally allowed discharges from a large refinery nearby. The effluent water would have been similarly tainted though the fault clearly did not lie in the power station cooling-water system.

In the trials at the Mercer warm-water fish farm, tasting tests found the shrimps and trout to be no different from any others and perfectly acceptable to consumers (Guerra *et al.* 1979a, b).

The use of heat exchangers to transport low grade heat is uneconomic and often impractical on the large scale (Hugenin 1976; see Langford 1983a). However, where the waste heat is in the form of steam or near-boiling water such devices may be both practical and economic.

One solution to the problems associated with contamination is to raise fish purely for ornamental purposes, for example the exotic goldfish and carps which can sell for extremely high prices (see Iles 1963; Parker and Krenkel 1969). In France there is at least one installation involved in this work at the present time.

(v) Other contaminants and potential problems
The other contaminants in cooling-water discharges have been discussed at length in the previous chapters (see Chapter 3) and clearly any organisms being cultured in a discharge would be exposed to such contaminants (Boyden and Romeril 1974).

Romeril and Davis (1976) found that eels and carp accumulated heavy metals when grown in the cooling water system of the Ratcliffe-on-Soar power station. The River Trent, from which the cooling water is abstracted, contained high concentrations of heavy metals from the effluents upstream. The highest concentration in eels was in the cold, raw Trent water. There was, however, no evidence of additional bioaccumulation in the fish from the heated water and, because of rapid growth, the tissue concentration was actually lower. Similar results were found for fish and shrimps cultured in the effluent of an Italian power station (Saroglia *et al.* 1981). On the other hand, Melard and Phillipart (1981) found heavy metals concentrated in the tilapia (*Satherodon niloticus*) cultured in the cooling-water discharge of a Belgian power station, though they were below the legally permitted levels.

Although molluscs, as filter feeders, would be expected to accumulate contaminants in heated effluent culture, the results of field studies show variable effects. Oysters (*Ostrea edulis*) and plaice (*Pleuronectes platessa*) at Hunterston contained levels of radio-nuclides well below the legal limits (see MAFF, 1972 *et seq.*) Oysters (*Crassostrea virginica*) cultured at the Maine Yankee power station grew faster and accumulated gamma-emitting radio-nuclides faster in the heated effluent than in cold water (Price *et al*. 1976; Hess *et al*. 1978). Again the ultimate levels were below the legal limits. Similar data were obtained for the accumulation of ^{65}Zn in the same species at the Humboldt Bay nuclear power station in California (Salo and Leet 1968). As we have already noted (p. 333) oysters (*C. gigas*) accumulated copper in the effluent plume of a Californian power station. The bioaccumulation of chlorinated hydrocarbons in culture systems is also of some concern, though again there is no evidence of levels harmful to man. Scott *et al*. (1983) found that *C. virginica* accumulated bromoform but depurated it rapidly in clean sea water. The effects were the same in both chlorinated and dechlorinated sea water.

Some 58 identifiable chlorinated compounds were identified in Asiatic clams (*Corbicula manilensis*) collected from near the cooling-water outfalls of coal-fired power stations. Some of the compounds were not associated with chlorinated effluents and originated from other, unidentified sources (Lee *et al*. 1983).

In the intensive systems used for culturing fish, the supply of oxygen may become a limiting factor, particularly in heated waters (Aston and Heathwood 1980; Aston 1981; Brown *et al*. 1984). Using eels as an example, the rate of oxygen consumption at 15°C is approximately 80 mg kg^{-1} h^{-1} and 150 mg kg^{-1} h^{-1} at 25°C. Thus to maintain a working concentration of oxygen of 5 mg litre^{-1} the rate of aeration needed at 25°C is 2·3 times that at 15°C. Aston's (1981) conclusion was that fish farms using heated effluents, even where cooling towers are present, would need aeration equipment to cope with oxygen demand at the highest temperatures and usual stocking rates. Barnett and Hardy (1984) considered that because of the higher oxygen requirements after feeding, short-term aeration in this period would be an advantage. Also, organisms which have low activity can be an advantage, for example turbot as compared with salmonids have low activity and oxygen consumption.

Aston and his colleagues (Aston *et al*. 1980) compared three commercial methods of aeration at a fish farm using heated effluent and recommended a simple choice of porous plastic tubing supplied with air from a low-pressure compressor. In seeming contradiction to these oxygen deficiency problems, problems have been reported in thermal discharges as a result of supersaturation of oxygen and nitrogen leading to symptoms of gas bubble disease (see p. 313) (Coutant and Genoway 1968; Malouf *et al.* 1972; Miller R. W. 1974; Rucker 1976; Goldberg 1978). Turbot (*S. maximus*) were killed by nitrogen supersaturation in a system heated by the effluent from Wylfa nuclear power station in North Wales. Levels of 118% saturation were the cause (Jones *et al.* 1981). Other reports are of salmonid mortalities in both fresh and brackish heated waters (see Barnett and Hardy 1984). In intensive systems gas concentrations can be maintained at acceptable levels by bubbling air from a similar system as already described above for alleviating oxygen deficiency. However, where fish are cultured using floating cages in an effluent canal or plume, artificial aeration is not usually practicable. Chamberlain and Strawn (1977) found that 99% of the fish reared in cages in a power station effluent canal near Galveston in Texas, died in 14 days as a result of gas bubble disease. Fish in cages at 3 m depth survived well. The problems of disease and parasite infestations in intensive aquaculture systems are well known (Roberts 1976, 1978). As we have seen, the effects of thermal discharges in open receiving waters can exacerbate the problems though some organisms appear to be less virulent in heated waters (Groberg *et al.* 1978; Sanders *et al.* 1978).

The major bacterial disease problems at warm water fish farms appear to have been caused by *Vibrio* spp. (Mihursky *et al.* 1970; Egusa 1981; Jones *et al.* 1981). Closed systems are most at risk and expensive dosing treatments may be necessary to control the diseases. The author noted that fish introduced into the heated, chlorinated water at a Midland power station already heavily infected with the trematode parasite *Gyrodactylus* spp., causing lesions, became free of the parasites after exposure for several weeks. The pathogenic amoeba *Naegleria fowleri* has also been recorded in heated fish-farm waters (De Jonckheere *et al.* 1983).

Barnett and Hardy (1984) list four advantages of using heated effluents from coastal power stations for aquaculture, in relation to disease control, namely:

—they are sited on open coasts, away from areas of heavy pollution;
—chlorination tends to suppress disease organisms;
—open, direct cooling systems provide large amounts of water which will flush the culture system of pathogens and parasites;
—higher temperatures will suppress cold water pathogens.

They do note, of course, that warm water pathogens may be favoured. We have already seen (p. 313) that in fresh waters heated effluents may have less advantageous effects on parasites and disease.

There is clearly still some considerable potential in the use of heated effluents for fish farming (Jones 1976, 1980), though, as yet, there are basic problems to overcome, particularly with regard to chlorination at coastal sites, and the effects of stocking densities and temperature on ammonia and oxygen (Aston and Brown 1978).

The choice of the site is the vital factor. A suitable site should include the following characteristics:

—A clean source of cooling water with low sediment concentrations, whether fresh, brackish or salt water.
—Multiple units of plant generating the heat to ensure continuous supply.
—Access to the heated cooling water before any contaminants enter.
—Easy access to cold, clean water for regulating the temperatures to the desired optimum.
—Preferably a topography which allows gravity feeding of the main water supplies to minimise pumping costs.

Before entering into a scheme any operator should carry out a careful market survey because of severe competition from the present commercial developments.

On ecological grounds, the use of a heated effluent for aquaculture may have serious disadvantages. A large-scale fish farm will produce large volumes of effluent itself and instead of only heat, this will carry a substantial load of organic matter and ammonia. This may be far worse for the receiving water than the original thermal discharge. Any treatment required is yet another cost for the farm operation (Barnett and Hardy 1984).

10.3.2 Other Uses of Waste Heat

The other systems for using waste heat have been thoroughly reviewed by several authors (Yee 1972, 1974; see IAEA 1974d; Beall *et al.* 1977; Lee and Sengupta 1977, 1982; Yarosh 1977; Majewski and Miller 1979; Marshall 1979; Olszewski *et al.* 1979; Howells *et al.* 1981; Langford 1983a; Barnett and Hardy 1984).

(i) Soil warming
Several trials have been carried out, in various countries, using heated effluents in networks of buried pipes to heat soils and enhance the growth of crops. Luckow and Reinken (1979) found that there was a significant increase in the yields of corn, sugar beet, winter wheat and pastureland in fields underheated by cooling water in Germany. Spring potatoes ripened 4

weeks earlier than in unheated fields and showed a 60% increase in yield, while soya beans, not normally grown in Germany yielded over $5\,t\,ha^{-1}$.

Yarosh (1977) and De Walle (1979) reviewed the work on soil warming in the USA and again showed that crops showed increased yields and earlier or double harvesting, particularly where frozen soils could be thawed and made workable early in the season. In the warmer regions, however, summer-planted crops did not benefit from the soil warming. Direct spraying of heated fresh water showed similar benefits. Barnett and Hardy (1984), discuss the use of heated sea water for soil warming but so far this has not been tried. In an assessment of the priority uses of waste heat, soil warming was ranked fifth (Olszewski *et al.* 1979). It is likely that the costs of installation and maintenance of such systems are too high to be offset by their biological advantages (see Barnett and Hardy 1984 for references).

(ii) Glasshouse and greenhouse heating and culture
Table 10.6 shows the crops under cultivation in heated greenhouses in Europe in 1987. At that time potted plants were the commonest crop (Aston 1988). At Drax power station in northern England a 20 acre (44 ha) greenhouse culture system has been in operation for some years using hydroponic culture and cooling water for heating. The system uses fan-assisted fin-tube heat converters and heating panels to maintain the required temperatures. The venture was based on experiments at another site (Statham *et al.* 1979). Tomatoes and lettuce were being grown commercially (Howells *et al.* 1981). Large-scale trials showed that the yields were similar to other systems but the costs of heating were much reduced.

In other countries similar ventures have been in operation since the 1960s and early 1970s (see Barnett and Hardy 1984; Aston 1988). For example in Romania, heat from a local power station has been used to heat a complex of over 80 ha of greenhouses for many years. The crops include cucumbers, eggplants, tomatoes and lettuces. Similar schemes are in use or under trial in Japan, North America, France and Czechoslovakia. In Japan, the heated water is pumped over the roofs and along the walls of greenhouses to maintain a temperature of 15–25°C. In other countries heat exchangers are used to transport high grade heat. These uses of waste heat were given lowest ranking by Olszewski *et al.* (1979).

Other uses of waste heat include heating animal shelters at farms, space heating particularly using dual-purpose, specially designed plant, heat for industrial processes, enhancement of sewage treatment, processes producing protein and desalination. Of these the system likely to best utilise heat directly is the combined heat and power system or total energy system,

TABLE 10.6
Plant species ranked in order of the number of projects in which they
are under culture. G = greenhouse culture, F = open field culture.
Total numbers include commercial as well as R and D projects (from
Aston 1988)

Species		No. of projects	
		Total	Commercial
Potted plants	G	6	6
Tomatoes	G	3	2
Asparagus	F	2	
Strawberries	G	2	
Young nursery plants	G	2	2
Salads	G	2	1
Cucumbers	G	1	1
Potatoes	F	1	
Cauliflowers	F	1	
Peppers	FG	1	
Lemons	FG	1	
Roses	F	1	1
Corn	F	1	
Rice	F	1	
Peach tree plants	F	1	
Green plants	FG	1	
Phytoplankton		1	
Cut flowers	FG	1	

some versions of which have been in use in colder countries for many years
(see Gay *et al.* 1976; Muir 1976; Majewski and Miller 1979; Marshall 1979;
Marecki 1988).

In the Soviet Union, over 1000 dual purpose heat and electricity
generating schemes were operating in 800 towns and cities in 1980
(Majewski and Miller 1979). Sweden and Finland also use these systems
extensively. Thermal efficiencies of up to 88% are claimed at the best plants.

These plants are specially designed to produce both electricity and high-
grade heat, either as steam or more recently hot water, which can be
transported to nearby buildings or industrial process plants for use. Figure
10.10 shows the conceptual layout of an integrated total energy scheme.

The major disadvantage of these schemes for electricity generation is
that the electrical output is not as great per unit of fuel used as in
conventional power stations but this can be offset by the use of the hot
water and, for fossil-fuelled plant, by reductions in the emissions of sulphur

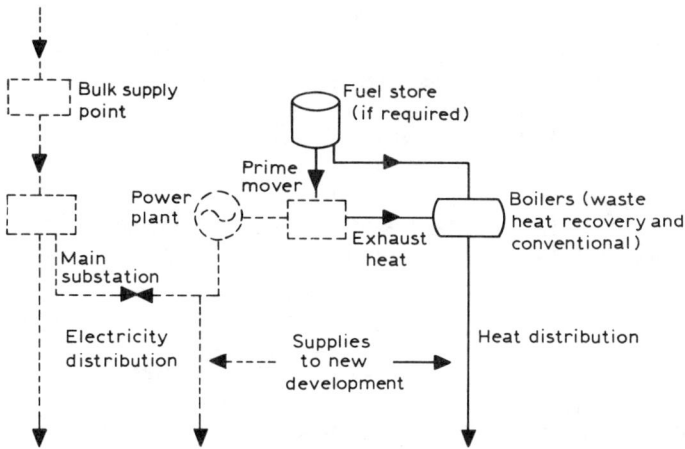

F_IG_. 10.10   Line diagram of an integrated total energy scheme.

dioxide as the exhaust gases are recycled to extract the heat. In countries with poor fresh-water supplies, the use of waste heat from a power station for desalination can be a great asset. The waste heat from a 2000 MW (thermal) plant could help produce $2 \cdot 7 \times 10^6 \, m^3 \, day^{-1}$ of fresh water (IAEA 1974d). Figure 10.11 shows the temperature ranges needed to supply the various systems in which waste heat can be used.

For the future, there is clearly much more work to do on the development of the uses of waste heat, though the economic benefits of any schemes are not yet universally proven. It must be borne in mind that, even making use of waste heat in the various systems, it is unlikely that, in another age of expansion of electricity demand and production, this will make a significant difference to the total amounts of heat to be discharged (Barnett and Hardy 1984).

10.4 THE CONTROL OF THERMAL DISCHARGES, CRITERIA AND THE LAW

It is abundantly clear from the published data that large-scale thermal discharges cause changes in the flora and fauna of natural water bodies, albeit mostly very localised. The changes to communities are clearly mostly caused by scour or biocides, except in the warmest effluents. The changes to physiological processes are, on the other hand, clearly related to temperature at sub-lethal levels. The extent to which effluent temperatures

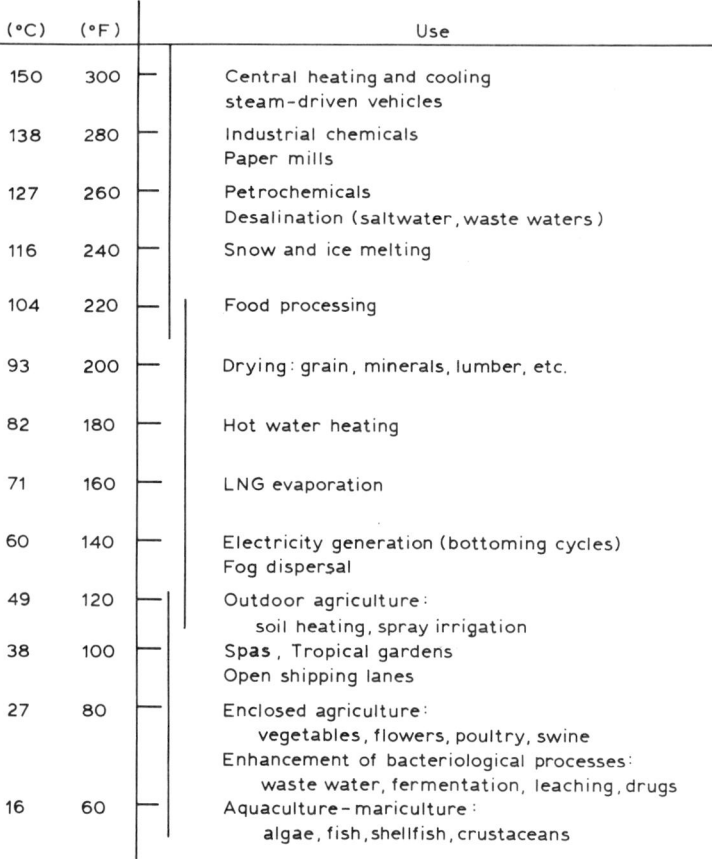

(°C)	(°F)	Use
150	300	Central heating and cooling steam-driven vehicles
138	280	Industrial chemicals Paper mills
127	260	Petrochemicals Desalination (saltwater, waste waters)
116	240	Snow and ice melting
104	220	Food processing
93	200	Drying: grain, minerals, lumber, etc.
82	180	Hot water heating
71	160	LNG evaporation
60	140	Electricity generation (bottoming cycles) Fog dispersal
49	120	Outdoor agriculture: soil heating, spray irrigation
38	100	Spas, Tropical gardens Open shipping lanes
27	80	Enclosed agriculture: vegetables, flowers, poultry, swine Enhancement of bacteriological processes: waste water, fermentation, leaching, drugs
16	60	Aquaculture–mariculture: algae, fish, shellfish, crustaceans

FIG. 10.11   Temperatures required for water to be used in various processes (from Majewski and Miller 1979).

require rigid control, particularly in waters where dilution is high and mixing rapid is arguable. Controls on outfall siting and design and the use of biocides are probably more important. In smaller bodies of water, controls on extreme high temperatures are much more significant.

There are different approaches to the control of polluting discharges in different countries and many countries have Federal and State or other local constraints which must be applied (Coulson and Forbes 1952; EPA 1972; McLoughlin 1976; Anon. 1979; Johnson 1979; Johnson and Courcelle 1989). In England and Wales, the control of water abstraction and effluents are covered by various Acts of Parliament, which allow the appropriate pollution control authority to grant consents for abstraction and the composition of any discharge, often based on its potential effects in its specific receiving water. The standards set for European waters and effluents are mostly dictated by EEC Directives with which the member countries are generally expected to comply. The control of pollution by the setting of Environmental Quality Objectives (EQOs) which consider the effects of any discharge on the total water quality of a receiving water has advantages and disadvantages to the potential polluter, but provided that the objectives are sufficiently ambitious in terms of improvement and clean water, the system is of overall benefit to the water body itself.

Langford (1983a) reviewed the constraints and the law relating to power-station effluents at that time, in various parts of the world. The conclusion was that none of the scientific data, from the in-situ studies indicated the need for rigid, and at times unnecessarily stringent, constraints on thermal discharges where mixing is rapid or the dilution is high (see Heimberger 1982). Effluent limits as low as 25°C and temperatures rises kept to 0·5–3°C are ecologically unnecessary and economically indefensible in many waters (House of Lords 1977). Leason (1974) argued that cooling water is not strictly a polluting discharge, but clearly this is not supported by the evidence presented in this volume, or by the law. In fact, prior to any state supported pollution control in the UK, legal action had been taken under Common Law against the discharge of water at other than normal temperatures to a stream (Wisdom 1979). As we have seen in Chapter 1, the potential hazards of heated discharges were noted early on in the history of the electricity industry, for example in the Electricity Act of 1919.

Various criteria have been established in different countries to control the temperatures of discharges. Attempts to create 'blanket' criteria, that is a single temperature value for a whole country or region which should not be exceeded, are not practical in many regions nor are they necessarily of benefit to the ecosystem in many places.

The seemingly rigid criteria set in the 1960s for the waters of the USA were subsequently modified to be more tailorable to specific conditions (Tarzwell 1970; Miller and Beck 1975; EPA 1976; Bugbee 1978; Jeter 1977) (see Table 10.7). These criteria were, in turn, also subsequently criticised (Thurston *et al.* 1979) for not considering all the factors involved in protecting the different species and ecosystems and for not being sufficiently stringent. Whilst this may be true for the dryer regions of the USA with small rivers and lakes, there can certainly be a case for relaxing standards set at the point of outfall in otherwise unpolluted waters where the receiving water offers huge dilution or rapid mixing.

TABLE 10.7
Temperature Criteria produced by US EPA for site-specific use

Freshwater Aquatic Life
For any time of year, there are two upper limiting temperatures for a location (based on the important sensitive species found there at that time):
1. One limit consists of a maximum temperature for short exposures that is time dependent and is given by the species-specific equation:

Temperature ($^\circ$C) = $1/b$ [$\log_{10}$(time in minutes) $- a$] $- 2$

Where:

a = intercept on the 'y' or logarithmic axis of the line fitted to experimental data which are available for some species from Appendix II-C, NAS, 1974.
b = slope of the line fitted to experimental data which are available for some species from Appendix II-C, NAS, 1974.
2. The second value is a limit on the weekly average temperature that:
   a. in the cooler months (mid-October to mid-April in the north and December to February in the south) will protect against mortality of important species if the elevated plume temperature is suddenly dropped to the ambient temperature, with the limit being the acclimation temperature minus 2°C when the lower lethal threshold temperature equals the ambient water temperature (in some regions this limitation may also be applicable in summer); or
   b. in the warmer months (April through October in the north and March through November in the south) is determined by adding to the physiological optimum temperature (usually for growth) a factor calculated as one-third of the difference between the ultimate upper incipient lethal temperature and the optimum temperature for the most sensitive important species (and appropriate life state) that normally is found at that location and time; or
   c. during reproductive seasons (generally April through June and September through October in the north and March through May and October through November in the south) meets site-specific requirements for successful migration, spawning, egg incubation, fry rearing, and other reproductive functions of important species. These local requirements should supersede all other requirements when they are applicable; or
   d. is a site-specific limit that is found necessary to preserve normal species diversity or prevent appearance of nuisance organisms.

Marine Aquatic Life
In order to assure protection of the characteristic indigenous marine community of a water body segment from adverse thermal effects:
1. the maximum acceptable increase in the weekly average temperature due to artificial sources is 1°C (1·8°F) during all seasons of the year, providing the summer maxima are not exceeded; and
2. daily temperature cycles characteristic of the water body segment should not be altered in either amplitude or frequency.

TABLE 10.7—*contd.*

Summer thermal maxima, which define the upper thermal limits for the communities of the discharge area, should be established on a site-specific basis. Existing studies suggest the regional limits expressed in the table:

	Short term maximum	Maximum true daily mean[a]
Sub-tropical Regions (south of Cape Canaveral and Tampa Bay, FL, and HI)	32·2°C (90°F)	29·4°C (85°F)
Cape Hatteras, NC to Cape Canaveral, FL	32·2°C (90°F)	29·4°C (85°F)
Long Island (south shore) to Cape Hatteras, NC	30·6°C (87°F)	27·8°C (82°F)

Baseline thermal conditions should be measured at a site where there is no unnatural thermal addition from any source, which is in reasonable proximity to the thermal discharge (within 5 miles) and which has similar hydrography to that of the receiving waters at the discharge.

[a] True daily mean = average of 24 hourly temperature readings.
From Thurston *et al.* (1979).

Krouse (1978) discussed the muddle of regulations, both Federal and State, which tried to cover the control of power station effluents in the USA. The evolution of issues were noted, for example, the drafting of legislation in 1971 and 1972 indicated that 'thermal' effects were paramount, whilst later regulations stressed the effects of entrainment and impingement. It is interesting to note that there are now few data to suggest that either are damaging to populations. Krouse (1978) also notes that much of the wording in the law is imprecise and ill-defined. From observations of this and subsequent environmental issues and their legislation it seems clear that poor legislation arises from poor science or the placing of political expediency above the scientific facts. As one example, the maximum temperature limit of 28°C on cyprinid waters and 21·5°C (Johnson and Courcelle 1989) in salmonid waters, in Europe can be impractical in the UK and in other countries, where, during hot dry summers, natural river temperatures in both salmonid and coarse fish reaches can reach 24–26°C (see Brooker *et al.* 1977; Koops 1978; Koops *et al.* 1980; Langford 1983a; Sweers 1983). Most fish inhabiting European rivers are able to tolerate temperatures up to 28°C or so without harm for long periods, though temperatures exceeding 30°C may not be tolerated for more than a few hours. However, there is clearly a case for ensuring that heated effluents do not heat the *whole* of a river reach for long periods, and it is essential that 'escape routes' are available for both migratory and non-migratory species if temperatures exceed 30°C. Blanket criteria do not usually take into account the vertical and horizontal stratification of thermal effluents, particularly in large bodies of water or the open sea which provide such escape routes or areas. The design and configuration of an effluent outfall is, therefore, a vital factor in minimising the effects of the effluent (Leason 1975, 1976; see Fischer *et al.* 1979).

The most stringent temperature criteria would need to be imposed on tropical waters where organisms are living already at high temperatures. However, as we have seen in various chapters, even in such waters, scour, chlorine or some other factor than heat may cause the largest disturbance to the ecosystem.

In essence, we have seen that most large thermal discharges are not simply vehicles for the disposal of heat but also have other properties which can affect ecosystems, either in a source water body or in a receiving water. Local, tailor-made constraints on the temperature, chemical composition and physical mode of discharge are, therefore, advisable to minimise the effects on the receiving water, where this is clearly necessary.

The importance of chlorine and its residuals in cooling waters is much greater than was envisaged in the early legislation, and there is still disagreement among American authorities as to the acceptable limit to protect ecosystems (EPA 1976; Capuzzo *et al.* 1977b; Lum 1978; Thurston *et al.* 1979; Seegert and Bogardus 1980; Mattice and Tsai 1983). The composition of the biocoenosis in a receiving water affected by a thermal discharge at any point in time will consist of:

—those indigenous species which can tolerate the temperature, chemical and physical changes caused by the effluent;
—those species which have survived entrainment through the system and colonised the discharge area;
—those species which have migrated into the area as a result of the changed conditions and have replaced species eliminated by the discharge,

The degree of ecological change and the area affected are the important elements for the ecologist to determine or to predict.

The decision on the amount of ecological disturbance which can be allowed or tolerated at any site must be a valued judgement based on the needs of the human population, economics, politics and the ecological value of the location.

10.5 THE FUTURE AND THE NEED FOR FURTHER INFORMATION

The prediction of the effects of future thermal discharges on specific aquatic ecosystems is clearly a complex exercise but one that will be very necessary as the demands for more electricity and more industrial development are met, particularly in the developing countries. The legal mechanisms for presenting ecological predictions exist in many countries. In the USA and more recently in Europe, the formal Environmental Impact Statement (Baker *et al.* 1977; Richkus 1977; Bendix and Graham 1978; Ward 1978; Landy 1979; see Langford 1983a; Effer 1984; Roberts and Roberts 1984; Jerrilov and Marionov 1987), will be the major vehicle. In the USA, however, the presentation of ecological information and prediction through these statements has been devalued by the bureaucratic complexity of the process and the masses of data presented (see Tucker 1978; Barnthouse *et al.* 1977). In many cases the huge amounts of data presented means that much of them will not be digested and scrutinised in detail and thus important ecological predictions could be submerged and lost. In this way the EIS procedure has become counter-productive. There is thus a need for simplified and summarised Statements. To some extent this is being realised in the UK and other countries at the present time where the EIS documents for new power stations are being produced in popularised versions. However, these cannot be expected to satisfy interested amateur or professional ecologists and thus access must be given to the more detailed information where relevant.

For some operating sites, the data on aquatic ecology have been edited into books which are of great use for the monitoring of the site, for students and to those concerned with the planning of future sites (e.g. Merriman and Thorpe 1976; Barnthouse *et al.* 1977; Hesse *et al.* 1982b).

There is also an urgent need for ecologists to study natural ecosystems over long periods and in as wide a range of habitats as possible so that we have a database from which to make predictions. In most siting studies these data are absent and the accuracy of the predictions is therefore diminished. It is in the long-term interests of governments, industrial managers and conservationists to provide the means to obtain the best data possible for predicting the effects of future developments and the long-term database is a major economic and ecological asset.

In conclusion, it seems clear that the prediction of the effects of thermal discharges from the new breed of power stations or other industries, need not be subject to the irresponsible extrapolation and exaggeration which was found in the 1960s and early 1970s. The research and surveys have not borne out the dire predictions of disaster much of which came from academic and political ambition. It is also clear that some of the legislation rushed through in some countries as a result of the high media and political profile of the issue was hasty and ill-conceived. Also, some of the so called science, particularly that used for the Environmental Impact Statements and consent procedures in the USA, has been of poor quality and has done both ecology and environmental protection a disservice. This work has mainly come from the less able academics and some of the opportunistic commercial consultant organisations.

There is no substitute, even in the field of applied ecology, site assessments or prediction, for sound objective science with high academic credibility (Howells and Gammon 1984). Whether this originates from the academic community, the consultant organisation or from scientists within the industry itself, is immaterial. The objective should be to allow the continued progress of civilisation and the improvement of man's living standards with a minimal disturbance to other ecosystems.

References and Bibliography

A large proportion of the publications listed are to be found in Proceedings of conferences or in collections of papers published in book form. This would make the full listing of each reference extremely cumbersome and so a shorter format is used.

As an example, Adams, A. P. and Lewis, B. G. (1978), is indicated as found in 'Gerhold, (Ed.)'. This is then identifiable as the Proceedings of a conference edited by Gerhold, R. M. published in 1978.

In the text, where a reference is made to a book, to a series of regular publications or to a proceedings, the reference is usually preceded by the word 'see'.

Abernethy, C. S. and Watson, D. G. (1976). Effects of temperature and acute irradiation on trout. In: *Pacific Northwest Laboratory, Annual Report for 1975, Part 2, Ecological Sciences*, Vaughan, D. E. (Ed.). US Energy Research and Development Administration, Richland, Washington, pp. 84–6.

Acker, R. F., Brown, B. F., DePalma, J. R. and Iverson, W. P. (Eds) (1972). *Proceedings of the Third International Congress on Marine Corrosion and Fouling*. National Bureau of Standards, Gaithersburg, Maryland.

Ackermann, W. C., White, G. F. and Worthington, E. B. (Eds) (1973). *Man-Made Lakes: their Problems and Environmental Effects*. Geophysical Monograph 17, American Geophysical Union, Washington, D.C.

Ackers, P. (1969). Modelling of heated water discharges, pp. 177–213, in Parker and Krenkel (Eds).

Adair, W. D. and Demont, D. J. (1971). Environmental responses to thermal discharges from Marshall Steam Station, Lake Norman, North Carolina. *Cooling Water Studies*. Edison Electric Co., Baltimore, Maryland and Johns Hopkins University.

Adair, W. D. and Hains, J. J. (1974). Saturation values of dissolved gases associated with the occurrence of gas-bubble disease in fish in a heated effluent, pp. 59–78 in Gibbons and Sharitz (Eds).

Adams, A. P. and Lewis, B. G. (1978). Microbiology of spray drift from cooling towers and cooling canals, pp. 31–6 in Gerhold (Ed.).

Adams, J. R. (1969). Ecological investigations related to thermal discharges. Pacific Coast Electrical Association, Engineering Operations Section, Annual Meeting, Los Angeles.

AEC (1973). Biological effects of thermal discharges. Annual progress report for 1973. Battelle Northwest Laboratories, Atomic Energy Commission, Washington, D.C.

Aho, J. M., Gibbons, J. W. and Esch, G. W. (1976). Relationship between thermal loading and parasitism in the mosquitofish, pp. 213–18 in Esch and McFarlane (Eds).

Alabaster, J. S. (1958). The behaviour of roach (*Rutilus rutilus*) in temperature gradients in large outdoor tanks. *Proc. Indo-Pacific Fish. Council*, **3**, 49–55.

Alabaster, J. S. (1962). The effect of heated effluents on fish. *Int. J. Air Wat. Pollut.*, **7**, 541–63.

Alabaster, J. S. (1967). The survival of salmon (*Salmo salar*) and sea-trout (*Salmo trutta*) in fresh and saline water at high temperatures. *Wat. Res.*, **1**, 717–30.

Alabaster, J. S. (1969). Effects of heated discharges on freshwater fish in Britain, pp. 354–74 in Krenkel and Parker (Eds) (b).

Alabaster, J. S. and Downing, A. L. H. (1966). A field and laboratory investigation of the effect of heated effluents on fish. *Fish. Inv. London, Series*, **1**, 6(4) 42 pp.

Alabaster, J. S., Garland, J. H. N. and Hart, I. C. (1971). Fisheries, cooling water discharges and sewage and industrial wastes, pp. 3–9 in CERL.

Alabaster, J. S. and Lloyd, R. (1980). *Water Quality Criteria for Freshwater Fish.* Butterworths, London and Boston.

Alden, R. W., Maturo, F. J. S. and Ingram, W. (1976). Interactive effects of temperature, salinity and other factors on coastal copepods., pp. 336–48 in Esch and McFarlane.

Alderdice, D. F. and Velsen, F. P. J. (1978). Relation between temperature and incubation time of eggs of chinook salmon (*Oncorhynchus tshawytscha*). *J. Fish. Res. Board. Can.*, **35**, 69–75.

Alderson, R. (1974). Sea-water chlorination and the survival and growth of the early development stages of plaice, *Pleuronectes platessa* L. and Dover sole, *Solea solea* (L). *Aquaculture*, **4**, 41–53.

All England Law Reports (1952). Vol. 1, p. 1326. Pride of Derby and Derbyshire Angling Association Ltd and Another v. British Celanese Ltd and others.

Allen, G. H., Boydstun, L. B. and Garcia, F. G. (1970). Reaction of marine fishes around warmwater discharge from an atomic steam-generating plant. *Prog. Fish Cult.*, **32** (January), 9–16.

Allen, J. R. M. and Wootton, R. J. (1982). The effect of ration and temperature on the growth of the three-spined stickleback, *Gasterosteus aculeatus*. L. *J. Fish Biol.*, **20**, 409–22.

Almatar, S. M. (1984). Effect of acute changes in temperature and salinity on the oxygen uptake of larvae of herring (*Clupea harengus*) and plaice (*Pleuronectes platessa*). *Mar. Biol.*, **80**, 117–24.

Alston, D. E., Lawrence, J. M., Bayne, D. R. and Campbell, F. F. (1978). Effects of thermal alteration on macro-invertebrate fauna in three artificial channels, pp. 569–79 in Thorp and Gibbons (Eds).

Ames, L. J., Felley, J. D. and Smith, M. H. (1979). Amounts of asymmetry in centrarchid fish inhabiting heated and non-heated reservoirs. *Trans. Am. Fish. Soc.*, **108**, 489–95.

Anderson, D. R. (1983). Chlorine-heavy metals interaction on toxicity and metal accumulation, pp. 811–26 in Jolley *et al.* (b).

Anderson, R. R. (1969). Temperature and rooted aquatic plants. *Ches. Sci.*, **10**(3,4), 157–9.

Anderson, T. P. and Lenat, D. R. (1978). Effect of power plant operation on the zooplankton community of Belews Lake, North Carolina, pp. 618–41 in Thorp and Gibbons (Eds).

Angelovic, J. W., White, J. C. and Davis, E. M. (1973). Interactions of ionizing radiation, salinity, and temperature on the estuarine fish (*Fundulus heteroclitus*), pp. 131–41 in Nelson and Evans (Eds).

Anon. (1979). New European standards for freshwater fish. Notes on Water Research, No. 20, Water Research Centre, Stevenage, UK.

Anon. (1988). Water treatment in the steel industry. *Water Services*, **92** (No. 1103), 21–2.

Ansell, A. D. (1963). The biology of *Venus mercenaria* in British waters, and in relation to generating station effluents. *Rep. Challenger Soc.*, **3**(xv).

Ansell, A. D., Lander, K. F., Coughlan, J. and Loosemore, F. A. (1964). Studies on the hard-shell clam, Venus mercenaria, in British waters. 1. Growth and reproduction in natural and experimental colonies. *J. Applied. Ecol.*, **1**, 63–82.

Appourchaux, M. (1952). Effects de la temperature de l'eau, sur la Faune et la Flore aquatiques. *L'eau*, **8**, 377.

Armstrong, F. A. J. and Scott, D. P. (1974). Photochemical dechlorination of water supply for fish tanks, with commercial water sterilizers. *J. Fish. Res. Board Can.*, **31**, 1881–5.

Arnaud, P. M., Bellan-Santani, D., Harmelin, J-G., Marinopoulos, J. and Zibrowis, H. (1981). Impact des rejets d'eau chaude de la centrale thermo-electrique EDF de Martigues-Ponteau (Mediterranee nord-occidentale) sur le zoobenthos des substrats durs superficiels, pp. 701–24 in EDF.

Arndt, H. E. (1968). Effect of heated water on a littoral community in Maine, in US Senate Public Works Committee on Thermal Pollution, 90th Congress, 2nd Session. Hearings before subcommittee on air and water.

Arnold, G. P. (1974). Rheotropism in fishes. *Biol. Revs*, **49**, 515–76.

Arthur, J. W. and Eaton, J. G. (1971). Chloramine toxicity to the amphipod *Gammarus pseudolimnaeus*, and the fathead minnow (*Pimphales promelas*). *J. Fish. Res. Board Can.*, **28**, 1841–5.

ASME (1975). *Cooling Tower Plume Modelling and Drift Measurement.* American Society of Mechanical Engineers, United States Engineering Center, New York.

Aston, R. J. (1968a). The effect of temperature on the life cycle, growth and fecundity of *Branchiura sowerbyi* (Oligochaeta: Tubificidae). *J. Zool. (Lond.)*, **154**, 29–40.

Aston, R. J. (1968b). The effect of Drakelow power stations on the dissolved oxygen content of the River Trent. CEGB Internal Report, RD/L/N 79/68, Leatherhead, UK.

Aston, R. J. (1973). Field and experimental studies on the effects of power station effluents on Tubificids (Oligochaeta, Annelida). *Hydrobiologia*, **42**(2–3), 225–42.

Aston, R. J. (1975). Fish culture in power station cooling water—a review. CEGB Internal Report, RD/L/M 494, Leatherhead, UK.

Aston, R. J. (1976). Aquaculture at power stations in the United States—a summary of information collected during a visit. CERL Internal Report, LM/BIOL/003, Leatherhead, UK.

Aston, R. J. (1981). The availability and quality of power station cooling water for aquaculture. *Proceedings of a Symposium on Aquaculture in Heated Effluents and Recirculation Systems*, pp. 39–58. European Inland Fisheries Advisory Commission, Stavanger, Norway, 28–30th May, 1980, Vol. 1, Berlin.

Aston, R. J. (1984). The case for temperature control in vermiculture. *Proc. Int. Conf. on Earthworms and Environment Management*. Cambridge, UK, 23–27 July.

Aston, R. J. (1988). Status of projects using reject heat for aquaculture and horticulture at power plants in the EEC. CEGB Internal Report RD/L/3303/R.88. Leatherhead, UK.

Aston, R. J. and Brown, D. J. A. (1973). Local and seasonal variation in populations of the leech *Erpobdella octoculata* (L) in a polluted river warmed by condenser effluents. *Hydrobiologia*, **47**(2–3), 347–66.

Aston, R. J. and Brown, D. J. A. (1975). Carp and eel culture in power station cooling-water. *Proc. 7th Coarse Fish. Conf. Liverpool University (March)*, pp. 72–80.

Aston, R. J. and Brown, D. J. A. (1978). Fish farming in heated effluents. *Proceedings of a Conference on Fish Farming and Wastes*, University College, London, January 4–5th, 1978 (cyclostyled).

Aston, R. J., Brown, D. J. A. and Macqueen, J. F. (1978). Temperature control and its advantages in carp culture using power station cooling water. CEGB Internal Report, RD/L/N/65/77, Leatherhead, UK.

Aston, R. J., Brown, D. J. A. and Milner, A. G. P. (1976). *Heated Water Farms at Inland Power Stations*. Central Electricity Generating Board, Newsletter No. 102, London.

Aston, R. J., Brown, D. J. A. and Milner, A. G. P. (1982). The effects of temperature on the culture of *Branchiura sowerbyi* (Oligochaets, Tubificidal) on activated sludge. *Aquaculture*, **29**, 137–45.

Aston, R. J. and Heathwood, A. (1980). The oxygen consumption of carp (*Cyprinus carpio*) in power station cooling water in relation to temperature and oxygen concentration. CEGB Internal Report, RD/L/N 167/79, Leatherhead, UK.

Aston, R. J. and Milner, A. G. P. (1976). Heated water farms at inland power stations. *Fishing Farming Int.*, June 1976, 41–4.

Aston, R. J. and Milner, A. G. P. (1980). A comparison of populations of *Asellus aquaticus* (L) above and below power stations in organically polluted reaches of the River Trent. *Freshwat. Biol.*, **10**, 1–14.

Aston, R. J. and Milner, A. G. P. (1981). Conditions required for the culture of *Branchiura sowerbyi* (Oligochaeta, Tubificidae), in activated sludge. *Aquaculture*, **26** (1981/1982), 155–60.

Aston, R. J. and Sadler, K. (1981). Macrophyte culture in cooling water. CEGB Internal Report, RD/L/2071N81, Leatherhead, UK.

Aston, R. J., Sadler, K. and Milner, A. G. P. (1981). The effects of temperature on the culture of *Branchiura sowerbyi*, (Oligochaeta, Tubificidae), on activated sludge. CEGB Internal Report, RD/L/211N81, Leatherhead, UK.

Aston, R. J., Sadler, K., Parry, M. and Milner, A. G. P. (1980). A comparison of aeration methods in relation to the oxygen requirements of eels at Hinkley Marine Farm. CEGB Internal Report, RD/L/N 50/80, Leatherhead, UK.

Audunson, T., Land, J. and Rye, H. (1975). Computations of the temperature response of stratified sill fjords to cooling-water discharges, pp. 113–38 in IAEA (a).

Baker, J. M. (Ed.) (1977). *Marine Ecology and Oil Pollution.* Applied Science Publishers, London.

Baker, M. S., Kaming, J. S. and Morrison, R. E. (1977). *Environmental Impact Statements: A Guide to Preparation and Review.* Practising Law Institute, New York.

Balligand, P., Grauby, A., Fourcy, A., de Cachard, M. B. and Dumont, M. (1976). Experience gained in France on heat recovery from nuclear plants for agriculture and pisciculture. *Nucl. Technol.*, **38**(1), 90–6.

Bamber, R. N. (1978). The effects of dumped pulverised fuel ash on the benthic fauna of the Northumberland Coast. PhD Thesis, University of Newcastle upon Tyne, UK.

Bamber, R. N. (1983). Pozzolanic aggregates of fly ash in the sea. *Mar. Biol.*, **77**, 151–4.

Bamber, R. N. (1984a). The benthos of a marine fly-ash dumping ground. *J. Mar. Biol. Assn. UK*, **64**, 211–26.

Bamber, R. N. (1984b). The autecology of *Cyathura carinata* (Kroyer), (Crustacea; Isopoda), in a cooling water discharge lagoon. CEGB Internal Report, TPRD/L/2655/N84, Leatherhead, UK.

Bamber, R. N. (1985). Coarse substrate benthos of Kingsnorth outfall lagoon, with observations on *Petricola pholadiformis* (Lamarck). CEGB Internal Report, TPRD/1/2759/N84, Leatherhead, UK.

Bamber, R. N. (1987a). The biology of the American ostracod *Sarsiella zostericola* Cushman in the vicinity of a British power station. *J. Micropalaeontol.*, **6**(1) 57–62.

Bamber, R. N. (1987b). The effects of acidic sea water on young carpet shell clams, *Venerupis decussata* (L) (Mollusca; Veneracea). *J. exp. Mar. Biol. Ecol.*, **108**, 241–301.

Bamber, R. N. (1990). Power station thermal effluents and marine crustaceans. *J. Therm. Biol.*, **15**(1) 91–6.

Bamber, R. N. and Henderson, P. A. (1981). Bradwell biological investigations; analysis of the benthic surveys of the River Blackwater up to 1975. CEGB Internal Report, RD/L/2042, R81, Leatherhead, UK.

Bamber, R. N. and Henderson, P. A. (1985). An examination of an 'exotic' species of fish recorded in association with heated discharges. CEGB Internal Publication, TRPD/D/L/2744/R84, Leatherhead, UK.

Bamber, R. N. and Spencer, J. F. (1984). The benthos of a coastal power station thermal discharge canal. *J. Mar. Biol. Assn.*, **64**, 603–23.

Banus, M. D. and Kolehmainen, S. E. (1976). Rooting and growth of red mangrove seedlings from thermally stressed trees, pp. 46–53, in Esch and McFarlane (Eds).

Baranava, V. V. (1980). Changes in the fatness of roach (*Rutilus rutilus*) under the influence of warm waste water from a district power station. *Hydrobiol. J.*, **16**(5), 62–6.

Barber, Y. M. (1972). Statement as presented before the Fourth Session of the Lake Michigan Enforcement Conference, Sherman House, Chicago, Illinois, September, 19–21.

Barica, J. (1975). Summerkill risk in prairie ponds and possibilities of its prediction. *J. Fish. Res. Board, Can.*, **32**, 1283–8.

Barnes, R. S. K. and Coughlan, J. (1970). Bradwell biological investigations: the environment in the vicinity of the barrier wall. CEGB Internal Report, RD/L/R 1711, Leatherhead, UK.

Barnes, R. S. K. and Coughlan, J. (1972). The bottom fauna of the Blackwater Estuary; The macrofauna of the area adjacent to Bradwell nuclear generating station. *The Essex Naturalist*, **33**(1), 1–32.

Barnes, R. S. K. and Green, J. (Eds) (1979). *The Estuarine Environmment.* Applied Science Publishers, London.

Barnett, P. R. O. (1971). Some changes in intertidal sand communities due to thermal pollution. *Proc. Roy. Soc. Lond. B*, **177**, 353–64.

Barnett, P. R. O. (1972). Effects of warm water effluents from power stations on marine life. *Ibid.*, **180**, 497–509.

Barnett, P. R. O. and Hardy, B. L. S. (1969). The effect of temperature on the benthos near Hunterston Generating Station, Scotland. *Ches. Sci.*, **10**, 255–6.

Barnett, P. R. O. and Hardy, B. L. S. (1984). Thermal Deformations, In: *Marine Ecology, Volume V, Ocean Management, Part 4, Pollution and Protection of the Seas, Pesticides, Domestic Wastes, and Thermal Deformations,* Kinne, O. (Ed.). Wiley—Interscience, New York, pp. 1769–1926.

Barnthouse, L. W., Klauda, R. J., Vaughan, D. S. and Kendall, R. L. (1977). Science, law and Hudson River power plants: A case study in environmental impact assessment. *Monogr. Amer. Fish. Soc.*, **4**, 1–355.

Baross, J. A. and Denning, O. (1983). Growth of 'black-smoker' bacteria at temperatures of at least 250°C. *Nature (Lond.)*, **803** (June), 423–6.

Barwick, D. H. and Lorenzen, W. E. (1984). Growth responses to changing environmental conditions in a South Carolina cooling reservoir. *Environ. Biol. Fish.*, **10**(4), 271–9.

Bass, M. L., Berry, C. R. and Heath, A. G. (1977). Histopathological effects of intermittent chlorine exposure on bluegill (*Lepomis macrochirus*) and rainbow trout (*Salmo gairdneri*). *Wat. Res.*, **11**, 731–5.

Bass, M. L. and Heath, A. G. (1977). Cardiovascular and respiratory changes in rainbow trout, *Salmo gairdneri* exposed intermittently to chlorine. *Wat. Res.*, **11**, 497–502.

BEA (1954). Changes in the dissolved oxygen content of river water used for cooling. Sub Committee (Effluents Testing), No. 8, British Electricity Authority, London, October.

Beall, S. E. (1973). Conceptual design of food complex using waste warm water for heating. *J. Environ. Qual.*, **2**, 207–15.

Beall, S. E. *et al.* (1977). Energy from cooling-water. *Indust. Water Engng*, **16**(6), 8–14.

Bean, R. M., Mann, D. C. and Neitzel, D. A. (1983). Organohalogens in chlorinated cooling waters discharged from nuclear power stations, pp. 383–90 in Jolley *et al.* (a).

Bean, R. M., Thomas, B. L. and Neitzel, D. A. (1985). Analysis of sediment from chlorination of power plant cooling-water, pp. 1357–70 in Jolley *et al.* (Eds).

Beauchamp, R. S. A. (1967). The effect of biological factors on the design and operation of power stations. In: *Biology and the Manufacturing Industries*, Brook, M. (Ed.). Institute of Biology and Academic Press, London, pp. 43–51.

Beauchamp, R. S. A. (1969a). The use of chlorine in the cooling-water systems of coastal power stations. *Ches. Sci.*, **10**(3–4), 280–1.

Beauchamp, R. S. A. (1969b). Some effects of the power industry on the aquatic environment in Britain, pp. 200–30 in Howells and Lauer (Eds).

Beauchamp, R. S. A., Ross, F. F. and Whitehouse, J. W. (1971). The thermal enrichment of aquatic habitats. *Proc. 5th Int. Wat. Pollut. Res. Conf. (July–August 1970)*. Pergamon Press, Oxford.

Beck, S. D. (1968). *Insect photoperiodism*. Academic Press, New York.

Becker, C. D. (1969). The food and feeding of juvenile chinook salmon in the Columbia River at Hanford, pp. 27–32 in AEC.

Becker, C. D. (1973). Columbia River thermal effects study: reactor effluent problems. *J. Water Pollut. Contr. Fed.*, **45**, 850–69.

Becker, C. D., Cushing, C. E., Gore, K. L., Baker, K. S. and McKenzie, D. H. (1979a). Synthesis and analysis of ecological information from cooling impoundments. Electric Power Research Institute, Report EA—1054, Volume 1, RP-880, Palo Alto, CA.

Becker, C. D., Cushing, C. E., Gore, K. L., Baker, K. S. and McKenzie, D. H. (1979b). Synthesis and analysis of ecological information from cooling impoundments. Volume 2, Appendix A, Study site-histories and data synopses, Electric Power Research Institute, EA-1054, RP-880, Palo Alto, CA.

Becker, C. D. and Genoway, R. G. (1979). Evaluation of the critical thermal maximum for determining thermal tolerance of freshwater fish. *Environ. Biol. Fish.*, **4**(3), 245–6.

Becker, C. D. and Thatcher, T. O. (1973). Toxicity of power plant chemicals to aquatic life. US Atomic Energy Commission Report No. Wash. 1249, US Government Printing Office, Washington, D.C.

Bedford, S. E. (1973). A survey of effluent disposal in the glass industry. British Glass Industry Research Association, Information Circular No. 108, Sheffield, UK.

Beitinger, T. L. (1976). Behavioural thermoregulation by bluegill exposed to various rates of temperature change, pp. 176–9 in Esch and McFarlane (Eds).

Beitinger, T. L. and Magnuson, J. J. (1975). Influence of social rank and size on thermoselection behaviour of bluegill (*Lepomis macrochirus*). *J. Fish. Res. Board Can.*, **32**(11), 2133–6.

Beitinger, T. L. and Magnuson, J. J. (1979). Growth rates and temperature selection of bluegill, *Lepomis macrochirus*. *Trans. Am. Fish. Soc.*, **108**, 378–82.

Bell, F. V. M. (1949). Timber structures, with particular reference to the maintenance of oil-loading jetties at Queen's Dock, Swansea. Institution of Civil Engineers (Marine and Waterways Division), Sess. 1948–9, pp. 3–11.

Bellanca, M. A. and Bailey, D. S. (1977). Effects of chlorinated effluents of aquatic ecosystem in the lower James River. *J. Water Pollut. Contr. Fed.* (April), 639–45.

Belter, W. G. (1975). Management of waste heat at nuclear power stations, its possible impact on the environment and possibilities of economic use, pp. 3–23 in IAEA (a).

Bender, R. S. and Profitt, M. A. (1974). Effects of thermal effluents on fish and invertebrates, pp. 438–47 in Gibbons and Sharitz (Eds).

Bendix, S. and Graham, H. R. (Eds) (1978). *Environmental Assessment: Approaching Maturity.* Ann Arbor, Michigan.

Bennett, D. H. (1971). Preliminary examination of body temperatures of largemouth bass (*Micropterus salmoides*) from an artificially heated reservoir. *Arch. Hydrobiol.*, **68**(3), 376–82.

Bennett, D. H. and Gibbons, J. W. (1972). Food of largemouth bass (*Micropterus salmoides*) from a South Carolina reservoir receiving heated effluent. *Trans. Am. Fish. Soc.*, **101**(4), 650–4.

Bennett, D. H. and Gibbons, J. W. (1974). Growth and condition of juvenile largemouth bass from a reservoir receiving thermal effluent, pp. 246–54 in Gibbons and Sharitz (Eds).

Bennett, D. H. and Gibbons, J. W. (1975). Reproductive cycles of largemouth bass (*Micropterus salmoides*) in a cooling reservoir. *Trans. Am. Fish. Soc.*, **1**, 77–82.

Bennett, D. H. and Goodyear, C. P. (1978). Response of mosquitofish to thermal effluent, pp. 498–510 in Thorp and Gibbons (Eds).

Bertsche, E. C. (1971). Treatment of water in nuclear fuel storage basins to control radioactivity release. *Proceedings of the 26th Industrial Waste Conference*, Purdue University Press, pp. 77–88.

Berg, J. D., Hoff, J. C., Roberts, P. V. and Matin, A. (1985). Disinfection resistance of *Legionella pneumophila* and *Escherichia coli* grown in continuous and batch culture, pp. 603–13 in Jolley *et al.* (Eds).

Bettelheim, J., Kyte, W. S. and Littler, A. (1981). Fifty years' experience of flue gas desulphurisation at power stations in the United Kingdom. *The Chemical Engineer* (June), 5–9.

Biesiadka, E., Kasprzak, K. and Kolasa, J. (1978). Effects of artificial rise of temperature on stagnant waters and their biocoenoces. *Int. Revue. ges. Hydrobiol.*, **63**(1), 41–56.

Biette, R. M. and Geen, G. H. (1979). Growth of underyearling sockeye salmon (*Oncorhynchus nerka*) under constant and cyclic temperatures in relation to live zooplankton size. *Can. J. Fish. Aquat. Sci.*, **37**, 203–10.

Bilyard, G. R. (1987). The value of benthic infauna in marine pollution monitoring studies. *Mar. Pollut., Bull.*, **18**(11), 581–5.

Birge, W. J. (1978). Aquatic toxicology of trace elements of coal and fly ash, pp. 219–40 in Thorp and Gibbons (Eds).

Blackie, J. R. *et al.* (1980). Environmental effects of deforestation. Freshwat. Biol. *Assn. Pub. No. 10*, Ambleside, UK.

Blake, N. J., Doyle, L. J. and Pyle, T. E. (1976). The macrobenthic community of a thermally altered area of Tampa Bay, Florida, pp. 296–301 in Esch and McFarlane (Eds).

Blank, L. W. (1984). Control of algal growth in cooling towers: Toxicity of low-level continuous chlorination. CEGB Internal Report, TPRD/L/2650/N84. Leatherhead, UK.

Blaylock, B. G. and Frank, M. L. (1978). The effect of ionizing radiation on the thermal tolerance of mosquitofish, pp. 785–93 in Thorp and Gibbons (Eds).

Block, C. J., Spotila, J. R., Standora, E. A. and Gibbons, J. W. (1984). Behavioural regulation of largemouth bass, *Micropterus salmoides* and Bluegill, *Lepomis macrochirus*, in a nuclear reactor cooling reservoir. *Environ. Biol. Fish.*, **11**(1), 41–52.

Block, R. M. (1974). Effects of acute cold shock on the channel catfish, pp. 109–18 in Gibbons and Sharitz (Eds).

Block, R. M., Burton, D. T., Gullans, S. R. and Richardson, L. N. (1978). Respiratory and osmoregulatory responses of white perch (*Morone americana*) exposed to chlorine and ozone in estuarine waters, pp. 351–60 in Jolley *et al.* (Eds) (c).

Block, R. M. and Helz, G. R. (Eds) (1976). *Proceedings of the Chlorination Workshop*, University of Maryland, Chesapeake Biological Laboratory, 15–18 March.

Block, R. M., Helz, G. R. and Davis, W. P. (1977). The fate and effects of chlorine in coastal waters. *Ches. Sci.*, **18**(1), 97–102.

Bloom, S. G., Levin, A. A. and Raines, G. E. (1974). Future uses of mathematical models to predict ecological effects from thermal discharges, pp. 174–89 in Gallagher (Ed.).

Board, P. A. (1972). The effects of temperature and other factors on the tunnelling of *Lyrodus pedicellatus* and *Teredo navalis*, pp. 797–805 in Acker *et al.* (Eds).

Boehlert, G. W. (1982). The effects of photoperiod and temperature on laboratory growth of juvenile *Sebastes diploproa* and a comparison with growth in the field. *US Fish. Bull.*, **79**, 789–94.

Bogardus, R. B., Teppen, T. C., Boies, D. B. and Horvath, F. J. (1978). Avoidance of monochloramine: Test tank results for rainbow trout, coho salmon, alewife, yellow perch, and spottail shiner, pp. 149–62 in Jolley *et al.* (Eds) (c).

Bohnsack, J. A. (1983). Resiliency of fish reef communities in the Florida Keys, following a January 1977 hypothermal fish kill. *Environ. Biol. Fish.*, **9**(1), 41–53.

Bongers, L. H., Burton, D. T., Liden, L. H. and O'Connor, T. P. (1978). Bromine chloride—An alternative biofouling control agent for cooling water treatment, pp. 735–52 in Jolley *et al.* (Eds) (c).

Bordet, F. (1980). Influence of the heating of water by a power station on plankton photosynthesis on the Seine at Porcheville. *Cahiers du Lasboratoire d'Hydrobiologie de Montereau*, **10**, 33–47 (Trans.).

Boreman, J., Goodyear, C. P. and Christensen, S. W. (1978). An empirical transport model for evaluating entrainment of aquatic organisms by power plants. Fish and Wildlife Service, Biological Services Program, FWS/OBS-78/90, US Department of the Interior, Washington, DC, 67 pp.

Bott, T. L. (1975). Bacterial growth rates and temperature optima in a stream with a fluctuating thermal regime. *Limnol. Oceanogr.*, **20**(2), 191–7.

Bott, T. R., Miller, P. C. and Patel, T. D. (1983). Biofouling in an industrial cooling system. *Proc. Biochem.*, **18**(1), 10–13.

Bourque, J. E. and Esch, G. W. (1974). Population ecology of parasites in turtles from thermally altered and natural aquatic communities, pp. 551–61 in Gibbons and Sharitz (Eds).

Bowen, M. (1976) Effects of a thermal effluent on the Ostracods of Par Pond, South Carolina, pp. 219–26 in Esch and McFarlane (Eds).

Bowers, A. B. and Naylor, E. (1964). Occurrence of *Atherina boyeri*, Risso, *Nature (Lond.)*, **202**, 318.

Bowles, R. R. and Merriner, J. V. (1978). Evaluation of ichthyoplankton sampling gear used in power plant entrainment studies, pp. 33–45 in Jensen (Ed.) (b).

Boyden, C. R. and Romeril, M. G. (1974). A trace metal problem in pond oyster culture. *Mar. Pollut. Bull.*, **5**(5), 74–8.

Boylen, C. W. and Brock, T. D. (1973). Effects of thermal additions from the Yellowstone geyser basins on the benthic algae of the Firehole River. *Ecology*, **54**(6), 1283–91.

Boytsov, M. P. (1974). Morphology of fingerlings in the Ivankovo Reservoir affected by hot water discharge from the Konakovo Hydroelectric Power Station. *Vopr. Ikhtiol. (USSR)*, **14**, 1046 (Trans.).

Bradford, J. M. and Burns, D. A. (1977). The effects of the Marsden 'A' thermal power station on the marine zooplankton. *N.Z. Oceanogr. Inst.*, **3**(9), 69–79.

Brady, D. K., Graves, W. L. and Geyer, J. C. (1969). Surface heat exchange at power plant cooling lakes. Edison Electric Institute Research, Project Report No. 5, EEI Publication Number 69-901, New York.

Brauer, G. A., Neill, W. H. and Magnusson, J. J. (1974). Effects of power plant on zooplankton distribution and abundance near plant's effluent. *Wat. Res.*, **8**, 485–9.

Bray, E. S. (1971). Observations on the reproductive cycle of the roach (*Rutilus rutilis*) with particular reference to the effects of heated effluents. *Proc. 5th Brit. Coarse Fish, Conf. University of Liverpool.* Janssen Services, London.

Bregman, J. (1969). Keynote address, pp. 3–14 in Krenkel and Parker (Eds) (b).

Brett, J. R. (1956). Some principles in the thermal requirements of fishes. *Qu. Rev. Biol.*, **31**(2), 75–87.

Brett, J. R. (1964). The respiratory metabolism and swimming performance of young sockeye salmon. *J. Fish. Res. Board Can.*, **21**, 1183–226.

Brett, J. R. (1970). Fishes, functional responses, Chap. 3 (Temperature), pp. 515–60 in Kinne (Ed.) (a).

Brett, J. R. (1979). Environmental factors and growth. In: *Fish Physiology*, Hoar, W. S., Randall, D. J. and Brett, J. R. (Eds). Academic Press, New York and London, pp. 599–675.

Brewer, W. H. (1866). Observations on the presence of living species in hot and saline waters of California. *Am. J. Sci.*, **41** (2nd Series) 391–4.

Briand, F. J. P. (1975). Effects of power plant cooling-systems on marine phytoplankton. *Mar. Biol.*, **33**, 135–46.

Brinkhurst, R. B. and Cook, D. G. (1974). Aquatic earthworms, (Annelida; Oligochaeta). In: *Pollution Ecology of Freshwater Invertebrates*, Hart, C. W. and Fuller, S. L. H. (Eds). Academic Press, New York and London, pp. 143–56.

Brisbin, I. L. (1974). Abundance and diversity of waterfowl inhabiting heated and unheated portions of a reactor cooling reservoir, pp. 579–93 in Gibbons and Sharitz (Eds).

Brock, T. D. (1975). Predicting the ecological consequences of thermal pollution from observations on geothermal habitats, pp. 599–621 in IAEA(a).

Brock, T. D. (1978). *Thermophilic Micro-organisms and Life at High Temperatures.* Springer, New York.

Brock, T. D. (1979). *Biology of Micro-organisms.* Prentice-Hall, Englewood Cliffs, N.J.

Brock, T. D. (1985). Life at high temperatures. *Science, NY,* **230** (4722), 132–8.

Brock, T. D. and Boylen, C. W. (1973). Presence of thermophilic bacteria in laundry and domestic hot water heaters. *Appl. Microbiol.*, **25**, 72–6.

Brock, T. D. and Yoder, C. D. (1971). Thermal pollution of a small river by a large university: bacteriological studies. *Proc. Indiana Acad. Sci.*, **80**, 183–8.

Brook, A. J. and Baker, A. L. (1972). Chlorination at power plants; Impact on phytoplankton productivity. *Science*, **176**, 1414–15.

Brook, I. M. (1975). Trophic relationships in a sea-grass community, *Thalassia testudinum* in Card Sound, Florida: Fish diets in relation to macrobenthic and cryptic faunal abundance. *Trans. Am. Fish. Soc.*, **106**(3), 219–29.

Brooke, L. T. (1975). Effect of different constant incubation temperatures on egg survival and embryonic development in Lake Whitefish (*Coregonus clupeaformis*). *Trans. Am. Fish. Soc.*, **104**, 555–9.

Brooker, M. P., Morris, D. L. and Hemsworth, R. J. (1977). Mass mortalities of adult salmon (*Salmo salar*) in the River Wye, 1976. *J. App. Ecol.*, **14**, 409–17.

Brooks, A. S. (1974). Phytoplankton entrainment studies at the Indian River estuary, Delaware, pp. 105–11 in Jensen (Ed.).

Brooks, A. S. and Liptak, N. E. (1979). The effect of intermittent chlorination on freshwater phytoplankton. *Wat. Res.*, **13**, 49–52.

Brooks, A. S. and Seegert, G. L. (1977). The effects of intermittent chlorination on rainbow trout and yellow perch. *Trans. Am. Fish. Soc.*, **106**(3), 278–86.

Brooks, A. S. and Seegert, G. L. (1978). A preliminary look at the effects of intermittent chlorination on selected warmwater fishes, pp. 95–110 in Jolley *et al.* (Eds) (c).

Brown, D. J. A. (1973). The effect of power station cooling-water discharges on the growth of coarse-fish fry. *Proc. 6th Brit. Coarse Fish Conf., Liverpool.* Janssen Services, London, pp. 191–202.

Brown, D. J. A. (1979). The distribution and growth of juvenile cyprinid fishes in rivers receiving power station cooling-water discharge. *Proc. 1st Brit. Freshw. Fish Conf.*, University of Liverpool, UK.

Brown, D. J. A. and Aston, R. J. (1975). Chlorine and chloramine concentrations in the cooling-water of Ratcliffe-on-Soar power station and the potential toxicity to fish. CEGB Internal Report, RD/L/N 134/75, Leatherhead, UK.

Brown, G. W. (1970). Predicting the effect of clearcutting on stream temperature. *J. Wat. Soil Conservat.*, **25**, 11–13.

Brown, G. W. and Brazier, J. R. (1972). Controlling thermal pollution in small streams. Report no. EPA R2-72083. Office of Research and Monitoring, Washington, US Environmental Protection Agency.

Brown, G. W. and Krygier, J. T. (1967). Changing water temperatures in small mountain streams. *J. Wat. Soil Conservat.*, **22**, 242–4.

Brown, G. W. and Krygier, J. T. (1970). Effects of clear cutting on stream temperature. *Wat. Resources Res.*, **6**, 1133–9.

Brown, J., Ray, N. J. and Ball, M. (1976). The disposal of pulverised fuel ash in water supply catchment areas. *Wat. Res.*, **10**, 1115–21.

Brown, J. A. G., Jones, A. and Matty, A. J. (1984). Oxygen metabolism of farmed turbot (*Scophthalmus maximus*), 1, The influence of fish size and water temperature on metabolic rate. *Aquaculture*, **36**, 273–81.

Brungs, W. A. (1971). Chronic effects of constant elevated temperature on the fathead minnow (*Pimephales promelas* Rafinesque). *Trans. Am. Fish. Soc.* **100**, 659–64.

Brungs, W. A. (1973). Effects of residual chlorine on aquatic life. J. Wat. Pollut. Contr. Fed., **45**, 2180–93.

Brungs, W. A. (1977). General considerations concerning the toxicity to aquatic life of chlorinated condenser effluent, pp. 109–13 in Jensen (Ed.).

Bryan, G. W. (1971). The effects of heavy metals (other than mercury) on marine and estuarine organisms. *Proc. Roy. Soc. (Lond. B)*, **177**, 389–410.

Bryan, G. W. and Hummerstone, C. G. (1973). Brown seaweed as an indicator of heavy metals in estuaries in south-west England. *J. Mar. Biol. Ass. UK*, **53**, 705–20.

Brylinski, J. M. (1981). Influence d'un enchauffement permanent des eaux par les rejets d'une centrale thermique sur le developpment de *Temora longicornis* (Copepoda, Calanoida), dans le port de Dunkerque, pp. 659–77 in EDF.

Buckland, F. (1883). *The Natural History of British Fishes*. Society for Promoting Christian Knowledge, Unwin Brothers, The Gresham Press, London.

Budenholzer, R. J., Hauser, C. G. and Oleson, (1971). Selecting heat rejection systems for future steam electric power plants. Westinghouse Enn., Nov., 168–75.

Bugbee, S. L. (1978). Implementation of Section 316, of the Federal Water Pollution Control Act, pp. 3–6 in Jensen (Ed.).

Buikema, A. L., Sherberger, S. R., Knauer, G. W., Newbern, L. A., Reading, J. T. and Cairns, J. (1978). Effect of simulated entrainment on the biology of a freshwater cladoceran, pp. 809–25 in Thorp and Gibbons (Eds).

Bulleid, M. J. (1974). A preliminary report on the effect of water temperature on the early growth of Rainbow trout (*Salmo gairdneri*). *J. Inst. Fish. Mgnt*, **5**(1), 16–22.

Bullimore, B., Dyrynda, P. E. J. and Bowden, N. (1978). The effects of falling temperature on the fauna of Swansea Docks. In: *Progress in Underwater Science, Vol. 3, Proc. 11th, Symp. Underwater Assoc.*, 1977. Gamble, J. C. and Yorke, R. A. (Eds). British Museum (Natural History), Pentech Press, London.

Bunting, D. L. (1974). Zooplankton. Thermal regulation and stress, pp. 50–5 in Gallaher (Ed.).

Burnett, J. M., McMillan, W., MacQueen, J. F., Moore, D. J. and Shepherd, J. G. (1974). The cooling of power stations, Paper No. 2.2.5, *Proc. 9th World Energy Conf.*, Detroit., WEC, London.

Burnett, N. C. and Fedyko, J. (1978). Prediction of leachate concentrations from ash and flue-gas desulphurisation sludge disposal sites. *Proceedings of the American Power Conference*, Vol. 40, pp. 986–93.

Burton, D. T., Capizzi, T. P., Margrey, S. L. and Wakefield, W. W. (1980–81). Effects of rapid changes in temperature on two estuarine crustaceans. *Mar. Environ. Res.*, **4**, 267–78.

Burton, D. T., Hall, L. W. and Margrey, S. L. (1980). Multifactorial chlorine, δT and exposure duration studies of spring power plant operations on three estuarine invertebrates, pp. 547–556 in Jolley *et al.* (c).

Burton, D. T., Hall, L. W., Jnr, Margrey, S. L. and Small, R. D. (1978). Interactions of chlorine, temperature change (δT), and exposure time on survival of striped bass (*Morone saxatilus* eggs and prolarvae. *J. Fish. Res. Board Can.*, **36**, 1108–13.

Burton, D. T. and Liden, L. H. (1978). Biofouling control alternatives to chlorine as a control agent for cooling water treatment, pp. 717–34 in Jolley *et al.* (Eds) (c).

Burton, D. T., Margrey, S. L. and Richardson, L. B. (1976). Effects of power plant entrainment temperatures on oxygen consumption patterns of amphipods and grass shrimp, pp. 93–9 in Esch and McFarlane (Eds).

Bush, R. M., Welch, E. B. and Mar, B. W. (1974). Potential effects of thermal discharges on aquatic systems. *Environ. Sci. Technol.*, **8**(6), 561–8.

Cabral, C. G., Agustines, A. A. and Chakroff, R. P. (1976). Effects of hot water discharge into Laguna Lake. *Technol. J.*, **1**(3), 24–32.

Cairns, J. (1956a). The effects of heat on fish. *Indust. Wastes*, May–June, 180–83.

Cairns, J. (1956b). The effects of increased temperatures on aquatic organisms. *Proceedings of the 10th Industrial Waste Conference, Purdue University Engineering Bulletin*, Vol. 40(1), p. 346.

Cairns, J. (1969). Ecological management problems caused by heated waste water discharges into the aquatic environment. *Proceedings of the Governors Conference on Thermal Pollution*, 18 July 1969, Michigan, Traverse City (mimeo).

Cairns, J. (1976). Heated waste water effects on aquatic ecosystems, pp. 32–38 in Esch and McFarlane (Eds).

Cairns, J. (1977). Effects of temperature changes and chlorination upon the community structure of aquatic organisms. In: *Towards a Plan of Action for Mankind, Volume 3. Biological Balance and Thermal Modifications*. Marois, M. (Ed.), Pergamon Press, Oxford and New York, pp. 129–44.

Cairns, J., Buikema, A. L., Heath, A. G. and Parker, B. C. (1978). Effects of temperature on aquatic organisms sensitivity to selected chemicals. Bulletin 106, Virginia Water Resources Research Center, Blacksburg, Virginia.

Cairns, J., Calhoun, W. F., McGinniss, M. J. and Straka, W. (1976a). Aquatic organisms response to severe stress following acutely sublethal toxicant exposure. *Wat. Resources Bull.*, **12**(6), 1233–43, American Water Resources Association.

Cairns, J. and Cherry, D. S. (1983). A site-specific field and laboratory evaluation of fish and Asiatic clam responses to coal-fired power plant discharges. *Wat. Sci. Technol.*, **15**, 10–37.

Cairns, J., Cherry, D. S. and Giattini, J. D. (1981). Correspondence between behavioural responses of fish in laboratory and field heated, chlorinated effluents, pp. 207–16 in Mitsch *et al.* (Eds).

Cairns, J., Crossman, J. S. and Dickson, K. L. (1970). The biological recovery of the Clinch River following a fly-ash pond spill. *Proceedings of the 25th Purdue Industrial Waste Conference*, 5–7 May, Purdue University Press.

Cairns, J., Dickson, K. L. and Crossman, J. S. (1972). The response of aquatic communities to spills of hazardous materials. *Proceedings of the National Conference on Hazardous Material Spills*, pp. 179–97.

Cairns, J., Dickson, K. L. and Lanza, G. (1979). Studies evaluating the biological effects of thermal additions to aquatic ecosystems. *Revista de Biologica*, **11**, 1–53.

Cairns, J., Heath, A. G. and Parker, B. C. (1975). Temperature influence on chemical toxicity to aquatic organisms. *J. Wat. Pollut. Contr. Fed.*, **47**(2), 267–80.

Cairns, J., Messenger, D. and Calhoun, W. F. (1976b). Invertebrate response to thermal shock following exposure to acutely sublethal concentrations of chemicals. *Arch. Hydrobiol.*, **77**, 164–75.

Cairns, J. and Messenger, D. (1974). An interim report on the effects of prior exposure to sublethal concentrations of toxicants upon the tolerance of snails to thermal shock. *Arch. Hydrobiol.*, **74**(4), 441–7.

Cairns, J. and Plafkin, J. L. (1975). Response of protozoan communities exposed to chlorine stress. *Archiv. Protistenkunde*, **117**, 47–53.

Calhoun, S. W., Zimmerman, E. G. and Beitinger, T. L. (1982). Stream regulation alters acute temperature preferenca of red shiners, *Notropis lutrensis. Can. J. Fish. Aquat. Sci.*, **39**, 360–3.

Cane, A., Key, G. and Rogers, A. (1981). A survey of the Trawsfynydd Lake Fishery, Summer 1979. CEGB Internal Report, NW/SSD/SR/60/81, Manchester, UK.

Cannon, T. C., Jinks, S. M., Kings, L. R. and Lauer, G. J. (1978). Survival of entrained ichthyoplankton and macro-invertebrates at Hudson River power plants, pp. 71–90 in Jensen (Ed.).

Capper, C. B. (1974). The protection of open recirculating cooling systems. *Effl. Wat. Treat. J.*, **125**, 577–83.

Capuzzo, J. M. (1979a). The effect of temperature on the toxicity of chlorinated cooling waters to marine animals—A preliminary review. *Mar. Pollut. Bull.*, **10**, 45–47.

Capuzzo, J. M. (1979b). The effects of halogen toxicants on survival, feeding and egg production of the rotifer, *Brachionus plicatilis. Estuar. Coastal Mar. Sci.*, **8**, 307–16.

Capuzzo, J. M., Davidson, J. H., Lawrence, S. A. and Libni, M. (1977a). The differential effects of free and combined chlorine on juvenile marine fish. *Estuar. Coastal Mar. Sci.*, **5**, 733–41.

Capuzzo, J. M., Goldman, F. C., Davidson, J. A. and Lawrence, S. A. (1977b). Chlorinated cooling waters in the marine environment: Development of effluent guidelines. *Mar. Pollut. Bull.*, **8**(7), 161–3.

Capuzzo, J. M. and Reynolds, W. W. (1980). Lobster behaviour in relation to power plants, pp. 227–240 in Hocutt *et al.* (Eds).

Carlson, D. M. (1974). Responses of planktonic cladocerans to heated waters, pp. 186–206 in Gibbons and Sharitz (Eds).

Carpenter, E. J., Anderson, S. J. and Peck, B. B. (1974a). Copepod and chlorophyll a concentrations in receiving waters of a nuclear power station and problems associated with their measurement. *Estuar. Coastal Mar. Sci.*, **2**, 83–8.

Carpenter, E. J., Peck, B. B. and Anderson, S. J. (1972). Cooling-water chlorination and productivity of entrained phytoplankton. *Mar. Biol.*, **16**, 37–40.

Carpenter, E. J., Peck, B. B. and Anderson, S. J. (1974b). Survival of copepods passing through a nuclear power station on North-eastern Long Island Sound, USA. *Mar. Biol.* **24**, 49–55.

Carpenter, E. J., Peck, B. B. and Anderson, S. J. (1974c). Summary of entrainment research at the Millstone Point nuclear power station, 1970 to 1972, pp. 31–35 in Jensen (Ed.) (b).

Carpenter, J. H. and Smith, C. A. (1978). Reactions in chlorinated seawater, pp. 195–208 in Jolley *et al.* (Eds) (c).

Carpenter, J. H., Smith, C. A. and Zika, R. G. (1980). Reaction products from the chlorination of seawater, pp. 379–86 in Jolley *et al.* (c).

Carr, A. P. and Blackley, M. W. L. (1986). Implications of sedimentological and hydrological processes on the distribution of radionuclides; The example of a salt marsh near Ravenglass, Cumbria. *Estuar. Coastal Shelf Sci.*, **22**, 529–43.

Carr, W. E. S. and Giesel, J. T. (1975). Impact of thermal effluent from a steam electric station on a marshland nursery area during the hot season. *Fish. Bull.*, **73**(1), 67–80.

Carrier, R. F. and Hannon, E. H. (1979). Entrainment: an annotated bibliography. Interim Report, Palo Alto, Electric Power Research Institute, EA-1049, April 1979.

Carstens, T. (1975). Trapping of heat in sill fjords, pp. 99–111 in IAEA(a).

Carter, S. R. (1978). Macroinvertebrate entrainment study at Fort Calhoun station, pp. 155–69 in Jensen (Ed.) (b).

Carter, S. R., Bazata, K. R. and Andersen, D. L. (1982). Macroinvertebrate communities of the channelized Missouri River near two nuclear power stations, pp. 147–85, in Hesse *et al.* (Eds) (b).

Casey, H. (1977). Origin and variation of nitrate nitrogen in the chalk springs, streams and rivers in Dorset and its utilisation by higher plants. *Prog. Wat. Technol.*, **8** (4 & 5) 225–35.

Casey, H. and Farr, I. S. (1982). The influence of within stream disturbance on dissolved nutrient levels during spates. *2nd Int. Symp. on Interactions between Sediments and Freshwater*, Queens University, Kingston, Canada. *Hydrobiologia*, **92**, 447–62.

Casterin, M. E. and Reynolds, W. W. (1982). Thermoregulatory behavior and diel activity of yearling winter flounder *Pseudopleuronectes americanus* (Walbaum). *Environ. Biol. Fish.*, **7**(2), 177–80.

Castenholz, R. W. and Wickstrom, C. E. (1975). Thermal streams, pp. 264–95 (Chapter 12) in Whitton (Ed.).

CEGB (1963). *Glossary of Power Station Terms*, Instructions Booklet No. 7. Personnel Department, CEGB, Newgate Street, London.

CEGB (1976). *Proceedings of the Seminar on Aquatic Environmental Studies. Dinorwig Pumped Storage Scheme.* Generation, Construction and Design Division, CEGB, Llanberis, North Wales.

CEGB Research (1987). Acid Rain, Issue No. 20. CEGB, London.

CEGB Research (1990) 30 Years of Marine Biology, Issue No. 23. CEGB, London.

Cember, H., Curtis, E. H. and Blaycock, B. G. (1978). Mercury bioconcentration in fish: temperature and concentration effects. *Environ. Pollut.*, **17**(4), 311–19.

CERL (1971). Symposium on Freshwater Biology and Electrical Power Generation. CEGB Internal Report, Parts 1 and 2, RD/L/M312, Leatherhead, UK.

CERL (1975). Symposium on Marine Science and Electricity Generation in Southampton Water, 27 June 1973, Fawley. CEGB Internal Report, LM/Biol/001, Leatherhead, UK.

Chamberlain, G. and Strawn, K. (1977). Submerged cage culture of fish in supersaturated thermal effluent. pp. 625–45. In: *Proceedings of the 5th Annual Meeting of the World Mariculture Society*, Avault, J. W. (Ed.). Lousiana State University. Baton Rouge.

Chang, S-L. (1978). Resistance of pathogenic *Naegleria* to some common physical and chemical agents. *Appl. Environ. Microbiol.*, **35**, 368–75.

Characklis, W. G. (1978). Biofouling film development and destruction in experimental systems, pp. 73–82 in Gerhold (Ed.).

Characklis, W. G., Trulear, M. G., Stathopoulos, N. and Chiang, L. C. (1980).

Oxidation and destruction of microbial films, pp. 349–68 in Jolley *et al.* (Eds) (c).

Cheremisinoff, N. P. and Cheremisinoff, P. N. (1983). *Cooling Towers. Selection, Design and Practice.* Ann Arbor Press. The Butterworth Group, Ann Arbor, Michigan.

Cherry, D. S. and Cairns, J. (1982). Biological monitoring. Part V. Preference and avoidance studies. *Wat. Research*, **16**(3), 263–301.

Cherry, D. S., Cairns, J., van Hassel, J. H., Specht, W. L., Lechleitner, R. A., Nicholson, R. B. and Whitaker, J. B. (1982a). Field and laboratory methods useful in predicting ecotoxicological effects of coal ash, pp. 1–31 in Jenkins and Schjodtzhansen (Eds).

Cherry, D. S., Dickson, K. L. and Cairns, J. (1975). Temperatures selected and avoided by fish at various acclimation temperatures. *J. Fish. Res. Board Can.*, **32**(4), 485–91.

Cherry, D. S., Dickson, K. L., Cairns, J. and Stauffer, J. R. (1977a). Preferred, avoided and lethal temperatures of fish during rising temperature conditions. ibid., **34**(2), 239–46.

Cherry, D. S., Guthrie, R. K. and Harvey, R. S. (1974). Temperature influence on bacterial populations in three aquatic systems. *Wat. Res.*, **8**, 149–55.

Cherry, D. S., Guthrie, R. K., Rodgers, J. H., Cairns, J. and Dickson, K. L. (1976). Responses of Mosquito fish (*Gambusia affinis*) to ash effluent and thermal stress. *Trans. Am. Fish. Soc.*, **105**(6), 686–94.

Cherry, D. S., Guthrie, R. K., Sherberger, F. F. and Larrick, S. R. (1979a). The influence of coal ash and thermal discharges on the distribution and bioaccumulation of aquatic invertebrates. *Hydrobiologia*, **62**(3), 257–67.

Cherry, D. S., Hoehn, R. C., Waldo, S. S., Willis, D. H., Cairns, J. and Dickson, K. L. (1977b). Field-laboratory determined avoidances of the spotfin shiner and the bluntnose minnow to chlorinated discharges. *Wat. Res. Bull.*, **13**(5), 1047–55.

Cherry, D. S., Larrick, S. R., Dickson, K. L. and Cairns, J. (1977c). Response of eurythermal and stenothermal fish species to chlorinated discharges. In: *Trace Substances in Environmental Health XL. A Symposium*, Hemphill, D. D. (Ed.). University of Missouri, Columbia, Missouri, pp. 413–18.

Cherry, D. S., Larrick, S. R., Dickson, K. L., Hoehn, R. C. and Cairns, J. (1977d). Significance of hypochlorous acid in free residual chlorine to the avoidance response of spotted bass (*Micropterus puntulatus*) and rosyface shiner (*Notropis rubellus*). *J. Fish. Res. Board Can.*, **34**(9), 1365–72.

Cherry, D. S., Larrick, S. R., Giattina, J. D., Cairns, J. and van Hassel, J. (1982b). Influence of temperature selection upon the chlorine avoidance of cold-water and warmwater fishes. *Can. J. Fish. Aquat. Sci.*, **39**(1), 162–73.

Cherry, D. S., Larrick, S. R., Giattina, J. D., Dickson, K. L. and Cairns, J. (1978). The avoidance response of the common shiner to total and combined residual chlorine in thermally influenced discharges, pp. 826–37 in Thorp and Gibbons (Eds).

Cherry, D. S., Larrick, S. R., Guthrie, R. K., Davis, E. M. and Sherberger, F. F. (1979b). Recovery of invertebrate and vertebrate populations in coal ash stressed drainage systems. *J. Fish. Res. Board Can.*, **36**, 1089–96.

Cherry, D. S., Rodgers, J. H., Graney, R. L. and Cairns, J. (1980). Dynamics and control of the Asiatic clam in the New River, Virginia. Bulletin 123 of the Virginia Water Resources Center, Blacksburg, Va.

Cherry, D. S., van Hassel, J. H., Ribbe, P. H. and Cairns, J. (1987). Factors influencing acute toxicity of coal ash to rainbow trout, (*Salmo gairdneri*) and bluegill sunfish (*Lepomis macrochirus*). *Wat. Res. Bull.*, **23**(2), 293–306.

Chiba, K. (1981). Present status of flow-through and recirculating systems and their limitations in Japan, pp. 41–51 in Tiews (Ed.).

Chow, W. and Kawaratani, R. K. (1983). Biofouling assessment and control; An Electric Power Research Institute overview, pp. 887–900 in Jolley *et al.* (b).

Christensen, S. W., Solomon, J. A. and Tyndall, R. L. (1984). Legionnaires disease bacteria in power plant cooling systems, in Oak Ridge Environmental Sciences Division Report, Publication No. 2261, Oak Ridge. TN, p. 12.

Christophersen, J. (1973). Micro-organisms, pp. 3–86 in Precht *et al.* (Eds).

Christy, E. J., Farlow, J. O., Bourque, J. E. and Gibbons, J. W. (1974). Enhanced growth and increased body size of turtles living in thermal and post-thermal aquatic systems, pp. 277–84 in Gibbons and Sharitz (Eds).

Christy, E. J. and Sharitz, R. R. (1975). Growth of aquatic herbs along a thermal gradient. *Proc. 2nd Thermal Ecology Symposium, Augusta*, April. US Atomic Energy Commission, Savannah River, GA.

Chu, T. Y. J. and Olem, H. (1980). Power industry wastes. *J. Wat. Pollut. Contr. Fed.*, **52**(6), 1433–45.

Chu, T-Y, J. and Olem, H. (1982). Power industry wastes. ibid., Literature Review Issue, **54**(6), 990–1002.

Chung, K. S. and Strawn, K. (1978). Stochastic approach to predict survival of estuarine animals exposed to hot discharge effluent, pp. 642–662 in Thorp and Gibbons (Eds).

Chung, K. S. and Strawn, K. (1982). Predicted survival of the bay anchovy (*Anchoa mitchilli*) in the heated effluent of a power plant on Galveston Bay, Texas. *Environ. Biol. Fish.*, **7**(1), 57–62.

Churchill, M. A. and Wojtalik, T. A. (1969). Effects of heated discharges on aquatic environment. The TVA Experience. American Power Conference Chicago, IL.

Clark, J. and Brownwell, W. (1973). Electric power plants in the coastal zone: environmental issues. American Littoral Soc. Spec. Publ. No. 7. Amer. Litt. Soc., NJ.

Clark, W. C., Shelbourn, J. E. and Brett, J. R. (1981). Effect of artificial photoperiod cycles, temperature and salinity on growth and smolting in underyearling coho (*Oncorhynchus kisutch*), chinook (*O. tscawytscha*) and sockeye (*O. nerka*) salmon. *Aquaculture*, **22**, 105–16.

Clarke, K. U. (1967). Insects and temperature, pp. 293–346 in Rose (Ed.).

Clement, L. J. (1972). The increase in the oxygen content of water in the River Maas, Amer and Donge due to the presence of Amer Power Station (Holland). *Elektrotechniek*, **50**(21), 838–42. CEGB Translation Series Number 6192, London.

Clugston, J. P. (1973). The effects of heated effluents from a nuclear reactor on species diversity abundance reproduction and movement of fish. PhD Thesis, University of Georgia, Athens.

Cole, R. A. (1977). Chlorination for the control of biofouling in thermal power plant cooling water systems, pp. 29–37 in Jensen (Ed.).

Cole, R. A. and Kelly, J. E. (1978). Zoobenthos in thermal discharge to Western Lake Erie. *J. Wat. Pollut. Contr. Fed.*, **50**(11), 2509–21.

Colson, E. W. (1974). Marine mammal investigations. In: *Environmental Investigations at Diablo Canyon*, Adams, J. R. and Hurley, J. F. (Eds). Pacific Gas and Electric Company, Department of Engineering Research, CA, pp. 253–8.

Constable, A. S. C. (1979). Dechlorination practices and control. Paper No. IWC-79-37, 40th Annual Meeting, International Water Conference, October, 30–31, Pittsburgh, PA.

Copeland, B. S., Laney, R. W. and Pendleton, F. C. (1974). Heat influences in estuarine ecosystems, pp. 423–37 in Gibbons and Sharitz (Eds).

Cootner, P. H. and Lof, G. O. (1965). *Water Demand for Stream Electric Generation.* Resources for the Future Incorporated, Johns Hopkins Press, Baltimore.

Corpe, W. A. (1972). Microfouling; The role of primary film-forming marine bacteria, pp. 598–609 in Acker *et al.* (Eds).

Corpe, W. A. (1978). Ecology of microbial attachment and growth on solid surfaces, pp. 57–66 in Gerhold (Ed.).

Cory, R. L. and Nauman, J. W. (1969). Epifauna and thermal additions. *Ches. Sci.*, **10**, 210.

Coste, M., Verrel, J. L., Bray, M. and Roche, O. (1978). Incidences du rechauffement des eaux de Seine sur la composition de la microflore diatomique benthique. *Cahiers du Laboratoire de Hydrobiologie, Montereau*, **6**, 27–44.

Cotter, A. J. R., Phillips, D. J. H. and Ahsanullah, M. (1982). The significance of temperature, salinity and zinc as lethal factors for the mussel, *Mytilus edulis* in a polluted estuary. *Mar. Biol.*, **68**, 135–41.

Coughlan, J. (1969). The littoral fauna of Milford Haven, near the outfall of Pembroke power station. CEGB Internal Report RD/L/N 27/69, Leatherhead, UK.

Coughlan, J. (1977a). Marine wood-borers in Southampton Water, 1951–1975. *Proc. Hants. Field Club Archaeol. Soc.*, **33**, 5–15.

Coughlan, J. (1977b). Review of wood-borer studies, Milford Haven, 1966–1976. CEGB Internal Report, RD/L/N 215/76, Leatherhead, UK.

Coughlan, J. (1978). Bradwell biological investigations; the bottom fauna of the Blackwater Estuary, 1875. CEGB Internal Report, RD/L/R 1969, Leatherhead, UK.

Coughlan, J. and Davis, M. H. (1983). Effect of chlorination on entrained plankton at several United Kingdom coastal power stations, pp. 1053–66 in Jolley *et al.* (b).

Coughlan, J. and Davis, M. H. (1985). Concentrations of chlorine around marine cooling-water outfalls: Prediction of a model, pp. 1459–68 in Jolley *et al.* (Eds).

Coughlan, J. and Fleming, J. M. (1978). A versatile pump-sampler for live zooplankton. *Estuaries*, **1**(2), 132–5.

Coughlan, J. and Whitehouse, J. W. (1977). Aspects of chlorine utilisation in the United Kingdom. *Ches. Sci.*, **18**(1), 102–11.

Coughtrey, P. J. (1983). *Ecological Aspects of Radionuclide Releases.* British Ecological Society Special Publication, Number 3, Blackwell, Oxford.

Coulson, H. J. W. and Forbes, V. A. (1952). In: *The Law of Waters and Land Drainage*, Hobday, S. R. (Ed.) (6th edn). Sweet and Maxwell, London.

Coutant, C. C. (1962). The effect of a heated water effluent upon the macroinvertebrate riffle fauna of the Delaware River. *Proc. Am. Acad. Sci.*, **36**, 58–71.

Coutant, C. C. (1967). Thermal pollution, biological effects. *J. Wat. Pollut. Contr. Fed.*, Annual Review.

Coutant, C. C. (1968a). Thermal pollution—biological effects. A review of the literature of 1967 on wastewater and water pollution control. *J. Wat. Pollut. Contr. Fed.*, **40**, 1047–52.

Coutant, C. C. (1968b). Effect of temperature on the development rate of bottom organisms, In: *Biological Effects of Thermal Discharges*, Annual Report of the Pacific NW Laboratories, US Atomic Energy Commission, Division of Biological Medicine, pp. 11–12.

Coutant, C. C. (1969a). Thermal pollution—biological effects. *J. Wat. Pollut. Contr. Fed. Annual Review.*, **41**, 1036–53.

Coutant, C. C. (1969b). Behavior of sonic-tagged chinook salmon and steelhead trout migrating past Hanford thermal discharges, pp. 21–6 in AEC.

Coutant, C. C. (1970a). Thermal pollution—biological effects. A review of the literature of 1969 on wastewater and water pollution control. *J. Wat. Pollut. Contr. Fed.*, **42**, 1025–57.

Coutant, C. C. (1970b). Biological aspects of thermal pollution—1. Entrainment and discharge canal effects. *CRC Crit. Rev. Environ. Contr.*, **1**, 341–81.

Coutant, C. C. (1971a). Thermal pollution—biological effects. A review of the literature of 1970 on water pollution control. *J. Wat. Pollut. Contr. Fed.*, **43**, 1292–334.

Coutant, C. C. (1971b). Effects of organisms of entrainment in cooling water steps towards predictability. *Nucl. Safety*, **12**, 600–7.

Coutant, C. C. (1973). Effects of thermal shock on the vulnerability of juvenile salmonids to predation. *J. Fish. Res. Board Can.*, **31**, 351–4.

Coutant, C. C. (1974). Evaluation of entrainment effect, pp. 1–12 in Jensen (Ed.).

Coutant, C. C. (1975). Temperature selection by fish. A factor in power plant impact assessments, pp. 575–95 in IAEA (a).

Coutant, C. C. (1976). How to put waste heat to work. *Environ. Sci. Technol.*, **10**(9), 868–71.

Coutant, C. C. (1977). Compilation of temperature preference data. *J. Fish. Res. Board Can.*, **34**, 740–5.

Coutant, C. C., Cox, D. K. and Moore, K. W. (1976). Further studies of cold-shock effects on susceptibility of young channel catfish to predation, pp. 154–8 in Esch and McFarlane (Eds).

Coutant, C. C. and Genoway, R. G. (1968). Final report on an exploratory study of interaction of increased temperature and nitrogen supersaturation on mortality of adult salmonids. Battelle Memorial Inst. Pacif. Northwest Labs, Richland, Washington.

Coutant, C. C. and Goodyear, C. P. (1972). Thermal effects. *J. Wat. Pollut. Contr. Fed.*, **44**, 1250–94.

Coutant, C. C. and Pfuderer, H. A. (1973). Thermal effects. *ibid.*, **45**, 1331–2593.

Coutant, C. C. and Pfuderer, H. A. (1974). Thermal effects. *ibid.*, **46**, 1476–540.

Coutant, C. C. and Talmage, S. S. (1975). Thermal effects. *ibid.*, **47**, 1656–710.

Coutant, C. C. and Talmage, S. S. (1976). Thermal effects. *ibid.*, **48**(6), 1486–544.

Coutant, C. C. and Talmage, S. S. (1977). Thermal effects. *ibid.* (June), 1369–425.
Coutant, C. C., Wasserman, C. S., Chung, M. S., Rubin, D. B. and Manning, H. (1978). Chemistry and biological hazard of a coal ash seepage stream. *ibid.*, **50**, 747–52.
Cowles, R. B. and Bogert, C. M. (1944). A preliminary study of the thermal requirements of desert reptiles. *Bull. Am. Mus. Nat. Hist.*, **83**, 265–96.
Cowx, I. G., Young, W. O. and Booth, P. J. (1987). Thermal characteristics of 2 regulated rivers in Mid-Wales, UK. *Regulated Rivers; Research and Management*, **1**(1), 85–91.
Cox, D. K. (1974). Effects of three heating rates on the critical thermal maximum of bluegill, pp. 158–63 in Gibbons and Sharitz (Eds).
Cragg-Hine, D. (1968). The macro-invertebrate fauna of the River Nene and Peterborough Power Station effluent channel. CEGB Internal Report, RD/L/N89/68, Leatherhead, UK.
Cragg-Hine, D. (1969a). The reproductive cycle of fishes in the effluent channel of Peterborough Power Station. CEGB Internal Report, RD/L/N157/69, Leatherhead, UK.
Cragg-Hine, D. (1969b). The feeding habits of fish in the effluent channel of Peterborough Power Station. CEGB Internal Report, RD/L/R-1556, Leatherhead, UK.
Cragg-Hine, D. (1971). Coarse fish populations in the electricity cut, Peterborough. *Proc. 5th British Coarse Fish. Conf.*, Liverpool, pp. 19–28. Janssen Services, UK.
Crane, A. M., Bahner, L. H. and Domey, R. G. (1980). Influences of selected tidal-activity dependent parameters on chlorine demand in an estuarine environment, pp. 407–14 in Jolley *et al.*
Cravens, J. B. (1982). Thermal effects. *J. Wat. Pollut. Contr. Fed.*, **54**(6), 812–29.
Cravens, J. B. and Harrelson, M. E. (1985). Thermal effects. *ibid.* (June), 649–57.
Cravens, J. B., Harrelson, M. E. and Talmage, S. S. (1983). Thermal effects. *ibid.* (June), 787–800.
Crawshaw, L. I. (1977). Physiological and behavioural reactions of fishes to temperature change. *J. Fish. Res. Board Can.*, **34**(5), 730–4.
Crawshaw, L. I., Lemons, D. E. and Russo, K. E. (1980). Crayfish behaviour in relation to power plants, pp. 241–60, in Hocutt *et al.* (Eds).
Crecelius, E. A. (1985). Fly-ash disposal in the ocean: An alternative worth considering, pp. 379–88 in Duedall *et al.* (Eds) (a).
Crema, R. and Pagliai, M. B. (1980). The structure of benthic communities in an area of thermal discharge from a coastal power station. *Mar. Pollut. Bull.*, **11**, 224–7.
Criepi (1977). *Thermal Effects on Marine Organisms; The State of the Art.* Bio-environment Laboratory, Central Research Institute of the Electric Power Industry, Abiko-city, Chiba, Japan.
Crisp, D. J. (Ed.) (1964). The effects of the severe winter of 1962–1963 on marine life in Britain. *J. Anim. Ecol.*, **33**, 165–210.
Crisp, D. J. (1965). The ecology of marine fouling, pp. 99–117. In: Goodman, G. T., Edwards, R. W. and Lambert, J. M. (Eds) *Ecology and the Industrial Society.* British Ecological Society Symposium No. 5. Blackwell, Oxford, pp. 99–117.

Crisp, D. J. (1972). Mechanisms of adhesion of fouling organisms, pp. 69–709 in Acker *et al.* (Eds).

Crisp, D. J. (1990). Water temperature in a stream gravel bed and implications for salmon incubation. *Freshwater Biology*, **23**, 601–12.

Crisp, D. J. and Molesworth, A. H. N. (1951). Habitat of *Balanus amphitrite* var. *denticulata* in Britain, *Nature (Lond.)*, **167**, 489.

Crisp, D. T. and Le Cren, E. D. (1970). The temperature of three different small streams in Northwest England. *Hydrobiologia*, **35**(2), 305–23.

Cross, F. A., Dean, J. A. and Osterberg, C. L. (1969). The effect of temperature, sediment, and feeding on the behavior of four radionuclides in a marine benthic amphipod. In: *Proceedings of the Second National Symposium on Radioecology*, Nelson, D. J. and Evans, F. C. (Eds), CONF-670503, USAEC (TID-4500), Washington, pp. 450–61.

Cumbie, P. M., Miskimen, T. A. and Rice, J. K. (1985). Environmental impacts of chlorine discharges: A utility industry perspective, pp. 63–71 in Jolley *et al.* (Eds).

Cummins, K. W. (1972). What is a river? A zoological description, pp. 33–52 in Oglesby *et al.* (Eds).

Curtis, E. J. C. (1969). Sewage fungus: its nature and effects. *Wat. Res.*, **3**, 289–311.

Cushing, D. H. (1976). The impact of climatic change on the fish stocks in the North Atlantic. *Geogr. J.*, **142**, 216–27.

Dahlberg, M. D. and Conyers, J. W. (1974). Winter fauna in a thermal discharge with observations on a macrobenthos sampler, pp. 414–22 in Gibbons and Sharitz (Eds).

Dale, H. M. and Gillespie, T. (1976). The influence of floating vascular plants on the diurnal fluctuations of temperature near the water surface in early spring. *Hydrobiologia*, **49**(3), 246–56.

Dandy, C. J. (1964a). An examination of the food and feeding habits of two species of fish in a cooling pond at Grove Road power station. CEGB Internal Report, RD/L/N 8/64, Leatherhead, UK.

Dandy, C. J. (1964b). An examination of the diet of a sample of trout collected from Lake Trawsfynnydd. CEGB Internal Report, RD/L/N 11/64, Leatherhead, UK.

Dandy, C. J. (1964c). A study of the feeding habits of the trout in Cavendish Dock (Roosecote power station). CEGB Internal Report, RD/L/N 19/64, Leatherhead, UK.

Danilievski, A. A. (1965). *Photoperiodism and Seasonal Development in Insects.* Oliver and Boyd, London.

Davies, B. R. and Walker, K. F. (Eds) (1986). *The Ecology of River Systems.* Monographia Biological, Vol. 60, Dr. W. Junk, Amsterdam.

Davies, I. (1966). Chemical changes in cooling water towers. *Int. J. Air Wat. Pollut.*, **10**, 853–63.

Davies, R. M. and Jensen, L. D. (1974). Effects of entrainment of zooplankton at three mid-Atlantic power plants, pp. 131–55 in Jensen (Ed.).

Davies, R. M., Hanson, C. H. and Jensen, L. D. (1976). Entrainment of estuarine zooplankton into a mid-Atlantic power plant. Delayed effects, pp. 349–57 in Esch and McFarlane (Eds).

Davis, M. H. (1977). Heavy metals in the common cockle, *Cerastoderma edulis* (*L*) from Southampton Water. MSc Thesis, Brunel University.

Davis, M. H. (1983a). The response of entrained phytoplankton to chlorination at a coastal power station (Fawley, Hampshire). CEGB Internal Report, TPRD/L/2470/N83, Leatherhead, UK.

Davis, M. H. (1983b). Preliminary observations on the effect of chlorination or the community structure of entrained phytoplankton. CEGB Internal Report, TPRD/L/2471/N83, Leatherhead, UK.

Davis, M. H. and Coughlan, J. (1978). Response of entrained plankton to low-level chlorination at a coastal power station, pp. 369–76 in Jolley *et al.* (Eds) (c).

Davis, M. H. and Coughlan, J. (1983). A model for predicting chlorine concentration within marine cooling circuits and its dissipation at outfalls, pp. 347–58 in Jolley *et al.* (a).

Davis, M. H. and Coughlan, J. (1984). Comparative studies of the effects on marine phytoplankton of three methods of power plant chlorination. 5th Conference on Water Chlorination: Environmental Impact and Health Effects. (unpublished poster).

Deacutis, C. F. (1978). Effect of thermal shock on predator avoidance by larvae of two fish species. *Trans. Am. Fish. Soc.*, **107**(4), 632–5.

De-Jonckheere, J. F. (1981). Pathogenic and nonpathogenic *Acanthamoeba* spp. in thermally polluted discharges and surface waters. *J. Protozool.*, **28**(1), 56–9.

De-Jonckheere, J. F., Mealrd, C. and Philippart, J. C. (1983). Appearance of pathogenic *Naegleria fowleri* (Amoebida, Vahilkampfiidae) in artificially heated water of a fish farm. *Aquaculture*, **18**, 73–8.

De-Jonckheere, J. and Van de Voorde, H. (1976). Differences in destruction of cysts of pathogenic and nonpathogenic *Naegleria* and *Acanthamoeba* by chlorine. *Appl. Environ. Microbiol.*, **31**, 294–7.

De-Jonckheere, J. F. and Van de Voorde, H. (1977). The distribution of *Naegleria fowleri* in man-made thermal waters. *Am. J. Trop. Med. Hyg.*, **26**(1), 10–15.

De-Jonckheere, J., Van Dijck, P. and Van de Voorde, H. (1975). The effect of thermal pollution on the distribution of *Naegleria fowleri*. *J. Hyg. Camb.*, **75**, 7–13.

Demont, D. J. and Miller, R. W. (1971). First reported incident of gas-bubble disease in the heated effluent of a steam generating station. *Proc. 25th Ann. Conf. S.E. Assoc. Game and Fish Comm.*, South Carolina, Charleston, pp. 392–9.

Dempsey, C. H. (1983). Entrainment of young fish in the cooling waters of Kingsnorth power station, Kent. CEGB Internal Report, TPRD/L/2428/N82, Leatherhead, UK.

de Nie, H. W. (1982). Effects of thermal effluents from the Bergum power station on the zooplankton in the Bergumermeer. *Hydrobiologia*, **95**, 337–50.

Descamps, B. (1977). Studies of the possibilities of utilising heated discharges in eel culture. *Pisciculture Francaise*, **49**, 25–36.

De Sylva, D. P. (1969). Theoretical considerations of the effects of heated effluents on marine fishes, pp. 229–93 in Krenkel and Parker (Eds) (b).

Devik, O. (1975). Waste heat and nutrient loaded effluents in the aquaculture. Report on the workshop held at Solstrand near Bergen, 24–28 February 1974, pp. 693–702 in IAEA (a).

Devik, O. (Ed.) (1976). *Harvesting Polluted Waters; Waste Heat and Nutrient-Loaded Effluents in the Aquaculture.* Environmental Science Research Series, Plenum Press, New York, London.

Devinney, J. S. (1980). Effects of thermal effluents on communities of benthic marine algae. *J. Environmental Management*, **11**, 225–42.

De Walle, D. R. (1979). Utilization and dissipation of waste heat by soil warming, in Lee and Sengupta.

Dickson, K. L., Cairns, J., Messenger, D. I., Plafkin, J. L. and Van der Schalie, W. (1977). Effects of intermittent chlorination on aquatic organisms and communities. *J. Wat. Pollut. Contr. Fed.*, **49**, 35–44.

Dickson, K. L., Hendricks, A. C., Crossman, J. S. and Cairns, J. (1974). Effects of intermittently chlorinated cooling tower blowdown on fish and invertebrates. *Am. Chem. Soc.*, **8**(9), 845–9.

Dizon, A. E., Neill, W. H. and Magnuson, J. J. (1977). Rapid temperature compensation of volitional swimming speeds and lethal temperatures in tropical tunas (Scombridae). *Environ. Biol. Fish.*, **2**(1), 83–92.

Donze, M. (1978). Measurements of the effect of heating on survival and growth of natural plankton populations. *Proc. Int. Ass. Theor. Appl. Limnol.*, **20**, 342–6.

Doudoroff, P. and Katz, M. (1950). Critical review of literature on the toxicity of industrial wastes and their components to fish. 1. Alkalis, acids and inorganic gases. *Sewage Industr. Wastes*, **22**(11), 1432–58.

Draley, J. E. (1977). Biofouling control in cooling towers and closed cycle systems, pp. 23–28 in Jensen (Ed.).

Dreesen, D. R., Gladney, E. S., Owens, J. W., Perkins, B. L., Wienke, C. L. and Wangen, L. E. (1977). Comparison of levels of trace elements extracted from fly-ash and levels found in effluent waters from a coal-fired plant. *Environ. Sci. Technol.*, **11**, 1017–19.

Dressel, D. M., Heinle, D. R. and Grote, M. G. (1972). Vital staining to sort live and dead copepods. *Ches. Sci.*, **13**, 156–9.

Dryer, W. and Benson, N. G. (1957). Observations on the influence of the New Johnsonville Steam Plant on fish and plankton populations. *Proc. Ann. Conf. S.E. Assoc. Game and Fish. Comm.*, Vol. 10, pp. 85–91.

DSIR (1964). Effects of polluting discharges on the Thames Estuary. Water Pollution Research Tech. Paper 11, Department of Scientific and Industrial Research, HMSO, London.

Duedall, I. W., Kester, D. R., Park, P. K. and Ketchum, B. H. (Eds) (1985a). *Energy Wastes in the Ocean*. Wiley-Interscience, London, New York.

Duedall, I. W., Kester, D. R., Park, P. J. and Ketchum, B. H. (1985b). Energy wastes in the marine environment: An overview, pp. 127–162 in Duedall *et al.* (Eds) (a).

Duever, M. J. and Abernethy, C. S. (1974). Synergistic effects of temperature and acute radiation. In: Battelle Pacific Northwest Laboratories, Annual report for 1973, BNWL-1850, Pt. 2, Richland, pp. 77–8.

Duma, R. J. (1978). The epidemiology of naturally acquired human, waterborne meningoencephalitis due to thermophilic pathogenic *Naegleria*, pp. 107–114 in Gerhold (Ed.).

Dunford, W. E. (1977). OTC Field fish spawning study. Ontario Hydro Research Division Report No. 80-45-K, Ontario, Canada.

Dunstall, T. G. (1985). Effects of entrainment on phytoplankton primary production at four thermal electric generating stations on the Laurentian Great Lakes. *Int. Revue ges. Hydrobiol.*, **70**(2), 247–57.

Dunster, H. J. (1978). Pollution resulting from the release of radioactive waste materials to the sea. *Mar. Pollut. Bull.*, **9**, 118–22.

Durrett, C. W. (1972). Density, distribution, production and drift of benthic fauna in a reservoir receiving thermal discharges from a steam electric generating station. MS Thesis, North Texas State University, Denton, Texas.

Durrett, C. W. and Pearson, W. D. (1975). Drift of macroinvertebrates in a channel carrying heated water from a power plant. *Hydrobiologia*, **46**(1), 33–43.

Dussart, B. (1955). La temperature des lacs et ses causes de variations. *Verh. int. Ver. Limnol.*, **12**, 78–96.

Dvorak, A. J. and Lewis, B. G. (Eds) (1978). Impacts of coal-fired power plants on fish, wildlife and their habitats. Biological Services Program, Fish and Wildlife Service, US Department of the Interior (FWS/OBS 78/29). US Government Printing Office, Washington, D.C., 259 pp.

Dyer, K. R. (1973). *Estuaries: a Physical Introduction*. Wiley-Interscience, London.

Eaton, R. A. (1976). Cooling tower fungi, pp. 359–87 in Jones, E. B. G.

Eaton, R. A. and Jones, E. B. G. (1971). New fungi on timber from water-cooling towers. *Nova Hedwigia*, **19**, 779–88.

Eble, A. F., Stolpe, N. R. and Evans, M. C. (1975). The use of thermal effluents of an electric generation station in New Jersey in aquaculture of the Great Malaysian prawn *Macrobrachium rosenbergii* and the Rainbow trout *Salmo gairdneri*. *Proc. Power Plant Waste Heat Utilization in Aquaculture Workshop*, November 6–7, Trenton, NJ.

EDF (1977). Influence des rejets thermiques sur le milieu vivant en mer et en estuaire. *Journees de la Thermo-ecologique*, Direction de l'Equipment, Electricité de France.

EDF (1981). Influence des rejets thermiques sur le milieu vivant en mer et en estuaire, *Journees de la Thermo-ecologique*, Direction de l'Equipment, Electricité de France.

Edgren, M. and Notter, M. (1980). Cadmium uptake by fingerlings of perch (*Perca fluviatilis*) by Cd-115 m at two different temperatures. *Bull. Environ. Contam. Toxicol.*, **24**, 647–51.

Edgren, M., Olsson, M. and Renberg, L. (1979). Preliminary results on uptake and elimination at different temperatures of p,p'-DDT and two chlorobiphenyls in perch from brackish water. *Ambio*, **VIII** (6), 270–2.

Edgren, M., Olsson, M. and Reutergardh, L. (1981). A one year study of the seasonal variations of sDDT and PCB levels in fish from heated and unheated areas near a nuclear power plant. *Chemosphere*, **10**(5), 447–52.

Edinger, J. E., Brady, D. K. and Graves, W. L. (1968a). The variation of water temperatures due to steam electric cooling operations. *J. Wat. Pollut. Contr. Fed.*, **40**(9), 1632–9.

Edinger, J. E., Duttweiler, D. W. and Geyer, J. C. (1968b). The response of water temperatures to meterological conditions. *Wat. Res.*, **4**(5), 1137–43.

Edinger, J. E. and Geyer, J. C. (1965). *Heat Exchange in the Environment*. Edison Electric Institute Research, New York.

Edington, J. M. (1966). Some observations on stream temperature. *Oikos*, **15**(11), 265–73.

Edsall, T. A. and Yocum, T. G. (1972). Review of recent technical information concerning the adverse effects of once-through cooling on Lake Michigan. US Fish and Wildlife Service, Bureau of Sport Fishing and Wildlife, Ann Arbor, Michigan (mimeo).

Edwards, R. W. and Brooker, M. P. (1982). *The Ecology of the Wye*. Monographie Biologicae, Vol. 50, Illies, J. (Ed.), Dr W. Junk, The Hague, Boston, London.

Edwards, R. W., Williams, P. E. and Williams, R. (1984). Ebbw, pp. 83–112 in Whitton.

EEI (1969). *Environmental Projects Related to Thermal Discharges*. Edison Electric Institute, New York.

Effer, W. R. (1984). Environmental information and methodologies required for the environmental assessment of Canadian power generation installations, pp. 283–309 in Roberts and Roberts.

Effer, W. R. and Bryce, J. B. (1975). Thermal discharge studies on the Great Lakes—the Canadian experience, pp. 371–87 in IAEA (a).

Egusa, S. (1981). Fish diseases and their control in intensive culture utilizing heated effluents or recirculating systems in Japan, pp. 33–9 in Tiews (Ed.).

Ehrlich, K. F., Hood, J. M., Muszynski, G. and McGowen, G. E. (1979). Thermal behaviour responses of selected California littoral fishes. *Fish Bull.*, **76**(4), 837–49.

EIFAC (1965). Water quality for European freshwater fish. Report on finely dissolved solids and inland fisheries. *Int. J. Air Water Pollut.*, **9**, 151–68.

EIFAC (1968a). Water quality criteria for European freshwater fish. Report on extreme pH values and inland fisheries. European Inland Fisheries Advisory Commission Tech. Paper No. 4, Rome.

EIFAC (1968b). Water quality criteria for freshwater fish. Report on water temperature and inland fisheries. ibid., No. 6, Rome.

EIFAC (1973). Water quality criteria for European freshwater fish. Report on chlorine and freshwater fish. ibid., No. 20, Rome.

Eiler, H. O. and Delfino, J. T. (1975). Comparative effects of two modes of cooling water discharge on Mississippi River biota and environmental ecosystems, pp. 685–92 in IAEA (a).

Eisenbud, M. (1973). *Environmental Radioactivity*. Academic Press, London, New York.

Eisler, R. (1982). *Trace Metal Concentrations in Marine Organisms*. Pergamon Press, Oxford.

Elder, H. W. and Hollingden, G. A. (1976). Flue gas desulphurization. In: *Proceedings of a National Conference on Health, Environmental Effects and Control Technology of Energy Use*, Environmental Protection Agency, Office of Research and Development, EPA-600/7-76-002. Washington, D.C., pp. 262–6.

Electricity Council (1973). *Electricity Supply in Great Britain. A Chronology: Organissation and Development* (2 vols). Electricity Council, London.

Electricity Council (1987). *Handbook of Electricity Supply Statistics*. The Electricity Council, London.

Elliott, J. M. (1971). Some methods for the statistical analysis of samples of benthic invertebrates. *Sci. Publ. Freshw. Biol. Assn*, No. 25. Ambleside, UK.

Elliott, J. M. (1972). Effect of temperature on the time of hatching in Baetis rhodani (Ephemeroptera: Baetidae). *Oecologia*, **9**, 47–51.

Elliott, J. M. (1976). The energetics of feeding, metabolism and growth of brown trout (*Salmo trutta L.*) in relation to body weight, water temperature and ration size. *J. Anim. Ecol.*, **45**, 923–33.

Elliott, J. M. (1978). Effect of temperature on the hatching time of eggs of Ephemerella ignita (Poda) (Ephemeroptera-Ephemerellidae). *Freshwat. Biol.*, **8**, 51–8.

Elliott, J. M. (1980). Some aspects of thermal stress in freshwater teleosts. In: *Stress and Fish*, Pickering, A. D. (Ed.). Academic Press, London and New York, pp. 209–45.

Elliott, J. M. and Tullett, P. A. (1978). A bibliography of samplers for benthic invertebrates. *Freshw. Biol. Assn Occasional Publ.*, No. 4.

Eloranta, P. V. (1983). Physical and chemical properties of pond waters receiving warm water effluent from a thermal power plant. *Wat. Res.*, **17**, 133–40.

Elser, H. J. (1965). Effect of warmed water discharge on angling in the Potomac River, Maryland, 1961–62. *Prog. Fish Cult.*, **27**, 79–86.

Emberton, H. and Turnpenny, A. W. H. (1981). Toxicity of chlorinated sea-water in relation to the culture of eels (*Anguilla anguilla* (L)) in heated effluents. CEGB Internal Report, RD/L/2112 N81, Leatherhead, UK.

Emberton, H. and Turnpenny, A. W. H. (1982). Experiments on the culture of eels (*Anguilla anguilla* (L.)) in raw and dechlorinated cooling water from a coastal power station. CEGB Internal Report, RD/L2144 N81, Leatherhead, UK.

Emmings, A. (1990). Water resource consequences following a nuclear event. *Atom*, **399** (January), 30–3.

Eng-Wilmot, D. L., Hitchcock, W. S. and Martin, D. F. (1977). Effect of temperature on the proliferation of *Gymnodinium breve* and *Gomphosphaeria aponina*. *Mar. Biol.*, **41**, 71–7.

Engel, R. E., Malzer, G. L. and Bergsrud, F. G. (1985). Saline cooling tower water for irrigation in Minnesota; crop and soil effects. *J. Environ. Qual.*, **14**(1), 32–6.

England, A. C., Fraser, D. W., Mallison, G. F., Mackel, D. C., Skaliy, P. and Gorman, G. W. (1982). Failure of *Legionella pneumophila* sensitivities to predict culture results from disinfectant-treated air-conditioning cooling-towers. *Appl. Environ. Microbiol.*, Jan., 240–4.

Englert, T. L., Lawler, J. P., Aydin, F. N. and Vachtsevanos, G. (1976). A model of striped bass population dynamics in the Hudson River. In: *Estuarine Processes, Vol. 1, Uses, Stresses and Adaptation to the Estuary*, Wiley, M. (Ed.). Academic Press, London and New York, pp. 137–50.

EPA (1972). *Water Quality Standards Criteria Digest*. US Environmental Protection Agency (August 1972), Washington, D.C.

EPA (1976). *Quality Criteria for Water*. US Environmental Protection Agency, Washington, D.C.

Eppley, R. W. (1972). Temperature and phytoplankton growth in the sea. *Fish Bull.*, **70**(4), 1063–85.

Eppley, R. W., Renger, E. H. and Williams, P. M. (1976). Chlorine reactions with sea-water constituents and the inhibition of photosynthesis of natural marine phytoplankton. *Estuar. Coastal Mar. Sci.*, **4**, 147–61.

Erickson, J. and Freeman, A. J. (1978). Toxicity screening of fifteen chlorinated and brominated compounds using four species of marine phytoplankton, pp. 307–10 in Jolley *et al.* (Eds) (c).

Ernst, E. J. (1970). Part II—Floras and faunas of the jetty and deeper areas, pp. 53–55 in Hechtel (Ed.).

Esch, G. W. and Hazen, T. C. (1978). Thermal ecology and stress: A case history for red-sore disease in largemouth bass, pp. 331–63 in Throrp and Gibbons (Eds).

Esch, G. W. and McFarlane, R. W. (Eds) (1976). Thermal Ecology, II. Technical Information Centre, Energy Research and Development Administration, ERDA Symposium Series (Conf. 750425), Springfield, Va.

Eure, H. E. and Esch, G. W. (1974). Effects of thermal effluent on the population dynamics of helminth parasites in largemouth bass, pp. 207–15 in Gibbons and Sharitz (Eds).

Evans, M. S. Hawkins, B. E. and Wurster, T. E. (1978). Effects of the Donald C. Cook nuclear power plant on zooplankton of southeastern Lake Michigan, pp. 125–39 in Jensen (Ed.).

Everich, D. and Gonzalez, J. G. (1977). Critical thermal maxima of two species of estuarine fish. *Mar. Biol.*, **41**, 141–5.

Evins, C. (1975). The toxicity of chlorine to some freshwater organisms. Water Research Centre, Tech. Report, T.W.8, Stevenage, Herts.

Fales, R. R. (1978). The influence of temperature and salinity on the toxicity of hexavalent chromium to the grass shrimp, (*Palaeomonetes pugio.* (Holthius)). *Bull. Environ. Contam. Toxicol.*, **20**(4), 447–50.

Falke, J. D. and Smith, M. H. (1974). Effects of thermal effluent on the fat content of the mosquito fish, pp. 100–8 in Gibbons and Sharitz (Eds).

Farrel, J. and Rose, A. H. (1967). Temperature effects on micro-organisms, in Rose (Ed.).

Farrell, J. R. and Tesar, M. A. (1982). Periphytic algae in the channelized Missouri River with special emphasis on apparent optimal temperatures, pp. 85–124 in Hesse *et al.* (Eds) (b).

Fava, J. A. and Seegert, G. L. (1983). Factors in the design of chlorine toxicological research, pp. 913–6 in Jolley *et al.* (b).

Fava, J. A., Rue, W. J., Chrostowski, P., Ferris, J. S. and Plugge, H. (1985). Conceptual approach to evaluate alternatives to chlorination for biofouling control, pp. 73–84 in Jolley *et al.* (Eds).

Fenton, P. J. and Norris, T. E. (1972). The economics and future of total energy and combined heat/power installations. *Proc. Thermo-Fluids Conference— Thermal Discharges.* Engineering and Ecology/4–7 December. Institute of Engineering, Sydney, Australia.

Ferens, M. C. and Murphy, T. M. (Jr) (1974). Effects of thermal effluent on populations of mosquito fish, pp. 237–45 in Gibbons and Sharitz (Eds).

Ferguson-Wood, E. J. (1967). *Microbiology of Oceans and Estuaries.* Elsevier Publishing Company, Amsterdam, London.

Fey, Von J. M. (1977). The heating of a mountain stream and the effects on the zoocoenosis demonstrated on the Lenne, Sauerland. *Arch. Hydrobiol. (Suppl. 53)*, **3**, 307–63.

Fickeisen, D. H. and Schneider, M. J. (Eds) (1976). Gas bubble disease. Proc. Workshop, Richland, Washington, 8–9 October, 1974, ERDA Tech. Inf. Centre, Oak Ridge, TN, Conf. 741033.

Filatova, T. N., Tsippert, M. R., Zair-Bek, I. A., Misyuk, V. A., Molitvina, V. P. and Uleskina, A. G. (1976). Thermal pollution of water bodies and the hydrologic aspects of the problem. *Soviet Hydrology: Selected Papers*, **15**(4), 257–64.

Fischer, H. B., List, E. J., Koh, R. C. Y., Imberger, J. and Brooks, N. H. (1979). *Mixing in Inland and Coastal Waters.* Academic Press, New York, London.

Fjerdingstad, E. (1975). Bacteria and fungi, pp. 129–53 in Whitton (Ed.).

Flemer, D. A. (1974). The effects of entrainment on phytoplankton at the Morgantown steam-electric station, Potomac River estuary, September, 5–8, 1972, pp. 163–175 in Jensen (Ed.).

Flemer, D. and Sherk, J. A. Jr (1977). The effects of steam electric station operation on entrained phytoplankton. *Hydrobiologia*, **55**(1), 33–44.

Fleming, J. M. (1970). An investigation into the growth of periphyton in the heated effluent from Trawsfynydd Nuclear Power Station, 1969. CEGB Internal Report, RD/L/N 199/70, Leatherhead, UK.

Fleming, J. M. and Coughlan, J. (1978). Preservation of vitally stained zooplankton for live/dead sorting. *Ches. Sci.*, **19**(2), 135–7.

Fliermans, C. B., Bettinger, G. E. and Fynsk, A. W. (1982). Treatment of cooling systems containing high levels of *Legionella pneumophila*. *Wat. Res.*, **16**, 903–9.

Fliermans, C. B., Cain, P. S. and Schmidt, E. L. (1975). Direct measurement of bacterial stratification in Minnesota lakes. *Arch. Hydrobiol.*, **76**(2), 248–55.

Fliermans, C. B., Gorden, R. W., Hazen, T. C. and Esch, G. W. (1978). Aeromonas distribution and survival in a thermally altered lake, pp. 93–100 in Gerhold (Ed.).

Flook, R. A. (1978). Problems associated with the re-use of purified sewage effluents for power station cooling purposes. *Prog. Wat. Technol.*, **10**(1/2), 105–11.

Flowers, B. (1976). *Nuclear Power and the Environment.* Royal Commission on Environmental Pollution, 6th Report, HMSO, London.

Foerster, J. W., Trainor, F. R. and Buck, J. D. (1974). Thermal effects on the Connecticut river. Phycology and Chemistry. *J. Wat. Pollut. Contr. Fed.*, **46**, 2138–52.

FORATOM (1987). *Nuclear Power in Western Europe.* Forum Atomique European, London.

Forbes, A. M. (1980). Decomposition and microbial colonisation of leaves in a stream modified by a coal ash effluent. *Hydrobiologia*, **76**, 263–7.

Forbes, A. M., Magnuson, J. J. and Harrell, D. M. (1981). Effects of habitat modifications from coal ash effluent on stream invertebrates; A synthesis *Trans. Am. Fish. Soc.*, Warmwater Fisheries Symposium, pp. 241–9.

Ford, R. F., Foreman, D. G., Grubbs, K. J., Kroll, C. D. and Watts, D. G. (1978). Effects of thermal effluent on benthic marine invertebrates determined from long-term simulation studies, pp. 546–68 in Thorp and Gibbons (Eds).

Forsyth, D. J. and McColl, R. H. S. (1974). The limnology of a thermal lake, Lake Rotowhero, New Zealand. II, General biology with emphasis on the benthic fauna of Chronomids. *Hydrobiologia*, **44**(1), 91–104.

Fox, J. L. and Corcoran, E. F. (1957). Thermal and osmotic counter measures against some typical marine fouling organisms. *Corrosion*, **14**, 31–2.

Fox, J. L. and Moyer, M. S. (1973). Some effects of a power plant on marine microbiota. *Ches. Sci.*, **14**(1), 1–10.

Fraizier, A. and Ancellin, J. (1975). Influence de la temperature sur la contamination d'especes marines par le der-59, pp. 51–63 in IAEA (c).

Fraley, J. J. (1979). Effects of elevated stream temperatures below a shallow reservoir on a cold water macroinvertebrate fauna, pp. 257–72 in Ward and Stanford (Eds) (a).

Franco, R. J. (1980). The successful use of non-chromate cooling water treatments. *Industr. Wat. Engng*, **17**(5), 14–18.

Frank, M. L. (1974). Relative sensitivity of different developmental stages of carp eggs to thermal shock, pp. 171–6 in Gibbons and Sharitz (Eds).

Freeman, R. F. and Sharma, R. K. (1977). *Survey of Fish Impingement at Power Plants in the United States, Vol. 11, Inland Waters*. Rep. no. ANL/ES-56. Argonne National Laboratory, Argonne, IL.

Fry, F. E. J. (1967). Responses of vertebrate poikilotherms to temperature, pp. 375–420 in Rose (Ed.).

Fry, F. E. J., Hart, J. S. and Walker, K. F. (1946). Lethal temperature relations for a sample of young speckled trout (*Salvelinus fontinalis*), University of Toronto Studies in Biology Series, Vol. 54, pp. 1–35.

Funnell, I. R. (1988). Infrared thermography in the electricity supply industry. In: *Applications of Thermal Imaging*, Burnay, S. G., Williams, T. L. and Jones, C. H. (Eds). Adam Hilger, IOP Publishing, Bristol and Philadelphia, pp. 73–103.

Gallagher, B. J. (Ed.) (1974). *Energy Production and Thermal Effects*. Ann Arbor Science Publishers, Ann Arbor, Michigan.

Gallaway, B. J. and Strawn, K. (1974). Seasonal abundance and distribution of marine fishes at a hot water discharge in Galveston bay, Texas. *Contrib. Mar. Sci.*, **18**, 71–137.

Gallaway, B. J. and Strawn, K. (1975). Seasonal and areal comparisons of fish diversity indices at hot-water discharge in Galveston Bay, Texas. *Contrib. Mar. Sci.*, 79–89.

Galloway, M. L. and Kilambi, R. V. (1988). Thermal enrichment of a reservoir and the effects on annulus formation and growth of largemough bass, *Micropterus salmoides. J. Fish Biol.*, **32**(3), 533–44.

Gallup, D. N. and Hickman, M. (1975). Effects of the discharge of thermal effluent from a power station on Lake Wabamun, Alberta, Canada, limnological features. *Hydrobiologia*, **6**, 45–69.

Gallup, D. N., Hickman, M. and Rasmussen, J. (1975). Effects of thermal effluent and macrophyte harvesting on the benthos of an Alberta lake. *Verh. int. Verein. theor. angew. Limnol.*, **19**, 552–61.

Gameson, A. L. H. (1957). Weirs and the aeration of rivers. *J. Inst. Wat. Res. Engrs*, **11**(6), 477–90, 72.

Gameson, A. L. H., Gibles, J. W. and Barrett, M. J. (1959). A preliminary temperature survey of a heated river. *J. Wat. Wat. Engng*, **63**, 13.

Gameson, A. L. H. and Truesdale, G. A. (1959). Some oxygen studies in streams. *J. Inst. Wat. Res. Engrs*, **9**, 571–94.

Gammon, J. R. (1976). Measurement of entrainment and predictions of impact on the Wabash and Ohio rivers, pp. 159–76 in Jensen (Ed.).

Gammon, K. M. (1969). Planning cooling water for power stations. *Proc. 4th Int. Conf. on Wat. Pollut. Res.*, Prague, Inst. Water. Res., Stevenage, UK, pp. 927–36.

Garey, J. F. (1980). A review and update of possible alternatives to chlorination for controlling biofouling in cooling-water systems of steam-electric generating stations, pp. 453–70 in Jolley *et al.*

Garside, E. T. (1970). Fishes, structural responses, pp. 561–616 in Kinne (Ed.) (a).

Gasparini, R. (1982). Water quality and the discharge of cooling water into rivers, lakes and coastal waters, pp. 15–30, in Jenkins and Schjodtzhansen (Eds).

Gaudy, R. (1981). Mortalite du zooplankton transitant dans les circuits de refroidissement de la centrale de Martiques-Ponteau, pp. 725–44 in EDF.

Gaufin, A. R. and Hern, S. (1971). Laboratory studies on the tolerance of aquatic insects to heated water. *J. Kansas Entomol. Soc.*, **44**, 240–5.

Gay, B., La Croix, M. J. B. and Ophel, I. L. (1976). Low grade heat: a resource in cold climates. *Proceedings of a Workshop at the Chalk River Nuclear Laboratories*, October 6–10, 1975. Atomic Energy of Canada, AECL 5322/1, Ontario, Canada.

Gehrs, C. W. (1974). Vertical movement of zooplankton in response to heated water, pp. 285–90 in Gibbons and Sharitz (Eds).

Gentile, J. H., Cardin, J., Johnson, M. and Sosnowski, S. (1976). *Power plants, Chlorine and Estuaries*. Environmental Research Laboratory, Office of Research and Development, US Environmental Protection Agency, Narragansett, Rhode Island, EPA-600 3-76-055.

Gentry, J. B., Garten, C. T. Jr, Howell, F. G. and Smith, M. H. (1975). Thermal ecology of dragonflies in habitats receiving reactor effluent, pp. 563–74 in IAEA (a).

Gerchakov, S. M. and Sallman, B. (1978). Biofouling and effects of organic compounds and micro-organisms on corrosion processes, pp. 67–72 in Gerhold (Ed).

Gerhold, R. M. (Ed.) (1978). Microbiology of power plant thermal effluents. Proceedings of a symposium. University of Iowa, Iowa City, Iowa.

GESAMP (1984). Thermal discharges in the marine environment. IMO/FAO/UNESCO/WMO/WHO/IAEA/UNEP, joint group of experts on the scientific aspects of marine pollution. Reports and studies No. 24.

Gessner, F. (1970). Plants, pp. 363–406 in Kinne (Ed.) (a).

Giattina, J. D., Cherry, D. S. and Larrick, S. R. (1981). Comparison of laboratory and field avoidance behaviour of fish in heated chlorinated water. *Tran. Am. Fish. Soc.*, **110**, 526–35.

Gibbons, J. W. (1976). Thermal alterations and the enhancement of species populations, pp. 27–31 in Esch and McFarlane (Eds).

Gibbons, J. W., Bennett, D. H., Esch, G. W. and Hazen, T. C. (1978). Effects of thermal effluent on body condition of largemouth bass. *Nature*, **274** (5670), 470–1.

Gibbons, J. W. and Sharitz, R. R. (Eds) (1974). *Thermal Ecology*. ERDA, Tech. Inf. Centre, US Atomic Energy Commission. Conf. 730505, Washington.

Gibbons, J. W. and Sharitz, R. R. (1981). Thermal ecology: environmental teachings of a nuclear reactor site. *Bioscience*, **31**, 293–8.

Gibbons, J. W., Sharitz, R. R., Howell, F. G. and Smith, M. H. (1975). Ecology of artificially heated streams, swamps and reservoirs on the Savannah River Plants; The Thermal Studies program of the Savannah River Ecology Laboratory, pp. 389–99 in IAEA (a).

Gibson, C. I., Thatcher, T. D. and Apts, C. W. (1976). Some effects of temperature, chlorine and copper on the survival and growth of the coon stripe shrimp, pp. 88–92 in Esch and McFarlane (Eds).

Gibson, C. I., Tone, F. C., Schirmer, R. E. and Blaylock, J. W. (1980).

Bioaccumulation and depuration of bromoform in five marine species, pp. 517–34 in Jolley *et al.*

Gibson, R. J. (1978). The behavior of juvenile Atlantic salmon (*Salmo salar*) and brook trout. (*Salvelinus fontinalis*) with regard to temperature and water velocity. *Trans. Am. Fish. Soc.*, **107**(5), 703–12.

Ginn, T. C. and O'Connor, J. W. (1976). Response of the estuarine amphipod Gammarus daiberi to chlorinated power plant effluent. *Estuar. Coastal Mar. Sci.*, **6**, 459–69.

Ginn, T. C., Poje, G. V. and O'Connor, J. M. (1978). Survival of planktonic organisms folowing passage through a simulated power plant condenser tube, pp. 91–102 in Jensen (Ed.) (b).

Ginn, C., Waller, W. T. and Lauer, G. J. (1976). Survival and reproduction of *Gammarus* spp. (Amphipoda) following short-term exposure to elevated temperatures. *Ches. Sci.*, **17**(1), 8–14.

Glassman, A. B. and Bennett, C. E. (1978). Response of the alligator to infection and thermal stress, pp. 691–702 in Thorp and Gibbons (Eds).

Glynn, P. W. (1965). Community composition, structure and interrelationships in the marine intertidal *Endocladia muricata–Balanus glandula* association in Monterey Bay, California. *Beaufortia*, **12**(148), 1–198.

Goldberg, E. D., Koide, M., Hodge, V., Flegal, A. R. and Martin, J. (1983). U.S. Mussel Watch; 1977–78 Results on trace metals and radionuclides. *Estuar. Coastal Shelf Sci.*, **16**, 69–93.

Goldberg, R. (1978). Some effects of gas-supersaturated seawater on Spisula solidissima and Argopecten irradians. *Aquaculture*, **4**, 282–7.

Goldman, J. C., Capuzzo, J. H. and Wong, G. T. F. (1978). Biological and chemical effects of chlorination at coastal power plants, pp. 291–306 in Jolley *et al.* (Eds) (c).

Goldman, J. C. and Davidson, J. A. (1977). Physical model of marine phytoplankton chlorination at coastal power plants. *Environ. Sci. Technol.*, **11**, 908–13.

Goldman, J. C. and Quimby, H. L. (1979). Phytoplankton recovery after power plant entrainment. *J. Wat. Pollut. Contr. Fed.*, **51**(7), 1816–23.

Gonzalez, J. G. (1974). Critical thermal maxima and upper lethal temperatures for the calanoid copepods *Acartia tona* and *A. clausi. Mar. Biol.*, **27**, 219–23.

Gonzalez, J. C. and Yevich, P. (1976). Response of an estuarine population of the blue mussel, *Mytilus edulis* to heated water from a steam generating plant. *Mar. Biol.*, **34**, 177–89.

Goodyear, C. P. (1977a). Assessing the impact of power plant mortality on the compensatory reserve of fish populations, pp. 186–95 in Van Winkle (Ed.).

Goodyear, C. P. (1977b). Mathematical methods to evaluate entrainment of aquatic organisms by power plants, Fish and Wildlife Service, Biological Services Program, Topical Brief, Fish and Wildlife Resources and Electric Power Generation, No. 3, FWS/OBS-76/20, 3, US Department of the Interior, Washington, D.C.

Gorden, R. W. and Fliermans, C. B. (1978). Survival and viability of *Escherichia coli* in a thermally altered reservoir, pp. 135–42 in Gerhold (Ed.).

Gordon, D. C. and Longhurst, A. R. (1979). The environmental aspects of a tidal power project in the upper reaches of the Bay of Hundy. *Mar. Pollut. Bull.*, **10**, 38–45.

Goryajnova, L. I. (1975). Effect of warm effluent from the Novorossiysk thermal power plant on zooplankton. *Hydrobiol. J.*, **11**(6), 19–23.

Goss, L. B. and Bunting, D. L. (1976). Thermal tolerance of zooplankton. *Wat. Res.*, **10**, 387–92.

Goss, L. B. and Cain, C. (1977). Power plant condenser and service water system fouling by Corbicula, The Asiatic clam, pp. 11–17 in Jensen (Ed.).

Gosse, Ph. (1982). Predicting the impact of power plant discharges on the water quality of a river with the aid of a mathematical mode, pp. 149–64 in Jenkins and Schodtzhansen (Eds).

Graham, T. P. (1974). Chronic malnutrition in four species of sunfish in a thermally loaded impoundment, pp. 151–7 in Gibbons and Sharitz (Eds).

Graney, R. L., Cherry, D. S., Rodgers, J. H. and Cairns, J. (1980). The influence of thermal discharges and substrate composition on the population structure and distribution of the Asiatic clam, *Corbicula fluminea* in the New River, Virginia. *The Nautilus*, **94**(4), 130–5.

Gras, R. A. (1982). Impact of thermal power plants on aquatic ecosystems; the French experience, pp. 59–66 in Jenkins and Schodtzhansen (Eds).

Grauby, A. (1978). Thermal release from nuclear power plants, their utilization for plant and fish production. 5th Int. Fair and Tech. Meeting of Nuclear Industries. Vol. D, Paper 3/8, Basel (3–8 Oct), IAEA, Vienna, Austria.

Gray, R. H., Genoway, R. G. and Barraclough, S. A. (1977). Behavior of juvenile chinook salmon (*Oncorhynchus tshawytscha*) in relation to simulated thermal effluent. *Trans. Am. Fish. Soc.*, **106**(4), 366–70.

Gray, R. H., Page, T. L. and Saroglia, M. G. (1983). Behavioral response of carp (*Cyprinus carpio*) and black bullhead (*Ictalurus melas*) from Italy to gas-supersaturated water. *Environ. Biol. Fish.*, **8**(2), 163–7.

Gray, R. H., Page, T. L., Saroglia, M. G. and Bronzi, P. (1982). Comparative tolerance to gas supersaturated water of carp, *Cyprinus* and black bullhead, *Ictalurus melas* from the USA and Italy. *J. Fish Biol.*, **20**, 223–7.

Grayum, M. M. (1973). Effects of thermal shock and ionizing radiation on primary productivity. In: *Radionuclides in Ecosystems*, Nelson, D. J. (Ed.). International Symposium on Radioecology, USAEC, Washington, pp. 639–44.

Grieve, J. A., Johnston, L. E., Dunstall, T. G. and Minor, J. (1978). A program to introduce site-specific chlorination regimes at Ontario hydro generating stations, pp. 77–84 in Jolley *et al.* (Eds) (c).

Grimas, U. (1961). The bottom fauna of natural and impounded lakes in northern Sweden (Ankaravattet and Blasjou), *Rep. Inst. Freshw. Res., Drottinholm*, **42**, 183–237.

Grimas, U. (1970). Warm water effluents in Sweden. *Mar. Pollut. Bull.*, **1**(10), 151–2.

Grimas, U. M. and Ehlin, U. (1975). Swedish studies on combination effects of thermal discharges in the aquatic environment, some aspects of power plant siting policy, pp. 69–79 in IAEA (c).

Grimes, C. B. and Mountain, J. A. (1971). Effects of thermal effluent upon marine fishes near the Crystal River Steam Electric Station. Florida Department of Natural Resources. Marine Research Lab. Professional Papers Series No. 17.

Groberg, W. J., McCoy, R. H., Pilcher, K. S. and Fryer, J. L. (1978). Relation of water temperature to infections of Coho salmon (*Oncorhynchus kisutch*), Chinook salmon (*O. tshawytscha*) and steelhead trout (*Salmo gairdneri*) with

Aeromonas salmonicida and *A. hydrophila. J. Fish. Res. Board Can.,* **35**(1), 1–7.

Groth, D. R. (1975). Chlorine-induced mortality in fish. *Trans. Am. Fish. Soc.,* **4**, 800–2.

Grove, R. (1983). Dispersion of chlorine at seven southern California coastal generating stations, pp. 333–46 in Jolley *et al.* (a).

Grove, R. S., Faeder, E. J., Ospital, J. and Bean, R. M. (1985). Halogen compounds discharged from a coastal power plant, pp. 1371–9 in Jolley *et al.* (Eds).

Guerin, J-P. and Reys, J-P. (1978). Influence d'une temperature elevee sur le rythme de ponte et la fecondite des populations mediterraneenes de *Scolelepis fuliginosa* (Annelide: Polychete) en elevage au laboratoire, pp. 341–48 in McLusky and Berry (Eds).

Guerra, C. R., Godfriaux, B. L., Eble, A. F., Farmanfarmaian, A., Pitman, R. and Campbell, P. (1976). Integration of thermal and food-processing wastes into a system for commercial culture of freshwater shrimp (Power plant waste heat utilization in aquaculture). Final report, Volume 1, NSF/RANN, Grant number 74-14079, AO1, Public Service Electric and Gas Company, Research and Development Department.

Guerra, C. R., Godfriaux, B. L. and Sheahan, C. J. (1979a). Utilization of waste heat from power plants by sequential culture of warm and cold weather species, pp. 2121–40 in Lee and Sengupta.

Guerra, C. R., Godfriaux, B. L. and Sheahan, C. J. (1979b). Venture analyses for a proposed commercial waste heat aquaculture facility. In: *Proc. 10th Annual Meeting. World Mariculture Society,* Avault, J. W. (Ed). Louisiana State University, Baton Rouge, pp. 28–38.

Guma'a, S. A. (1978). The effects of temperature on the development and mortality of eggs of perch, *Perca fluviatilis. Freshwat. Biol.,* **8**, 221–7.

Gundersen, K. and Bienfang, P. (1972). Thermal pollution: Use of deep, cold, nutrient-rich sea-water for power plant cooling and subsequent aquaculture in Hawaii. *Marine Pollut. Sea Life,* FAO, Rome, pp. 513–16.

Gurney, J. D. and Cotter, I. A. (1966). *Cooling Towers.* Maclaren, London.

Gurtz, M. E. and Weiss, C. M. (1974a). Effect of thermal stress on phytoplankton productivity in condenser cooling water, pp. 490–507 in Gibbons and Sharitz (Eds).

Gurtz, M. E. and Weiss, C. M. (1974b). Response of phytoplankton to thermal stress, pp. 177–85 in Jensen (Ed.).

Guthrie, R. K. and Cherry, D. S. (1978). Microbial interactions in thermally influenced waters, pp. 15–22 in Gerhold (Ed.).

Guthrie, R. K., Cherry, D. S. and Ferebee, R. N. (1974a). A comparison of thermal loading effect on bacterial populations in polluted and non-polluted aquatic systems. *Wat. Res.,* **8**, 143–8.

Guthrie, R. K., Cherry, D. S. and Rodgers, J. W. (1974b). The impact of ash basin effluent on biota in the drainage system. *Proc. 7th Mid Atlantic Ind. Waste Conf.,* 12–14 November, Drexel University, Philadelphia, PA.

Guthrie, R. K., Cherry, D. S., Singleton, F. L. and Harvey, R. S. (1978). The effects of coal basin effluent and thermal loading on bacterial populations of flowing streams. *Environ. Pollut.,* **17**(4), 297–302.

Gutknecht, J. (1965). Uptake and retention of cesium-137 and zinc-65 by seaweeds. *Limnol. Oceanogr.,* **10**(1), 58–66.

Haag, R. W. and Gorham, P. R. (1977). Effects of thermal effluent on standing crop and net productions of *Elodea canadensis* and other submerged macrophytes in Lake Wabamum, Alberta. *J. Appl. Ecol.*, **14**, 835–51.

Haag, W. R. and Lietzke, M. H. (1980). A kinetic model for predicting the concentrations of active halogen species in chlorinated saline cooling, pp. 415–26 in Jolley *et al.*

Hadderingh, R. H. (1978). Mortality of young fish in the cooling water system of Bergum Power Station. *Proc. Int. Assoc. Theoret. Appl. Limnol.*, **20**, 347–52.

Hair, J. R. (1971). Upper lethal temperature and thermal shock tolerances of the Opossum shrimp *Neomysis awatschensis* from the Sacramento–San Joaquim estuary, California. *Calif. Fish Game*, **57**, 17–27.

Hall, G. E. (Ed.) (1971). *Reservoir Fisheries and Limnology*. Spec. Pub. No. 8, American Fish Society, Washington, D.C.

Hall, L. W. (1980). Blue crab behaviour in relation to power plants, pp. 207–26 in Hocutt *et al.* (Eds).

Hall, L. W., Burton, D. T. and Margrey, S. L. (1979). The influence of acclimation temperature on the interactions of chlorine, elevated temperature and exposure duration for grass shrimp, *Palaeomonetes pugio*. *Trans. Am. Fish. Soc.*, **106**, 626–31.

Hall, L. W., Burton, D. T., Margrey, S. L. and Graves, W. C. (1983a). Predicted mortality of Chesapeake Bay organisms exposed to simulated power plant chlorination conditions at various acclimation temperatures, pp. 1005–18 in Jolley *et al.* (b).

Hall, L. W., Helz, G. R. and Burton, D. T. (1981). *Power Plant Chlorination. A Biological and Chemical Assessment*. Electric Power Research Institute, Research Report, RP. 1312-1, Palo Alto, CA.

Hall, L. W., Hocutt, C. H. and Stauffer, J. R. (1978). Implication of geographical location on temperature preference of white perch, *Morone americana*. *J. Fish. Res. Board, Can.*, **35**, 1464–8.

Hall, L. W., Margrey, S. L., Graves, W. C. and Burton, D. T. (1983b). Avoidance responses of juvenile Atlantic menhaden *Brevoortia tyrannus* subjected to simultaneous chlorine and δT conditions, pp. 983–92 in Jolley *et al.* (b).

Hamer, P., Jackson, J. and Thurston, E. F. (Eds) (1961). *Industrial Water Treatment Practice*. Butterworths and Imperial Ltd, London.

Hamilton, D. H. (1978). Chlorine application for the control of condenser fouling, pp. 687–94 in Jolley *et al.* (Eds).

Hamilton, D. H., Flemer, D. A., Keefe, C. W. and Mihursky, J. A. (1970). Power plants: Effects of chlorination on estuarine primary production. *Science*, **167**(3941), 197–8.

Hannah, L. (1979). *Electricity before Nationalisation*. Macmillan Press, London.

Hannon, E. H. (1978). A document collection of electricity utility studies related to steam—electric power station cooling system effects on water quality and aquatic biota. Atomic Ind. Forum Research Project, EA-872 (877-1) (Interim Report), Electricity Power Research Institute, Palo Alto, CA.

Harden-Jones, F. R. (1968). *Fish Migration*. Edward Arnold, London.

Harleman, D. R. F. (1969). Mechanics of condenser water discharge from thermal power plants, pp. 144–65 in Parker and Krenkel (Eds).

Harmsworth, R. V. (1974). Artificial cooling lakes as unique aquatic ecosystems, pp. 56–66 in Gallagher (Ed.).

Harrison, F. L. (1977). Quarterly progress report on chemical effluents in surface waters from nuclear plants. Laurence Livermore Laboratory, Livermore, CA.

Harrison, F. L. and Rice, D. W. Jr (1978). Copper sensitivity of adult Pacific oysters, pp. 301–15 in Thorp and Gibbons (Eds).

Harrison, S. J. and Phizacklea, A. P. (1987). Vertical temperature gradients in muddy intertidal sediments in the Forth estuary, Scotland. *Limnol. Oceanog.*, **32**(4), 954–63.

Hartnoll, R. G. (1978). The effect of salinity and temperature on the post-larval growth of the crab, *Rhithropanopeus harrisii*, pp. 349–58 in McLusky and Berry (Eds).

Hartwell, I. S. and Hoss, D. E. (1979). Thermal shock resistance of spot (*Leistomus xanthurus*) after acclimation to constant or cycling temperature. *Trans. Am. Fish. Soc.*, **108**, 397–400.

Hartwig, E. O. and Valentine, R. (1983). Bromoform production in tropical open ocean waters; Ocean thermal energy conversion chlorination, pp. 311–32 in Jolley *et al.* (Eds) (a).

Harvey, R. S. (1974). Temperature effects on the sorption of radionuclides by aquatic organisms, pp. 28–42 in Gibbons and Sharitz (Eds).

Hathaway, E. S. (1927). The relation of temperature to the quantity of food consumed by fishes. *Ecology*, **8**(4), 428–34.

Hatle, S. and Lampar, M. (1975). Exploitation of waste heat, pp. 731–40 in IAEA (a).

Hauser, J., Eppel, D. and Tanzer, F. (1980). Analysis of thermal impact in tidal rivers and estuaries. *Wat. Res.*, **14**(10) 1409–19.

Hawes, F. B. (1970). Thermal problems—old hat in Britain. CEGB Newsletter No. 83, Central Electricity Generating Board, London.

Hawes, F. B., Coughlan, J. and Spencer, J. F. (1975). Environmental effects of the heated discharges from Bradwell Nuclear Power Station and of the cooling systems of other power stations, pp. 423–37 in IAEA (c).

Hawkes, H. A. (1962). Biological aspects, pp. 311–432 in Klein.

Hawkes, H. A. (1969). Ecological changes of applied significance from waste heat, pp. 15–53 in Parker and Krenkel (Eds).

Haymes, G. T. (1980). Fish Spawning in Chlorine Plumes at Three Thermal Generating Stations. Ontario Hydro Research Division, Report No. 80-123-K, Ontario, Canada.

Hazen, T. C. and Fliermans, C. B. (1979). Distribution of *Aeromonas hydrophila* in natural and man-made thermal effluents. *Appl. Environ. Microbiol.*, **38**(1), 166–8.

Heath, A. G. (1977). Toxicity of intermittent chlorination to freshwater fish, Influence of temperature and chlorine form. *Hydrobiologia*, **56**(1), 39–47.

Heath, A. G. (1978). Influence of chlorine form and ambient temperature on the toxicity of intermittent chlorination to freshwater fish, pp. 123–34 in Jolley *et al.* (Eds) (c).

Hechtel, G. J. (Ed.) (1970). Biological effects of thermal pollution. Northport. Stony Brook Technical Report, Marine Sciences Research Center, State University of New York, New York.

Hedgpeth, J. W. and Gonor, J. J. (1969). Aspects of the potential effect of thermal alteration on marine and estuarine benthos, pp. 80–118 in Krenkel and Parker (Eds) (b).

Heimberger, B. E. (1982). Zero liquid discharge control in United States coal-fired steam electric stations, pp. 101–18 in Jenkins and Schodtzhansen (Eds).

Hein, M. K. and Koppen, J. D. (1979). Effects of thermally elevated discharges on the structure and composition of estuarine periphyton diatom assemblages. *Estuar. Coastal Mar. Sci.*, **9**, 385–401.

Heinle, D. R. (1969). Temperature and zooplankton. *Ches. Sci.*, **10**, 186–209.

Heinle, D. R. (1976). Phytoplankton and zooplankton collection problems and relevance of entrainment to the ecosystem, pp. 101–18 in Jensen (Ed).

Heinle, D. R., Millsaps, H. S., Jr and Millsaps, C. V. (1974). Zooplankton investigations at Morgantown, pp. 157–62 in Jensen (Ed).

Hellawell, J. M. (1978). *Biological Surveillance of Rivers—A Biological Monitoring Handbook*. NERC and Water Research Centre, Stevenage, UK.

Hellawell, J. M. (1986). *Biological Indicators of Freshwater Pollution and Environmental Management*. Elsevier Applied Science, Amsterdam, London.

Helz, G. R., Dotson, D. A. and Sigled, A. C. (1983). Chlorine demand; Studies concerning its chemical basis, pp. 181–90 in Jolley *et al.* (a).

Helz, G. R., Sigled, A. C. and Hill, C. A. (1980). Mechanisms of chlorine degradation in estuarine waters, pp. 387–94 in Jolley *et al.*

Helz, G. R., Sugam, R. and Sigled, A. C. (1984). Chemical modifications of estuarine water by a power plant using continuous chlorination. *Environ. Sci. Technol.*, **18**(3), 192–9.

Henderson, P. A., Holmes, R. H. A. and Bamber, R. N. (1988). Size-selective mortality in the sand smelt *Atherina boveri* Risso, and its role in population regulation. *J. Fish Biol.*, **33**, 221–33.

Henderson, P. A., Turnpenny, A. W. H. and Bamber, R. N. (1984a). Long term stability of a sand smelt (*Atherina presbyter*, Cuvier), population subject to power station cropping. *J. Appl. Ecol.*, **21**, 1–10.

Henderson, P. A., Whitehouse, J. W. and Cartwright, G. (1984b). The growth and mortality of larval herring *Clupea harengus* in the River Blackwater estuary, 1978–80. *J. Fish Biol.*, **24**, 613–22.

Henessey, R. A. S. (1971). *The Electric Revolution*. Oriel Press, London, 183 pp.

Hergenrader, G. L., Harrow, L. G., King, R. G., Cada, G. F. and Schlesinger, A. B. (1982). Larval fishes in the Missouri River and the effects of entrainment, pp. 185–225, in Hesse *et al.* (Eds) (b).

Hergott, S. J., Jenkins, D. and Thomas, J. F. (1978). Power plant cooling-water chlorination in northern California. *J. Wat. Pollut. Contr. Fed.*, **50**(11), 2590–601.

Hess, C. T., Smith, C. W. and Price, A. H. (1978). Use of heated reactor effluent for culturing shellfish. *Proceedings 10th National Shellfish Sanitation Workshop*, Hunt Valley, Maryland, June 29–30, 1977, US Atomic Energy Commission, Washington, D.C.

Hess, C. T., Smith, C. W., Price, A. H. and Darling, I. C. (1979). Using heated effluent from an 835 MW nuclear power station for shellfish aquaculture, pp. 2229–41 in Lee and Sengupta.

Hesse, L. W., Bliss, Q. P. and Zuerlin, G. J. (1982a). Some aspects of the ecology of adult fishes in the channelized Missouri River with special reference to the effects of two nuclear power stations, pp. 225–78, in Hesse *et al.* (Eds) (b).

Hesse, L. W., Hergenrader, G. L., Lewis, H. S., Reetz, S. D. and Schlesinger, A. B. (Eds) (1982b). The Middle Missouri River. A collection of papers on the biology with special reference to power station effects. The Middle Missouri River Study Group. P.O. Box 934, Norfolk, NE 68701, USA.

Hestagen, I. H. (1979). Temperature selection and avoidance in the sand giby, *Pomatoschistus minutus* (Pallas), collected at different seasons. *Environ. Biol. Fish.*, **4**(4), 369–77.

Hetherington, J. A. (1976). Radioactivity in surface and coastal waters of the British Isles (1974). Fish. Radiobiol. Lab. Tech. Rep. FRL 11, MAFF, Lowestoft.

Hetherington, J. A. and Harvey, B. R. (1978). Uptake of radioactivity by marine sediments and implications for monitoring metal pollutants. *Mar. Pollut. Bull.*, **9**(4), 102–6.

Hettler, W. F. and Clements, L. C. (1978). Effects of acute thermal stress on marine fish embryos and larvae, pp. 171–90 in Jensen (Ed.) (b).

Hickman, M. (1974). Effects of the discharge of thermal effluent from a power station on Lake Wabumum, Alberta, Canada—The epipelic and epipsammic algal communities. *Hydrobiologia*, **45**(2–3), 199–215.

Hickman, M. (1982). The removal of a heated water discharge from a lake and the effect upon an epiphytic algal community. *Hydrobiologia*, **87**, 21–32.

Hickman, M. and Klarer, D. M. (1974). The growth of some epiphytic algae in a lake receiving thermal effluent. *Arch. Hydrobiol.*, **74**, 403–26.

Hickman, M. and Klarer, D. M. (1975). The effect of the discharge of thermal effluent from a power station on the primary productivity of epiphytic algal community. *Brit. Phycol. J.*, **10**, 81–91.

Hill, B. J. (1977). The effect of heated effluent on egg production in the estuarine prawn Upogebia africana (Ortmann). *J. Exp. Mar. Biol. Ecol.*, **29**, 291–302.

Hillbricht-Ilkowska, A. and Zdanowski, B. (1978). Effect of thermal effluent and retention time on lake functioning and ecological efficiencies in plankton communities. *Internationale Revue der Gesamten Hydrobiologie*, **63**(5), 609–17.

Hillman, R. E. (1980). Behavior of bivalve molluscs, pp. 309–26 in Hocutt *et al.* (Eds).

Hillman, R. E., Davis, N. W. and Wennemer, J. (1977). Abundance, diversity and stability in shore-zone fish communities in an area of Long Island Sound affected by the thermal discharge of a nuclear power station. *Estuar. Coastal Mar. Sci.*, **5**, 355–81.

Hindle, E. (1932). Some new thermophilic organisms. *J. Roy. Microsc. Soc.*, **L11**, 123–33.

Hindley, P. D. and Miner, R. M. (1972). Evaluating water surface heat exchange co-efficients. *Journal of the Hydraulics Division, Proc. Am. Soc. Civil Engineers*, **HY8**, 1411–26.

Hines, A. H. (1979). Effects of a thermal discharge on reproductive cycles in Mytilus edulis and Mytilus californianus (Mollusca; Bivalvia). *Fishery Bull.*, **77**(2), 498–503.

Hirayama, K. and Hirano, R. (1970a). Influence of high temperature and residual chlorine on marine phytoplankton. *Mar. Biol.*, **7**, 205–13.

Hirayama, K. and Hirano, R. (1970b). Influences of high temperature and residual chlorine on the marine planktonic larvae. *Bull. Fac. Fish. Nagasaki Univ.*, **29**, 83–90.

Hoadley, A. W., Ajello, G. and Masterson, N. (1975). Preliminary studies of fluorescent pseudomonads capable of growth at 41°C in swimming pool water. *Appl. Microbiol.*, **29**(4), 527–31.

Hoagland, K. E. and Turner, R. D. (1980). Range extensions of teredinids (shipworms) and polychaetes in the vicinity of a temperate-zone nuclear generating station. *Mar. Biol.*, **58**, 55–64.

Hocutt, C. H. (1973). Swimming performance of three warmwater fishes exposed to a rapid temperature change. *Ches. Sci.*, **14**(1), 11–16.

Hocutt, C. H. and Edinger, J. E. (1980). Fish behaviour in flow fields, pp. 143–82 in Hocutt *et al.* (Eds).

Hocutt, C. H., Stauffer, J. R. Jr, Edinger, J. E., Hall, L. W. and Morgan, R. P. II. (Eds) (1980). *Power Plants; Effects on Fish and Shellfish Behaviour.* Academic Press, New York, London.

Hodson, P. V., Spry, D. J. and Blunt, B. R. (1980). Effects of rainbow trout (*Salmo gairdneri*), of a chronic exposure to waterborne selenium. *Can. J. Fish. Aquat. Sci.*, **37**, 233–40.

Hofer, R., Forstner, H. and Rettenwamder, R. (1982). Duration of gut passage and its dependence on temperature and food consumption in roach, *Rutilus rutilus* L.; laboratory and field experiments. *J. Fish Biol.*, **20**, 290–9.

Hoffman, G. L. and Bauer, O. N. (1971). Fish parasitology in reservoirs—A review, pp. 495–512 in Hall (Ed.).

Hokanson, K. E. F. (1977). Temperature requirements of some percids and adaptations to the seasonal temperature cycle. *J. Fish. Res. Board Can.*, **34**, 1524–50.

Hokanson, K. E. F., Kleiner, C. F. and Thorslund, T. W. (1977). Effects of constant temperatures and diel temperature fluctuations on specific growth and mortality rates and yield of juvenile rainbow trout, *Salmo gairdneri. ibid.*, **34**, 639–48.

Holme, N. A. and McIntyre, A. D. (1971). *Methods for Study of Marine Benthos.* I.B.P. Handbook No. 16, Blackwell, Oxford, Edinburgh.

Holmes, D. W. *et al.* (1974). Pond and cage culture of channel catfish in Virginia. *J. Tenn. Acad. Sci.*, **49**, 74–8.

Holmes, N. J. (1970a). Marine fouling in power stations. *Mar. Pollut. Bull.*, **1**(7), 105–6.

Holmes, N. J. (1970b). The design of chlorination schedules for reducing mussel fouling in power station cooling systems. CEGB Internal Report, RD/L/R 1686, Leatherhead, UK.

Holzwarth, G., Balmer, R. G. and Soni, L. (1984a). The fate of chlorine in recirculating cooling towers, Field Results. *Wat. Res.*, **18**(11), 1429–35.

Holzwarth, G., Balmer, R. G. and Soni, L. (1984b). The fate of chlorine and chloramine in cooling towers, Henry's Law constants for flashoff. *ibid.*, **18**(11), 1421–7.

Homer, M. (1976). Seasonal abundance, biomass, diversity and trophic structure of fish in a salt-marsh tidal creek affected by a coastal power plant, pp. 259–67 in Esch and McFarlane (Eds).

Hooper, W. C. *et al.* (1978). Rearing of brook trout and lake trout in the thermal effluent of a coal-fired generating station. *Trans. Can. Electr. Assoc.* (*Eng. Oper. Div.*), **17**(4), paper 78-1-223.

Hoornbeek, F. K., Sawyer, P. J. and Sawyer, E. S. (1982). Growth of winter flounder (*Pseudopleuronectes americanus*), and smooth flounder, (*Liopsetta putnami*), in heated and unheated water. *Aquaculture*, **28**, 363–73.

Horoszewicz, L. (1983). Reproductive rhythm in tench *Tinca tinca* (L), in fluctuating temperatures. *Aquaculture*, **32**(1/2), 79–92.

Horwitz, R. J. (1981). The direct and indirect impacts of entrainment on estuarine communities—the transfer of impacts between trophic levels, pp. 185–98 in Mitsch *et al.* (Eds).

Hose, J. E., Hunt, W. and Stoffel, R. J. (1983a). Physiological responses of a marine fish exposed to chlorinated seawater at concentrations near its avoidance threshold. *Mar. Environ. Res.*, **8**, 241–54.

Hose, J. E., Stoffel, R. J. and Zerba, K. E. (1983b). Behavioral responses of selected marine fishes to chlorinated seawater. Mar. Environ. Res., **9**, 37–59.

Hose, J. E., King, T. D., Zerba, K. E., Stoffel, R. J., Stephens, J. S., Jr and Dickinson, J. A. (1983c). Does avoidance of chlorinated seawater protect fish against toxicity? Laboratory and field observations, pp. 967–82 in Jolley *et al.* (b).

Hoss, D. E., Coston, L. C., Baptist, J. P. and Engel, D. W. (1975). Effects of temperature, copper and chlorine on fish during simulated entrainment in power plant condenser cooling systems, pp. 519–27 in IAEA (a).

Hoss, D. E., Hettler, W. F. and Coston, L. C. (1974). Effects of thermal shock on larval estuarine fish—Ecological implications with respect to entrainment in power plant cooling systems. In: *The Early Life History of Fish*, Blaxter, J. H. S. (Ed.). Springer-Verlag, Berlin-Heidelberg-New York, pp. 357–71.

Hostgaard-Jensen, P., Klitgaard, J. and Pedersen, K. M. (1977). Chlorine decay in cooling water and discharge into seawater. *J. Wat. Pollut. Contr. Fed.*, 1832–41.

House of Lords (1977). EEC Environment Policy Select Committee on the European Communities. R/2005/76, Water Standards for Freshwater Fish, HMSO, London.

Houston, A. H. (1982). Thermal effects upon fishes, NRCC associate committee on scientific criteria for environmental quality. National Research Council of Canada, Publication No. 18566, Ottawa, Canada.

Howell, F. G. and Gentry, J. B. (1974). Effects of thermal effluents from nuclear reactors on species diversity of aquatic insects, pp. 562–71 in Gibbons and Sharitz (Eds).

Howells, G. D. (1977). In and out of hot water. *Mar. Pollut. Bull.*, **8**(11), 245–8.

Howells, G. D. (1980). The effects of power station operation on the marine environment. In: *Seminari Internazionali Sull' Inquinamento Marino*, Della Croce, N. (Ed.). Instituto di Scienze Ambientali Marine, Universita di Genoa, Italia, pp. 137–50.

Howells, G. D. (1983). The effects of power stations cooling water discharges on aquatic ecology. *J. Inst. Wat. Pollut. Cont.*, **82**(1), 10–17.

Howells, G. D., Aston, R. J., Rippon, J. E. and Sadler, K. (Eds) (1980). Exploitation of Reject Heat and Organic Wastes, Report of a Seminar held at CERL on 6th November 1979. CEGB Internal Report, RD/L/N 80/80, Leatherhead, UK.

Howells, G. D. and Gammon, K. M. (1984). The role of research in meeting environmental assessment needs for power station siting, pp. 310–329 in Roberts and Roberts.

Howells, G. D. and Langford, T. E. (1982). Effects of power station cooling water discharges on marine organisms in temperate waters. CEGB Internal Report, TPRD/L/2286/N82, Leatherhead, UK.

Howells, G. D., Taylor, M. T. and Aston, R. J. (1981). Power Station Reject Heat for Biomass Production. CEGB Internal Report, RD/L/217 N81, Leatherhead, UK, 20 pp.

Howells, G. P. and Lauer, G. (Eds) (1969). Hudson River Ecology. *Proceedings of the Second Symposium held at Sterling Forest, Tuxedo, New York*, October 1969. Department of Environmental Conservation, New York State.

Howland, E. B. and Pope, D. H. (1983). Distribution and seasonality of *Legionella pheumophila* in cooling towers. *Curr. Microbiol.*, **9**, 319–24.

Howles, L. R. (1984). Nuclear Station Achievement 1983. *Nucl. Engng Int.*, May, 36–8.

Hrs-Brneko, M. (1978). The relationship of temperature and salinity to larval development in mussels (*Mytilus galloprovincalis* L.), pp. 359–66 in McLusky and Berry (Eds).

Hubbert, M. K. (1971). Energy resources for power production, pp. 13–43 in IAEA.

Hugenin, J. E. (1976). Heat exchangers for use in the culturing of marine organisms. *Ches. Sci.*, **17**(1), 61–4.

Huh, H. T. (1980). Effects of thermal effluents on marine biota in costal waters of Korea. *Acta Oceanographica Taiwanica*, **11**, 1–9.

Humpesch, U. H. and Elliott, J. M. (1980). Effect of temperature on the hatching time of eggs of three *Rithrogena* spp. (Ephemeroptera) from Austrian streams and an English stream and river. *J. Anim. Ecol.*, **49**, 643–61.

Humphris, T. H. (1977). The use of sewage effluent as power station cooling water. *Wat. Res.*, **11**, 217–23.

Humphris, T. H. and Rippon, J. E. (1978). The effect of chlorine on nitrifying bacteria. CEGB Internal report, RD/L/N 164/77, Leatherhead, UK.

Hunt, F. R. (1971). Power station site selection in England and Wales, pp. 647–57 in IAEA.

Hunt, G. J. (1985). Radioactivity in surface and coastal waters of the British Isles, 1983. Ministry of Agriculture Fisheries and Food, Directorate of Fisheries Research, Aquat. Environ. Monit. Rep., MAFF Direct. Fish. Res., Lowestoft, UK.

Hunt, G. J. (1989). Radioactivity in surface and coastal waters of the British Isles, 1988. Aquat. Environ. Monit. Rep. MAFF Direct. Fish. Res., Lowestoft, UK.

Hutchinson, G. E. (1957). *A Treatise on Limnology, Vol. 1*. John Wiley, New York.

Hutchinson, V. H. (1976). Factors influencing thermal tolerances of individual organisms, pp. 10–26 in Esch and McFarlane (Eds).

Hynes, H. B. N. (1960). *The Biology of Polluted Waters*. Liverpool University Press, Liverpool.

Hynes, H. B. N. (1970). *The Ecology of Running Waters*. Liverpool University Press, Liverpool.

IAEA (1966). *Disposal of Radioactive Wastes into Seas, Oceans and Surface Waters*. International Atomic Energy Agency, Vienna.

IAEA (1969). *Environmental Contamination by Radioactive Materials.* International Atomic Energy Agency, Vienna.

IAEA (1971). *Environmental Aspects of Nuclear Power Stations.* Proceedings Series, International Atomic Energy Agency, Vienna.

IAEA (1972). *Thermal Discharges at Nuclear Power Stations: Their Management and Environmental Impacts.* Report of a Panel Meeting, 23–27th Oct. 1982. International Atomic Energy Agency, Vienna.

IAEA (1974a). *Environmental Surveillance around Nuclear Installations,* Vol. 1. International Atomic Energy Agency, Vienna.

IAEA (1974b). *ibid.,* Vol. 2.

IAEA (1974c). *Population Dose Evaluation and Standards for Man and his Environment.* International Atomic Energy Agency, Vienna.

IAEA (1974d). Thermal discharges at nuclear power stations: Their management and environmental impacts. Technical Report Series, Number 155, International Atomic Energy Agency, Vienna.

IAEA (1975a). *Environmental Effects of Cooling Systems at Nuclear Power Stations.* International Atomic Energy Agency, Vienna.

IAEA (1975b). *Impact of Nuclear Releases into the Aquatic Environment.* International Atomic Energy Agency, Vienna.

IAEA (1975c). *Combined Effects of Radioactive, Chemical and Thermal Releases into the Environment.* International Atomic Energy Agency, Vienna.

IAEA (1976). *Effects of Ionizing Radiation on Aquatic Organisms and Ecosystems.* International Atomic Energy Agency, Technical Report Series 172, Vienna.

IAEA (1979). *Methodology for Assessing Impacts of Radioactivity on Aquatic Ecosystems.* International Atomic Energy Agency, Technical Report Series, No. 190, Vienna.

IAEA (1989). Nuclear share of electricity production rises in 1988. *IAEA Newsbriefs,* **4**(4) (May), 35.

IAWPR (1982). *Proceedings of the International Conference on Coal fired Power Plants and the Aquatic Environment,* IAWPR, IUPAC, Nordforsk, Copenhagen, 16–18 August, Denmark.

Icanberry, J. and Adams, J. R. (1974). Zooplankton survival in cooling water systems of four thermal power plants on the California coast. Interim report March 1971–Jan. 1972, pp. 13–22 in Jensen (Ed).

Iles, R. (1963). Cultivating fish for food and sport in power station water. *New Scientist,* 31 January, 324.

Ilus, E. and Keskitalo, J. (1987). Aquatic macrophytes in the sea area around the Louisla nuclear power station, south coast of Finland. *Aqua Fennica,* **16**(2), 111–23.

Industrial Water Society (1981). Facts and theories on Legionnaires disease, based on the proceedings of a conference held in May 1981. I.W.S. Intelligence Report, Tamworth, UK.

Ingersoll, C. G. and Claussen, D. L. (1984). Temperature selection and critical thermal maxima of the fantail darter, *Etheostoma flabellare* and johnny darter, *E. nigrum* related to habitat and season. *Environ. Biol. Fish.,* **11**(2), 131–8.

Inman, G. W. Jr and Johnson, J. D. (1978). The effect of ammonia concentration on the chemistry of chlorinated seawater, pp. 235–52 in Jolley *et al.* (Eds) (c).

Ito, H. (1980). Observations on the upper lethal temperatures of five species of sea-snails, Family Buccinidae. *Bulletin of the Hokkaido, Reg. Fisheries Research Laboratories*, **45**, 57–63.

Itzkowitz, N., Schubel, J. R. and Woodhead, M. J. (1983). Responses of summer flounder, *Paralichthys dentatus* embryos to thermal shock. *Environ. Biol. Fish.*, **8**(2), 125–35.

IUPDEE (1979). Environmental problems posed by the siting of large capacity thermal power stations on the coast. Report on the Joint Group of Experts for the study of thermal Power Station Cooling Problems, Warsaw Congress (June 79). International Union of Producers and Distributors of Electrical Energy, Brussels.

Iwanski, M. L. and Chu, T. J. (1985). Power industry wastes. J. Wat. Pollut. Contr. Fed., (June), *Literature Review Issue*, 599–609.

Jacobson, P. M. (1976). Oxygen balance in the condenser cooling water system of the Connecticut Yankee Plant, pp. 35–60 in Merriman and Thorpe (Eds).

Janssen, J. and Giesey, J. P. (1984). A thermal effluent as a sporadic cornucopia: effects on fish and zooplankton. *Environ. Biol. Fish.*, **II**(3), 193–203.

Jarman, R. T. and de Turville, C. M. (1981). Effects of Pembroke power station on the temperatures in Milford Haven, Wales. *Wat. Sci. Technol.*, **13**(7), 295–304.

Jaske, R. T. and Goebel, J. B. (1967). Effects of dam construction on temperatures of Columbia River. *J. Am. Water Works Assoc.*, **59**, 935–43.

Jenkins, S. H. and Schjodtzhansen, P. (Eds) (1982). *Cooling-Water Discharges from Coal-Fired Power Plants; Water Pollution Problems*. Pergamon Press, Oxford.

Jenner, H. A. (1980). The biology of the mussel *Mytilus edulis* in relation to fouling problems in industrial cooling water systems. *La tribune du CEBEDEAU*, **33**, 1–8.

Jenner, H. A. (1982). Physical methods in the control of mussel fouling in seawater cooling systems. *ibid.*, **35**, 163–75.

Jenner, H. A. (1983). Control of mussel fouling in the Netherlands; experimental and existing methods. In: *Proceedings of the Symposium on Condenser Macrofouling Control Technologies—The state of the Art*. EPRI, Hyannis, Massachussets, June, pp. 407–21.

Jennison, B. L. (1978). Effects of thermal effluents on reproduction in a sea anemone, pp. 470–83 in Thorp and Gibbons (Eds).

Jensen, L. D. (Ed.) (1974). Entrainment and intake screening. Proc. 2nd Entrainment and Screening Workshop, Rep. No. 15, Edison Electric Institute, New York.

Jensen, L. D. (Ed.) (1976). Third National Workshop on Entrainment and Impingement. Section 316 (b)—research and compliance. Ecological Analysts, New York.

Jensen, L. D. (Ed.) (1977). *Biofouling Control Procedures: Technology and Ecological Effects. Pollution Engineering and Technology*, 5. Marcel Dekker, New York, Basel.

Jensen, L. D. (1978a). Microbiological and ecological issues associated with condenser cooling entrainment, pp. 11–14 in Gerhold (Ed.).

Jensen, L. D. (Ed.) (1978b). Fourth National Workshop on Entrainment and Impingement. E.A. Communications, Ecological Analysts, New York.

Jensen, L. D., Davies, R. M., Brooks, A. S. and Meyers, C. D. (1969). *The Effects of Elevated Temperature upon Aquatic Invertebrates*. Edison Electric Institute Publication No. 69-900, Report No. 4, New York.

Jensen, L. D., Davies, R. M., Smith, R. A. and Brooks, A. S. (1974). Entrainment of planktonic organisms into cooling water systems of three mid-Atlantic thermal power plants, pp. 95–104 in Jensen (Ed.).

Jerrilov, A. and Marionov, U. (1987). *Environmental Impact Assessment: a Practical Approach.* IVL Swedish Environmental Research Institute, Stockholmm IVL, Rept. No. 859.

Jeter, C. R. (1977). An approach to thermal water quality standards, Paper 11-A-3 in Lee and Sengupta (Eds).

Jirka, G. (1974). Submerged diffusers for cooling water discharges, pp. 85–91 in Gallagher (Ed.).

Jobling, M. (1981). Temperature tolerance and the final preferendum; rapid methods for the assessment of optimum growth temperatures. *J. Fish Biol.*, **19**, 439–55.

Jobson, H. E. (1978). Thermal model for evaporation from open channels. International Association for Hydraulic Research (cyclostyled).

Johnson, P. B. and Hasler, A. D. (1977). Winter aggregations of carp (Cyprinus carpio) as revealed by ultrasonic tracking. *Trans. Am. Fish. Soc.*, **106**(6), 556–9.

Johnson, R. G. (1965). Temperature variation in the infaunal environment of a sand flat. *Limnol. Oceanogr.*, **10**(1), 114–20.

Johnson, S. P. (1979). *The Pollution Control Policy of the European Communities.* Graham and Trotman, London.

Johnson, S. P. and Courcelle, G. (1989). *The Environmental Policy of the European Communities.* International Environmental Law and Policy Series. Graham and Trotman, London.

Jokiel, F. L. and Coles, S. L. (1974). Effects of heated effluent on hermatypic corals at Kahe Point, Oahu. *Pacific Sci.*, **28**(1), 1–18.

Jolley, R. L. (1985). Basic issues in water chlorination: A chemical perspective, pp. 19–38 in Jolley *et al.* (Eds).

Jolley, R. L. and Carpenter, J. H. (1983a). A review of the chemistry and environmental fate of reactive oxidant species in chlorinated water, pp. 3–48 in Jolley *et al.* (a).

Jolley, R. L. and Carpenter, J. H. (1983b). Review of analytical methods for reactive oxidant species in chlorinated waters, pp. 611–52 in Jolley *et al.* (a).

Jolley, R. L., Brungs, W. A., Cumming, R. B. and Jacobs, V. A. (Eds) (1980). *Water Chlorination; Environmental Impact and Health Effects, Vol. 3.* Ann Arbor Science, Butterworth Group, Ann Arbor, Michigan.

Jolley, R. L., Brungs, W. A., Cotruvo, J. A., Cumming, R. B., Mattice, J. S. and Jacobs, V. A. (Eds) (1983a). *Water Chlorination; Chemistry and Water Treatment, Vol. 4, Book 1.* Ann Arbor Science, The Butterworth Group, Ann Arbor, Michigan.

Jolley, R. L., Brungs, W. A., Cotruvo, J. A., Cumming, R. B., Mattice, J. S. and Jacobs, V. A. (Eds) (1983b). *Water Chlorination; Environment, Health and Risk, Vol. 4, Book 2.* Ann Arbor Science, The Butterworth Group, Ann Arbor, Michigan.

Jolley, R. L., Bull, R. J., Davies, W. P., Katz, S., Roberts, M. H. and Jacobs, V. A. (Eds) (1985). *Water Chlorination: Chemistry, Environmental Impacts and Health Effects, Vol. 5.* Lewis Publishers, Chelsea, Michigan.

Jolley, R. L., Cumming, R. B., Pitt, W. W., Taylor, F. G., Thompson, J. E. and

Hartman, S. J. (1978a). Ecological impact of chloro-organics produced by chlorination of cooling-tower waters, pp. 85–92 in Gerhold (Ed.).

Jolley, R. L., Gorchev, H. and Hamilton, D. H. (Eds) (1978c). *Water Chlorination: Environmental Impacts and Health Effects, Vol. 2.* Ann Arbor Science, Ann Arbor, Michigan.

Jolley, R. L., Pitt, W. W. Jr, Taylor, F. G. Jr, Hartmann, S. J., Jones, G. Jr and Thompson, J. E. (1978b). An experimental assessment of halogenated organics in waters from cooling towers and once-through systems, pp. 695–706 in Jolley *et al.* (Eds) (c).

Jones, A., Brown, J. A. G., Douglas, M. T., Thompson, S. J. and Whitfield, R. J. (1981). Progress toward developing methods for the intensive farming of turbot (*Scopthalmus maximus*) in cooling water from a nuclear power station, pp. 481–96 in Tiews (Ed.).

Jones, C. (1976). Cooling water set to go commercial. *Fish Farm. Int.,* **3**(4), 14–15.

Jones, C. (1980). Eels from Somerset. *ibid.,* **7**(1), 24–6.

Jones, E. B. G. (1976). *Recent Advances in Aquatic Mycology.* Paul Elek, London.

Jones, J. G. (1975). Heterotrophic micro-organisms, pp. 141–54 in Whitton (Ed.).

Jones, J. G. (1977). The study of aquatic microbial communities, pp. 1–25 in Skinner and Shewan (Eds).

Jones, J. R. E. (1964). *Fish and River Pollution.* Butterworths, London.

Jones, R. E. (1978). Heavy metals in the estuarine environment. Technical Report TR 73, April, Water Research Centre, Stevenage, UK.

Jude, D. J. (1976). Entrainment of fish larvae and eggs on the Great Lakes, with special reference to the D. C. Cook nuclear plant on Lake Michigan, pp. 177–200 in Jensen (Ed.).

Kaeding, L. R. and Kaya, C. M. (1978). Growth and diets of trout from contrasting environments in a geothermally heated stream: The Firehole River of Yellowstone National Park. *Trans. Am. Fish. Soc.,* **107**(3), 432–38.

Kalin, R. J. (1970). Part III—Ecology of the microbenthos, pp. 77–85 in Hechtel (Ed.).

Kamath, P. R., Bhat, I. S. and Ganguly, A. K. (1971). Environmental behaviour of discharged radioactive effluents at Tarapur atomic power station, pp. 475–94 in IAEA.

Kamath, P. R., Bhat, I. S. and Ganguly, A. K. (1975). Seasonal features of thermal abatement of shoreline discharges at nuclear sites, pp. 217–27 in IAEA, (a).

Karadi, G. W., Krizek, R. J. and Csallory, S. C. (Eds) (1971). *Pumped Storage Development and its Environmental Effects.* American Water Resources Association. University of Wisconsin, Milwaukee (19–24 September 1971).

Karas, P. and Neuman, E. (1981). First year growth of Perch (*Perca fluviatilis* L.) and Roach (*Rutilus rutilus* (L.)) in a heated Baltic bay. *Rep. Inst. Freshwat. Res., Drott.,* **59**, 48–63.

Kastendiek, J., Schroeter, S. C. and Dixon, J. (1981). The effect of the seawater cooling system of a nuclear generating station on the growth of mussels in experimental populations. *Mar. Pollut. Bull.,* **12**(12), 402–7.

Kasweck, K. L. (1978). Genetic variability of *Escherichia coli* in thermally stressed reactor effluent waters, pp. 143–148 in Gerhold (Ed.).

Katz, B. M. (1977). Chlorine dissipation and toxicity presence of nitrogenous compounds. *J. Wat. Pollut. Contr. Fed.,* **49**(7), 1627–35.

Kaya, C. M. (1977). Reproductive biology of rainbow and brown trout in a geothermally heated stream, the Firehole River of Yellowstone National Park. *Trans. Am. Fish. Soc.*, **16**(4), 354–61.

Kaya, C. M. (1978). Thermal resistance of rainbow trout from a permanently heated stream, and of two hatchery strains. *Prog. Fish Cult.*, **40**(4), 138–42.

Kedl, R. J. and Coutant, C. C. (1976). Survival of juvenile fishes receiving thermal and mechanical stresses in a simulated power plant condenser, pp. 394–400 in Esch and McFarlane (Eds).

Keenan, J. H. (1941). *Thermodynamics*. John Wiley, New York, London.

Keeney, R. L. (1980). *Siting Energy Facilities*. Academic Press, New York, London.

Kelley, R. B. (1974). Large scale water cooling via floating spray devices, pp. 92–8 in Gallagher (Ed.).

Kelso, J. R. M. (1974). Influence of a thermal effluent on movement of Brown bullhead (*Ictalurus nebulosus*) as determined by ultrasonic tracking. *J. Fish. Res. Board Can.*, **31**, 1507–13.

Kelso, J. R. M. (1976). Movement of yellow perch (*Perca flavescens*) and white sucker (*Catostomus commersoni*) in a nearshore Great Lakes habitat, subject to a thermal discharge. *ibid.*, **33**, 42–53.

Kelso, J. R. M. and Leslie, J. K. (1979). Entrainment of larval fish by the Douglas Point Generating Station, Lake Huron, in relation to seasonal succession and distribution, *ibid.*, **36**, 37–41.

Kelso, J. R. M. and Minns, C. K. (1975). Summer distribution of the nearshore fish community near a thermal generating station as determined by acoustic census. *ibid.*, **32**, 1409–18.

Kemp, H. T., Little, R. L., Holdman, V. L. and Darby, R. L. (1973). *Water Quality Data Book, Vol. 5, Effects of Chemicals on Aquatic Life*. US Environmental Protection Agency, Office of Research and Development, Washington, D.C.

Kemp, W. M., Smith, W. H. B., McKellar, H. N., Lehman, M. E., Homer, M., Young, D. L. and Odum, H. T. (1977). Energy cost-benefit analysis applied to power plants near Crystal River, Florida. In: *Ecosystem Modelling in Theory and Practice—An Introduction with Case Histories*. Hall, C. A. S. and Day, J. W. (Eds). John Wiley, London, pp. 508–43.

Kennedy, V. S. and Mihursky, J. A. (1967). Bibliography on the effects of temperature in the aquatic environment. University of Maryland Nat. Resources Inst., Contribution No. 326, May.

Kerambrun, P. (1983). Effect of thermal pollution on marine organisms. *Oceanis*, **9**(8), 627–51 (Fr.).

Kerr, J. E. (1953). Studies of fish preservation at the Contra Coasta Steam Plant of the Pacific Gas and Electric Company. *Fish. Bull.*, No. 92, California Fish and Game.

Kerr, N. M. (1976). Farming marine flatfish using waste heat from sea-water cooling. *Energy World*, October, 2–10.

Keser, M., Larson, B. R., Vadas, R. L. and McCarthy, W. (1978). Growth and ecology of *Spartina alterniflora* in Maine after a reduction in thermal stress, pp. 420–33 in Thorp and Gibbons (Eds).

Kestin, J. (Ed.) (1980). *Sourcebook on the Production of Electricity from Geothermal Energy*. US Dept. of Energy, DOE/RA/403-1, Washington, D.C.

Key, D. and Davidson, P. E. (1981). A Review of the Development of the Solent

Oyster Fishery, 1972–1980. Laboratory Leaflet No. 52, Ministry of Agriculture Fisheries and Food, Directorate of Fisheries Research, Lowestoft, UK.

Khalanski, M. (1975). Etudes realisées en France sur les consequences ecologiques de la refrigeration des centrales thermiques circuit ouvert, pp. 461–76 in IAEA (a).

Khalanski, M. (1977). Influence du fonctionnement d'une centrale thermique sur la production primaire planctonique du port de Dunkerque, pp. 101–44 in EDF.

Khalanski, M. (1978). Perturbations ecologigues liées a l'implantation de centrales thermiques de grande puissance sur le littoral. *Oceanis*, **4**, 152–95.

Khalanski, M. (1981). Structure et production du phytoplancton du port de Dunkerque incidence du fonctionnement de la centrale thermique, pp. 621–49 in EDF.

Khalanski, M. and Bordet, F. (1980). Effects of chlorination on marine mussels, pp. 557–68 in Jolley *et al.*

Khalanski, M. and Bordet, F. (1981). Modalite d'action du chlore sur la moule marine, pp. 745–59 in EDF.

King, J. R. and Mancini, E. R. (1976). Effects of power plant cooling water entrainment on the drifting macroinvertebrates of the Wabash River (Indiana) pp. 368–72 in Esch and McFarlane (Eds).

King, R. G. (1978). Entrainment of Missouri River fish larvae through Fort Calhoun station, pp. 45–56 in Jensen (Ed.).

King, T. L., Zimmerman, E. G. and Beitinger, T. L. (1985). Concordant variation in thermal tolerance and allozymes of the red shiner, *Notropis lutensis*, inhabiting tailwater sections of the Brazos River, Texas. *Environ. Biol. Fish.*, **13**(1), 49–57.

Kinne, O. (Ed.) (1970a). *Marine Ecology, Vol. 1, Environmental Factors, Part 1*, Wiley-Interscience, New York.

Kinne, O. (1970b). General introduction, pp. 321–46 in Kinne (Ed.).

Kinne, O. (1970c). Invertebrates, pp. 407–514 in Kinne (Ed.).

Kinne, O. (Ed.) (1975). *Marine Ecology, Vol. II, Physiological Mechanisms, Part 2.* John Wiley, New York.

Kirk, R. G. (1979). Marine fish and shellfish culture in the member states of the European Economic Community. *Aquaculture*, **16**, 95–122.

Kititsina, L. A. (1973). Effect of hot effluent from thermal and nuclear power plants on invertebrates in cooling ponds. *Hydrobiol. J.*, **9**(5), 67–79.

Klein, L. (1962). *River Pollution, Vol. 2, Causes and Effects.* Butterworths, London.

Knight-Jones, E. W. and Morgan, E. (1966). Responses of marine animals to changes in hydrostatic pressure. In: *Oceanography and Marine Biology. An Annual Review, Vol. 4*, Barnes, H. (Ed.). George Allen and Unwin, London, pp. 267–300.

Knights, B. (1987). Agonistic behaviour and growth in the European eel, *Anguilla anguilla* L. in relation to warm-water aquaculture. *J. Fish Biol.*, **31**, 265–76.

Knutson, K. M., Berguson, S. R., Rastetter, D. L., Mischuk, M. W., May, F. B. and Kuhl, G. M. (1976). Seasonal pumped entrainment of fish at the Monticello, MN, Nuclear Power Installation, Department of Biological Sciences, St Cloud State University, St Cloud, Minnesota.

Knutzen, J. (1981). Effects of decreased pH on marine organisms. *Mar. Pollut. Bull.*, **12**(1), 25–9.

Kolehmainen, S. E., Martin, F. D. and Schroeder, P. B. (1975). Thermal studies on tropical marine ecosystems in Puerto Rico, pp. 409–21 in IAEA (a).

Kolehmainen, S. E., Morgan, T. and Castro, R. (1974). Mangrove root communities in a thermally altered area in Guyanilla bay, Puerto Rico, pp. 371–90 in Gibbons and Sharitz (Eds).

Koops, F. B. J. (1972). Report on plankton investigations in the cooling circuit of Flevo power station near Lelystad. *N.V. tot, Keuring van Electrotechnische Materialen, Arnhem*, Nederland, No. IV, 7984–72.

Koops, F. B. J. (1974). Investigation of hydrobiological effects due to cooling water discharges of electric power plants. *Hydrobiol. Bull.*, **7**(3), 86–95.

Koops, F. B. J. (1975). Plankton investigations near Flevo power stations. *Verh. int. Ver. Limnol.*, **19**, 2207–13.

Koops, F. B. J. (1976). Hydrobiological cooling-water research in the Netherlands. Centre Belge d'Etude et de Documentation des Eaux (Aout-Septembre), 393/4, 318–20.

Koops, F. B. J. (1978). Some problems in finding good standards for cooling water. *Proc. Int. Assoc. Theoret. Appl. Limnol.*, **20**, 353–6.

Koops, F. B. J., Donze, M. and Hadderingh, R. H. (1980). Optimizing the use of cooling water to reduce its impact on the aquatic environment. *Proc. 11th World Energy Conf.*, Munich, Vol. 3, WEC, Detroit, London, pp. 424–39.

Koppe, P. (1974). Water pollution and the capacity of waters for waste heat. Paper 2.2-2, in *Proc. 9th World Energy Conf.*, WEC, Detroit, London.

Korneev, A. (1976). Warm water pond-fish culture in the USSR, *Fish. Gaz.*, 26.

Kostylev, V. A. and Yesipova, M. A. (1982). The effects of effluents of the Kursk thermal power station on zooplankton. *Hydrobiol. J.*, **17**(5), 26–8.

Krajewski, W. F., Kraszewski, A. K. and Grenney, W. J. (1982). A graphical technique for river water temperature predictions. *Ecolog. Modell.*, **17**, 209–24.

Kreh, T. W. and Derwort, J. E. (1976). Effects of entrainment through Oconee Nuclear Station on carbon-14 assimilation rates of phytoplankton, pp. 331–5 in Esch and McFarlane (Eds).

Krenkel, P. A. and Parker, F. L. (1969a). Engineering aspects, sources and magnitude of thermal pollution, pp. 10–52 in Krenkel and Parker (Eds).

Krenkel, P. A. and Parker, F. L. (Eds) (1969b). *Biological Aspects of Thermal Pollution*. Vanderbilt University Press, Portland, Oregon.

Krouse, C. A. (1978). The power plant regulatory programme; A congressional dilemma, pp. 7–10, in Jensen (Ed.).

Kurtz, J. B., Bartlett, C. L. R., Newton, U. A., White, R. A. and Jones, N. L. (1982). *Legionella pneumophila* in cooling water systems: Report of a survey of cooling towers in London and a pilot trial of biocides. *J. Hyg., Camb.*, **88**, 369–79.

Kutty, M. N. and Sukumaran, N. (1975). Influence of upper and lower temperature extremes on the swimming performance of *Tilapia Mossambica. Trans. Am. Fish. Soc.*, **105**(4), 755–61.

Kyser, J. M., Paddock, R. A. and Policastro, A. J. (1975). Analysis of three years' complete field temperature data from different sites of heated surface discharges into Lake Michigan, pp. 249–309 in IAEA (a).

Kyte, W. S. (1986). Possible emission-control technologies for coal-fired power stations in the UK. Paper to *The Problem of Acid Emission—An Opportunity for British Industry.* Institution of Chemical Engineers, University of Birmingham, 22–23 September.

Kyte, W. S., Bettelheim, J. and Cooper, J. R. P. (1983). Sulphur oxides control options in the UK electric power generation industry. In: *Effluent Treatment in the Process Industries.* European Federation of Chemical Engineering, London Publication Series, No. 31, pp. 39–47.

Lackey, J. B. (1974). Entrainment studies at Turkey Point on Biscayne Bay: Have thermal effects affected the plankton of Biscayne Bay? pp. 187–91 in Jensen (Ed.).

Lamb, T. J. (1972). Marine fouling control by electrolytic hypochlorite, pp. 995–1004 in Acker *et al.* (Eds).

Landry, A. M. and Strawn, K. (1973). Annual cycle of sport fishing activity at a warmwater discharge into Galveston Bay, Texas. *Trans. Am. Fish Soc.,* **102**(3), 573–7.

Landy, M. (1979). *Environmental Impact Statement Glossary.* IFI Plenum Press, New York, London.

Langford, T. E. (1963). The coarse fishes of the River Ancholme, Lincs. Report to Lincolnshire River Board, Boston, Lincs (cyclostyled).

Langford, T. E. (1966). Fishery biology as applied by a River Board. Year Book of the River Authorities Association, pp. 3–16.

Langford, T. E. (1967). Preliminary observations on the macro-invertebrate faunas of rivers, in relation to heated effluents from power stations. CEGB Internal Report, RD/L/N 124/67, Leatherhead, UK.

Langford, T. E. (1970). The temperature of a British river upstream and downstream of a heated discharge from a power station. *Hydrobiologia,* **35**(3–4), 353–75.

Langford, T. E. (1971a). The distribution, abundance and life-histories of stoneflies (Plecoptera) and mayflies (Ephemeroptera) in a British river, warmed by cooling-water from a power station. *ibid.,* **38**(2), 339–77.

Langford, T. E. (1971b). The biological assessment of thermal effects in some British rivers, pp. 9–40 in CERL.

Langford, T. E. (1972). A comparative assessment of thermal effects in some British and North American rivers, pp. 319–51, in Oglesby *et al.* (Eds).

Langford, T. E. (1974). Ecology and cooling water from power stations—A review of recent biological research in Britain, Paper 2.2-4, in *Proc. 9th World Energy Conf.,* WEC, Detroit, London.

Langford, T. E. (1975). The emergence of insects from a British river warmed by power station cooling-water. Part 11. The emergence patterns of some species of Ephemeroptera, Trichoptera, and Megaloptera, in relation to water temperature and river flow upstream and downstream of the cooling-water outfalls. *Hydrobiologia,* **47**(1), 91–133.

Langford, T. E. (1977). Biological problems with the use of sea-water for cooling. *Chem. Ind.* 16 July, 612–16.

Langford, T. E. (1983a). *Electricity Generation and the Ecology of Natural Waters.* Liverpool University Press, Liverpool.

Langford, T. E. (1983b). Records of fishes from the thermal discharge canal at Kingsnorth Power Station. CEGB Internal Report, TPRD/C/2477/N83, Leatherhead, UK.

Langford, T. E. (1983c). The ecological effects of thermal discharges on tropical and sub-tropical marine ecosystems. Paper to GESAMP. (cyclostyled).

Langford, T. E. (1987). The effects of a thermal discharge on the growth and feeding of bass, *Dicentrarchus labrax*, in the Medway estuary, England. CEGB Internal Report, TPRD/L3126/R 87, Leatherhead, UK.

Langford, T. E. (1988). Ecology and cooling water from power stations. *Atom* **385** (November) 4–7.

Langford, T. E. and Aston, R. J. (1972). The ecology of some British rivers in relation to warm water discharges from power stations. *Proc. Roy. Soc. Lond.*, **181**, 45–57.

Langford, T. E. and Daffern, J. R. (1975). The emergence of insects from a British river warmed by power station cooling-water. Part 1. The use and performance of insect emergence traps in a large spate river and the effects of various factors on total catches upstream and downstream of the cooling-water outfalls. *Hydrobiologia*, **46**(1), 71–114.

Langford, T. E. and Howells, G. D. (1976). The use of biological monitoring in the freshwater environment by the electrical industry in the UK. In: *Use of Biological Monitoring in the Freshwater Environment*, Alabaster, J. S. (Ed.). Academic Press, London, pp. 115–24.

Langford, T. E., Milner, A. G. P. and Fleming, J. M. (1979). The movements and distribution of some common bream (*Abramis brama* L.) in the vicinity of power station intakes and outfalls in British rivers as observed by ultra-sonic tracking. CEGB Internal Report, RD/L/N 145/78, Leatherhead, UK.

Langford, T. E. and Sherwood, S. (1970). A comparison of the numbers of trout and perch caught in nets set in the warm and cold areas of a power station cooling water reservoir—Trawsfynydd. CEGB Internal Report, RD/L/N 133/70, Leatherhead, UK.

Langridge, J. and McWilliam, J. R. (1967). Heat responses of higher plants, pp. 231–86 in Rose (Ed.).

Lanza, G. R., Lauer, G. J., Ginn, T. C., Storm, P. C. and Zubarik, L. (1975). Biological effects of simulated discharge plume entrainment at Indian Point nuclear power station, Hudson River estuary, USA, pp. 95–124 in IAEA (c).

Larimore, R. W. and McNurney, J. M. (1979). *Evaluation of a Cooling Lake Fishery, Vol. 2, Lake Sangchris Ecosystem Modelling*. Electric Power Research Institute, EA-1148, RP, 573, Palo Alto, CA.

Larimore, R. W., McNurney, J. M., Swadener, S. D., Buckler, J., Waite, S. W. and Coutant, L. W. (1979a). *Evaluation of a Cooling Lake Fishery, Vol. 3, Fish population studies*. Electric Power Reearch Institute, EA-1148, Palo Alto, CA.

Larimore, R. W., McNurney, J. M., Swadener, S. D., Buckler, J., Waite, S. W. and Coutant, L. W. (1979b). *Evaluation of a Cooling-Lake Fishery, Vol. 4*, Fish food resources studies, Final report. Electric Power Research Institute, EA-1148, Palo Alto, CA.

Larrick, S. R., Cherry, D. S., Dickson, K. L. and Cairns, J. Jr (1978). The use of various avoidance indices to evaluate the behavioural response of the golden shiner to components of total residual chlorine, pp. 135–48 in Jolley *et al.* (Eds).

Larrick, S. R., Clark, J. R., Cherry, D. S. and Cairns, J. Jr (1981). Structural and functional changes of aquatic heterotrophic bacteria to thermal, heavy metal and fly ash effluents. *Wat. Res.*, **15**, 875–80.

Larson, G. L., Hutchins, F. E. and Lamperti, L. P. (1977). Laboratory determination of acute and sublethal toxicities of inorganic chloramines to early life stages of Coho salmon (*Oncorhynchus kisutch*). *Trans. Amer. Fish. Soc.*, **106**(3), 268–77.

Larson, G. L. and Schlesinger, D. A. (1978). Toward an understanding of the toxicity of intermittent exposures of total residual chlorine to freshwater fishes, pp. 111–22 in Jolley *et al.* (Eds).

Lauer, G. J., Waller, W. T., Bath, D. W., Meeks, W., Heffner, R., Ginn, T., Zubarik, L., Bibko, P. and Storm, P. C. (1974). Entrainment studies on Hudson River organisms, pp. 77–92 in Jensen (Ed.).

Lavis, M. E. and Smith, K. (1972). 'Reservoir storage and the thermal regime of rivers with special reference to the River Lune, Yorkshire. *Sci. Total Environ.*, **1**, 81–90.

Lawler, J. P. (1976). Physical measurements; Their significance in the prediction of entrainment effects, pp. 59–92 in Jensen (Ed.).

Lawler, Matusky, and Skelly—Engineers (1979). *Ecosystem Effects of Phytoplankton and Zooplankton Entrainment.* Electric Power Research Institute, EA-1038 Interim Report, April 1979. Palo Alto, CA.

Lawler, J. P., Morris, R. A., Goldwyn, G., Abood, K. A. and Englert, T. L. (1974). Hudson River striped bass model, pp. 83–94 in Jensen (Ed.).

Lawrie, J. P. (1976). An assessment of the heated water installation at the Furnace salmon hatchery (1971–75). *Rep. Salm. Res. Trust Ireland*, **20**, 58–65.

Leason, D. B. (1974). Planning aspects of cooling-towers. *Atmos. Environmt*, **8**, 307–12.

Leason, D. B. (1975). The planning of coastal power stations, pp. 5–16 in CERL.

Leason, D. B. (1976). Future power stations—What will they be like?'. Paper to Ann. Conf. of Royal Inst. Chartered Surveyors, Edinburgh, 1975, CEGB, London.

Lee, G. F. (1979). Persistence of chlorine in cooling-water from an electric generating station. *Proc. ASCEJ Environ. Engng Div.*, **105**(E.E.4), 757–73.

Lee, N. E., Haag, W. R. and Jolley, R. L. (1983). Cooling water pollutants; Bioaccumulation by *Corbicula*, pp. 851–71 in Jolley *et al.* (Eds) (b).

Lee, S. S. and Sengupta, S. (1977). Waste heat management and utilisation, Vols 1, 11 and 111. Proceedings of a Conference, 9–11 May 1976, Miami Beach, Florida, Department of Engineering, University of Miami.

Lee, S. S. and Sengupta, S. (1982). *Proc. 3rd Conf. on Waste Heat Management and Utilization.* EPA-600/59 82-008. US Environ. Protect. Agency, Washington, D.C.

Leggett, W. C. (1976). The American shad (*Alosa sapidissima*) with special reference to its migration and population dynamics in the Connecticut River, pp. 169–226 in Merriman and Thorpe (Eds).

Leitheiser, R. M., Ehrlich, K. F. and Thum, A. B. (1978). Comparison of a high-volume pump and conventional plankton nets for collecting fish larvae entrained in power plant cooling systems. *J. Fish. Res. Board Can.*, **36**, 81–4.

Leland, H. V., Luoma, S. N. and Wilkes, D. J. (1977). Heavy metals. *J. Wat. Pollut. Contr. Fed.*, June, 1340–68.

Lemkuhl, D. M. (1972). Change in thermal regime as a cause of reduction of benthic fauna downstream of a reservoir. *J. Fish. Res. Board Can.*, **29**(9), 1329–32.

Lenat, D. R. (1978). Effects of power plant operation on the littoral benthos of Belews lake, North Carolina, pp. 580–96 in Thorp and Gibbons (Eds).

Leopold, L. B., Langbien, W. B. *et al.* (1964). *Fluvial Processes in Geomorphology.* Freeman, San Francisco.

Lewin, J. (1975). *British Rivers.* Allen and Unwin, London.

Lewis, B. G. (1964). Water flow and marine fouling in culverts: A review of literature up to 1962. CEGB Internal Report, RD/L/M60, Leatherhead, UK.

Lewis, R. M. (1966). Effects of salinity and temperature on survival and development of larval Atlantic menhaden *Brevoortia tyrannus. Trans. Am. Fish. Soc.,* **95**(4), 423–6.

Lewis, W. M., Jr (1974). Evaluation of heat distribution in a South Carolina reservoir receiving heated water, pp. 1–27 in Gibbons and Sharitz (Eds).

Leynaud, G. and Allardi, J. (1975). Incidences d'un rejet thermique en milieu fluivial sur les mouvements des populations ictyologiques, pp. 401–7 in IAEA (a).

Lietzke, M. H. (1978). A kinetic model for predicting the composition of chlorinated water discharged from power plant cooling systems, pp. 707–16 in Jolley *et al.* (Eds).

Lindberg, S. E. and Hutchinson, T. C. (Eds) (1987). *Heavy Metals in the Environment,* Vols 1 and 2. CEP Consultants, Edinburgh.

Linsley, Kraeger and Associates (1980). *Proceedings; Workshop on Water Supply for Electric Energy.* Electric Power Research Institute, SW-79-237, Palo Alto, CA.

Locan, D. T. and Maurer, D. (1975). Diversity of marine invertebrates in a thermal effluent. *J. Wat. Pollut. Contr. Fed.,* **47**(3), 515–23.

Loftus, M. E. (1978). Respiratory oxygen demand of estuarine cooling waters in Indian River, Delaware, pp. 37–56 in Gerhold (Ed.).

Loi, T-n and Wilson, B. J. (1979). Macroinfaunal structure and effects of thermal discharges in a mesohaline habitat of Chesapeake Bay, near a nuclear power plant. *Mar. Biol.,* **55**, 3–16.

Lowe-McConnell, R. H. (Ed.) (1966). *Man Made Lakes.* Institute of Biology Symposia No. 15, Academic Press, London, New York.

Luckow, H. and Reinken, G. (1979). The Agrotherm research project, pp. 117–28 in Lee and Sengupta (Eds).

Lucu, C., Pavicic, J., Skreblin, M. and Mastrovic, M. (1980) Toxicological effects of biocide slimicide C-30 on some marine invertebrates. *Mar. Pollut. Bull.,* **11**, 294–6.

Lum, J. (1978). Status of regulation development for cooling water discharges, pp. 823–29 in Jolley *et al.* (Eds).

Luning, K. (1980). Control of algal life-history by daylength and temperature, pp. 915–40 in Price *et al.* (Eds).

Lutz, P. and Merle, G. (1982). Discontinuous mass chlorination of natural draft cooling towers, pp. 197–214 in Jenkins and Schodtzhansen (Eds).

Macan, T. T. (1957). The Ephemeroptera of a stony stream. *J. Anim. Ecol.,* **26**, 317–42.

Macan, T. T. (1958a). The temperature of a stony stream. *Hydrobiologia,* **12**, 89–106.

Macan, T. T. (1958b). Causes and effects of short emergence periods in insects. *Verh. int. Ver. Limnol.,* **13**, 845–9.

Macan, T. T. (1960) A key to the British fresh- and brackish-water gastropods. *Sci. Publ. Freshwater Biol. Ass. no. 13,* Ferry House, Ambleside.

Macan, T. T. (1963). *Freshwater Ecology.* Longmans, Green, London.

Macan, T. T. (1970). *Biological Studies of the English Lakes*. Longmans, Green, London.

Macan, T. T. (1974). Freshwater invertebrates. In: *The Changing Flora and Fauna of Britain*, Hawksworth, D. L. (Ed.). Systematics Association Special Volume No. 6, Academic Press, London, New York, pp. 143–55.

Macan, T. T. and Maudsley, R. (1966). The temperature of a moorland fishpond. *Hydrobiologia*, **27**, 1–22.

Macinnes, J. R. and Calabrese, A. (1978). Response of embryos of the American oyster, *Crassostrea virginica* to heavy metals at different temperatures, pp. 195–202 in McLusky and Berry (Eds).

Macintosh, D. J. (1978). Some responses of tropical mangrove fiddler crabs to high environmental temperatures, pp. 49–56 in McLusky and Berry (Eds).

Mackichan, K. A. (1967). Diurnal temperature variations of three Nebraska streams. US Geological Survey Professional Papers, 575. B, 223–34.

Macqueen, J. F. (1978). Background water temperatures and power station discharges. *Adv. Wat. Resources*, **1**(4), 195–203.

Macqueen, J. F. (1980). Concentration of contaminants discharged with power station cooling water. *ibid.*, 165–72.

Macqueen, J. F. and Howells, G. (1978). Waste-heat disposal—A cool look at warm water. *CEGB Res.*, **7**, 33–44.

Madding, R. P., Tokar, J. V. and Marmer, G. J. (1975). A comparison of aerial infra-red and in-situ thermal plume measurement techniques, pp.163–184 in IAEA (a).

MAFF (1972 *et seq.*). Annual Technical Reports of the Fisheries Radiological Laboratory. Ministry of Agriculture, Fisheries and Food, Lowestoft, UK.

Magnuson, J. J., Crowder, L. B. and Medvick, P. A. (1977). Temperature as an ecological resource. *The American Zoologist*, **19**, 331–43.

Majewski, W. and Miller, D. C. (Eds) (1979). Predicting effects of power plant once-through cooling on aquatic ecosystems. Contribution to the International Hydrobiological Programme, No. 20, UNESCO Paris.

Major, R. L. and Mignell, J. L. (1966). Influence of rocky Reach Dam and the temperature of the Okagon River on the upstream migration of sockeye salmon. *Fishery Bull.*, **66**(1), 131–47.

Malmgren-Hansen, A. and Dahl-Madsen, K. I. (1982). Modelling the consequences of cooling water discharge from the 'Vendyssel' power plant, pp. 139–58 in IAWPR.

Malouf, R., Keck, D., Maurer, D. and Epifanio, C. (1972). Occurrence of gas-bubble disease in three species of bivalve molluscs. *J. Fish. Res. Board Can.*, **29**(5), 588–9.

Mance, G. (1987). *Pollution Threat of Heavy Metals on Aquatic Environments*. Elsevier Applied Science Publishers, London.

Mangarella, P. A. and Van Dusen, E. S. (1973). Submerged thermal discharges from ocean-sited power plants. ASME 73-WA/Oct-13, 8 pp.

Mangum, D. C. *et al.* (1973). Methods for controlling marine fouling in intake systems. US Department of Interior, June, NTIS Accessions No. PB 221 909, Washington, D.C.

Mann, K. H. (1965). Heated effluents and their effects on the invertebrate fauna of rivers. *Proc. Soc. Water Treat. Exam.*, **14**, 45–50.

Marcello, R. A. Jr and Fairbanks, R. R. (1976). Gas bubble disease of Atlantic menhaden, *Brevoortia tyrannus*, at a coastal nuclear plant, pp. 75–80 in Fickeisen and Schneider (Eds).

Marciak, Z. (1977). Influence of thermal effluents from an electric power plant on the growth of bream in the Konin Lake complex. *Tocz. Nauk, roln (H)*, **97**, 41–3.

Marcy, B. C. (1971). Survival of young fish in the discharge canal of a nuclear power plant. *J. Fish. Res. Board Can.*, **28**, 1057–60.

Marcy, B. C. (1976a). Fishes of the Lower Connecticut river and the effects of the Connecticut Yankee Plant, pp. 61–114 in Merriman and Thorpe (Eds).

Marcy, B. C. (1976b). Plankton, fish eggs and larvae of the Lower Connecticut River and the effects of the Connecticut Yankee Plant including entrainment, pp. 115–40 in Merriman and Thorpe (Eds).

Marcy, B. C., Beck, A. D. and Ulanowicz, R. E. (1978). Effects and impacts of physical stress on entrained organisms, pp. 135–88 in Schubel and Marcy (Eds).

Marcy, B. S. and Jacobson, P. M. (1976). Early life-history studies of American Shad in the Lower Connecticut River and the effects of the Connecticut Yankee Plant, pp. 141–68 in Merriman and Thorpe (Eds).

Marcy, B. C., Kranz, V. R. and Barr, R. P. (1980). Ecological and behavioural characteristics of fish eggs and young influencing their entrainment, pp. 29–74 in Hocutt *et al.* (Eds).

Marecki, J. (1988). *Combined Heat and Power Generating Systems.* Institution of Electrical Engineers, London.

Margrey, S. L., Burton, D. T. and Hall, L. W. (1981). Seasonal temperature and power chlorination effects on estuarine invertebrates. *Archiv. Environ. Contam. Toxicol.*, **10**(6), 747–53.

Markowski, S. (1959). The cooling water of power stations: A new factor in the environment of marine and freshwater invertebrates. *J. Anim. Ecol.*, **28**, 243–58.

Markowski, S. (1960). Observations on the response of some benthonic organisms to power station cooling water. *ibid.*, **29**, 349–57.

Markowski, S. (1962). Faunistic and ecological investigations in Cavendish dock, Barrow-in-Furness. *J. Anim. Ecol.*, **31**, 43–52.

Markowski, S. (1966). The diet and infection of fishes in Cavendish Dock, Barrow-in-Furness. *J. Zool. (London)*, **150**, 183–97.

Marmer, G. J., Tokar, J. V. and Madding, R. P. (1975). Comparison of thermal scanning and in situ techniques for monitoring thermal discharges. *Wat. Resources Bull.*, **11**(6), 1157–80.

Marshall, K. C. (1972). Mechanism of adhesion of marine bacteria to surfaces, pp. 625–32 in Acker *et al.* (Eds).

Marshall, W. (1979). *Combined Heat and Electrical Power Generation in the United Kingdom.* Department of Energy (Energy Pap. 35), HMSO, London.

Martin, D. B. and Arneson, R. D. (1978). Comparative limnology of a deep-discharge reservoir and surface-discharge lake on the Madison River, Montana. *Freshwat. Biol.*, **8**, 33–42.

Martin, H. C. (Ed.) (1986). *Proc. Int. Symp. on Acid Precipitation.* Muskako, Ontario, 15–20 September 1985. Parts 1 and 2. D. Reidel, Dordrecht.

Martin, M. *et al.* (1977). Copper toxicity experiments in relation to abalone deaths observed in a power plant's cooling-waters. *Calif. Fish. Game*, **63**(2), 95–100.

Martin Marietta Corporation (1977). Summary of current findings: Calvert Cliffs nuclear power plant aquatic monitoring program. Maryland power plant siting program, Maryland.

Martin, W. J. and Gentry, J. B. (1974). Effect of thermal stress on dragonfly nymphs, pp. 133–45 in Gibbons and Sharitz (Eds).

Mason, I. L. (1939). Studies on the fauna of an Algerian hot spring. *J. Exp. Biol.*, **16**, 487–98.

Masnik, M. T. (1979). The effects of thermal effluents on the populations of shipworms (Teredinidae; Mollusca) in the vicinity of a nuclear power station, pp. 301–21 in Lee and Sengupta (Eds).

Masse, H., Modot, C. and Mace, A-M. (1978). Influence de la temperature sur la reproduction et la survie de quelques Nassariidae (Mollusca: Gasteropda), pp. 367–74 in McLusky and Berry (Eds).

Masselchein, W. J. (1983). Nuclear discharges and the quality of river water. H.20. **16**(23), 524–33.

Masselchein, W. J. and Genot, J. (1982). Impact of nuclear power plants of the PWR type on river water quality (case report of the River Meuse). *Wat. Sci. Technol.*, **14** (4/5), 199–214.

Massengill, R. R. (1976a). Benthic fauna. 1965–1967 versus 1968–1972, pp. 39–54 in Merriman and Thorpe (Eds).

Massengill, R. R. (1976b). Entrainment of zooplankton at the Connecticut Yankee Plant, pp. 55–60 in Merriman and Thorpe (Eds).

Mathur, D., Robbins, T. W. and Purdy, E. J. Jr (1980). Assessment of thermal discharges on zooplankton on Conowingo Pond, Pennsylvania. *Can. J. Fish. Aquat. Sci.*, **37**, 937–44.

Matthews, W. J. (1986). Geographic variation in thermal tolerance of a widespread minnow, *Notropis lutrensis* of the North American mid-west. *J. Fish Biol.*, **28**, 407–17.

Mattheeuws, A., Genin, M., Detollenaere, A., Micha, J-C. and Mine, Y. (1981). Etude de la reproduction du gardon (*Rutilus rutilus*) et des effets d'une elevation provoquee de la temperature en Meuse sue cette reproduction. *Hydrobiologia*, **85**, 271–82.

Mattice, J. S. (1976). Effect of temperature on growth, mortality, reproduction and production of adult snails, pp. 73–80 in Esch and McFarlane (Eds).

Mattice, J. S. (1985). Chlorination of power plant cooling waters, pp. 39–62 in Jolley *et al.* (Eds).

Mattice, J. S. and Tsai, S. C. (1983). Total residual chlorine as a regulatory tool, pp. 901–12 in Jolley *et al.* (b).

Mattice, J. S. and Zittel, H. E. (1976). Site-specific evaluation of power plant chlorination. *J. Water Pollut. Contr. Fed.*, **48**(10), 2284–308.

McCain, J. (1975). Fouling community changes induced by the thermal discharge of a Hawaiian power plant. *Environ. Pollut.*, **9**, 63–72.

McCauley, R. W. (1958). Thermal relations of geographic races of *Salvelinus. Can. J. Zool.*, **36**, 655–62.

McCauley, R. W. (1977a). Laboratory methods for determining temperature preference. *J. Fish. Res. Board Can.*, **34**, 749–52.

McCauley, R. W. (1977b). Seasonal effects on temperature preference in Yellow Perch *Perca flavescens.* US EPA Ecological Research Series, EAPA 600/3/77-088.

McCauley, R. W. and Casselman, J. M. (1981). The final preferendum as an index of the temperature for optimum growth in fish. *Proc. World Symp. Aquaculture in Heated Effluents and Recirculation Systems*, Vol. 11. Heeveman, Berlin, pp. 82–93.

McCauley, R. W., Elliott, J. R. and Read, L. A. A. (1977). Influence of acclimation temperature on preferred temperature in the rainbow trout, *Salmo gairdneri. Trans. Am. Fish. Soc.*, **106**(4), 362–5.

McCauley, R. and Huggins, N. (1976). Behavioural thermoregulation by rainbow trout in a temperature gradient, pp. 171–5 in Esch and McFarlane (Eds).

McCauley, R. W. and Scott, D. P. (1960). Removal of free chlorine from running water by sodium thiosulphate. *J. Fish. Res. Board Can.*, **17**(4), 601.

McColl, R. H. S. and Forsyth, D. J. (1973). The limnology of a thermal lake, Lake Rotowhero, New Zealand. I. General description and water chemistry. *Hydrobiologia*, **43**(3–4), 313–32.

McDonald, D. B. and Bernhard, H. F. (1978). Effects of open and closed cycle condenser cooling modes on bacterial indicator organisms, p. 149 in Gerhold (Ed.).

McFadden, J. T. (1977). An argument supporting the reality of compensation in fish populations and a plea to let them exercise it, pp. 153–79 in Van Winkle (Ed.).

McFarlane, R. W. (1976). Fish diversity in adjacent ambient, thermal, and post-thermal freshwater streams, pp. 268–71 in Esch and McFarlane (Eds).

McGuire, H. E. (1977). The effect of liquid waste discharges from steam generating facilities, Richland, Battelle Pacific N.W. Laboratories, B.N.W.L. 2393, Washington, D.C.

McKelvey, K. K. and Brooke, M. (1959). *The Industrial Cooling Tower*, Elsevier, Amsterdam, London.

McKenney, C. L. and Dean, J. M. (1974). Effects of acute exposure to sub-lethal concentrations of cadmium on the thermal resistance of the mummichog, pp. 43–53 in Gibbons and Sharitz (Eds).

McLaren, I. A. (1963). Effects of temperature on growth of zooplankton and the adaptive value of vertical migration. *J. Fish. Res. Board Can.*, **20**(3), 685–727.

McLarney, W. O., Engstrom, D. G. and Todd, J. H. (1974). Effects of increasing temperature on social behaviour in groups of bullhead (*Ictalurus natalis*). *Environ. Pollut.*, **7**, 111–19.

McLean, R. I. (1972). Chlorine tolerance of the colonial hydroid *Bimeria franciscana. Ches. Sci.*, **13**(3), 229–30.

McLean, R. I. (1973). Chlorine and temperature stress on estuarine invertebrates. *J. Wat. Pollut. Contr.*, **45**(5), 837–41.

McLeese, D. W. (1956). Effects of temperature, salinity and oxygen on the survival of the American lobster. *J. Fish. Res. Board Can.*, **13**(2), 247–72.

McLoughlin, J. (1976). *The Law and Practice Relating to Pollution Control in the Member States of the European Communities: A Comparative Survey*. Graham and Trotman, London.

McMahon, J. W. and Docherty, A. E. (1975). Effects of heat enrichment on species succession and primary production in freshwater plankton, pp. 529–45 in IAEA (a).

McMahon, R. F. (1975). Effects of artificially elevated water temperatures on the growth, reproduction and life-cycle of a natural population of *Physa virgata*, Gould. *Ecology*, **56**, 1167–75.

McNae, W. (1968). A general account of the flora and fauna of mangrove swamps and forests in the Indo-West-Pacific Region, In: *Advances in Marine Biology*, *Vol. 6*, Russell, F. S. and Yonge, M. (Eds). Academic Press, London, New York, pp. 74–270.

McNaught, D. C. (1976). Recovery of entrained zooplankton populations. A model to predict impact on Great Lakes communities, pp. 93–100 in Jensen (Ed.).

McNeely, D. L. and Pearson, D. (1974). Distribution and condition of fishes in a small reservoir receiving heated waters. *Trans. Am. Fish Soc.*, **3**, 518–30.

Medvick, P. A. and Miller, J. M. (1979). Behavioural thermoregulation in three Hawaiian reef fishes. *Environ. Biol. Fish.*, **4**(1), 23–8.

Melard, Ch. and Phillippart, J. C. (1981). Pisciculture intensive du tilapia *Sarotherodon nioticus* dans les effluents thermiques d'une central nucleaire en Belgique, pp. 673–658 in Tiews (Ed.).

Meldrim, J. W. and Fava, J. A. (1977). Behavioural avoidance of estuarine fishes to chlorine. *Ches. Sci.*, **18**, 154–7.

Meldrim, J. W. and Gift, J. J. (1971). Temperature preference, avoidance and shock experiments with estuarine fishes. Ichthyological Associates Bulletin, 7, Ithaca, New York.

Mellanby, K. (1939). Low temperature and insect activity. *Proc. Roy. Soc.*, B N 849, **127**, 473–87.

Mellanby, K. (1940a). The activity of certain arctic insects at low temperatures. *J. Anim. Ecol.*, **9**(2), 296–301.

Mellanby, K. (1940b). Temperature coefficients and acclimatization. *Nature, Lond.*, **146**, 165.

Mellanby, K. (1940c). Temperature acclimatization in amphibia. *J. Physiol.*, **98**, 4.

Mellanby, K. (1980). *The Biology of Pollution*. Institute of Biology, Studies in Biology N38 (2nd edn), Edward Arnold, London.

Menasveta, P. (1981). Lethal temperature of marine fishes of the Gulf of Thailand. *J. Fish Biol.*, **18**, 603–7.

Merriman, D. and Thorpe, L. M. (Eds) (1976). The Connecticut River Ecological Study, The Impact of a Nuclear Power Plant. *Am. Fish. Soc. Monograph*, No. 1, 252 pp.

Mesarovic, M. M. (1975). Waste heat disposal from steam-electric plants with reference to the stochastic nature of some environmental conditions and to thermal pollution control regulations, pp. 311–29 in IAEA (a).

Middaugh, D. P., Couch, J. A. and Crane, A. M. (1977). Response of early life history stages of the striped bass, *Morone saxatilis*, to chlorination. *Ches. Sci.*, **18**, 141–53.

Middaugh, D. P., Dean, J. M., Dowey, R. G. and Floyd, G. (1978). Effect of thermal stress and total residual chlorine on early life stages of the Mummichog, *Fundulus heteroclitus*. *Mar. Biol.*, **46**, 1–8.

Mihursky, J. A., McErlean, A. J. and Kennedy, V. S. (1970). Thermal pollution, aquaculture and pathobiology in aquatic systems. *J. Wildl. Dis.* 6 Oct., Proc. Annual Conf. 347–55.

Milanov, T. (1973). Cooling problems in thermally polluted recipients. *Nordic Hydrology*, **4**, 237–55.

Miller, D. C. and Beck, A. D. (1975). Development and application of criteria for marine cooling waters, pp. 639–56 in IAEA (a).

Miller, D. S. and Brighouse, B. A. (1984). Thermal Discharges—A guide to power and process plant cooling-water discharges into rivers, lakes and seas. British Hydromechanics Research Association, London.

Miller, M. C., Hater, G. R., Federle, T. W. and Reed, J. P. (1976). Effects of a power plant operation on the biota of a thermal discharge channel, pp. 251–8 in Esch and McFarlane (Eds).

Miller, P. C. (1974). Potential use of vegetation to enhance cooling in holding ponds, pp. 610–27 in Gibbons and Sharitz (Eds).

Miller, R. W. (1974). Incidence and cause of gas-bubble disease in a heated effluent, pp. 79–93 in Gibbons and Sharitz (Eds).

Milner, A. G. P. (1984). The invertebrate faunas of cooling water systems at two British power stations. Thesis for M.I. Biol., Institute of Biology, London.

Ministry of Health (1949). *Prevention of River Pollution* (Hobday). HMSO, London.

Minns, C. K., Kelso, J. R. M. and Hyatt, W. (1978). Spatial distribution of nearshore fish in the vicinity of two thermal generating stations, Nanticoke and Douglas Point, on the Great Lakes. *J. Fish. Res. Board Can.*, **35**, 885–92.

Mitchell, R. (Ed.) (1972). *Water Pollution. Micro-biology.* Wiley-Interscience, London, New York.

Mitchell, R. (1974). Aspects of the ecology of the lamellibranch *Mercenaria mercenaria* (L), in British Waters. *Hydrobiol. Bull.*, **8**(1/2), 124–38.

Mitchell, S. J. and Cech, J. J., Jr (1983). Ammonia-caused gill damage in channel catfish (*Ictalurus punctatus*) confounding effects of residual chlorine. *Can. J. Fish. Aquat. Sci.*, **40**, 242–7.

Mitsch, W. J., Bosserman, R. W. and Klopatek, J. M. (Eds) (1981). *Energy and Ecological Modelling, Developments in Ecological Modelling, 1.* Elsevier Scientific Publishing Company, Amsterdam, London, New York.

Mitton, J. B. and Koehn, R. (1975). Genetic organisation and adaptive response of allozymes to ecological variables in *Fundulus heteroclitus. Genetics*, **79**, 97–111.

Modern Power Station Practice (1971), *Vol. 11, Planning and Layout.* CEGB, London, and Pergamon Press, Oxford.

Moller, B. and Dahl-Madsen, K. I. (1982). Biological monitoring of thermal effects of cooling water discharges from Danish power plants, pp. 89–100 in Jenkins and Schodtzhansen (Eds).

Moller, H. (1978a). *Effects of Power Plant Cooling on Aquatic Biota—An Indexed Bibliography.* Institut Fur Meereskunde an der Christian-Abrechts Universitat Kiel, No. 58. Kiel, Germany.

Moller, H. (1978b). Ecological effects of cooling water of a power plant at Kiel Fjord. *Sonderdruck aus*, **26** (1977/8), H3–4, 117–30.

Monn, M. G. *et al.* (1979). A survey of capital costs of closed-cycle cooling-systems for steam electric power plants. *Proc. 41st Ann. Mtg American Power Conf.*, Chicago, IL. American Power Assoc., Washington, D.C.

Moodie, G. E. E. (1985). Gill raker variation and the feeding niche of some temperate and tropical freshwater fishes. *Environ. Biol. Fish.*, **13**(1), 71–6.

Moore, C. J. and Frisbie, C. M. (1972). A winter sport fishing survey in a warm-water discharge of a steam-electric station on the Patuxent River, Maryland. *Ches. Sci.*, **13**(12), 110–15.

Moore, C. J., Fuller, S. L. H. and Burton, D. T. (1975). A comparison of food habits of white perch (*Morone americana*) in the heated effluent canal of a steam electric station and in an adjacent river system. *Environ. Lett.*, **8**(4), 315–23.

Moore, C. J., Stevens, G. A., McErlean, A. J. and Zion, H. H. (1973). A sport fishing survey in the vicinity of a steam-electric station on the Patuxent estuary, Maryland. *Ches. Sci.*, **14**(3), 160–70.

Moore, D. J. and James, K. W. (1973). Water temperature surveys in the vicinity of power stations with special reference to infra-red techniques. *Wat. Res.*, **7**, 807–20.

Morawska, B. (1984). The effect of water temperature elevation on incipient and cumulative fecundity of batch spawning tench, *Tinca tinca*. *Aquaculture*, **42**, 273–88.

Morgan, R. P., II (1980). Biocides and fish behaviour, pp. 75–102 in Hocutt *et al.* (Eds).

Morgan, R. P. and Carpenter, E. J. (1978). Biocides, pp. 95–134 in Schubel and Marcy (Eds).

Morgan, R. P. and Prince, R. D. (1977). Chlorine toxicity to eggs and larvae of five Chesapeake Bay fishes. *Trans. Am. Fish. Soc.*, **106**(4), 380–5.

Morgan, R. P. and Prince, R. D. (1978). Chlorine effects on larval development of striped bass (*Morone saxatilis*), white perch (*M. americana*) and blueback herring (*Alosa aestivalis*), ibid., **107**(4), 636–41.

Morgan, R. P. and Stross, R. G. (1969). Destruction of phytoplankton in the cooling water supply of a steam electric station. *Ches. Sci.*, **10**(3–4), 165–71.

Morita, R. Y. (1966). Marine psychrophilic bacteria. In: *Oceanography and Marine Biology, An Annual Review, Vol. 4*. Barnes, H. (Ed.). George Allen and Unwin, London, pp. 105–22.

Morris, A. W. and Bale, A. J. (1975). The accumulation of cadmium, copper, manganese and zinc by *Fucus vesiculosus* in the Bristol Channel. *Estuar. Coastal Sci.*, **3**, 153–63.

Morris, J. C. and Isaac, R. A. (1983). A critical review of kinetic and thermodynamic constants for the aqueous chlorine–ammonia system, pp. 49–62 in Jolley *et al.* (Eds)(a).

Morrison, B. R. S. (1989). The growth of juvenile Atlantic salmon *Salmo salar* L. and brown trout *Salmo trutta* L. in a Scottish river subject to a cooling-water discharge. *J. Fish. Biol.*, **35**(3), 539–56.

Morton, S., Bartlett, C. L. R., Bibby, L. F., Hutchinson, D. N., Dyer, J. V. and Dennis, P. J. (1986). Outbreak of legionnaires disease from a cooling-water system in a power station. *Brit. J. Industr. Med.*, **43**, 630–5.

Moser, R. E. (1981). FGD options offer environmental trade offs. *Hydrocarbon Processing*, October, 88–92.

Moss, J. L., Boonyaratpalin, S. and Shelton, W. L. (1978). Movement of three species of fishes past a thermally influenced area in the Coosa River, pp. 534–45 in Thorp and Gibbons (Eds).

Mount, D. I. (1971). Thermal standards in the United States of America, pp. 195–99 in IAEA.

Mountford, K., Mullen, R. S. and Shippen, R. S. (1974). Laboratory simulation of power plant effects; response of some estuarine phytoplankters to time–temperature combinations, pp. 193–98 in Jensen (Ed.).

Muir, N. (1976). Swedish study looks at district heating for small towns. *Energy Int.*, **13**(4), 30–2.

Muller, K. (1963). Diurnal rhythm in 'organic drift' of *Gammarus pulex*. *Nature (London)*, **198**, 806–7.

Muller, R. and Fry, F. E. J. (1976). Preferred temperature of fish; a new method, *J. Fish. Res. Board Can.*, **33**, 1815–17.

Muller-Fuega, A. (1978). Centrales nucleaires et ressources marines au Japon. Publication de l'Association por le Developpment de l'Aquaculture, No. 6, 33, 240, St. Andre de Cubzac, France.

Mulstay, R. (1971). Winter survey of polychaete fauna, pp. 91–103 in Studies on the Effects of a Steam–Electric Generating Plant on the Marine Environment at Northpoint, New York. Tech. Rep. No. 9. Marine Sciences Research Centre. State University of New York.

Murarka, I. P., Porcella, D. B. and Brockeen, R. W. (1981). Ecosystem models to assess impacts of power plant cooling systems, pp. 165–72 in Mitsch *et al.* (Eds).

Murphy, J. C., Garten, C. T., Smith, M. H. and Standora, E. A. (1976). Thermal tolerance and respiratory movement of bluegill from two populations tested at different levels of acclimation temperature and water hardness, pp. 145–47 in Esch and McFarlane (Eds).

Murphy, T. M. and Brisbin, I. L. (1974). Distribution of alligators in response to thermal gradients in a reactor cooling reservoir, pp. 313–21 in Gibbons and Sharitz (Eds).

Murray, A. J. (1979). Metals, organochlorine pesticides and PCB residue levels in fish and shellfish landed in England and Wales during 1974. Aquatic Environment Monitoring Report, Number 2, Ministry of Agriculture, Fisheries and Food, Directorate of Fisheries Research, Lowestoft, UK.

Murray, S. A. (1980). Periphyton responses to chlorination and temperature, pp. 641–50 in Jolley *et al.*

Murray, S. A. (1983). Haematological responses of bluegill to chlorination and temperature tests to an operating power plant, pp. 993–1004 in Jolley *et al.* (b).

Myers, E. P. and Ditmars, J. D. (1985). Ocean thermal energy conversion; environmental effects, pp. 163–94 in Duedall *et al.* (Eds).

Naiman, R. J. (1975). Food habits of the Amargosa pupfish in a thermal stream. *Trans. Am. Fish. Soc.*, **104**(3), 536–8.

Nair, N. B. and Saraswathy, M. (1971). The biology of the wood-boring teredinid molluscs. *Adv. Mar. Biol.*, **9**, 335–509.

Nakatani, R. E. (1969). Effects of heated discharges on anadromous fishes, pp. 294–317 in Krenkel and Parker (Eds).

Nakatani, R. E., Miller, D. and Tokar, J. V. (1971). Thermal effects and nuclear power stations in the U.S.A., pp. 561–72 in IAEA.

Nash, C. E. (1968). Power stations as sea farms. *New Scientist*, 14 November, 367–9.

Nash, C. E. (1969). Thermal aquaculture. *Sea Frontier*, **15**(5), 268–76.

Nash, C. E. (1970). Marine fish farming. *Mar. Pollut. Bull.*, **1**(2), 28–30.

Nash, C. E. (1974). Residual chlorine retention and power plant fish farms. *Prog. Fish. Cult.*, **36**(2), 92–5.

Nauman, S. W. and Cory, R. L. (1969). Thermal additions and epifaunal organisms at Chalk Point, Maryland. *Ches. Sci.*, **10**, 218.

Nawrocki, S. S. (1977). A study of fish abundance in Niantic Bay, with particular reference to the Millstone point nuclear power plant. M.S. Thesis, University of Connecticut, Storrs, loc. cit. Schubel and Marcy (Eds) (1978), p. 210.

Naylor, E. (1959). The fauna of a warm dock. *Proc. XVth Int. Congr. Zool.*, Sect. 3, pp. 259–62.

Naylor, E. (1965a). Biological effects of heated effluent in docks at Swansea, S. Wales. *Proc. Zool. Lond.*, **144**(2), 253–68.

Naylor, E. (1965b). Effects of heated effluents upon marine and estuarine organisms. In: *Advances in Marine Biology, 3*, Russell, F. S. and Yonge, M. (Eds). Academic Press, London, New York, pp. 63–103.

Nebeker, A. V. (1971a). Effect of temperature at different altitudes on the emergence of aquatic insects from a single stream. *J. Kans. Entomol. Soc.*, **44**, 26–35.

Nebeker, A. V. (1971b). Effect of high winter water temperatures on adult emergence of aquatic insects. *Wat. Res.*, **5**, 777–83.

Nebeker, A. V., Stevens, D. G. and Stroud, R. K. (1979). Temperature and oxygen–nitrogen gas ratios affect fish survival in air-supersaturated water. *ibid.*, **13**, 299–303.

Neel, J. K. (1963). Impact of reservoirs. In: *Limnology in North America.* Frey, D. G. (Ed.) Madison, Wisconsin, pp. 575–93.

Negus, C. L. (1966). A quantitative study of growth and production of unionid mussels in the River Thames at Reading. *J. Anim. Ecol.*, **35**, 513–32.

Neill, W. H. and Brauer, G. (1970). Ecological responses of fishes and fish-food organisms to heated effluents: Case study of Lake Monona, Wisconsin. Annual Report to the Wisconsin Utilities Association and Office of Water Resources Research. Department of the Interior Laboratory of Limnology, University of Wisconsin, Madison, Wisconsin.

Neill, W. H. and Magnusson, J. J. (1974). Distributional ecology and behavioural thermoregulation of fishes in relation to heated effluent from a power plant at Lake Monona, Wisconsin. *Trans. Am. Fish Soc.*, **10.3**(4), 663–710.

Neilsen, T. K. and Rasmussen, J. (1982). On the definition of a mixing zone, pp. 161–4 in Jenkins and Schodtzhansen (Eds).

Nelson, D. H. (1974). Growth and developmental responses of larval toad populations to heated effluent in a South Carolina Reservoir, pp. 264–76 in Gibbons and Sharitz (Eds).

Nelson, D. J. and Evans, F. C. (Eds) (1973). Radionuclides in ecosystems. *Proc. 3rd Nat. Symp. on Radioecology, Vol. 2*, 10–12 May 1971. USAEC Conf. 670503, Oak Ridge, Tennessee.

Nelson, D. J., Kaye, S. V. and Booth, R. S. (1972). Radionuclides in river systems, pp. 367–88 in Oglesby *et al.* (Eds).

Nelson-Smith, A. (1970). The problem of oil pollution of the sea. In: *Advances in Marine Biology, Vol. 8.* Russell, F. S. and Yonge, M. (Eds) Academic Press, London, New York, pp. 215–306.

Neuman, E. (1977). Activity and distribution of benthic fish in some Baltic archipelagos with special reference to temperature. *Ambio*, Special Report, **5**, 47–55.

Neumann, E. (1979a). Catch–temperature relationship in fish species in a brackish heated effluent. *Rep. Inst. Freshw. Res. Drottingholm*, **58**, 88–106.

Neuman, E. (1979b) Activity of perch, *Perca fluviatilis L.* and roach, *Rutilus rutilus L.* in a Baltic bay, with special reference to temperature. *ibid.*, **58**, 107–25.

Neuman, E. (1979c). Fishery biology investigations at the Oskarshamn power plant, 1962–1978, National Swedish Environmental Protection Board Bulletin. A summary. SNV PM, 1154E.

Neuman, E. (1982). Thermal discharges and fish fauna in Sweden, pp. 67–88 in Jenkins and Schodtzhansen (Eds).

Newell, R. C. (1970). *The Biology of Intertidal Animals.* Logos Press, London.

Ney, J. J. and Schumacher, P. D. (1978). Assessment of damage to fish larvae by entrainment sampling with submersible pumps. *Environ. Sci. Technol.*, **12**(6), 715–16.

Nichols, B. L., Anderson, R., Banta, W., Forman, E. J. and Boutwell, S. H. (1980). Evaluation of the effects of the thermal discharge on the submerged aquatic vegetation and associated fauna in the vicinity of the C.P. Crane generating station. Maryland Power Plant Siting Program, Contract P. 33-79-02, Annapolis, USA.

Nilsen, G. and Kallquist, T. (1975). Thermal effects on marine biota in experimental systems, pp. 499–518 in IAEA (a).

Noel, J. and Schwartzbrod, J. (1981). The influence of a discharge of warm water on the survival of *Salmonella.* Cahiers du Laboratoire d'Hydrobiologie de Montereau, No. 12, pp. 7–12.

Nordlie, K. J. and Arthur, J. W. (1981). Effects of elevated water temperature on insect emergence in outdoor experimental channels. *Environ. Pollut. (Series A)*, **25**(1), 53–65.

North, W. J. (1969). Biological effects of heated water discharge at Morro Bay, California. *Proc. Int. Seaweed Symp.*, 6.

Norton, M. G. (1985). Colliery-waste and fly-ash dumping off the north-eastern coast of England, pp. 423–48 in Duedall *et al.* (Eds).

Nuclear Regulatory Commission (1977). The environmental effects of using coal for generating electricity. US Department of Commerce, National Technical Information Service, Report No. PB-267 237, Washington, D.C.

Nugent, R. S. (1970). The effects of thermal effluent on some macro-fauna of a sub-tropical estuary. PhD Dissertation, University of Miami, Florida.

Nursall, J. R. and Gallup, D. N. (1971). The responses of the biota of Lake Wabamun, Alberta to thermal effluent. In: *Proc. Int. Symp. on Identity and Measurement of Environmental Pollutants*, Ottawa, Canada, pp. 295–304.

Nuttall, N. (1990). Deadly zebras of the deep. *The Times*, 15 March, p. 38.

Nyman, L. (1972). Some effects of temperature on eel (*Anguilla*) behavior. *Rep. Inst. Freshw. Res., Drottingholm*, **52**, 90–102.

Nyman, L. (1975). Behaviour of fish influenced by hotwater effluents as observed by ultrasonic tracking. *Rep. Inst. Freshw. Res. (Swed.)*, **54**, 63–75.

Obeng, L. E. (Ed.) (1969). *Man-made lakes: The Accra Symposium.* Ghana Universities Press.

Oden, B. J. (1979). The freshwater littoral meiofauna in a South Carolina reservoir receiving thermal effluents. *Freshwat. Biol.*, **9**, 291–304.

Odum, H. T. (1974). Energy cost-benefit models for evaluating thermal plumes, pp. 628–49 in Gibbons and Sharitz (Eds).

OECD (1988). *The Radiological Impact of the Chernobyl Accident in OECD Countries*. Nuclear Energy Agency, Paris.

Oehme, F. W. (Ed.) (1978). *Toxicity of Heavy Metals in the Environment, Parts I and II*. Marcel Dekker, New York.

Ogawa, H. (1979). Modelling of power plant impacts on fish populations. *Environ. Managemt*, **3**(4), 321–30.

Oglesby, R. T., Carlson, C. A. and McCann, J. A. (Eds) (1972). *River Ecology and Man*. Academic Press, New York, London.

Olla, B. L. and Samet, C. (1978). Effects of elevated temperature on early embryonic development of the tautog, *Tautoga onitis*. *Trans. Am. Fish. Soc.*, **107**, 820–4.

Olla, B. L., Studholme, A. L., Bejda, A. J., Samet, C. and Martin, A. D. (1975). The effect of temperature on the behaviour of marine fishes; A comparison among Atlantic mackerel, *Scomber scombrus*, bluefish, *Pomatomus saltatrix* and tautog, *Tautoga onitis*, pp. 299–306 in IAEA (c).

Olszewski, M., Suffern, J. S., Coutant, C. C. and Cox, D. K. (1979). An overview of waste heat utilization research at Oak Ridge National Laboratory, pp. 2299–320 in Lee and Sengupta.

Oppenheimer, C. H. (1963). (Ed.) *Marine Microbiology*. C. C. Thomas Publications, Springfield, IL.

Oppenheimer, C. H. (1970). Bacteria, fungi and blue-green algae, pp. 347–62 in Kinne (Ed).

Opresko, D. M. and Hannon, E. H. (1979). *Chemical Effects of Power Plant Cooling-Waters: an Annotated Bibliography*. Electric Power Research Institute, EA-1072 RP-877 (ORNL/EIS-134), Palo Alto, CA.

Orrison, L. A., Cherry, W. B., Tyndall, R. L., Fliermans, C. B., Gough, S. B., Lambert, M. A., McDougal, L. K., Bibb, W. F. and Brenner, D. J. (1983). *Legionella oakridgensis* unusual new species isolated from cooling tower water. *Appl. Environ. Microbiol.*, **45**(2), 536–45.

Osborne, L. L. and Davies, R. M. (1987). The effects of a chlorinated discharge and a thermal outfall on the structure and composition of the aquatic macroinvertebrate communities in the Sheep River, Alberta, Canada. *Wat. Res.*, **21**(8), 913–21.

Osborne, M. C. and Lum, J. (1978). EPA activities on condenser biofouling control. US Environmental Protection Agency (cyclostyled).

Oseid, D. and Smith, L. L. (1972). Swimming endurance and resistance to copper and Malathion of bluegills treated by long-term exposure to sublethal levels of hydrogen sulphide. *Trans. Am. Fish. Soc.*, **101**(4), 620–5.

Osterberg, C. L. (1985). Nuclear power wastes and the ocean, pp. 127–62 in Duedall *et al*. (Eds).

Othmer, D. F. and Roels, O. A. (1973). Power, fresh water, and food from cold, deep sea water. *Science*, **182**, 121–5.

Ott, F. S. and Forward, R. B. (1976). The effect of temperature on phototaxis and geotaxis by larvae of the crab, *Phithropanopeus harrisi* (Gould). *J. Exp. Mar. Biol. Ecol.*, **23**, 97–107.

Ottendorfer, L. J. (1975). Impact of thermal discharges on the ecosystem of the Austrian Danube. Present situation and future development, pp. 785–92 in IAEA (a).

Otto, R. G. (1976). Thermal effluents, fish and gas-bubble disease in southwestern Lake Michigan, pp. 121–9 in Esch and McFarlane (Eds).

Otto, R. G., Kitchel, M. A. and O'Hara Rice, J. (1976). Lethal and preferred temperatures of the Alewife (*Alosa pseudoharengus*) in Lake Michigan. *Trans. Am. Fish. Soc.*, **1**, 97–106.

Overrein, L. N., Seip, H. M. and Tollan, A. (1982). Acid precipitation: Effects on forests and fish. Final Report of the SNSF project 1972–80. Norwegian Forest Research Institute, Aas, Norway.

Owens, M. and Edwards, R. W. (1964). A chemical survey of some English rivers. *Proc. Soc. Wat. Treat. Exam.*, **13**, 134–44.

Owens, M., Knowles, G. and Clark, A. (1969). The prediction of the distribution of dissolved oxygen in rivers. *Adv. Wat. Pollut. Res., Proc. 4th Int. Conf., Prague*, pp. 125–37.

Page-Jones, R. M. (1971). The automatic detection of low-levels of dissolved free chlorine in fish farming experiments using seawater effluents. *Prog. Fish. Cult.*, **33**, 99–102.

Paller, M. H., Lewis, W. M., Heidinger, R. C. and Wawronowicz, L. J. (1983). Effects of ammonia and chlorine on fish in streams receiving secondary discharges. *J. Wat. Pollut. Contr. Fed. Conference Preview Issue* (August), 1087–97.

Palmegiano, G. and Saroglia, M. G. (1981). Winter shrimp culture in thermal effluents, pp. 297–302 in Tiews (Ed.).

Palmer, C. J., Culley, M. B. and Claridge, P. N. (1979). A further occurrence of *Atherina boyeri Risso*. 1810 in North-Eastern Atlantic waters. *Environ. Biol. Fish.*, **4**(1), 71–5.

Pannell, J. P. M., Johnson, A. E. and Raymont, J. E. G. (1962). An investigation into the effects of warmed water from Marchwood power station into Southampton Water. *Proc. Inst. Civil Engrs*, **23**, 35–62.

Parker, E. D., Hirshfield, M. F. and Gibbons, J. W. (1973). Ecological comparisons of thermally affected aquatic environments. *J. Wat. Pollut. Contr. Fed.*, **45**(4), 726–33.

Parker, F. L. and Krenkel, P. A. (Eds) (1969). *Engineering Aspects of Thermal Pollution*. Vanderbilt University Press, Portland, Oregon.

Parkin, R. B. and Stahl, J. B. (1981). Chironomidae (Diptera) of Baldwin Lake, Illinois—a cooling reservoir. *Hydrobiologia*, **76**(1/2), 119–27.

Patel, B., Balani, M. C., Patel, S., Panday, V. K. and Soman, S. D. (1975). Impact of thermal and radioactive effluents on a tropical nearshore system, pp. 17–33 in IAEA (c).

Patrick, R. (1969). Some effects of temperature on freshwater algae, pp. 161–85 in Krenkel and Parker (Eds).

Patrick, R. (1974). Effects of abnormal temperatures on algal communities, pp. 335–49 in Gibbons and Sharitz (Eds).

Patten, B. G. (1980). Short term thermal resistance of hexagrammid eggs and planktonic larvae from Puget Sound. *Trans. Am. Fish. Soc.*, **109**, 427–32.

Paul, J. F. and Lick, W. J. (1974). A numerical model for thermal plumes and river discharges. *Proc. 17th Conf. Great Lakes*, Ann Arbor, Michigan, pp. 445–55.

Pearson, W. D. and Franklin, D. R. (1968). Some factors affecting drift rates of *Baetis* and *Simuliidae* in a large river. Ecology, **49**, 75–81.

Peck, B. B. and Warren, R. S. (1978). Nitrate reductase activity and primary productivity of phytoplankton entrained through a nuclear power station on north-eastern Long Island sound, pp. 392–409 in Thorp and Gibbons (Eds).

Pentreath, R. J. (1980). *Nuclear Power, Man and the Environment.* The Wykham Science Series, Taylor and Francis, London.

Perkins, E. J. (1974). *The Biology of Estuaries and Coastal Waters.* Academic Press, London, New York, San Francisco.

Perry, L. G. and Tranquilli, J. A. (1984). Age and growth of largemouth bass in a thermally altered reservoir, as determined from otoliths. *North Am. J. Fish. Managmnt*, **4**, 321–30.

Peters, D. S., Boyd, M. T. and DeVane, J. C., Jr (1976). The effect of temperature, salinity and food availability on the growth and food conversion efficiency of post-larval pinfish, pp. 106–13 in Esch and McFarlane (Eds).

Peters, D. S., Kjelson, M. A. and Boyd, M. T. (1974). The effect of temperature on food evacuation rate in the pinfish (*Lagodon rhomboides*), spot (*Lepistomus zanthurus*) and silverside (*Menidia menidia*). *Proc. 26th Ann. Conf. of the S.E. Assoc. of Game and Fish.* Commissioners 1972, pp. 637–43.

Peterson, R. E. and Seo, K. K. (1977). Thermal aquaculture. In: *Proc. 8th Annual Meeting, World Mariculture Society,* Avault, J. W. (Ed.). Louisiana State University, Baton Rouge, pp. 491–503.

Peterson, R. H. (1976). Temperature selection of juvenile Atlantic salmon, (*Salmo salar*) as influenced by various toxic substances. *J. Fish. Res. Board Can.,* **33**, 1722–30.

Peterson, R. H. and Sutterlin, A. M. (1979). Temperature preference of several species of *Salmo* and *Salvelinus* and some of their hybrids. *ibid.,* **36**, 1137–40.

Petts, G. E. (1984). *Impounded Rivers. Perspectives for Ecological Management.* Wiley-Interscience, Chichester, UK.

Philbin, T. W. and Philipp, H. D. (1971). Thermal effects studies in New York State, pp. 575–88 in IAEA.

Philipart, J-Cl. and Ruwet, J-Cl. (1982). Ecology and distribution of tilapias, pp. 15–59 in Pullin and Lowe-McConnell (Eds).

Phillips, D. J. H. (1980). *Quantitative Aquatic Biological Indicators. Their Use to Monitor Trace Metal and Organochlorine Pollution.* Applied Science Publishers, London.

Pickering, P. and Davis, M. H. (1983). Effects of entrainment on the numbers and diversity of estuarine diatoms: Fawley power station. CEGB Internal Report, TPRD/L/2457/N 83, Leatherhead, UK.

Pipe, E. J. (1972). The Central Electricity Generating Board and Regional Water Authorities. *Proc. Conf. on Management of National and Regional Water Resources,* Institution of Civil Engineers, London, 3–4 Oct.

Plumb, R. H., Simmons, L. L. and Collins, M. (1980). Assessment of intermittently chlorinated discharges using chlorine half-life, pp. 435–44 in Jolley *et al.* (Eds).

Poje, G. V., Riordan, S. A. and O'Connor, J. M. (1981). Power plant entrainment simulation utilizing a condenser tube simulator, Final Report, NUREG/CR-2091. Office of Nuclear Regulatory Research, US Nuclear Regulatory Commission, Washington, D.C.

Poje, G. V., Riordan, S. A. and O'Connor, J. M. (1983). Power plant chlorination; Immediate and persistent effects of sub-lethal concentrations on an estuarine crustacean, pp. 1039–52 in Jolley *et al.* (b).

Polgar, T. T., Bongers, L. H. and Krainak, G. M. (1976). Assessment of power plant effects on zooplankton in the near field, pp. 358–68 in Esch and McFarlane (Eds).

Polikarpov, G. G. (1966). *Radioecology of Aquatic Organisms. The Accumulation and Biological Effect of Radioactive Substances.* North-Holland Publishing Co., Reinhold Book Division, Amsterdam.

Polivannaya, M. F. (1974). Food of perch and ruffe in the zone of effluent from the Kurakhovka State Regional Electric Power Station. *Hydrobiol. J. (USSR)*, **10**, 84–94.

Polivannaya, M. F. and Sergeyeva, O. A. (1971a). Zooplankton dynamics and biology of some dominant species in cooling ponds of the southern Ukraine. In: *Simpoz. povliyan, podogr. vod TEPP.* (copy 2), Borok.

Polivannaya, M. F. and Sergeyeva, O. A. (1971b). Biology of common species of Cladocera in the Kurakhovka cooling pond. *Hydrobiol. J. (USSR)*, **7**, 6.

Porter, W. P. and Tracy, C. R. (1974). Modelling the effects of temperature changes on the ecology of the garter snake and leopard frog, pp. 594–609 in Gibbons and Sharitz (Eds).

* Powell, A. H. (1963). The effect of normal operation of a generating station on the dissolved oxygen content of the circulating water, CEGB Internal Report No. 4/63, London.

Power, M. E. and Todd, J. H. (1976). Effects of increasing temperature on social behaviour in territorial groups of pumpkinseed sunfish (*Lepomis gibbosus*). *Environ. Pollut.*, **10**, 217–23.

Precht, J., Christophersen, J., Hensel, H. and Larcher, W. (Eds) (1973). *Temperature and Life.* Springer-Verlag, Berlin, Heidelberg, New York.

Prentice, E. F. (1969). Gull predation in a reactor discharge plume. In: *Biological Effects of Thermal Discharges.* Annual Prog. Rept. for 1968. AEC Research and Development Dept., BNWL-1050, Battelle Northwest.

Preston, A. (1971). The radiological consequences of releases from nuclear facilities to the aquatic environment, pp. 3–23 in IAEA.

Preston, A. (1974). The radiological consequences of releases from nuclear facilities to the aquatic environment, pp. 3c–21 in IAEA (b).

Price, A. H., Hess, C. T. and Smith, C. W. (1976). Observations of *Crassostrea virginica*, cultured in the heated effluent and discharged radionuclides of a nuclear power reactor. *Proc. Nat. Shellfisheries Assn*, **66**, 1–15.

Price, J. H., Irvine, D. E. G. and Farnham, W. F. (1980). *The Shore Environment, Vol. 2, Ecosystems.* The Systematics Association, Special Volume No. 17(a), Academic Press, New York, London.

Prince, E. D. and Mengel, L. J. (1981). Aggregation of spottail shiners in the heated discharge of a nuclear power station. *Trans. Am. Fish. Soc.*, **110**(2), 221–5.

Pringle, B. H., Hissong, D. E., Katz, E. L. and Mulawka, S. T. (1968). Trace metal accumulation in estuarine molluscs. *ASCE, J. Sanit. Engng Div.*, **94**, 455–75.

Profitt, M. A. (1969). Effects of heated discharges upon aquatic resources of White River, at Petersburg, Indiana. Indiana Water Resources Center, Report No. 3, Indiana.

PSEG (1977). Integration of thermal and food processing residuals into a system for commercial culture of freshwater shrimp (power plant waste heat utilization in aquaculture). *Pub. Serv. Elec. Gas Comp. Rep. Final Report*, January, Vol. 1.

Pullin, R. S. V. and Lowe McConnell, R. H. (Eds) (1982). The Biology and Culture of Tilapias, ICLARM Conference Proceedings 7. International Center for Living Aquatic Resources Management, Manila.

Punzi, V. L. and Patel, R. D. (1985). Predicting chlorine compounds in powerplant cooling-tower systems, pp. 1469–87 in Jolley *et al.* (Eds).

Quinn, T., Esch, G. W., Hazen, T. C. and Gibbons, J. W. (1978). Long-range movement and homing bass (*Micropterus salmoides*) in a thermally altered reservoir. *Copeia*, 3(3), 542–5.

Rachyunas, L. A. (1973). Feeding of fishes in the cooling ponds of the Lithuanian State Power Plant. *Hydrobiol. J.*, 9, 13–18. (USSR), *Hydrobiol. J.*, 9, 21.

Ramade, A. (1981). Impact du transit sur la production photoplanctonique et certains parametres de la biomasse dans les circuits de refroidissement de la centrale de Martiques-Ponteau, pp. 589–607 in EDF.

Raney, E. C. and Menzel, B. W. (1969). *Heated Effluents and Effects on Aquatic Life with Emphasis on Fishes (A bibliography)*. Cornell University Water Research and Marine Science Centre, Philadelphia Electric Co., and Ichthyological Association BLU, No. 2.

Rankin, J. S., Buck, J. D. and Foerster, J. W. (1974). Thermal effects on the microbiology and chemistry of the Connecticut River—A summary, pp. 350–5 in Gibbons and Sharitz (Eds).

Rasmussen, J. B. (1982). The effect of thermal effluent before and after macrophyte harvesting on standing crop and species composition of benthic macroinvertebrate communities in Lake Wabamur, Alberta. *Can. J. Zool.*, 60(12), 3196–205.

Ray, D. L. (Ed.) (1959). *Marine Boring and Fouling Organisms*. University of Washington Press, Seattle.

Ray, J. and Skulec, S. (1983). Waste heat and water from atomic power plants and environmental protection. *Ochrana ovzdusi.*, 15(3), 33–8.

Raymont, J. E. G. (1976). The introduction of new species in habitats of heated effluents, pp. 185–209 in Devik (Ed.).

Raymont, J. E. G. and Carrie, B. G. A. (1964). The production of zooplankton in Southampton Water. *Int. Revue ges. Hydrobiologie*, 49(2), 185–232.

Rebhun, M. and Engel, G. (1988). Re-use of waste-water for industrial cooling systems. *J. Wat. Pollut, Contr. Fed.*, 60(2), 237–41.

Reetz, S. D. (1982). Phytoplankton studies in the Missouri River near Fort Calhoun station and Cooper nuclear stations, pp. 71–84 in Hesse *et al.* (Eds) (b).

Reeve, M. R. and Cosper, E. (1970). The acute thermal effects of heated effluents on the copepod, *Acartia tonsa* from a sub-tropical bay and some problems of assessment. FAO Technical Conference on Marine Pollution and its Effects on Living Resources and Fishing, pp. 1–5.

Reintjes, J. W. and Pached, A. L. (1966). The relation of menhaden to estuaries. American Fisheries Society, Special Publication, No. 3, pp. 50–8.

Relini, G. and Oliva, G. D. (1972). Biological studies on fouling problems in Italy, pp. 757–66 in Acker *et al.* (Ed.).

Repsys, A. J. and Rogers, G. D. (1982). Zooplankton studies in the channelized Missouri River, pp. 125–46 in Hesse *et al.* (Eds) (b).

Reutter, J. M. and Herdendorf, C. E. (1974). Laboratory estimates of the seasonal final temperature preferenda of some Lake Erie fish. *Proc. 17th Conf. Great Lakes Res.*, p. 59.

Reymeysen, J. *et al.* (1979). Dry cooling towers. UNIPEDE, Warsaw. Paper 2 O/D.2, June 1979.

Reynolds, J. E. III and Wilcox, J. R. (1985). Abundance of West Indian manatees (*Trichechtus manatus*) around selected Florida power plants following winter cold fronts 1982–1983. *Bull. Mar. Sci.*, **36**(3), 413–22.

Reynolds, W. M. (1977). Fish orientation behavior; An electronic device for studying simultaneous responses to two variables. *J. Fish. Res. Board Can.*, **34**, 300–4.

Reynolds, W. W. and Casterlin, M. E. (1976). Thermal preferenda and behavioural thermoregulation in three centrarchid fishes, pp. 185–90 in Esch and McFarlane (Eds).

Reynolds, W. W. and Casterlin, M. E. (1978). Complementary of thermoregulatory rhythms in *Micopterus salmoides* and *M. dolomieui*. *Hydrobiologia*, **60**(1), 89–91.

Reynolds, W. W. and Casterlin, M. E. (1979). Behavioral thermoregulation and activity in *Homarus americanus* Milne Edwards. *Comp. Biochem. Physiol.*, **64A**, 25–8.

Reynolds, W. W. and Thompson, D. A. (1974). Responses of young Gulf grunnion, *Leuresthes sardina*, to gradients of light, turbulence and oxygen. *Copeia*, **3**, 144–9.

Rheinheimer, G. (1974). *Aquatic Microbiology*. Wiley-Interscience, London.

Rice, R. T. and Baptist, J. P. (1974). Ecological effects of radioactive emissions from nuclear power plants. In: *Human and Ecologic Effects of Nuclear Power Plants*, pp. 373–439.

Rice, T. R. (1963). Accumulation of radionuclides by aquatic organisms. In: Studies of the fate of certain radionuclides in Estuarine and other aquatic environments, Sabo, J. and Bedrosian, P. H. (Eds). US Public Health Service Publication No. 999-R-3, pp. 35–50.

Richards, F. P. and Ibara, R. M. (1977). The preferred temperatures of the brown bullhead, *Ictalurus nebulosus* with reference to its orientation to the discharge canal of a nuclear power plant. *Trans. Am. Fish. Soc.*, **107**(2), 288–94.

Richards, F. P., Reynolds, W. W. and McCauley, R. W. (1977). Temperature preference studies in environmental impact assessments: An overview with procedural recommendations. *J. Fish. Res. Board Can.*, **34**, 728–61.

Richkus, W. A. (1975). The response of juvenile alewives to water currents in an experimental chamber. *Trans. Am. Fish. Soc.*, **104**(3), 494–8.

Richkus, W. A. (1977). Aquatic impact assessment at Calvert Cliffs. *Record of the Maryland Power Plant Siting Act*, **6**(1), 1–7.

Ricker, W. E. (1975). Computation and Interpretation of biological statistics of fish populations. *Bull. Fish. Res. Board Can.*, **191**.

Ringler, N. H. and Hall, J. D. (1975). Effects of logging on water temperature and dissolved oxygen in spawning beds. *Trans. Am. Fish. Soc.*, **104**, 111–21.

Rippon, J. (1971) The bacteria of aquatic environments in power stations, pp. 95–114 in CERL.

Rippon, J. E. (1979). *U.K. Biofouling Control Practices*. CEGB (cyclostyled paper), Leatherhead, UK.

Rippon, J. E. and Wood, M. J. (1970). The association of bacterial growth with stress-corrosion cracking. CEGB Internal Report, RD/L/N 122/70, Leatherhead, UK.

Ritchie, J. (1927). Report on prevention of growth of mussels in sub-marine shafts and tunnels at Westbank Electric Station, Portobello. *Trans. R. Scot. Soc. Arts.*

Roberts, M. H. and Gleeson, R. A. (1978). Acute toxicity of bromochlorinated seawater to selected estuarine species with a comparison to chlorinated seawater toxicity. *Mar. Environ. Res.*, **1**(1), 19–30.

Roberts, R. D. and Roberts, T. M. (1984). *Planning and Ecology.* Chapman and Hall, London.

Roberts, R. J. (1976). Bacterial diseases of farmed fishes. In: *Microbiology in Agriculture*, Skinner, F. A. and Carr, J. G. (Eds). Academic Press, London, pp. 52–62.

Roberts, R. J. (Ed.) (1978). *Fish Pathology.* Bailliere Tindall, London.

Robertson, R. F. S. (1977). Utilisation of waste heat from electricity generating stations. AECL Report 5689, Whiteshell Nuclear Research Establisment.

Robson, A. (1984). The radiological impact of electricity generation by UK coal and nuclear systems. *Sci. Tot. Environ.*, **35**, 417–30.

Roessler, M. A. (1971). Environmental changes associated with a Florida power plant. *Mar. Pollut. Bull.*, **2**(6), 87–90.

Roessler, M. A., Bach, S., Josselyn, M., Hixon, R. and Brook, I. (1979). Thermal effluent discharges from a power plant into Cord Sound Florida— observations before and after. *Environ. Conservat.*, **6**(2), 127–37.

Roessler, M. A., Beardsley, G. L., Rehrer, R. and Garcia, J. (1975). Effects of thermal effluents on the fishes and benthic invertebrates of Biscayne Bay—Card Sound, Florida. Technical Report of the University of Miami, Florida.

Rogers, A. (1977). Dinorwic pumped-storage scheme. Aquatic environmental studies programme. 1. Introduction and general appraisal. CEGB NW/SSD/SR 82/77, London.

Rogers, G. D. (1978). Entrainment of crustacean zooplankton through Foort Calhoun station, pp. 141–54 in Jensen (Ed.) (b).

Romberg, G. P., Spigarelli, S. A., Prepejchal, W. and Thommes, M. M. (1974). Fish behaviour at a thermal discharge into Lake Michigan, pp. 296–312 in Gibbons and Sharitz (Eds).

Romeril, M. G. (1971). Preliminary observations on trace metal accumulation in the hard shell clam, *Mercenaria mercenaria.* CEGB Internal Report, RD/L/N/31/71, Leatherhead, UK.

Romeril, M. G. (1972). Trace metals in the common cockle, *Cerastoderma edule.* CEGB Internal Report, RD/L/N 179/72, Leatherhead, UK.

Romeril, M. G. (1974). Trace metals in sediments and bivalve mollusca in Southampton Water and the Solent. *Rev. Int. Oceanogr. Med.*, **xxxiii**, 31–47.

Romeril, M. G. (1975). Trace metal studies in Southampton Water and the Solent, pp. 69–82 in CERL.

Romeril, M. G. (1976a). Iron concentrations in mussels transplanted into Southampton Water and the Solent. CEGB Internal Report, LM/BIOL/012, Leatherhead, UK.

Romeril, M. G. (1976b). Further observations on trace metal concentrations in the tissues of the hard shell clam, *Mercenaria mercenaria* in Southampton Water. CEGB Internal Report, RD/L/N 11/76, Leatherhead, UK.

Romeril, M. G. (1979). The occurrence of copper, iron and zinc in the hard shell clam *Mercenaria mercenaria* and sediments of Southampton Water. *Estuar. Coastal Mar. Sci.*, **9**, 423–34.

Romeril, M. G. and Davis, M. H. (1976). Trace metal levels in eels grown in power station cooling-water. *Aquaculture*, **8**, 139–49.

Roosenburg, W. J. (1969). Greening and copper accumulation in the American Oyster, *Crassostrea virginica*, in the vicinity of a steam electric generating station. *Ches. Sci.*, **10**, 241–52.

Rose, A. H. (Ed.) (1967). *Thermobiology*. Academic Press, London, New York.

Rosenweig, W. D., Minnigh, H. A. and Pipes, W. O. (1983). Chlorine demand and inactivation of fungal propagules. *Appl. Environ. Microbiol.*, **45**(1), 182–6.

Ross, A. J. and Smith, C. A. (1974). Effect of temperature on survival of *Aeromonas liquefaciens, Aeromonas salmonicida, Chondrococcus columnaris* and *Pseudomonas fluorescens. Progr. Fish Cult.*, **36**(1), 51–62.

Ross, F. F. (1954). Changes in the dissolved oxygen content of river water used for direct cooling. British Electricity Authority, Sub-committee No. 8, (Effluents Testing), London, 12 pp.

Ross, F. F. (1959). The operation of thermal power stations in relation to streams. *J. Inst. Sewage Purif.*, **16**, 16–29.

Ross, F. F. (1970). Warm water discharges into rivers and the sea. CEGB Newsletter No. 86, Central Electricity Generating Board, London.

Ross, F. F. and Whitehouse, J. W. (1973). Cooling-towers and water quality. *Wat. Res.*, **7**, 623–31.

Ross, M. J. and Siniff, D. B. (1982). Temperatures selected in a power plant thermal effluent by adult yellow perch (*Perca flavescens*) in winter. *Can. J. Fish. Aquat. Sci.*, **39**, 346–51.

Rossie, J. P. (1974). Dry cooling towers, pp. 99–112 in Gallagher (Ed.).

Rothwell, R. (1971). The calefaction of Lake Trawsfynydd, pp. 115–30 in CERL.

Round, F. E. (1973). *The Biology of the Algae*. Edward Arnold, London.

Rucker, R. R. (1972). Gas bubble disease of salmonids—a critical review. US Fish and Wildlife Service Technical Paper 58.

Rucker, R. R. (1976). Gas bubble disease: mortalities of coho salmon. *Oncorhynchus kisutch*, in water with constant total gas pressure and different oxygen-nitrogen ratios. *Fish. Bull.*, **73**(4), 915–18.

Rulifson, R. A. (1977). Temperature and water velocity effects on the swimming performances of young-of-the-year striped mullet (*Mugil cephalus*), spot (*Leiostomus zanthurus*), and pinfish (*Lagodon rhomboides*). *J. Fish. Res. Board Can.*, **34**, 2316–22.

Ruttner, F. (1963). *Fundamentals of Limnology* (Trans. by Frey, D. G. and Frey, F. E. J.), University of Toronto Press, Toronto, Canada.

Sadler, K. (1979). Effects of temperature on the growth and survival of the European eel, *Anguilla anguilla* L. *J. Fish Biol.*, **15**, 499–507.

Sadler, K. (1980). Effect of the warm-water discharge from a power station on fish populations in the River Trent. *J. Appl. Ecol.*, **17**(2), 349–57.

Saenger, P., Robson, J. and Ewald, J. (1980a). Cooling water and the Gladstone environment. Report of the Queensland Electricity Generating Board, Brisbane, Australia.

Saenger, P., Stephenson, W. and Moverley, J. (1980b). The estuarine macrobenthos of the Calliope River and Auckland Creek, Queensland. *Memoranda of the Queensland Museum*, **20**(1), 143–61.

Saenger, P., Stephenson, W. and Moverley, J. (1982). Macrobenthos of the cooling water discharge canal of the Gladstone power station, Queensland. *Austr. J. Mar. Freshw. Res.*, **33**, 1083–95.

Sager, D. R. and Cofield, C. R. (1984). Differential accumulation of selenium among axial muscle, reproductive and liver tissues of four warmwater fish species. *Water Resources Bull.*, **20**(3), 359–63.

Salo, E. O. and, Leet, W. L. (1968). The concentration of zinc-65 by oysters maintained in the discharge canal of a nuclear power plant. *Proc. Second Symposium on Radioecology*, University of Michigan. Ann Arbor, Michigan.

Sanders, J. E., Pilcher, K. S. and Fryer, J. L. (1978). Relation of water temperature to bacterial kidney disease in Coho salmon (*Oncorhynchus kisutch*) sockeye salmon (*O. nerka*) and steelhead trout (*Salmo gardneri*). *J. Fish. Res. Board Can.*, **35**, 8–11.

Sanders, J. G. and Ryther, J. H. (1980). Impact of chlorine on the species composition of marine phytoplankton, pp. 631–40 in Jolley *et al.*

Sandholm, M., Oksanen, H. E. and Pesonen, L. (1973). Uptake of selenium by aquatic organisms. *Limnol. Oceanogr.*, **18**(3), 496–9.

Sandstrom, O. (1985). Recipient Monitoring at Forsmark Nuclear Power Station. Environmental Quality Laboratory, SNV 1915, National Swedish Environmental Protection Board, Solna, Sweden.

Sappo, G. B. (1975). Effect of hot effluent discharge from the Konokovo Power Plant on the growth of bream in the Ivan'kobo Reservoir. *Hydrobiol. J. Ichthyol.*, **11**(6), 43–7.

Sappo, G. B. (1976). The formation of local population of bream (*Abramis brama orientalis*) in the heated water zone of the Konakovo Power Station. *ibid.*, **16**(1), 35–45.

Sarker, A. L. (1977). Feeding ecology of the bluegill, *Lepomis macrochirus*, in two heated reservoirs of Texas. *Trans. Am. Fish. Soc.*, **106**(6), 596–601.

Saroglia, M. G., Quierazza, G. and Scarano, G. (1981). Water quality criteria for aquaculture in thermal effluents, heavy metals and residual anti-fouling products, pp. 99–112 in Tiews (Ed.).

Saroglia, M. G. and Scarano, G. (1983). Risk from residual chlorine in cooling waters used for aquacultural purposes, pp. 1437–46 in Jolley *et al.* (b).

Sarvala, J. (1979). Effect of temperature on the duration of egg, nauplius and copepodite development of some freshwater benthic Copepoda. *Freshwat. Biol.*, **9**, 515–34.

Sastry, A. N. (1976). Effects of constant and cyclic temperature regimes on the pelagic larval development of a brachyuran crab, pp. 81–7 in Esch and McFarlane (Eds).

Sastry, A. N. (1978). Physiological adaptation of *Cancer irroratus* larvae to cyclic temperatures, pp. 57–66 in McLusky and Berry (Eds).

Saunders, R. L. (1976) Heated effluent for the rearing of fry for farming and release, pp. 213–36 in Devik (Ed.).

Schmidt, W. (1915). *Annalen der Hydrographic and Maritimen Meteorologic*, pp. 111–24 and 169–78.

Schneider, M. J., Genoway, R. G. and Barraclough, S. A. (1974). Preliminary studies on the effects of fatigue on thermal tolerance of rainbow trout, pp. 177–85 in Gibbons and Sharitz (Eds).

Schroeder, L. A. and Callaghan, W. M. (1981). Thermal tolerance and acclimation of two species of *Hydra. Limnol. Oceanogr.*, **26**(4), 690–6.

Schubel, J. R. (1974). Effects of exposure to time-excess temperature histories typically experienced at power plants on the hatching success of fish eggs. *Estuar. Coastal Mar. Sci.*, **2**, 105–16.

Schubel, J. R. and Auld, A. H. (1974). Hatching success of blueback-herring and striped bass eggs with various time vs temperature histories, pp. 164–70 in Gibbons and Sharitz (Eds).

Schubel, J. R., Coutant, C. C. and Woodhead, P. N. J. (1978). Thermal effects of entrainment, pp. 19–94 in Schubel and Marcy (Eds).

Schubel, J. R. and Koo, T. S. Y. (1976). Effects of various time-excess temperature histories on hatching success of blueback herring, American shad and striped bass eggs, pp. 165–70 in Esch and McFarlane (Eds).

Schubel, J. R. and Marcy, B. C. (Eds) (1978). *Power Plant Entrainment. A Biological Assessment*. Academic Press, London, New York.

Schubel, J. R., Smith, C. F. and Koo, T. S. (1977). Thermal effects of power plant entrainment on survival of larval fishes: A laboratory assessment. *Ches. Sci.*, **16**(3), 290–8.

Schwartz, F. J. (1964). Effects of winter water conditions on fifteen species of captive marine fishes. *Am. Midl. Nat.*, **71**, 434–44.

Schwartzbrod, J. and Noel, J. (1980). Recherche des salmonelles dans un plan d'eau utilise pour le rejet des eaux d'une centrale thermique. *Techniques et Sciences Municipales*, **75**(3), 105–9.

Scott, D. L. (1973). *Pollution in the Electric Power Industry. Its Control and Costs.* Lexington Books, Toronto, London.

Scott, G. I. (1983). Physiological effects of chlorine produced oxidants, dechlorinated effluents and trihalomethanes on marine invertebrates, pp. 827–42 in Jolley *et al.* (Eds) (b).

Scott, G. I. and Middaugh, D. P. (1978). Seasonal chronic toxicity of chlorination to the American oyster, *Crassostrea virginica*. (G), pp. 311–28 in Jolley *et al.* (Eds).

Scott, G. I., Middaugh, D. P., Crane, A. M., McGlothlin, N. P. and Watabe, N. (1980). Physiological effects of chlorine-produced oxidants and uptake of chlorination by-products in the American oyster, *Crassostrea virginica* (Gmelin), pp. 501–16 in Jolley *et al.*

Scott, G. I., Middaugh, D. P. and Klingensmith, S. (1983). Bioconcentration of bromoform by American oysters *Crassostrea virginica*, (G), exposed to chlorinated and dechlorinated seawater with notes on survival and feeding, pp. 1029–38 in Jolley *et al.* (Eds) (b).

Scott, G. I., Oswald, E. O., Sammons, T. I., Baughman, D. S. and Middaugh, D. P. (1985). Interactions of chlorine-produced oxidants, salinity and a protistan parasite in affecting lethal and sub-lethal physiological effects in the Eastern or American oyster, pp. 463–81 in Jolley *et al.* (Eds).

Seaman, C. V., Hill, L. O., Vignon, B. V., Stanford, T. B. and Hunter, M. D. (1983). Halogenated organic study at selected Tennessee Valley Authority fossil-fuelled power plants, pp. 373–82 in Jolley *et al.* (a).

Seegert, G. L. and Bogardus, R. B. (1980). Ecological and environmental factors to be considered in developing chlorine criteria, pp. 961–72 in Jolley *et al.*

Seegert, G. L. and Brooks, A. S. (1978a). The effects of intermittent chlorination on Coho salmon, alewife, spottail shiner and rainbow smelt. *Trans. Am. Fish. Soc.*, **107**(2), 346–53.

Seegert, G. L. and Brooks, A. S. (1978b). Dechlorination of water for fish culture: Comparison of the activated carbon, sulfite reduction, and photochemical methods. *J. Fish. Res. Board Can.*, **35**, 88–92.

Seegert, G. L., Brooks, A. S. and Latimer, D. L. (1977). The effects of a 30 minute exposure of selected Lake Michigan fishes and invertebrates to residual chlorine, pp. 91–9 in Jensen (Ed.).

Serns, S. L. and Strawn, K. (1975). Age and growth of Bluegill (*Lepomis macrochirus*) in two heated Texas reservoirs. *Trans. Am. Fish. Soc.*, **104**(3), 506–12.

Shafland, P. L. and Pestrak, J. M. (1982). Lower lethal temperatures for fourteen non-native fishes in Florida. *Environ. Bio. Fish.*, **7**(2), 149–56.

Shapiro, M. A., Kard, M. H., Keleti, G., Sykora, J. L. and Martinez, A. J. (1980). The role of free-living amoebae occurring in heated effluents as causative agents of human disease, pp. 135–49 in Jenkins and Schodtzhansen (Eds).

Sharitz, R. R., Gibbons, J. W. and Gause, S. C. (1974). Impact of production-reactor effluents on vegetation in a southeastern swamp forest, pp. 356–62 in Gibbons and Sharitz (Eds).

Shelbourne, J. E. (1970). Marine fish cultivation: priorities and progress in Britain. In: *Marine Aquaculture*, McNeil, W. J. (Ed.). Oregon State University Press, pp. 15–36.

Sherberger, F. F., Benfield, E. F., Dickson, K. L. and Cairns, J., Jr (1977). Effects of thermal shocks on drifting aquatic insects: A laboratory simulation. *J. Fish. Res. Board Can.*, **34**(4), 529–36.

Shrode, J. B., Zerba, K. E. and Stephens, J. S., Jr (1982). Ecological significance of temperature tolerance and preference of some inshore California fishes. *Trans. Am. Fish. Soc.*, **111**, 45–51.

Shubert, L. E. (Ed.) (1981). *Algae as Ecological Indicators*. Academic Press, London.

Shuter, B. J., Wismer, D. A., Regier, H. A. and Matuszek, J. E. (1985). An application of ecological modelling; Impact of thermal effect on a smallmouth bass population. *Trans. Am. Fish. Soc.*, **114**(5), 631–51.

Siler, J. R. and Clugston, J. P. (1975). Largemouth bass under conditions of extreme stress. In: *Black Bass biology and management*, Stroud, R. H. and Clepper, H. (Eds). Sport Fishing Institute, Washington, D.C., pp. 333–41.

Simmons, C. N., Armitage, B. J. and White, J. C. (1974). An ecological extraction of heater water discharges on phytoplankton blooms in the Potomac River. *Hydrobiologia*, **45**(4), 441–66.

Skinner, F. A. and Shewan, J. M. (Eds) (1977). *Aquatic Microbiology*. The Society for Applied Bacteriology, Symposium Series Number 6, Academic Press, London, New York.

Smedile, E. and Parisi, V. (1975). Effect of entrainment in power station cooling systems studied using periphytic communities, pp. 127–41 in IAEA (c).

Smith, K. (1972). River water temperatures—an environmental review. *Scot. Geog. Mag.*, **88**, 211–20.

Smith, K. (1975). Water temperature variations within a major river system. *Hydrology*, **6**, 153–69.

Smith, R. A., Brooks, A. S. and Jensen, L. D. (1974a). Effects of condenser entrainment on algal photosynthesis at mid-Atlantic power plants, pp. 113–22 in Jensen (Ed.).

Smith, R. A. and Jensen, L. D. (1974). Effects of condenser destruction on dissolved oxygen levels in the James River, pp. 123–29 in Jensen (Ed.).

Smith, R. L. and Rhodes, D. (1983). Body temperature of the salmon shark, *Lamna ditropis. J. Mar. Biol. Assn*, **63**, 243–4.

Smith, W. H. B., McKellar, H., Young, D. L. and Lehman, M. E. (1974b). Total metabolism of thermally affected coastal systems on the west coast of Florida, pp. 475–89 in Gibbons and Sharitz (Eds).

Snow, C. E. and Sladek, W. A. (1978). The effects of chlorination on in-situ coliform and faecal Streptococcus bacteria in power-plant cooling water and the receiving stream, pp. 125–33 in Gerhold (Ed.).

Solski, A. (1974). The influence of discharged heated waters from the power station at Skawinka on microflora of Vistula River. *Polski Archiwum Hydrobiologii*, **21**(1), 75–82.

Soon Kil Yi (1987). Effects of cooling water system of a power plant on marine organisms, Effects on benthic organisms. *Bull. Kor. Fish. s*, **20**(5), 391–407 (Ko).

Southwood, J. R. E. (1966). *Ecological Methods*. Chapman and Hall, London.

Spencer, J. F. (1970). Diurnal and seasonal temperature changes in the littoral soil on Pwllcrochan Flats, Milford Haven, in relation to Pembroke Power Station. CEGB Internal Report, No. RD/L/R 1641, Leatherhead, UK.

Spencer, J. F. (1975). Temperature studies related to the dispersal of the cooling water from Fawley power station, pp. 17–22 in CERL.

Spencer, J. F. (1977). Temperature studies in the Blackwater Estuary, 1974–75. CEGB Internal Report, RD/L/R/1962, Leatherhead, UK.

Spencer, J. F. (1982). A preliminary study of residual and organo-chlorine levels at Kingsnorth power station. CEGB Internal Report, TPRD/1/2313, No. 82, Leatherhead, UK.

Spigarelli, S. A. (1975). Behavioural responses of Lake Michigan fishes to a nuclear power plant discharge, pp. 479–98 in IAEA (c).

Spigarelli, S. A., Goldstein, R. M., Prepejchal, W. and Thommes, M. M. (1982). Fish abundance and distribution near three heated effluents to Lake Michigan. *Can. J. Fish. Aquat. Sci.*, **39**, 305–15.

Spigarelli, S. A., Romberg, G. P., Prepejchal, W. and Thommes, M. M. (1974). Body-temperature characteristics of fish at a thermal discharge on Lake Michigan, pp. 119–32 in Gibbons and Sharitz (Eds).

Spigarelli, S. A. and Smith, D. W. (1976). Growth of salmonid fishes from heated and unheated areas of Lake Michigan—measured by RNA and DNA ratios, pp. 100–5 in Esch and McFarlane (Eds).

Spigarelli, S. A. and Thommes, M. M. (1979). Temperature selection and estimated thermal acclimation by rainbow trout (*Salmo gairdneri*) in a thermal plume. *J. Fish. Res. Board Can.*, **36**, 366–76.

Spigarelli, S. A., Thommes, M. M. and Beitinger, T. L. (1977). The influence of body weight on heating and cooling of selected Lake Michigan fishes. *Comp. Biochem. Physiol.*, **56A**, 51–7, 222.

Spigarelli, S. A., Thommes, M. M., Prepejchal, W. and Goldstein, R. M. (1983). Selected temperatures and thermal experience of Brown Trout, *Salmo trutta* in a steep thermal gradient in nature. *Environ. Biol. Fish*, **8**(2), 137–49.

Spotila, J. R. (1974). Behavioral thermoregulation of the American alligator, pp. 322–34 in Gibbons and Sharitz (Eds).

Sprague, J. B. (1963). Resistance of four freshwater crustaceans to lethal high temperatures and low oxygen. *J. Fish. Res. Board Can.*, **20**, 387.

Sprague, J. B. and Drury, D. E. (1969). Avoidance reactions of salmonid fish to representative pollutants. In: *Adv. Water Poll. Res., Proc. 4th Int. Conf.*, Pergamon Press, Oxford, pp. 169–79.

Sprules, W. M. (1947). An ecological investigation of stream insects in Algonquin Park, Ontario. *University of Toronto, Stud. Biol. Ser.*, **56**, 1–81.

Spurr, G. and Scriven, R. A. (1975). United Kingdom experience of the physical behaviour of headed effluents in the atmosphere and in various types of aquatic systems, pp. 227–47 in IAEA (a).

Squires, L. E., Rushford, S. R. and Brotherson, J. D. (1979). Algal response to a thermal effluent. Study of a power station on the Provo River, Utah, USA. *Hydrobiologia*, **63**, 17–32.

Stangenberg, M. and Pawlaczyk, M. (1960). The influence of warm water influx from a power station upon the formation of biocoenotic communities in a river. *Zesz-nauk. Politech. Wr. Wroclaw*, No. 40, *Inzyn Sanit.*, **1**, 67–106; *Water Poll. Abst.*, **35**(3), 579.

Stasko, A. B. and Pincock, D. G. (1977). Review of underwater telemetry, with emphasis on ultrasonic techniques. *J. Fish. Res. Board Can.*, **34**, 1261–85.

Statham, J., Drakes, G. D. and Whitaker, D. K. (1979). The use of reject heat in power station condenser cooling water for crop production in greenhouses. Final report on the Eggborough experimental project. CEGB Internal Report, NER/SSD/R 398, Harrogate, UK.

Stauffer, J. R. (1980). Influence of temperature on fish behaviour, pp. 103–42 in Hocutt *et al.* (Eds).

Stauffer, J. R., Cherry, D. S., Dickson, K. L. and Cairns, J., Jr (1975a). Laboratory and field temperature preference and avoidance data of fish related to the establishment of standards. In: *Fisheries and Energy Symposium*, Saila, S. B. (Ed.). Lexington Books, D. C. Heath and Company, Lexington, pp. 119–39.

Stauffer, J. R., Dickson, K. L. and Cairns, J., Jr (1974). A field evaluation of the effects of heated discharges on fish distribution. *Wat. Resources Bull.*, **10**(5), 860–75.

Stauffer, J. R., Dickson, K. L., Cairns, J., Calhoun, W. F., Masnik, M. T. and Myers, R. H. (1975b). Summer distribution of fish species in the vicinity of a thermal discharge, New River, Virginia. *Arch. Hydrobiol.*, **76**(3), 287–301.

Stauffer, J. R., Dickson, K. L., Heath, A. G., Lane, G. W. and Cairns, J. (1975c). Body temperature change of bluegill sunfish subjected to thermal shock. *Prog. Fish. Cult.*, **37**(2), 90–2.

Stauffer, J. R. and Edinger, J. E. (1980). Power plant design and fish aggregation phenomena, pp. 9–28 in Hocutt *et al.* (Eds).

Stauffer, J. R., Jr, Kaesler, R. L., Cairns, J. Jr and Dickson, K. L. (1980). Selecting groups of fish to optimize acquisition of information on thermal discharges. *Wat. Resources Bull.*, **16**(6), 1097–101.

Stevens, D. E. and Finlayson, B. J. (1978). Mortality of young striped bass entrained at two power plants in the Sacramento-San Joaquim delta, California, pp. 57–70 in Jensen (Ed.).

Stevens, A. R., Tyndall, R. L., Coutant, C. C. and Willaert, E. (1977). Isolation of the etiological agent of primary amoebic meningoencephalitis from artificially heated waters. *Appl. Environ. Microbiol.* (Dec.), 701–45.

Stewart, J. E., Horner, G. W. and Arie, B. (1972). Effects of temperature, food and starvation on several physiological parameters of the lobster, *Homarus americanus. J. Fish. Res. Bd Canada.*, **29**, 439–42.

Stewart, L. (1968). The movements of salmon in relation to variations in air and water temperatures. Lancashire River Authority Fisheries Department Report, Lancaster.

Stiles, C. D. and Blake, N. J. (1976). Seasonal distribution of a podocopid ostracod in a thermally altered area of Tampa Bay, Florida, pp. 227–34 in Esch and McFarlane (Eds).

Stober, Q. J., Dinnel, P. A., Hurlburt, E. F. and DiJulio, D. H. (1980). Acute toxicity and behavioural responses of coho salmon (*Oncorhynchus kisutch*) and shiner perch (*Cymatogaster aggregata*) to chlorine in heated sea-water. *Wat. Res.*, **14**, 347–54.

Stober, Q. J. and Hanson, C. H. (1974). Toxicity of chlorine and heat to pink (*Oncorhynchus gorbuscha*) and Chinook salmon (*O. tschawytscha*). *Trans. Am. Fish. Soc.*, **3**, 569–75.

Stock, J. M. and Strachan, A. R. (1977). Heat as a marine fouling control process at coastal electric generating stations, pp. 55–62 in Jensen (Ed.).

Stockner, J. G. (1968). The ecology of a diatom community in a thermal stream. *Brit. Phycol. Bull.*, **3**(3), 501–14.

Storr, J. F. (1974). Plankton entrainment by the condenser systems of nuclear power stations of Lake Ontario, pp. 291–5 in Gibbons and Sharitz (Eds).

Storr, J. F. and Schlenker, G. (1974). Response of perch and their forage to thermal discharges in Lake Ontario, pp. 363–70 in Gibbons and Sharitz (Eds).

Straney, D. D., Briese, L. A. and Smith, M. H. (1974). Bird diversity and thermal stress in a cypress swamp, pp. 572–78 in Gibbons and Sharitz (Eds).

Stratton, C. L. and Lee, G. F. (1975). Cooling towers and water quality. *J. Wat. Pollut. Contr. Fed.*, **47**, 1901–12.

Straughan, D. (1980a). Impact of Southern California Edison's operations on intertidal solid substrates in King Harbor. Report No. 80-RD-95. Institute for Marine and Coastal Studies, University of Southern California, Los Angeles.

Straughan, D. (1980b). The impact of shoreline thermal discharge on rocky intertidal fauna. Report No. 81-RD-3, Institute for Marine and Coastal Studies, University of Southern California, Los Angeles.

Straughan, D. and Straughan, I. R. (1972). Marine biological survey of intertidal and shallow sub-tidal reef and sandy areas at HECO plant, Kahe, Hawaii, A report to URS, San Mateo for H.E.C.O. Institute for Marine and Coastal Studies, University of Southern California, Los Angeles.

Strawn, K. (1970). Beneficial uses of warm water discharges in surface waters. In: *Electric Power and Thermal Discharges*. Eisenbud, M. and Gleason, G. (Eds). Gordon and Breach, New York.

Streffer, C. (1975). Interaction mechanisms of radioactive, chemical and thermal releases from the nuclear industry; Methodology for considering co-operative effects, pp. 3–10 in IAEA (c).

Strickland, J. B. (1969). Remarks on the effects of heated discharges on marine zooplankton, pp. 73–7 in Krenkel and Parker (Eds).

Stroud, R. H. and Douglas, P. A. (1968). Thermal pollution of water. *Bull. Sport. Fish. Inst.*, **191**, 1–8.

Stuart, T. J. and Stanford, J. A. (1978). A case of thermal pollution limited primary productivity in a southwestern USA reservoir. *Hydrobiologia*, **58**(3), 199–211.

Studt, J. F. and Blake, N. J. (1976). Chronic in-situ exposure of the bay scallop to power plants effluent, pp. 235–42 in Esch and McFarlane (Eds).

Stuntz, W. E. and Magnuson, J. J. (1976). Daily ration, temperature selection and activity of bluegill, pp. 180–84 in Esch and McFarlane (Eds).

Sugam, R. and Helz, G. R. (1980). Seawater chlorination; A description of chemical speciation, pp. 427–34 in Jolley *et al.*

Summers, J. K. and Polgar, T. T. (1981). A procedure for determining and evaluating potential power plant entrainment impact *or* a modeller revisits his assumptions, pp. 199–207 in Mitsch *et al.* (Eds).

Sverdrup, H. U., Johnson, M. W. and Fleming, R. H. (1963). *The Oceans: Their Physics, Chemistry and Biology*. Prentice Hall, Englewood Cliffs, N.J.

Swale, E. M. F. (1964). A study of the phytoplankton of a calcareus river. *J. Ecol.*, **52**, 433–66.

Swartzman, G. L., Haar, R. T., McKenzie, D. H. and Zaret, T. (1981). Evaluation of the usefulness of ecological simulation models in power plant cooling systems, pp. 173–84 in Mitsch *et al.* (Eds).

Swedish State Power Board (1974). Cooling Water-Effects on the Environment. Report of the Environment Protection Foundation, Solna, Sweden.

Sweers, H. E. (1974). Two methods to measure the heat dissipation of discharged cooling water; a phenomenological approach. *Electrotechnik*, **52**(11), 615–18.

Sweers, H. E. (1980). Thermal characteristics of the river Rhine. *N.V. Kema. Arnhem*, **13**(4), 71–7.

Sweers, H. E. (1983). Heat discharge regulations and the problem of variability of natural water temperature. *Electrotechniek*, **61**, 678–85.

Sykes, G. and Skinner, F. A. (Eds) (1971). *Microbial Aspects of Pollution*, Society for Applied Bacteriology, Symposium Series No. 1. Academic Press, London, p. 289.

Sylvester, J. R. (1972). Effect of thermal stress on predator avoidance in sockeye salmon. *J. Fish. Res. Board Can.*, **29**, 601–3.

Sylvester, J. R. (1975a). Critical thermal maxima of three species of Hawaiian estuarine fish: a comparative study. *J. Fish. Biol.*, **7**, 257–62.

Sylvester, J. R. (1975b). Biological considerations on the use of thermal effluents for finfish aquaculture. *Aquaculture*, **6**, 1–10.

Szybalski, W. (1967). Effects of elevated temperatures on DNA and on some polynucleotides; Denaturation, renaturation, and cleavage of glycosidic and phosphate ester bonds, pp. 73–113 in Rose (Ed.).

Tagatz, M. E. (1961). Tolerance of striped bass and American shad to changes of temperature and salinity. United States Fish and Wildlife Service, Special Scientific Report, Fisheries No. 388, Washington, D.C., 8 pp.

Tagatz, M. E. (1969). Some relations of temperature acclimation and salinity to thermal tolerance of the blue crab, *Callinectes sapidus. Trans. Am. Fish. Soc.*, **4**, 713–16.

Talmage, S. S. and Coutant, C. C. (1980). Thermal effects. *J. Wat. Pollut. Contr. Fed.*, **52**(6), 1575–1616.

Talmage, S. S. and Opresko, D. M. (1981). Literature review: responses of fish to thermal discharges. Electric Power Research Institute, E. A.—1840, Palo Alto, CA.

Tansey, M. R. and Fliermans, C. B. (1978). Pathogenic species of thermophilic and thermotolerant fungi in reactor effluents of the Savannah River Plant, pp. 633–90 in Thorp and Gibbons (Eds).

Tarzwell, C. M. (1970). Thermal requirements to protect aquatic life. *J. Wat. Pollut. Contr. Fed.*, **42**(5), part 1, 823–8.

Teleki, G. C. (1976). The incidence and effect of once-through cooling on young-of-the-year fishes of Long Point Bay, lake Erie; A preliminary assessment, pp. 387–93 in Esch and McFarlane (Eds).

Templeton, W. L. and Coutant, C. C. (1971). Studies on the biological effects of thermal discharges from nuclear reactors to the Columbia River at Hanford, pp. 591–612 in IAEA.

Tenessen, K. J. and Miller, J. L. (1978). Effects of thermal discharge on aquatic insects in the Tennessee valley. Environmental Protection Agency, EPA-600/7-78-128, Washington, D.C.

Tetra-Tech Inc. (1980). *Methodology for Evaluation of Multiple Power Plant Cooling System Effects, Vols I–VI*. Electric Power Research Institute, RP-878-1, Palo Alto, CA.

Thatcher, T. O. (1974). Combined effects of mercury and temperature on the mortality of rainbow trout, pp. 54–8 in Gibbons and Sharitz (Eds).

Thatcher, T. O. (1978). The relative sensitivity of Pacific northwest fishes and invertebrates to chlorinated sea water, pp. 341–50 in Jolley *et al.* (Eds).

Thatcher, T. O., Schneider, M. J. and Wolf, E. G. (1976). Bioassays on the combined effects of chlorine, heavy metals and temperature on fishes and fish food organisms. Part 1. Effects of chlorine and temperature on juvenile brook trout (*Salvelinus fontinalis*). *Bull. Environ. Contam. Toxicol.*, **15**(1), 40–8.

Thomas, P., Bartos, J. M. and Brooks, A. S. (1980). Comparison of the toxicities of monochloramine and dichloramine to rainbow trout, *Salmo gairdneri* under various time conditions, pp. 581–8 in Jolley *et al.*

Thompson, T. J. (1971). Role of nuclear power in the United States of America, pp. 91–118 in IAEA.

Thomson, J. M. (1963). Mortality threshold of fish in fly-ash suspension. *Austr. J. Sci.*, **25**, 414–15.

Thorhaug, A. (1974). Effect of thermal effluents on the marine biology of southeastern Florida, pp. 518–31 in Gibbons and Sharitz (Eds).

Thorhaug, A. (1979). Biological effects of power plant thermal effluents in Card Sound, Florida. *Environ. Conservat.*, **6**(2), 127–37.

Thorhaug, A. (1980). Biological effects of thermal effluents in the marine environment: tropics and sub-tropics with a guideline (February) (unpublished typescript only).

Thorhaug, A. (1987). Large scale seagrass restoration in a damaged estuary. *Mar. Pollut. Bull.*, **18**(8), 442–6.

Thorhaug, A., Blake, N. and Schroeder, P. B. (1978). The effect of heated effluents from power plants on seagrass (*Thalassia*) communities, quantitatively comparing estuaries in the subtropics to the tropics. *Aquaculture*, **9**, 181–7.

Thorhaug, A. and Roessler, M. A. (1977). Seagrass community dynamics in a subtropical estuarine lagoon. *Aquaculture*, **12**, 253–77.

Thorhaug, A., Segar, D. and Roessler, M. A. (1974). Impact of a power plant on a subtropical estuarine environment. *Mar. Pollut. Bull.*, **5**(10), 166–9.

Thorp, J. H. and Gibbons, J. W. (Eds) (1978). *Energy and Environmental Stress in Aquatic Systems.* Technical Information Centre, US Department of Energy.

Thulin, J. (1981). Diseases and parasites of fish at the Barseback nuclear power station, 1980. *Naturvardsverket, Meddelande, SNV pm*, 1429 (cyclostyled).

Thurston, R. V., Russo, R. C., Fetterholf, C. M., Jr, Edsall, T. A. and Barber, Y. M., Jr (Eds) (1979). *A Review of the EPA Red Book: Quality Criteria for Water.* Water Quality Section, American Fisheries Society, Bethesda. MD. 313 pp.

Tiews, K., (Ed.) (1981). *Aquaculture in Heated Effluents and Circulation Systems, Vol. 11. Proceedings of a World Symposium, Stavanger,* 1980, Heenemann, Berlin.

Tilley, L. J. *et al.* (1978). Response of the Asiatic Clam, *Corbicula fluminea,* to gamma radiation. *Health Phys.*, **35**(11), 704–7.

Tilly, L. J. (1974). Respiration and net productivity of the plankton community in a reactor cooling reservoir, pp. 462–74 in Gibbons and Sharitz (Eds).

Tinker, J. (1971). The smug and silver Trent. *New Scientist and Science Journal,* 10th June, 614–18.

Tinsman, J. C. and Maurer, D. L. (1974). Effects of a thermal effluent on the American oyster, pp. 223–36 in Gibbons and Sharitz (Eds).

Tinsman, J. C., Tinsman, S. G. and Maurer, D. (1976). Effects of a thermal effluent on the reproduction of the American oyster, pp. 64–72 in Esch and McFarlane (Eds).

Todd, R. D. and Bender, J. F. (1982). Water quality characteristics of the Missouri River near Fort Calhoun and Cooper Nuclear Stations, pp. 39–70 in Hesse *et al.* (Eds).

Tongiorgi, P., Tosi, L. and Balsamo, M. (1986). Thermal preferences in upstream migrating glass eels of *Anguilla anguilla* (L). *J. Fish Biol.*, **28**, 501–10.

Trace Metal Data Institute (1981). Water reclamation from cooling tower blowdown, using reverse osmosis. Bulletin 603, TMDI El Paso, Texas, 82 pp.

Trefethen, P. (1972). Man's impact on the Columbia River, pp. 77–98 in Oglesby *et al.* (Eds).

Trembley, F. J. (1960). Research projects on effects of condenser discharge water on aquatic life, Progress Report, 1956–59. Institute of Research, Lehigh University, Bethlehem, PA.

Trembley, F. J. (1965). Effects of cooling-water from steam-electric power plants on stream biota. In: *Biological Problems in Water Pollution,* US Dept. Health

Education and Welfare, 999 WP-25, Government Printing Office, Washington, D.C., pp. 334–45.

Trent River Authority. Annual Reports 1964–1974. Trent River Authority, Nottingham.

Trotzky, H. M. and Gregory, P. W. (1974). The effects of water flow manipulation below a hydro-electric power dam on the bottom fauna of the Upper Kennebec river, Maine. *Trans. Am. Fish. Soc.*, **103**(2), 318–24.

Truchan, J. W. (1977). Toxicity of residual chlorine to freshwater fish: Michigan's experience, pp. 79–89 in Jensen (Ed.).

Tsai, C. (1975). Effects of sewage treatment plants on fish: A review of literature. Chesapeake Research Consortium Inc. Pub. No. 36. University of Maryland, CEES. Contrib. No. 637.

Tucker, W. S. (1978). Power plant siting under the N.E.P.A. process. Success or failure, pp. 251–63 in Bendix and Graham (Eds).

Turner, A. and Thayer, T. A. (1980). Chlorine toxicity in aquatic ecosystems, pp. 607–30 in Jolley *et al.*

Turner, A. and Chu, A. (1983). Chlorine toxicity as a function of environmental variables and species tolerance, pp. 927–46 in Jolley *et al.* (b).

Turnpenny, A. W. H. (1981). An analysis of mesh sizes required for screening fishes at water intakes. *Estuaries*, **4**(4), 363–8.

Turnpenny, A. W. H. (1984). Critical swimming speeds of O-group pout (*Trisopterus luscus, L.*) in relation to body size and water temperature. CEGB Internal Report, TPRD/L/2657/N84, Leatherhead, UK.

Turnpenny, A. W. H. (1988). The behavioural basis of fish exclusion from coastal power station cooling water intakes. CEGB Internal Report, RD/L/3301/R88, Leatherhead, UK.

Turnpenny, A. W. H. and Henderson, P. A. (1981). Application of Leslie matrix models for assessing effects of power stations on fish populations. CEGB Internal Report, RD/L 2115N81, Leatherhead, UK.

Turnpenny, A. W. H., Langford, T. E. and Aston, R. J. (1985). Power stations and fish. CEGB Research, **17**, 27–39.

Turnpenny, A. W. H., Utting, N. J., Milner, R. S. and Riley, J. D. (1983). The effect of fish impingement at Sizewell 'A' power station on North Sea fish stocks— Sizewell 'B' Public Inquiry Support Document. CEGB, London.

Turoboyski, L. (1973). Investigations on influence of heated waters from the power station at Skawina on Skawinka and Vistula Rivers, Poland. *Polskie Archiwum Hydrobiologii*, **20**(3), 443–60.

Tyndall, R. L. (1980). Control of *Legionella pheumopila*, pathogenic free-living amoebae and pathogenic fungi in cooling waters, pp. 29–38 in Jolley *et al.*

Tyndall, R. L., Kuhl, G. and Bechrold, J. (1983). Chlorination as an effective treatment for controlling pathogenic *Naegleria* in cooling waters of an electric power plant, pp. 1097–1103 in Jolley *et al.* (b).

Tyndall, R. L., Willaert, E., Stevens, A. R. and Coutant, C. C. (1978). Isolation of pathogenic *Naegleria* from artificially heated waters, pp. 117–23 in Gerhold (Ed.).

Udey, L. R. (1978). Effect of temperature on the infectious process in Pacific salmonids, pp. 103–7 in Gerhold (Ed).

Udey, L. R., Fryer, J. L. and Pilcher, K. S. (1975). Relation of water temperature to ceratomyxosis in rainbow trout, (*Salmo gairdneri*) and coho salmon, (*Oncorhynchus kisutch*). *J. Fish. Res. Board Can.*, **32**, 1545–51.

UKAEA (undated). *Glossary of Atomic Terms*. United Kingdom Atomic Energy Authority, London.

UNIPEDE (1977). Expert group on cooling water problems. Survey of legislation and rules on water quality in common use in a number of European countries. Study Document 20/D. 4.

United Nations (1976). The demand for water: Procedures and methodologies for projecting water demands in the context of regional and national planning. UN Natural resources, Water Series No. 3.

Urbistondo, R., Roman, J. and 7 others (1974). Comparison of two types of ecological discontinuities in the River Tagus; Reservoir and nuclear plant. Paper 2.2.10, *Proceedings of the 9th World Energy Conference*, Detroit. WEC, London.

USAEC (1971). *Thermal Effects and US Nuclear Power Stations*. USAEC Division of Reactor Development and Technology, Washington, D.C.

US Department of the Interior (1953). Laboratory investigations of 81 fly ashes. Bureau of Reclamation, Engineering Laboratory Branch, Concrete Laboratory Report No. C.680 (11th September).

USSR, State Committee on the Utilisation of Atomic Energy (1986). *The Accident at the Chernobyl Nuclear Power Plant and its Consequences*. Compiled for the IAEA experts meeting, Part 1, General Material, Vienna.

Utting, N. J. and Millican, P. F. (1977). Further experiments to determine the effect of entrainment through Bradwell Nuclear Power Station on the larvae of *Ostrea edulis* L. CEGB Internal Report, RD/L/N 130/77, Leatherhead, UK.

Uziel, M. S. and Hannon, E. H. (1979). *Impingement—an Annotated Bibliography*. Electric Power Research Institute, EA 1050, Palo Alto, CA.

Vaas, K. F. and Sachlan, F. (1955). Limnological studies on diurnal fluctuations in shallow ponds in Indonesia. *Verh. Int., Ver. Limnol.*, **12**, 309–19.

Vadas, R. L., Keser, M. and Rusanowski, P. C. (1976a). Influence of thermal loading on the ecology of intertidal algae, pp. 202–12 in Esch and McFarlane (Eds).

Vadas, R. L., Keser, M., Rusanowski, P. C. and Larson, B. R. (1976b). The effects of thermal loading on the growth and ecology of a northern population of *Spartina alterniflora*, pp. 54–63 in Esch and McFarlane (Eds).

Vadas, R. L., Keser, M. and Larson, B. (1978). Effects of reduced temperatures on previously stressed populations of an intertidal alga, pp. 434–51 in Thorp and Gibbons (Eds).

Vaillancourt, G. and Couture, R. (1975). Influence de l'apport thermique originaire de la centrale nucleaire Gentilly sur la temperature de l'eau et sur les Gasteropodes, pp. 449–58 in IAEA (a).

van Densen, W. L. T. and Hadderingh, R. H. (1982). Effects of entrapment and cooling water discharge by the Bergum power station on the 0+ fish in the Bergumermeer. *Hydrobiologia*, **95**, 351–68.

van Eeden, W. N. (1975). Power station effluent control and the re-use of ash-water for cooling water treatment. *J. Wat. Pollut. Contr.*, **74**(2), 211–15.

Vanderhorst, J. R. (1982). *Effects of Chlorine on Marine Benthos*. Electric Power Research Institute, EA-2696, RP 1224-4, Palo Alto, CA.

Vanderhorst, J. R., Bridge, J. R. and Fellingham, G. W. (1983). Long-range chlorination in open microcosms, interpretations, pp. 797–810 in Jolley *et al.* (Eds) (b).

van Weers, A. W. (1975). The effect of temperature on the uptake and retention of Co and Zn by the common shrimp *Crangon crangon* (L), pp. 35–47 in IAEA (c).

Van Winkle, W. (Ed.) (1977). *Proceedings of the Conference on Assessing the Effects of Power-Plant-Induced Mortality on Fish Populations*. Pergamon Press, Oxford.

Venkataramiah, A., Lakshmi, G. J., Best, C. M., Gunter, G., Hartiwig, E. D. and Valentine, R. (1983). Effects of chlorinated discharges on marine animals, pp. 947–66 in Jolley *et al.* (b).

Verlaque, M., (1977). Le peuplement algal au voisinage de la central thermique de Martiques-Ponteau, (Golfe de Fos, France); Esquisse de la distribution des peuplements phytobenthiques superficiels, pp. 215–32 in EDF.

Verlaque, N., Giraud, G. and Boudouresque, C. F. (1981). Le peuplement algal de Martigues-Ponteau (Golfe de Fos, France); Etude de la zone de decollement de la tach thermique, pp. 679–99 in EDF.

Vernberg, F. J. (1978). Multiple factor and synergistic stresses in aquatic systems, pp. 726–47 in Thorp and Gibbons (Eds).

Vernberg, F. J. and Vernberg, W. B. (1974). Synergistic effects of temperature and other environmental parameters on organisms, pp. 94–9 in Gibbons and Sharitz (Eds).

Vesey, G. and Langford, T. E. (1985). The biology of the black goby, *Gobius niger*, in the English south-coast bay. *J. Fish Biol.*, **27**, 417–29.

Verstraete, W., Voet, J. P. and Varstaan, H. (1975). Shifts in microbial groups of river water upon passage through cooling systems. *Environ. Pollut.*, **8**, 275–81.

Vogel, J. and Hoff, F. (1981). *Atmospheric Effects of Cooling Lakes.* EA, 1762, RP, 578, Electric Power Research Institute, Palo Alto, CA.

Wackenuth, E. C. and Levine, G. (1977). Experience in the use of bromine chloride for antifouling at steam electric generating stations, pp. 55–62 in Jensen (Ed.).

Waite, T. D., Jorden, R. N. and Kawaratani, R. (1978). Evaluation of alternative chemical treatments for biofouling control in electric power facilities, pp. 753–74 in Jolley *et al.* (Eds).

Walker, J. H. and Lawson, J. D. (1977). Natural stream temperature variations in a catchment. *Wat. Res.*, **2**, 373–7.

Walling, D. E. and Foster, I. D. L. (1975). Variations in the natural chemical concentration of river water during flood flows, and the lag effect; some further comments. *J. Hydrol.*, **26**, 237–44.

Walters, J. (1977). An ecological study of *Hydrobia ulvae*, (Pennant), in the Medway estuary, including the possible effects of power station warm water effluent. PhD Thesis, Department of Biological Sciences, City of London Polytechnic, London.

Ward, D. V. (1978). *Biological Environmental Impact Studies: Theory and Methods.* Academic Press, New York, London.

Ward, G. H., (1982). Thermal plume area calculation, *Proc. Am. Soc. Chem. Engng, Energy Division*, **108**(EY 2), 104–15.

Ward, J. V. (1974). A temperature stressed stream ecosystem below a hypolimnial release mountain reservoir. *Arch. Hydrobiol.*, **74**(2), 247–75.

Ward, J. V. (1976). Effects of thermal constancy and seasonal temperature displacement on community structure of stream macroinvertebrates, pp. 302–7 in Esch and McFarlane (Eds).

Ward, J. V. and Stanford, J. A. (Eds) (1979a). *The Ecology of Regulated Streams.* Plenum Press, New York, London, 398 pp.

Ward, J. V. and Stanford, J. A. (1979b). Ecological factors controlling stream zoobenthos with emphasis on thermal modification of regulated streams, pp. 35–56 in Ward and Stanford (Eds).

Warinner, J. E. and Brehmer, M. L. (1966). The effects of thermal effluents on marine organisms. *Int. J. Air Wat. Pollut.*, **10**(4), 277–89.

Waslenchuk, D. (1983). Concentration, reactivity and fate of copper, nickel, and zinc associated with a cooling water plume in estuarine waters, 11. The particulate phases. *Environ. Pollut. (Series B)*, **5**, 59–70.

Water Research Centre (1976). *Thermal Pollution. A Literature Survey Covering the period 1968–1975.* WRC Reading List No. 34, Stevenage, Herts.

Waters, T. F. (1972). The drift of stream insects. *Ann. Rev. Entomol.*, **17**, 253–72.

Watt, J. D. and Thorne, D. J. (1965). Composition and pozzolanic properties of pulverised fuel ashes. 1. Composition of fly ashes from some British power stations and properties of their component particles. *J. Appl. Chem.*, **15**, 585–604.

Waugh, G. D. (1964). Observations on the effects of chlorine on the larvae of oysters (*Ostrea edulis* (L)) and barnacles (*Elminius modestus* (Darwin)). *Ann. Appl. Biol.*, **54**, 423–40.

Webb, P. W. (1978). Temperature effects on acceleration of rainbow trout, *Salmo gairdneri. J. Fish. Res. Board Can.*, **35**, 1427–32.

Webster, K. E., Forbes, A. M. and Magnuson, J. J. (1981). Behavioral responses of stream macroinvertebrates to habitat modification by a coal ash effluent. *Trans. Am. Fish. Soc.*, Warmwater Stream Symposium, pp. 408–14.

WEC (1974). *Proceedings of the 9th World Energy Conference*, Detroit, 22–27 September, 1974, Divisions 1–4 (4 Vols), WEC, London.

Welch, M. O. and Ward, C. H. (1978). Primary productivity: Analysis of variance in a thermally enriched aquatic system, pp. 381–91 in Thorp and Gibbons (Eds).

Welch, P. S. (1952). *Limnology.* McGraw-Hill, New York, Toronto, London.

Wellings, F. M., Lewis, A. L., Amuso, P. T. and Chang, S. L. (1977). *Naegleria* and water sports. *Lancet*, **i**, 199–200.

Wetzel, D. L. (1974). Zooplankton entrainment. In: *Operational thermal monitoring program of Lake Michigan, near Keewaunee Nuclear Power Plant.* January to December, 1974. Industrial Biotest Laboratories Inc., Chicago.

Wheeler, A. (1969). *Key to the Fishes of Northern Europe.* Frederick Warne, London.

Wheeler, A. and Maitland, P. S. (1973). The scarcer freshwater fishes of the British Isles, I, Introduced species. *J. Fish Biol.*, **5**, 49–68.

Whipple, R. T. P. (1963). Considerations on the siting of outfalls for the sea disposal of radioactive effluent in tidal waters. *Int. J. Air Wat. Pollut.*, **7**, 889–906.

White, G. C. (1972). *Handbook of Chlorination.* van Nostrand Reinhold, New York, Cincinnati, Toronto, London, Melbourne.

White, J. C., Hammond, R. A., Wooding, N. H., Jr and Brehmer, M. L. (1976). Temperature as a growth accelerator in the Spot (Teleost; Sciaenidae), pp. 113–17 in Esch and McFarlane (Eds).

White, J. W., Woolcott, W. S. and Kirk, W. L. (1977). A study of the fish community in the vicinity of a thermal discharge in the James River, Virginia. *Ches. Sci.*, **18**(2), 161–71.

Whitehouse, J. W. (1971). Some aspects of the biology of Lake Trawsfynydd; a power station cooling pond. *Hydrobiologia*, **38**(2), 253–88.

Whitehouse, J. W. (1975). Chlorination of cooling water: A review of literature on the effects of chlorine on aquatic organisms. CEGB Internal Report, RD/L/M 496, Leatherhead, UK.

Whitehouse, J. W. (1978). Continuous low-level chlorination for marine fouling control at power stations in the United Kingdom, pp. 361–368 in Jolley *et al.* (Eds).

Whitehouse, J. W. (1989). Studies of biofouling on plastics packing using miniature cooling towers. *Waterline*, (April) 19–30.

Whitehouse, J. W. and Aston, R. J. (1964). A survey of the fauna above and below nine power station outfalls. CEGB Internal Report, RD/L/N 12/64, Leatherhead, UK.

Whitehouse, J. W. and Key, D. (1978). Bradwell Biological Investigations; Fauna of the intertidal zone at Bradwell, Essex. CEGB Internal Report, RD/L/N 19/78, Leatherhead, UK, 12 pp.

Whitehouse, J. W., Khalanski, M., Saroglia, M. G. and Jenner, H. A. (1984). The control of biofouling in marine and estuarine power stations. CEGB/EDF/ENEL/KEMA, Leatherhead, UK.

Whiting, D., Langford, T. E. and Foster, D. J. (1976). Analyses of angling competition catch-data from the River Trent—with special reference to the location of power stations. CEGB Internal Report, SSD/MID/N13/76, Leatherhead, UK.

Whitton, B. A. (Ed.) (1975). *River Ecology*. Blackwell, London.

Whitton, B. A. (1984). *Ecology of European Rivers*. Blackwell, Oxford.

Widdows, J. (1976). Physiological adaptation of *Mytilus edulis* to cyclic temperature. *J. Comp. Physiol.*, **105**, 115–28.

Wilcox, J. R. (1979). Waste heat utilisation in aquaculture. Some futuristic and plausible schemes. Paper X-A-35 in Lee and Sengupta (Eds).

Wilde, E. W. and Tilly, L. J. (1981). Structural characteristics of algal communities in thermally altered artificial streams. *Hydrobiologia*, **76**(1/2), 57–63.

Wilkonska, H. and Zuromska, H. (1977). Changes in the species composition of fry in the shallow littoral of heated lakes of the Konin lakes complex. *Rocz. Nauk. roln. (H)*, **97**, 132–4.

Willaert, E. and Stevens, A. R. (1976). Isolation of pathogenic amoebae from thermal discharge water. *Lancet*, **ii**, 741.

Williams, D. D. and Hynes, H. B. N. (1976). The recolonization mechanisms of stream benthos. *Oikos*, **27**, 265–72.

Williams, G. E. and Strawn, K. (1980). Effects of power plants on penaeid shrimps, pp. 261–308 in Hocutt *et al.* (Eds).

Williams, R. B. and Murdoch, M. B. (1973). Effects of continuous low-level gamma radiation on sessile marine invertebrates. IAEA-SM-158/34. International Atomic Energy Agency, Vienna, pp. 551–63.

Wills, D. (1972). Physical Aspects of Cooling Water Discharges, 1963—Oct. 1971. C. E. Bib. 217, CEGB, London.

Wisconsin Electric Power Company (1976). Point Beach nuclear plant: Final Report WPDES, Intake Monitoring Studies Permit Number W1-0000957, Wisconsin Electric Power Company, Madison, Wisconsin.

Wisdom, A. S. (1979). *The Law of Rivers and Watercourses*, 4th edn. Shaw and Sons, London.

Witt, A. (1971). The evaluation of environmental alteration by thermal loading and acid pollution in the cooling reservoir of a steam electric station. NTIS, PB-197, Missouri Water Resources Research Abstract 4, W71-04019.

Wojtalik, T. A. and Waters, T. F. (1970). Some effects of heated water on the drift of two species of stream invertebrates. *Trans. Am. Fish. Soc.*, **99**(4), 782–8.

Wolfe, D. A. and Coborn, C. B., Jr (1970). Influence of salinity and temperature on the accumulation of caesium-137 by an estuarine clam under laboratory conditions. *Health Phys.*, **18**, 499–505.

Wolters, W. R. and Coutant, C. C. (1976). The effect of cold shock on the vulnerability of young bluegill to predation, pp. 162–4 in Esch and McFarlane (Eds).

Wong, G. T. F. (1980a). Some problems in the determination of total residual chlorine in sea-water. *Wat. Res.*, **14**(1), 51–60.

Wong, G. T. F. (1980b). The effect of temperature on the dissipation of chlorine, pp. 395–406 in Jolley *et al.*

Wong, G. T. F. and Davidson, J. A. (1977). The fate of chlorine in seawater. *Wat. Res.*, **11**, 971–8.

Wong, S. L., Clark, B., Kirby, M. and Kosciuw, R. F. (1978). Water temperature fluctuations and seasonal periodicity of *Cladophora* and *Potamogeton* in shallow rivers. *J. Fish. Res. Board Can.*, **35**, 866–70.

Wood, E. J. F. (1967). *Microbiology of Oceans and Estuaries*. Elsevier Publishing Company, London, New York.

Wood, H. C. (1868). Notes on some algae from a Californian hot spring. *Queckett J. Microsc. O.S.C. Science*, **8**, 250.

Wood, M. J. (1979). The effects of passage through a power station on micro-organisms entrained in estuarine cooling water. CEGB Internal Report, No. RD/L/N/180/79, Leatherhead, UK.

Woodhead, P. M. J. (1964). The death of fish and sub-littoral in the North Sea and the English Channel during the winter of 1962–63. *J. Anim. Ecol.*, **33**, 169–73.

Woodhead, P. M. J., Parker, J. H., Duedall, I. W. and Ayer, F. A. (1980) Environmental compatibility and engineering feasibility for utilisation of FGD waste in artificial fishing reef construction, pp. 695–700 in Proceedings of a Symposium on Flue-gas Desulphurisation, State University of New York, Stony Brook, New York.

Woods Hole Oceanographic Institute (1952). *Marine Fouling and its Prevention*. US Naval Institute, Annapolis.

Woodyard, J. P. and Sanning, D. E. (1978). The environmental impact of FGD sludge disposal. *Proceedings of the 71st Annual Meeting of the Air Pollution Control Association*, June 25–30. Houston, Texas.

WPCD (1982). Characterization and treatability of drainage samples from coal piles at steam electric power stations. Economic and Technical Report, EPS 3-WP-82-4, Water Pollution Control Directorate, Environmental Protection Service, Canada.

Wrenn, W. B. (1976a). Temperature preference and movement of fish in relation to a long, heated discharge channel, pp. 191–4 in Esch and McFarlane (Eds).

Wrenn, W. B. (1976b). Preliminary assessment of larval fish entrainment, Colbert steam plant, Tennessee River, pp. 381–386 in Esch and McFarlane (Eds).

Wroblewsteigo, A. (1977). Bottom fauna of the heated lakes. *Monographie Fauny Polski*, Tom 7, Krakow, Poland.

Wurtz, C. B. (1969). The effects of heated discharges on freshwater benthos, pp. 199–213 in Krenkel and Parker (Eds).

Wurtz, C. B. and Dolan, T. (1960). A biological method used in the evaluation of effects of thermal discharge in the Schuylkill River. *Proc. 15th Ind. Waste Conf. Purdue*, Purdue University, pp. 461–72.

Wurtz, C. B. and Renn, C. E. (1965). Water temperatures and aquatic life. Publication No. 65-901, Edison Electric Institute (June), Johns Hopkins University, Baltimore, Maryland.

Wurtzbaugh, W. A. and Davis, G. E. (1977). Effects of temperature and ration level on the growth and food conversion efficiency of *Salmo gairdneri*, Richardson. *J. Fish Biol.*, **11**, 87–8.

Yagi, H. and Ceccaldi, H. J. (1984). The combined effects of temperature and salinity on the metamorphosis and larval growth of the common prawn, *Palaemon serratur* (Pennant), (Crustacea, Decapoda, Palaeomonidae). *Aquaculture*, **37**, 73–85 (Fr.).

Yardley, D., Avise, J. C., Gibbons, J. W. and Smith, M. H. (1974). Biochemical genetics of sunfish, iii. Genetic subdivision of fish populations inhabiting heated waters, pp. 255–63 in Gibbons and Sharitz (Eds).

Yarosh, M. M. (1977). Waste energy utilization—needs and effects, pp. 267–282. In: *Proceedings of the World Conference Towards a Plan of Actions for Mankind, Vol. 3. Biological Balance and Thermal Modifications*. Marois, M. (Ed.). Institut de la View, Paris, 1974, Pergamon Press, Oxford.

Yee, W. C. (1972). Thermal aquaculture: engineering and economics. *Environ. Sci. Technol.*, **6**(3), 232–6.

Yee, W. C. (1974). Thermal aquaculture; potential and problems, pp. 174–89 in Gallagher (Ed.).

Yoder, C. O. and Gammon, J. R. (1976). Seasonal distribution and abundance of Ohio River fishes at the J. M. Stuart Electric Generating Station, pp. 284–95 in Esch and McFarlane (Eds).

Yongue, W. H., Jr, Berrent, B. L. and Cairns, J., Jr (1979). Survival of *Euglena gracilis* exposed to sublethal temperature and hexavalent chromium. *J. Protozool.*, **26**(1), 122–5.

Yost, F. E. and Talmage, S. S. (1981). *Ecological Investigations at Power Plant Cooling Lakes, Reservoirs and Ponds; An Annotated Bibliography*. Electric Power Research Institute, EA-1874, RP 877, Palo Alto, CA.

Yost, F. E. and Uziel, M. S. (1981). *Impingement and Entrainment; An Updated Bibliography*. Electric Power Research Institute, EA-1855, RP 877, Palo Alto, CA.

Young, D. L. (1974). Studies of Florida Gulf Coast salt marshes receiving thermal discharges, 532–50 in Gibbons and Sharitz (Eds).

Young, J. S. and Frame, A. B. (1976). Some effects of a power plant effluent on estuarine epibenthic organisms. *Int. Rev. ges. Hydrobiologie*, **61**(1), 37–61.

Young, J. S. and Gibson, C. I. (1973). Effect of thermal effluent on migrating menhaden. *Mar. Pollut. Bull.*, **4**(6), June, 94–5.

Young, G. K., Tseng, M. T. and Taylor, R. S. (1971). Estuary water temperature sensitivity to meteorologic conditions. *Wat. Resources Res.*, **3**(5), 1173–81.

Zawisza, J. and Backiel, T. (1972). Some results of fishery biological investigations of heated lakes. *Verh. int. Verein. Limnol.*, **18**, 1190–7.

Zeeman, S. I. and Grunewald, R. (1978). Size-fractionated primary productivity in Lake Michigan near the Kewaunee nuclear power plant, pp. 364–81 in Thorp and Gibbons (Eds).

Zeiman, J. C. and Ferguson-Wood, E. J. (1975). Effects of thermal pollution on tropical-type estuaries, with emphasis on Biscayne Bay, Florida. In: *Tropical Marine Pollution*, Ferguson-Wood, E. J. and Johannes, R. E. (Eds). Elsevier Scientific Publishing Co., New York, pp. 75–98.

Zeitoun, I. H. (1977). The effect of chlorine toxicity on certain blood parameters of adult rainbow trout (*Salmo gairdneri*). *Environ. Biol. Fish.*, **1**(2), 189–95.

Zeitoun, I. H. (1978). The recovery and haematological rehabilitation of chlorine-stressed adult rainbow trout (*Salmo gairdneri*). *Environ. Biol. Fish.*, **3**(4), 355–9.

Zgarheva, N. N. and Mordukhai-Boltovskoy, F. D. (1980). The effect of heated waters of the Konakovo district power plant on the phytophilous fauna of pondweed in Ivankovo reservoir. *Hydrobiol. J.*, **15**(6), 35–40.

Zillich, J. A. (1972). Toxicity of combined chlorine residuals to freshwater fish. *J. Wat. Pollut. Contr. Fed.*, **44**(2), 212–20.

Zimmerman, R. C., Saponja, D. and Nancarrow, D. R. (1974). Characteristics of non-thermal liquid effluents from two coal-fired power plants in Alberta, Canada, Paper 2.2-1, *Proc. 9th World Energy Conf.*, Detroit, WEC, London.

Zobell, C. E. (1946). *Marine Microbiology*. Chronica Botanica, Watham, MA.

Index